# Topics in
# Current Physics

**43**

# Topics in Current Physics  Founded by Helmut K. V. Lotsch

Volumes 1–38 are listed on the back inside cover

# Structure and Dynamics of Surfaces II

## Phenomena, Models, and Methods

Edited by W. Schommers
and P. von Blanckenhagen

With Contributions by
J. Als-Nielsen   E. Bauer   H. van Beijeren
G. Benedek   P. von Blanckenhagen
G. Doyen   W. Hanke   L. Miglio
A. Muramatsu   I. Nolden   W. Schommers

With 167 Figures

Springer-Verlag Berlin Heidelberg New York
London Paris Tokyo

Dr. Wolfram Schommers
Dr. Peter von Blanckenhagen

Kernforschungszentrum Karlsruhe, Institut für Nukleare Festkörperphysik, Postfach 3640, D-7500 Karlsruhe, Fed. Rep. of Germany

ISBN-13: 978-3-642-46593-2    e-ISBN-13: 978-3-642-46591-8
DOI: 10.1007/978-3-642-46591-8

Library of Congress Cataloging-in-Publication Data. (Rev. for vol. 2) Structure and dynamics of surfaces. (Topics in current physics ; 41, 43) Bibliography: p. : v. 1. Includes index. 1. Surfaces (Physics) 2. Surface chemistry. I. Schommers, W. (Wolfram), 1941–. II. Blanckenhagen, P. (Peter), 1936–. III. Series. QC 173.4.S 94 S 76   1987   530.4′1   86-6654

© Springer-Verlag Berlin Heidelberg 1987
Softcover reprint of the hardcover 1st edition 1987

2153/3150-543210

# Preface

Structural and dynamical properties of surfaces are of considerable importance for the elucidation of surface phenomena. For example, the *adsorption* of particles and *chemical reactions* at surfaces are influenced by these properties. Also the *electronic* surface states of electronic device materials (e.g., Schottky barrier diodes and metal-insulator-semiconductor devices) depend critically on the structure of the semiconductor surface and the overlayers.

The treatment *Structure and Dynamics of Surfaces* is divided into two parts. The first volume (Topics Curr. Phys., Vol. 41) deals with recent developments in surface-structural research using atomic beam scattering, scanning tunneling microscopy, surface channeling, high-resolution electron microscopy, etc. The theory of vibrations in the *harmonic* approximation of surface and adsorbate layers is given. New investigations by high-resolution electron loss spectroscopy are reported. The molecular-dynamics method is introduced and discussed in connection with *anharmonic* surface effects.

This second volume is a direct continuation and deals with both *ordered* and *disordered* surfaces and with *anharmonic* as well as *harmonic* dynamical surface effects. The mean-square amplitudes of the particles are significantly larger at the surface of the crystal than in the bulk and, therefore, *structural disorder* and *anharmonic effects* are expected to be more significant at the surface than in the bulk. In this connection it should be emphasized that also *liquid surfaces* are of fundamental importance for current surface-physics problems as well as for chemistry and biology.

As in the previous volume, also in this Topics volume the problems are discussed at the microscopic level. Novel experimental techniques and theoretical developments, which have been applied successfully during the last decade, are integrated. The contributions include the *background* of the subject, *typical results*, and in most cases also trends of *future developments*. First, some introductory remarks concerning surface phenomena and their analysis by scattering experiments are given and then the discussion of *Structure and Dynamics of Surfaces I* will be continued by the following eight reviews: Study of surface phonons by means of the Green's function method; layer growth and surface diffusion; phase transitions on single-crystal surfaces and in chemisorbed layers; solid and liquid surfaces studied by synchrotron x-rays; statistical mechanics of the liquid surface and the effect of premelting; the roughening transition; structural and dynamical aspects of adsorption and desorption; many-body description of surface elementary excitations and its application to semiconductors.

The editors would like to thank each of the contributors for excellent cooperation. Many thanks, too, to the founder of Topics in Current Physics, Dr. H.K.V. Lotsch, for his encouragement.

Karlsruhe, March 1987 *W. Schommers · P. von Blanckenhagen*

# Contents

# List of Contributors

*Als-Nielsen, Jens*
Risø National Laboratory, DK-4000 Roskilde, Denmark

*Bauer, Ernst*
Physikalisches Institut, Technische Universität Clausthal und SFB 126,
Göttingen-Clausthal, Leibnizstraße 4,
D-3392 Clausthal-Z., Fed. Rep. of Germany

*Beijeren, Henk van*
Instituut voor Theoretische Fysica, Rijksuniversiteit Utrecht,
Princetonplein 5, P.O. Box 80.006, NL-3508 TA Utrecht, The Netherlands

*Benedek, Giorgio*
Dipartimento di Fisica dell' Università and Gruppo Nazionale di Struttura della
Materia del CNR, Via Celoria 16, I-20133 Milano, Italy

*Blanckenhagen, Peter von*
Institut für Nukleare Festkörperphysik,
Kernforschungszentrum Karlsruhe GmbH,
Postfach 3640, D-7500 Karlsruhe 1, Fed. Rep. of Germany

*Doyen, Gerold*
Institut für Physikalische Chemie der Universität München,
Theresienstraße 37, D-8000 München 2, Fed. Rep. of Germany

*Hanke, Werner*
Physikalisches Institut der Universität Würzburg,
D-8700 Würzburg, Fed. Rep. of Germany

*Miglio, Leonida*
Dipartimento di Fisica dell' Università and Gruppo Nazionale di Struttura della
Materia del CNR, Via Celoria 16, I-20133 Milano, Italy

*Muramatsu, Alejandro*
Max-Planck-Institut für Festkörperforschung, Heisenbergstraße 1,
D-7000 Stuttgart 80, Fed. Rep. of Germany, also at
Physikalisches Institut der Universität Würzburg,
D-8700 Würzburg, Fed. Rep. of Germany

*Nolden, Irmgard*
Instituut voor Theoretische Fysica, Rijksuniversiteit Utrecht,
Princetonplein 5, P.O. Box 80.006, NL-3508 TA Utrecht, The Netherlands

*Schommers, Wolfram*
Institut für Nukleare Festkörperphysik,
Kernforschungszentrum Karlsruhe GmbH,
Postfach 3640, D-7500 Karlsruhe 1, Fed. Rep. of Germany

# 1. Introduction: Surface Phenomena and Their Analysis by Scattering Experiments

P. von Blanckenhagen and W. Schommers

With 10 Figures

In this introductionary chapter a synopsis of several surface phenomena is presented and methods are discussed for analyzing surface phenomena by elastic and inelastic scattering experiments using different probes. It is related to the chapters in this volume and intended to supplement them. Inelastic atom and neutron scattering are considered in more detail because they are not treated in the other contributions to this volume and to the first volume [1.1]. Indications of future developments of the different methods will be compiled in Sect. 1.2 followed by comments on the temperature dependence of surface properties and a summary of the subsequent chapters.

## 1.1 Surface Phenomena

In the introduction of [1.1] some surface phenomena were already discussed as examples to show up the relevance of the structure and the dynamics of surfaces. Here a somewhat more general summary of surface phenomena is given and the role of scattering experiments is emphasized.

Many surface *structures* have been analyzed experimentally till now; bare surfaces as well as adsorbed layers have been studied by low-energy electron diffraction and other methods of analysis [1.1–5]. In contrast to this the experimental study of the surface *dynamics* is still in its infancy.

### 1.1.1 Reconstruction and Relaxation

Surface reconstruction, that is a change of the size or shape of the structural unit cell parallel to the surface with respect to the bulk unit cell, has been observed at metal surfaces [1.6] as well as at semiconductor surfaces [1.7–9]. Adsorbed species may induce or destroy the reconstruction. On the other hand, substrate reconstructions can profoundly influence the behavior of an adsorbed layer. It has been demonstrated that substrate reconstruction plays a critical role in epitaxial layer growth on semiconductor surfaces [1.7].

Surface relaxation, desribed by a change of the interlayer distances vertical to the surface compared to the interlayer distances in the bulk, is in general a multilayer phenomenon. This fact was found in 1980 by a theoretical consideration of metal surfaces [1.12,13]. After that, evidence for *multilayer relaxation* was obtained for several different surfaces by analysis of low-energy electron diffraction (LEED) data with adapted models [1.14] and by high-energy ion scattering experiments [1.15]. Recently, a new type of relaxation − a rippled relaxation − has resulted from a LEED analysis of the NiAl(110) surface where the Al sites are displaced approximately 0.22 Å above the Ni sites [1.16]. A remarkable temperature dependence of the relaxation of surface layers of a noble-gas model crystal was found by molecular-dynamics calculations using a realistic pair potential [1.17]. Such an effect corresponds to the enhanced thermal expansion of interlayer distances vertical to the surface as detected by LEED experiments on an Xe single crystal [1.18]. But the temperature dependence of surface relaxations has still to be proven experimentally.

Recent reviews on reconstructions and relaxations have been given in [1.6b,c] and remarks concerning the related theories, for example, in [1.9–13].

### 1.1.2 Phase Transitions

Phase transitions such as order-order or order-disorder transitions have been observed on bare surfaces as well as on adsorbed surface layers. They may be reversible as well as irreversible. They occur spontaneously by temperature variation or can be induced by small coverages of adsorbates [1.6,19,20]. In cases of irreversible phase transitions the presence of a small amount of certain adsorbates are important for the transitions between different surface structures. The following example will illustrate these findings [1.21]: The $(1 \times 5)$ structure on the Ir(100) surface corresponds to the thermodynamic equilibrium

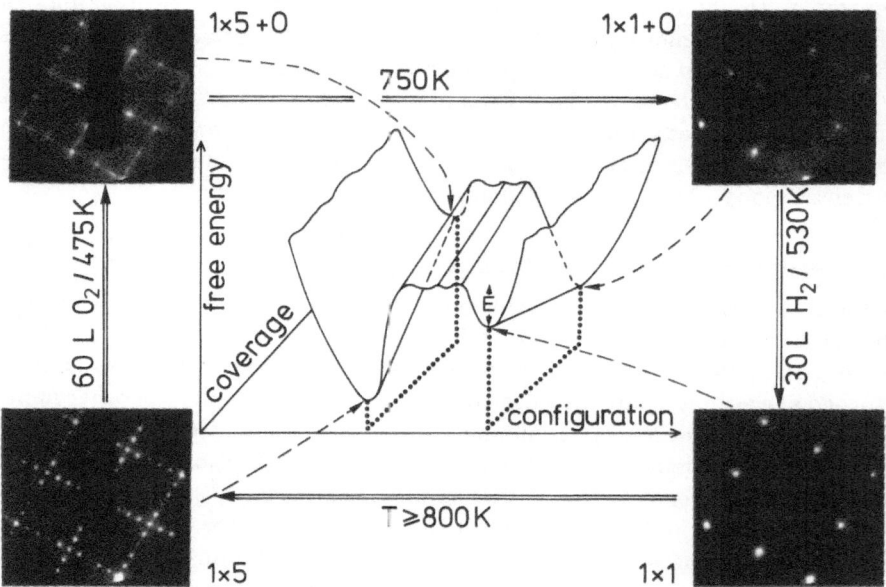

**Fig. 1.1.** Surface structure reconstruction on Ir(100): $(1 \times 1) \rightarrow (1 \times 5)$; Schematic behavior of the surface free energy as a function of the surface configuration and coverage (with oxygen). The respective LEED pattern are displayed for different minima [1.21]

of the bare surface (Fig. 1.1). By the adsorption of oxygen the reconstruction is removed and the metastable $(1 \times 1)$ structure is obtained on the bare surface after the oxygen has been removed by short-time exposure to hydrogen. Then an increase of the temperature causes the formation of the reconstructed $(1 \times 5)$ structure. The special treatment of the surface at certain temperatures and the related free energies are shown in Fig. 1.1. The transition from the metastable $(1 \times 1)$ to the stable $(1 \times 5)$ structure depends on temperature and time. By measurement of the time dependence of diffraction spot intensities at different temperatures the activation energy for the transition was obtained.

Besides reconstruction there are other phase transitions, which have been analyzed by scattering experiments. An extensive review on the theory of surface phase transitions as well as on experiments on bare surfaces and on chemically adsorbed layers is given in Chap. 4. Phase transitions in physisorbed layers are discussed in [1.22–25], and in [1.26] an example for order-disorder transitions on the surface of an alloy crystal has been reported. In Chap. 7 a review on theoretical and experimental work on the roughening transition is given and the effect of premelting is considered in Chap. 6. The theories on phase transitions have not only been proved by experiments but also by Monte Carlo simulations and molecular-dynamics calculations [1.27–31].

The characteristic temperature dependence of spatial and temporal fluctuations near the transition temperature can be studied by scattering experiments. The profiles of diffraction peaks have to be analyzed carefully in order to separate the Bragg intensity from diffuse scattering due to the critical fluctuations in space. Hence, a good instrumental resolution is needed for such studies [1.20]. The time-dependent critical fluctuations on surfaces have not yet been studied by inelastic scattering experiments.

A fundamental problem, the solution of which requires a detailed analysis of correlation functions by means of scattering experiments, consists of understanding the connection of the finite size behavior to first- and second-order phase transitions. A consequence of the finite-size effect in two-dimensional systems would be, for example, a shift of the maximum of thermodynamic functions like the susceptibility from the ideal transition temperature $T_c$ to temperatures below $T_c$, and a broadening of the temperature interval in which the change of the order parameter takes place [1.32]. Both phenomena have been observed, for example, for a reconstructive phase transition on W(001) [1.33].

### 1.1.3 Defects

Defects at surfaces and overlayers constitute a loss of structural order and symmetry; they may be associated with localized electronic states. Therefore defects may affect electronic and chemical properties, transport properties, thermodynamic and mechanical properties [1.34,35]. Defects can be analyzed by diffraction experiments and by imaging experiments. Whereas imaging, like

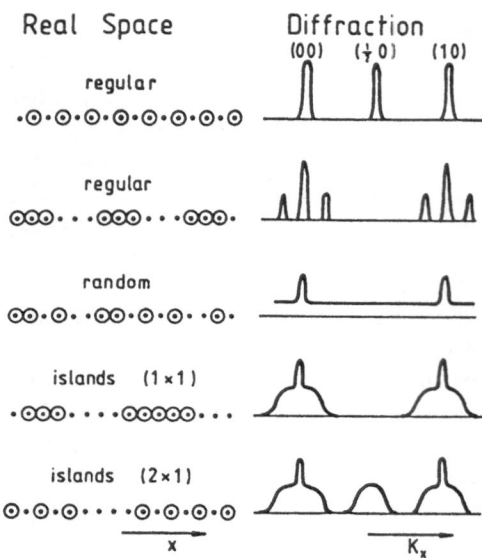

Real Space

regular

·◎·◎·◎·◎·◎·◎·◎·◎·◎

regular

◎◎◎···◎◎◎···◎◎◎·

random

◎◎·◎··◎◎·◎·◎··◎·

islands  (1×1)

·◎◎◎····◎◎◎◎◎···

islands  (2×1)

◎·◎·◎····◎·◎·◎·◎

x

Diffraction

(00)    (½0)    (10)

Kx

**Fig. 1.2.** Analysis of defects at surfaces by diffraction experiments: Schematic representation of diffraction patterns for different defect arrangements (half of the lattice sites are occupied) [1.35]. $K_x$ is the momentum transfer in x-direction

transmission electron microscopy, provides information on single features with a resolution down to 0.2 nm, the diffraction experiments yield average features with a maximum transfer width up to about 200 nm [1.35]. Different surface defects may have quite different effects on the diffraction pattern. The classification of surface defects and their influence on the diffraction spot profile is shown schematically in Fig. 1.2. These idealized profiles will be obtained after deconvolution with the response function of the instrument. New high-resolution instruments have been developed in order to identify and qualify surface defects by electron scattering. From the half width of spot profiles an average size of a periodic unit-like terrace or domain can be derived. However, such detailed studies of surface defects on real surfaces (using electrons, He atoms or x-rays as probes) are still in their infancy.

### 1.1.4 Phonons and Mean Vibrational Amplitudes

Surface phonons and their dispersion relations were predicted by model calculations for noble gases and ionic crystals as already in the early 1970s. But the first measurement of a phonon dispersion relation extending up to the zone boundary was not reported until 1981 [1.36a]. Branches of acoustical phonon dispersion relations have meanwhile been measured for several insulator and metal surfaces [1.36–43]. The experimental and theoretical results on ionic crystals are reviewed in Chap. 2. For LiF(100) a part of an optical phonon branch was also detected [1.42]. However, no complete set of dispersion relations has so far been determined experimentally. Also dispersion relations for adsorbed layers like oxygen on nickel(111) [1.41] and noble gases on graphite(0001) [1.39] were measured. A Kohn anomaly in the surface phonon dispersion relation for

Pt(111) has been observed recently [1.43]. Phonon lifetimes and phonon soften-ing related to continuous phase transitions are interesting problems for further research with improved instruments. The theory of surface phonons can de-scribe most of the experimental results ([1.40,41] and Chap. 2). However, the use of realistic potentials (e.g., pseudo potentials) for the determination of the dynamics of metal surfaces is just at the beginning [1.44]. Also the consid-eration of anharmonicities, which are obviously much more important at the surface than in the bulk, is restricted to a few molecular-dynamics calculations [1.1,45].

Mean vibrational amplitudes of atoms located at surfaces are generally up to about a factor two larger than of atoms located in the bulk [1.47]. The attenuation of scattered beams due to the vibrations of scattering atom is

$$I/I_0 \sim e^{-2W} \ , \tag{1.1}$$

here 2W is the Debye-Waller factor which is proportional to the thermal mean-square displacement $\langle u^2 \rangle$. In the limit of high temperatures $\langle u^2 \rangle$ can be ex-pressed by a surface Debye temperature $\theta_s$ as follows

$$\langle u^2 \rangle = \frac{3\hbar T_s}{M k_B \theta_s^2} \ , \tag{1.2}$$

where $T_s$ is the temperature of the surface. Examples for experimental and theoretical surface Debye temperatures have been compiled in Table 1.1. The reduction of the Debye temperature at the surface is approximately of the same

**Table 1.1.** Effective Debye temperature at the surface of some typical materials (all tem-peratures are given in units of [K])

| Surface | Method | $\Delta T_s$ | $\theta_{exp}^s$ | $\theta_{th}^s$ | $\theta_{exp}^b$ | Reference |
|---------|--------|--------------|------------------|-----------------|------------------|-----------|
| graphite | LEED | 100–600 | 325 | 424 | 530 | [1.54] |
| NaCl(100) | H-atom scattering | 300–700 | 240 | | 318 | [1.55] |
| Cu(100) | He-atom scattering | 70–773 | 270* | 229 | 350 | [1.56a] |
| Cu(111) | He-atom scattering | 70–773 | 360* | 237 | | |
| Cu(110)** | He-atom scattering | 70–423 | 350* | 236 | | |

$\Delta T_s$: Analyzed temperature range.
$\theta_{exp}^s$: Experimental surface Debye temperature.
$\theta_{exp}^b$: Experimental bulk Debye temperature.
$\theta_{th}^s$: Theoretical Debye temperature from model calculations.
* $\theta_{exp}^s$ is enhanced due to the Armand-effect (Sect. 1.2.6).
** Strong deviations from the linear temperature dependence (according to (1.2)) occur at $T_s > 423$ K, see also [1.56b]

order for the different types of crystals. It is reproduced qualitatively by model calculations. Usually lattice dynamics calculations are performed in the frame of the quasi-harmonic approximation [1.46]. But these model calculations do not consider the temperature-dependent effects due to the anharmonicity of the particle interactions at surfaces as indicated, for example, by the enhanced thermal expansion vertical to the (111) surface of Xe crystals detected by LEED experiments [1.18]. It would be very elucidating to perform similar temperature dependent studies also on other types of materials (e.g., metals).

### 1.1.5 Bound Resonant States

Bound resonant states appear as structure in the angular-dependent intensity distribution of scattered atoms when the atoms are captured at the surface due to the attractive atom surface potential [1.47]. The resonance condition for selective adsorption can be fulfilled in a purely elastic way. The "elastic resonance" occurs at selected wave vectors and affects the intensity of the diffraction peaks as well as the probability of inelastic transitions. The resonance condition can also be satisfied after creation or annihilation of a phonon.

Bound resonant states have been observed in the He-atom scattering on ionic and metallic crystals as well as on graphite crystals. From such data the parameters can be derived for the attractive part of the particle-surface interaction potential and models for adsorption processes can be proved [1.47,48,88,89].

### 1.1.6 Vibrations of Adsorbed Species

Vibrational spectroscopy of adsorbates is a powerful means for identifying the adsorbed species, their bound state, their concentration, and for detecting the change of these properties during chemical reactions on surfaces [1.49–53]. In connection with LEED experiments vibrational spectroscopy yields supplementary information about the surface structure and the properties of adsorption sites. If it is possible to observe all of the vibration modes, then the point group symmetry of the adsorption site can be established. A coverage-dependent vibrational frequency points to the occurrence of coupled vibrational behavior within the adsorbed layer, and a shift of intra-molecular vibrational frequencies from the gas-phase value is indicative of chemical bonding changes due to chemisorption at the surface. Vibrational frequencies can be measured with high resolution by different spectroscopic techniques, such as electron energy loss spectroscopy (EELS) and infrared-reflection-absorption spectroscopy (IRAS). From angular-dependent EELS data informations on the orientation of adsorbed molecules can be derived. Experiments with extreme energy resolution may yield lifetimes of bound states and thus information about energy dissipation processes.

### 1.1.7 Diffusion, Adsorption and Desorption

For several surface phenomena (such as epitaxial layer growth or catalysis) surface diffusion as well as adsorption and desorption processes are important.

Diffusion on metallic surfaces has been studied extensively at the microscopic scale by means of the field ion microscope (FIM). Elementary steps of diffusion processes were analyzed (for example, particle exchange and cluster formation [1.57]). Different methods have been used for the study of the temperature dependence of diffusion coefficients up to the melting point. For all metals studied as yet (e.g., for Cu, Ni, W and Mo) an enhanced surface diffusion was observed near the melting point (Chap. 3).

The migration of atoms at the surface by diffusion processes can be described by statistical mechanical models using space- and time-dependent correlation functions [1.1,58]. For the confirmation of such models the quasielastic scattering law has to be measured by inelastic scattering experiments (Sect. 1.2).

Adsorption and desorption processes have been studied among others by volumetric and spectroscopic methods as well as by scattering experiments [1.3,59,60]. Whereas the activation energies for adsorption and desorption can be derived from temperature-dependent rate experiments more detailed information on the adsorption-desorption mechanism can be obtained from the analysis of time-of-fight distributions of desorbed molecules [1.60].

By molecular-beam scattering studies of reactive scattering processes on surfaces insight into the kinetics of elementary surface reactions can be obtained [1.61]. A review on structural and dynamical aspects of adsorption and desorption, in particular on recent theoretical developments is given in Chap. 8.

### 1.1.8 Segregation

Segregation takes place on surfaces of alloys and results in a change of the relative concentration of the constituents with respect to the concentration in the bulk [1.62]. Due to this effect the phase diagram for alloy surfaces shows systematic deviations from the bulk phase diagram (Fig. 1.3). The segregation has been recognized as a very general phenomenon having important implications on catalysis, corrosion, and microelectronics. Grain-boundary segregation influences the strength of alloys [1.63,64]. The structure of segregated overlayers and the combined action of ordering and segregation deserve more attention in theory and experiment (e.g., in connection with scattering experiments) [1.63].

### 1.1.9 Interfaces and Superlattices

The structural and electronic properties of interfaces are of considerable interest for the design of microelectronic devices. For a detailed understanding of the behavior of metal-semiconductor contacts (Schottky barriers) the interplay between the crystallographic structure, the material composition and the electronic properties have to be studied [1.65–67].

Semiconductor superlattices can now be grown by thin-film techniques such as molecular beam epitaxy (MBE) or chemical vapor deposition (CVD) having designed electronic structure and nearly defect-free interfaces in a lattice-matched case such as $GaAs-Ga_{1-x}Al_xAs$. Several novel phenomena have been

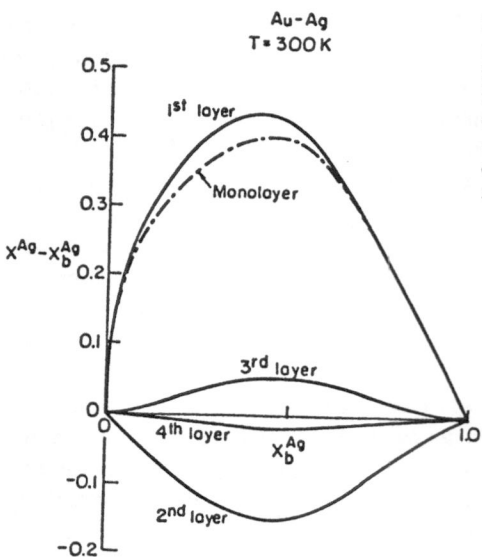

Au-Ag
T = 300 K

**Fig. 1.3.** Segregation on the (111) surface of an Ag-Au alloy: Surface excess of Ag as a function of the bulk composition is shown for the layers 1 to 4 as obtained from model calculations [1.62]. $X^{Ag}$ is the atom fraction of Ag at the surface, $X_b^{Ag}$ is the atom fraction of Ag in the bulk. The segregation within the monolayer approximation is also shown

found in such engineered structures such as, for example, multibarrier tunneling and the anomalous quantized Hall effect [1.68]. Also metal-metal superlattices have been grown with a regularity of a few tens of a nanometer by sputtering techniques. Artificial metal-metal layering can influence a number of physical properties like superconductivity and magnetism [1.69]. The lattice dynamics of superlattices has been considered in [1.70].

### 1.1.10 Magnetism and Electronic States

Surface magnetism has been studied extensively for model systems by Monte Carlo simulations [1.28]; for example, a magnetic surface reconstruction, i.e. the formation of an antiferromagnetic sturcture in the outermost layer of a ferromagnetic crystal, has been found by this method [1.71]. Experimental results on the magnetism on surfaces and at interfaces were obtained by magnetization measurements, Mössbauer spectroscopy, ferromagnetic resonance and LEED [1.72]. Recently, spin-polarized electron scattering has become a very promising method of studying magnetic surface properties [1.73–75]. First results on the Stoner continuum in Fe have been obtained by inelastic scattering of polarized electrons [1.75a].

Electronic surface states are known from band-structure calculations for crystals including free surfaces [1.76,77]. These surface states became accessible experimentally with the development of angular-resolved photoelectron spectroscopy; this method was improved dramatically after synchrotron radiation sources became available to surface science [1.78,79]. With the development of the inverse photoemission spectroscopy, the electronic states in the energy range between the vacuum level and the Fermi level are also accessible [1.80].

## 1.2 Scattering Methods for Surface Analysis

Most of the surface phenomena mentioned in Sect. 1.1 are related to structural and dynamical properties of surfaces. These properties can be analyzed by scattering experiments using different probes, such as electrons, He atoms, neutrons, x-rays or ions. The probes are complementary to each other; each has its own area of application. Typical properties of these probes are given in Table 1.2 (energy-wavelength relations, penetration depths, surface sensitivities, etc.). Topographic and microscopy methods which allow the analysis of local features of surfaces such as scanning tunnel electron microscopy (STM) [1.81, 84b], field ion microscopy (FIM) [1.57, 98] and high-resolution electron microscopy [1.84, 99] are not considered. The comparison does not take into account methods which are used mainly in chemical analysis (e.g., secondary ion-mass spectroscopy (SIMS) [1.100], and light spectroscopy as applied in the measurement of vibration frequencies of adatoms [1.84] and in the analysis of chemical reactions [1.101] taking place on surfaces). In Table 1.2 key references are also given to the appropriate literature describing methods of scattering experiments and their application.

**Table 1.2.** Properties of different probes for surface analysis by scattering experiments

| Properties | Electrons | He atoms | Neutrons | X-rays | Ions |
|---|---|---|---|---|---|
| | | | Probes | | |
| Application: | | | | | |
| elastic scattering | × | × | × | × | × |
| inelastic scattering | × | × | × | | |
| Wavelength-energy relation ($\lambda[\text{Å}]$, E[eV]) | $\lambda = \frac{12.3}{\sqrt{E}}$ | $\lambda = \frac{0.143}{\sqrt{E}}$ | $\lambda = \frac{0.286}{\sqrt{E}}$ | $\lambda = \frac{12398}{E}$ | |
| Surface sensitivity [monolayers] | $10^{-3}$–$10^{-4}$ | $10^{-3}$ | $10^{-3}$ | $10^{-3}$ | $10^{-2}$ |
| Typical sample surface area [mm$^2$] | 5 | 5 | $(10^6$–$10^9)^*$ | $10(10^6)^*$ | 1 |
| Analyzing depth [number of layers] | 3–5 | 1 | $\geq 1^{**}$ | $\geq 1$ | 1–10 |
| Transfer width [nm] | 20 | 20 | 100 | 1000 | |
| Energy resolution [meV] | 5 | 0.3 | 0.1 | – | – |
| Sample: (s: single crystal; e: exfoliated graphite substrate) | s | s | e | e,s | s |
| Maximum ambient pressure at the sample [mbar] | $10^{-4}$ | $10^{-4}$ | $10^3$ | $10^3$ | $10^{-6}$ |
| References | [1.50,52, 53,81–87] | [1.36,37,47, 81,88–90] | [1.91–93, 106] | [1.94,117] | [1.84, 95–97] |

\* exfoliated substrate
\*\* adsorbed layers on exfoliated substrates or powders

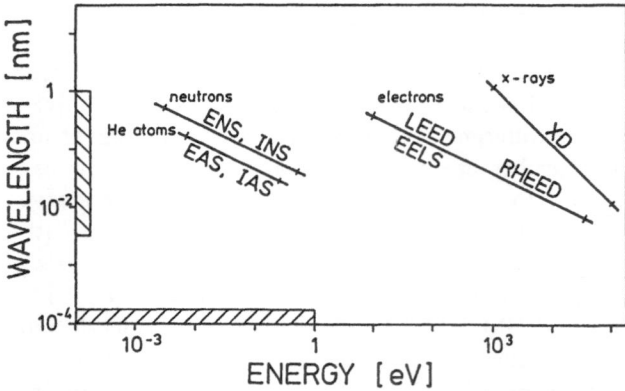

**Fig. 1.4.** Wavelength as a function of energy for different probes as used in scattering experiments. The parameter ranges useful for surface analysis are indicated. The regions for typical distances and excitation energies at surfaces are hatched on the axes. (XD: x-ray diffraction; LEED: low-energy electron diffraction; RHEED: reflection high-energy electron diffraction; EELS: electron energy loss spectroscopy; ENS: elastic neutron scattering; INS: inelastic neutron scattering, EAS: elastic He-atom scattering; IAS: inelastic He-atom scattering)

In the following we will discuss the applicability of different probes to scattering experiments on surfaces. For structure analysis by diffraction the wavelength of the probe must be of the order of the interatomic distances, and the analysis of surface dynamics requires probes with energies of the order of the excitation energies or at least a spectrometer resolution of such value. In Fig. 1.4 useful ranges of wavelengths and of probe energies are indicated. All probes presented can be used for structure analysis and inelastic studies but x-rays can be used only for structure analysis.

Due to their relatively large mass thermal He atoms and neutrons have a rather low velocity. Hence, their energy can be sorted out by time-of-flight measurements. It depends on the problem under investigation whether the time-of-flight spectrometer or the crystal spectrometer is better suited for inelastic scattering experiments. The relatively large momentum of thermal He atoms and neutrons as well as of electrons makes the exploration of excitations up to large wave vectors (of the order of the reciprocal lattice vector) possible.

## 1.2.1 Electron Scattering

Electrons are the most versatile probes in surfaces analysis so far. Their wavelength can be adapted to the interatomic distances. In addition, the high energy resolution realized in modern spectrometers allows spectroscopic studies to be made from the meV range, corresponding to phonon excitation energies, up to the eV range, corresponding to vibrational energies of adsorbed molecules. A serious drawback in the application of low-energy electrons is their strong interaction with the target atoms so that the scattered intensity is strongly contaminated by multiply scattered electrons. On the other hand, the strong interaction is the reason for the relatively short penetration depth which is for 100 eV electrons only of the order of 0.5 nm [Ref. 1.1, Fig. 7.3].

Two methods are used in the surface-structure analysis with electrons. These are LEED (mostly applied with normal incidence of 25–500 eV electrons) [1.81–85] and reflection high-energy electron diffraction (RHEED) (using 5–15 keV electrons with grazing incidence) [1.85, 86].

Only recently it has become possible to measure the phonon dispersion relations by electron energy loss spectroscopy (EELS) [1.41]. This technique is now being extended to inelastic scattering of spin polarized electrons [1.75]. Typical results from LEED are shown in Fig. 1.5 together with vibrational spectra measured by EELS for CO overlayers on Ni(111) and Pt(111) surfaces. Despite the similar intralayer arrangement of CO molecules the vibrational spectra are rather different due to the different positioning of the CO molecules with respect to the substrate atoms on both crystals.

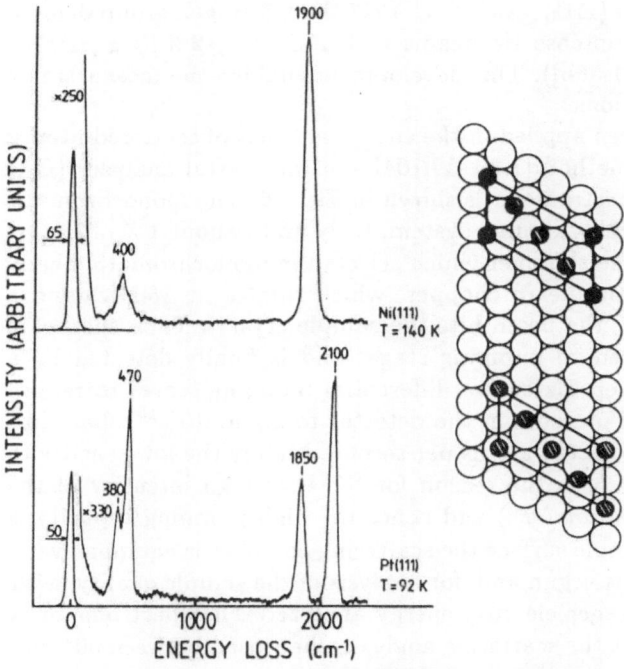

**Fig. 1.5.** Electron energy loss spectra of Ni(111) and Pt(111) surfaces, each covered with half a monolayer of CO which is ordered in a c(4 × 2) structure, as derived from LEED experiments. On the nickel surface the vibration spectrum indicates only a single CO species on a site of high symmetry. The only possibility of positioning the two-dimensional CO lattice on the surface which is consistent with the single type of adsorption site is to place all CO molecules into twofold bridges. By similar reasoning, half of the CO molecules must occupy on-top sites on the Pt(111) surfaces as shown in the picture [1.50]

## 1.2.2 Scattering of He Atoms

The wavelengths and the energies of thermal He beams are appropriate for the study of the structure and the dynamics of surfaces. Due to the large diameter of these probes they interact mainly with atoms belonging to the outermost layer. Since their energy is relatively low ($\lesssim 0.1 \, \text{eV}$) the scattering is entirely non-destructive. Although factorization of the scattering cross-section cannot be performed as completely as for neutrons (e.g., the surface corrugation leads to an intensity modulation) He-atom scattering can be used, for example, for the measurement of the scattering law of adsorbed atoms in order to analyze surface diffusion. This has been shown by model calculations in [1.102]. While He atoms have been used for the analysis of ordered structures for more than 10 years, only now they became also a tool for the study of disordered surfaces by monitoring of very low coverages of admolecules and allowing the investigation of the migration of adsorbed species [1.103].

The early work on inelastic scattering of He atoms was hampered by insufficient energy resolution ($\Delta E_0 \gtrsim 10\,\%$). In 1977 the Göttingen group demonstrated the production of intense He beams with $\Delta E_0/E_0 \gtrsim 2\,\%$ (e.g., $\Delta E_0 \sim$ 0.4 meV at $E_0 = 20 \, \text{meV}$ [1.36b]). This development enabled the measurement of phonon dispersion relations.

Two methods have been applied in the energy analysis of scattered atoms: the time-of-flight (TOF) method [1.36–39,104] and the crystal analyzer (CA) method [1.105]. A TOF spectrometer is shown in Fig. 1.6. The monochromatic beam is produced by a nozzle-skimmer system. Only up to about 1 % of the He atoms pass the orifice of the skimmer which selects the monochromatic beam. After passing the slit of the beam chopper, which rotates at 150 Hz, and a differential pumping stage, the beam hits the sample crystal. Then the beam passes through four differential pumping stages and is finally detected by a magnetic mass spectrometer. Extensive differential pumping serves to reduce the He partial background pressure in the detector to about $10^{-15}$ mbar, corresponding to about 100 detected atoms per second. Mainly the low sensitivity of the detector ($10^{-3} - 10^{-4}$) is the reason for the very high intensity of the incoming beam ($6 \times 10^{19}$ atoms/sr s) and hence the high pumping capacity is needed. To prepare the sample surface the scattering chamber is equipped with a target heater, an ion sputter gun and, for analysis of the sample quality, with a LEED system and an Auger electron energy analyzer. The spectrometer is designed in such a way that the scattering angle can be varied between 50° and 200°, but for most experiments it was fixed at 90°.

Typical TOF spectra measured with a spectrometer for He atoms scattered on an LiF(100) surface are shown in Fig. 1.7. The peaks 1, 4 and 6 are attributed to surface phonons (Rayleigh mode). Peak 5 is due to bulk phonons and the peaks 2 and 3 are due to incoherent scattering. From the parameters (energy and momentum transfer) of such peaks the phonon dispersion relation is obtained. Several of such data sets are discussed in comparison with results from model calculations in [1.40] and in Chap. 2.

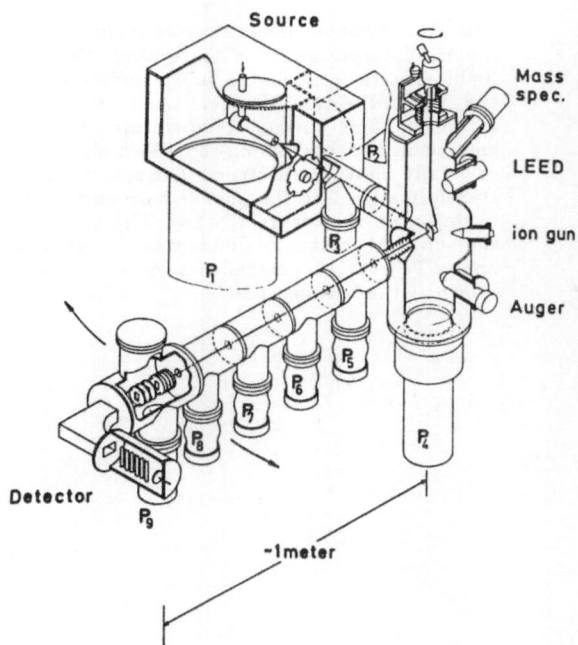

Source

Mass spec.

LEED

ion gun

Auger

Detector

$P_9$

~1meter

**Fig. 1.6.** Time-of-flight spectrometer for He-atom scattering experiments [1.36d]: The He-atom beam is produced in the source chamber, passes the chopper chamber and hits the target in a chamber which is equipped with instruments for preparation and analysis of surfaces (ion sputter gun, Auger electron spectrometer, LEED and mass spectrometer). The scattered atoms reach the detector chamber after they have passed four differential pumping stages

The first atomic beam spectrometer with a crystal analyzer was built by the Ottawa group, which used an LiF crystal as energy analyzer for the scattered atoms [1.105]. But due to its design the full capacity of the crystal spectrometer cannot be used as in the case neutron spectroscopy as yet (with the free choice of constant wave vector and constant energy scans). Whereas in a TOF spectrometer the relative energy resolution of the analyzer is given by the ratio of the time uncertainties (pulse width, time-of-flight through the sensitive detector volume and path differences due to the finite size of the sample surface) to the total flight time the energy resolution of the crystal analyzer $\Delta E_A$ is given by angular uncertainties in the scattering plane and by the mosaic width of the analyzer crystal. For both types of spectrometers the energy width of the incident beam $\Delta E_0$ has to be considered in order to get the final energy resolution $\Delta E$. For the CA spectrometer described above $\Delta E$ is about 1 meV at $E_0 = 22.6$ meV.

### 1.2.3 Neutron Scattering
Neutrons are available with appropriate wavelengths and energies for the study of the structure and dynamics. Multiple scattering is negligible in most applications due to the weak interaction of neutrons with the nuclei of the sample. The nuclear interaction between the neutron and the nucleus has the form of a $\delta$-function and hence the absolute determination of the scattering cross-section is straightforward.

13

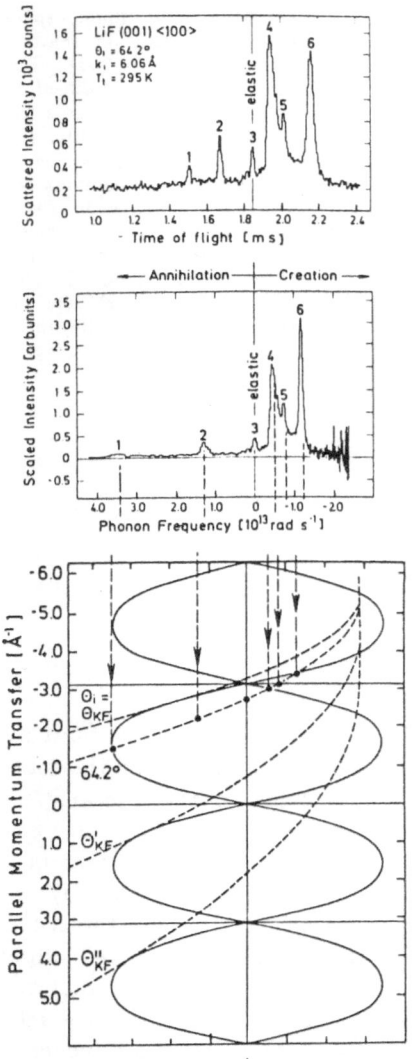

**Fig. 1.7.** Phonon inelastic scattering of He atoms: Interpretation of a typical time-of-flight spectra from an LiF(001) surface along the [100] direction. Incident angle $\theta_i$, wave factor $k_i$ and surface temperature $T_t$ are entered in the diagram on-top which shows an originally measurd spectrum. The spectra after transformation into the phonon frequency scale is shown on the diagram below. The traces of time-of-flight spectra for different final scattering angles $\theta_{KF}$ in the extended zone diagram is shown at the bottom [1.104]

The poor surface sensitivity of neutrons and the relatively low flux of the available neutron sources limits the application of neutrons in surface science so far to samples with very large surface areas (adsorbed layers on exfoliated graphite and pure powders).

Neutrons interact with matter in two ways: (1) they interact with the nuclei of atoms with a scattering amplitude which varies from nucleus to nucleus and is independent from the momentum and energy transfer; (2) neutrons have a magnetic moment and, therefore, they interact also with magnetic fields as, for example, caused by atomic electrons. These interactions make the neutrons

14

a useful tool for the study of the structure and dynamics of adsorbed layers including magnetic structures and excitations.

Only exploratory surface scattering experiments could be performed on single crystals [1.106] and superlattices as yet [1.93]. But several different adsorbed layers on exfoliated graphite or on metal powders and surfaces of bare powders have been studied with neutrons. In these experiments substrate and bulk data, respectively, have to be subtracted in order to get the scattered intensity due to the surface.

Adsorbed layers were investigated in diffraction experiments, by incoherent quasi-elastic scattering and by inelastic scattering [1.84b, 91–93, 107]. From the line width of quasi-elastically scattered neutrons the diffusion constant of $CH_4$ adsorbed on exfoliated graphite has been derived as a function of coverage $\theta$ and temperature [1.108]. For systems with continuous diffusion the macroscopic diffusion equation is valid and the so-called incoherent scattering law $S_s(\mathbf{Q}, \omega)$ becomes a Lorentzian

$$S_s(\mathbf{Q}, \omega) = \frac{DQ^2}{\pi[(DQ^2)^2 + \omega^2]} \, , \qquad (1.3)$$

$\hbar\mathbf{Q}$ and $\hbar\omega$ denote, respectively, the momentum and energy transfers in the collision between the probe and the scattering system. From the width of Lorentzian ($\Delta E = 2\hbar DQ^2$) the diffusion coefficient D can be determined. Typical Lorentzian scattering laws at T = 80 K and for a $CH_4$ coverage of $\theta = 0.9$ for different momentum transfers Q are presented in Fig. 1.8a. The shapes of the lines are well described by Lorentzians, (1.3), and their widths are proportional to $Q^2$ as shown for different $\theta$ in Fig. 1.8b. At $\theta = 0.9$ the diffusion constant is $2.2 \times 10^{-5}\,\text{cm}^2/\text{s}$. This value, which is typical for liquids, means that under the conditions in question the adsorbed $CH_4$ molecules form a two-dimensional liquid.

In connection with (1.3) let us still note the following: In general, we have to distinguish between the *coherent* scattering law $S(\mathbf{Q}, \omega)$ and the *incoherent* scattering law $S_s(\mathbf{Q}, \omega)$. $S(\mathbf{Q}, \omega)$ is given in terms of a correlation function $G(\mathbf{r}, t)$ (formulated in space and time) by [1.109]

$$S(\mathbf{Q}, \omega) = \frac{1}{2\pi} \int d\mathbf{r}\, dt \exp[i(\mathbf{Q} \cdot \mathbf{r} - \omega t)] G(\mathbf{r}, t) \, . \qquad (1.4)$$

$S_s(\mathbf{Q}, \omega)$ is connected to a correlation function $G_s(\mathbf{r}, t)$ by

$$S_s(\mathbf{Q}, \omega) = \frac{1}{2\pi} \int d\mathbf{r}\, dt \exp[i(\mathbf{Q} \cdot \mathbf{r} - \omega t)] G_s(\mathbf{r}, t) \, . \qquad (1.5)$$

In the classical limit, the G functions have the following meaning: $G(\mathbf{r}, t)$ represents the probability of finding a particle at position $\mathbf{r}$ at time t if there was *any* particle at the origin $\mathbf{r} = 0$ at t = 0. On the other hand, $G_s(\mathbf{r}, t)$ represents the probability of finding a particle at position $\mathbf{r}$ at time t if the

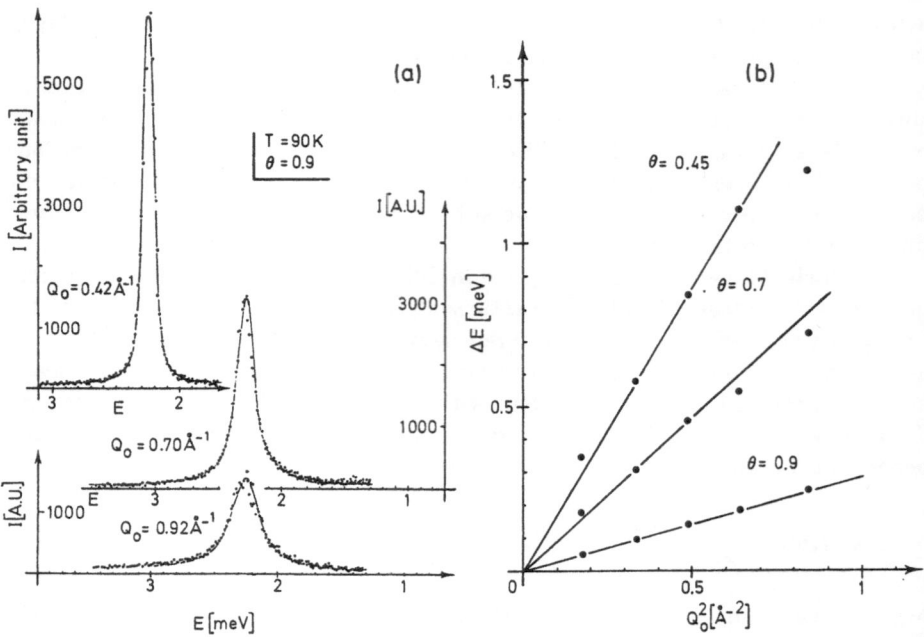

**Fig. 1.8a,b.** Surface diffusion studied by neutron scattering: **(a)** Incoherent quasi-elastic neutron scattering spectra (difference of the spectra from $CH_4$-covered and bare-exfoliated graphite) obtained at $T = 90\,K$ and coverage of $\theta = 0.9$ for different momentum transfers $Q_0$. **(b)** Full width at half maximum of the quasi-elastic spectra as a function of the square of the momentum transfer $Q_0$ for different coverages [1.108]

*same* particle was at $r = 0$ at $t = 0$. In other words, $G(r, t)$ describes the *collective* behavior of the system and $G_s(r, t)$ the *single-particle* behavior.

In the classical limit (1.5) may be expressed by

$$S_s(Q, \omega) = \frac{1}{2\pi} \int dt \exp(-i\omega t) \exp\left[-\frac{Q^2}{6}\langle r^2(t)\rangle\right][1 + 0(Q^4)] \;, \qquad (1.6)$$

where $\langle r^2(t)\rangle$ is the mean-square displacement as a function of time. The neglect of the $Q^4$ and higher-order terms correspond to the Gaussian approximation for $G_s$. From (1.6) one obtains the following relation

$$\frac{\omega^2 S_s(Q, \omega)}{Q^2} = \frac{k_B T}{m} f(\omega) + \left(\frac{Q k_B T}{2m}\right)^2 \int_0^\infty d\omega' f(\omega') f(\omega - \omega') + 0(Q^4) \;, (1.7)$$

where $f(\omega)$ is the Fourier transform of the velocity autocorrelation function. In the limit $/Q/\to 0$ the expression $\omega^2 S_s(Q, \omega)/Q^2$ is directly proportional to the frequency spectrum $f(\omega)$. In the case of a harmonic solid $f(\omega)$ is the frequency spectrum of the normal modes, the phonon density of states. It should be emphasized that in the limit $\omega \to 0$ the frequency spectrum $f(\omega)$ is directly proportional to the diffusion coefficient, see (6.65).

### 1.2.4 X-Ray Scattering

X-rays have wavelengths comparable to the interatomic distances but their energies are in the 10 to 100 keV region and up to now the energy resolution of x-ray spectrometers was insufficient for inelastic scattering experiments to study the surface dynamics. Diffraction experiments with a very high momentum resolution can be performed if a synchrotron radiation source is available. Several high-resolution experiments have been performed in order to study phase transitions in two-dimensional systems [1.117]. X-ray diffraction experiments on bare single crystal surfaces became possible after the grazing incidence diffraction method had been introduced successfully ([1.94] and Chap. 5).

### 1.2.5 Scattering of Ions

Ion scattering is a powerful real space technique in surface crystallography. It is based on the effects of *channeling, shadowing* and *blocking*. Also vibrational amplitudes can be measured by the application of these methods [1.95,96]. Whereas high-energy ion scattering or Rutherford backscattering spectroscopy, (RBS, ion energy ~1 MeV) determines surface atomic positions relative to subsurface positions, low-energy ion scattering spectroscopy (ISS, ion energy ~1 keV) determines the relative positions of surface atoms and from these the lateral surface structure.

The difference between ISS and RBS is depicted in Fig. 1.9. RBS is characterized by a thin shadow cone (radius ~0.1 Å); atoms D and B are shadowed by atoms C and A. On the other hand, ISS is characterized by a thick shadow cone (radius 1 Å); atom C in the first layer shadows atom E in the same layer and atom D is invisible in the surface sensitive ISS. From that it is clear that RBS is a tool for analyzing interatomic distances perpendicular to the surface and for the analysis of interfaces. ISS, using low-energy alkali-ions (whose neutralization probability is very small in contrast to noble gas ions), became a

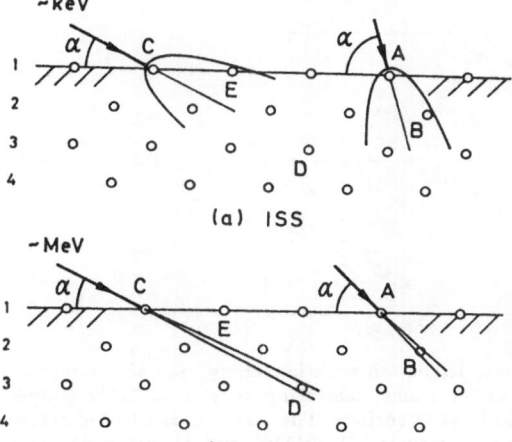

Fig. 1.9a,b. Comparison of shadow cones for low- and high-energy ion scattering for surface structure analysis [1.97]: (a) ISS – low energy ion scattering spectroscopy (ion energy ~1 keV); (b) RBS – Rutherford backscattering (ion energy ~1 MeV)

17

useful tool in the analysis of surface reconstructions. An example of the application of ISS with 2000 eV Na$^+$ ions is given in Fig. 1.10; it shows intensity distributions obtained for the unreconstructed Ni(110)-(1 × 1) and the reconstructed Ni(110)-(2 × 1) surface [1.111]. The two dominating peak positions in Fig. 1.10a are explained in the lower insert. The second peak in Fig. 1.10a is due to the atoms of the second layer. After the reconstruction took place (by adding less than 1 mbar s oxygen) the pattern of the Ni(110)-(2 × 1) structure was observed by LEED and the ion intensity distribution, as shown in Fig. 1.10b, was measured. Clearly, the first slope is shifted to smaller angles

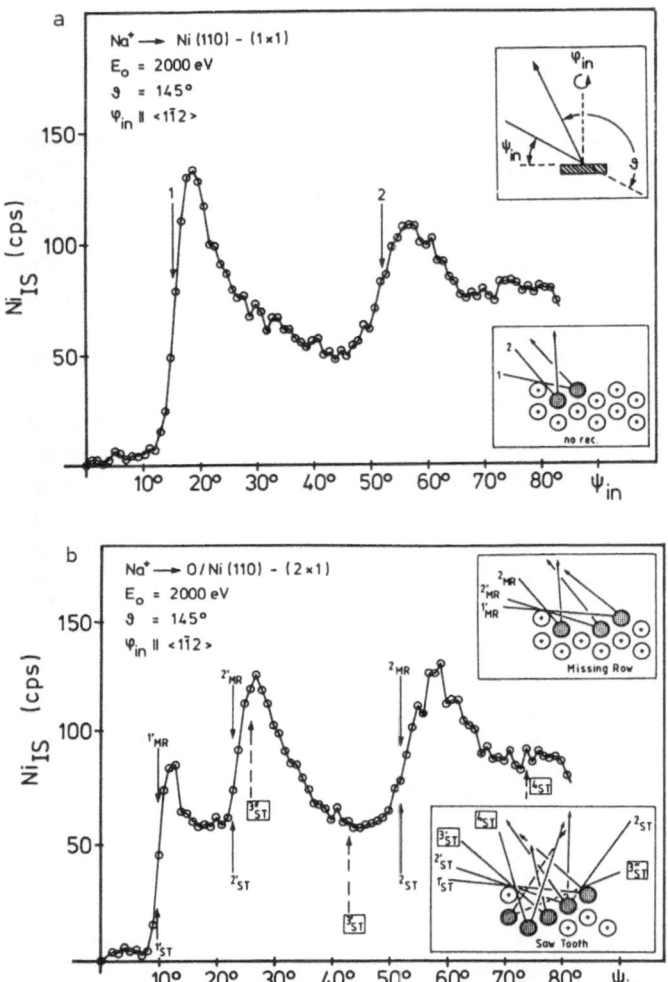

**Fig. 1.10a,b.** Determination of surface reconstruction with low energy ion scattering: **(a)** Intensity pattern of Na$^+$ backscattered at Ni atoms; scattering angle $\delta = 145°$; incident plane in the [1$\bar{1}$2] azimuth of the Ni(110)-(1 × 1) surface. The insert shows the scattering geometry. **(b)** Na$^+$ backscattering at the reconstructed O/Ni(110)-(2 × 1)-surface [1.111]

$\psi_{in}$ and a new peak appears at $\psi_{in} = 24°$. From these data it can be concluded that the saw-tooth model is not true and the Ni(110)-(2 × 1) surface structure is well described by the missing row model.

## 1.2.6 Comparison of Different Scattering Methods

The methods for scattering experiments on surfaces are not yet fully developed and additional probes may be used for surface analysis.

As electron scattering suffers from strong multiple scattering, low-energy positrons have been tested for use in diffraction experiments [1.112]. By change of sign of the charge from $e^-$ to $e^+$ the exchange term in the interaction potential can be eliminated and elastic scattering is weaker for positrons than for electrons. However, the inelastic mean free path is shorter for positrons and, consequently, multiple scattering is less important for positrons. But the positron sources are relatively complicated and, hence, this method has not been fully developed as yet.

Due to strong multiple scattering the evaluation of electron scattering data is very complicated. Only for scattering on noble gases and other crystals with large unit cells [1.83a] can the kinematic approximation and the factorization of the cross-section into scattering amplitude and scattering law be applied. Both advantages are given in the case of x-ray and neutrons, however, these probes are not very surface sensitive and large bulk contributions have to be subtracted from the scattering data. By the application of the grazing incident diffraction this problem could be overcome in several recent x-ray experiments ([1.94] and Chap. 5). Atoms are very surface sensitive and prove the structure and dynamics of the outermost layer, but the probe-target interaction is more complicated, e.g. due to the size of the probes and, hence, generally the scattering cross-section cannot be factorized as for neutron and x-ray scattering.

Atoms and electrons are the only probes which at present are useful in surface phonon analysis of bare surfaces. For atoms the energy resolution is by more than a factor of ten better than for electrons. Therefore, these probes are particularly suited for the study of quasi-elastic scattering as, for example, caused by surface diffusion and critical fluctuations.

The application of the He-atom scattering is restricted at large momentum transfers and at large energy transfers by two effects [1.47]:

1) The *Armand effect* considers that the assumption of single atom-atom scattering is not justified owing to the finite size of the incident atom. The correlation between displacements of neighboring surface atoms have to be taken into account. This effect is important at short phonon wavelengths and leads to an exponential decrease of the inelastic scattering cross-section with increasing momentum transfer [1.113]. He atoms are scattered by the charge density due to the surface atoms at an appreciable distance above the surface. The corrugation induced by optical modes and modes polarized parallel to the surface is relatively small at this distance. Hence, He atoms are less sensitive than electrons for this modes and for phonons near the zone boundary.

2) The *Levi effect* considers the role of the collision time on the inelastic scattering probability. The cross-section decreases with increasing excitation energy if the collision time $t_c$ becomes large relative to the mean vibration period $t_0$ because the forces exerted on the particle through the vibrating atom start to cancel [1.113]. Principally this reduction of the cross-section could be avoided by using higher incoming energies. However, then multiphonon scattering will cause a large background. This problem may be the reason why, up to now, high-frequency adsorbate modes could not be detected by He-atom scattering.

By use of the full capability of crystal analyzers in atomic beam spectrometers it would be possible to optimize the scan direction in the $\hbar\omega(q)$ area in analogy to neutron crystal spectroscopy. In TOF spectroscopy the trace of the spectra in the $\hbar\omega(q)$ plane changes with scattering angle but cannot be varied independently as desirable (Fig. 1.7). In electron scattering experiments (EELS) with presently available spectrometers the trace is nearly perpendicular to the q axis.

For electron scattering the "adiabatic approximation" holds in the energy range in question, i.e. $t_c \ll t_0$. Electron energy loss spectroscopy as used for the measurement of vibrational frequencies is based on dipole scattering. Large differential scattering cross sections are provided by inelastic scattering associated with the adsorbed species. However, dipole scattering is restricted to small momentum transfers parallel to the surface and can therefore not be applied to the analysis of phonon dispersion relations. For these experiments the so-called "impact scattering" is employed, where the electrons are scattered from the localized core potentials. The differential scattering cross-section for inelastic impact scattering processes is significantly smaller than that for dipole scattering. Furthermore, the impact scattering is strongly energy dependent [1.50]. As in low-energy electron diffraction multiple scattering contributes also considerably to the inelastically scattered intensity. Only improvements of the experimental techniques and optimization of the incoming energy enabled the first results on phonon dispersion relations by electron scattering experiments [1.41].

By a detailed study of the shape of the diffraction spots (using electrons, He atoms or x-rays as probes) surface defects and their kinetics can be analyzed. But such experiments require a high instrumental resolution. The width of the instrumental resolution function depends on the instrumental parameters like energy width and angular width of the scattered beam. It corresponds to a spatial resolution which would be achieved in the case of a perfect crystal surface. In atom and electron scattering experiments the transfer width is typical 10–20 nm [1.35,115]. The best values were obtained with special electron diffractometers [1.34,35,85].

## 1.2.7 Conclusion and Outlook

In future the analysis of the temperature dependence of structural and dynamical surface properties will need more attention as already indicated in the preceding sections. The study of phase transitions on surfaces requires detailed temperature-dependent inelastic scattering experiments. Despite the enormous progress in the development of methods for the analysis of the structure and dynamics of surfaces there is still a large potential for further improvements. In the following an outlook is given for possible future developments of some methods for surface studies by the scattering of different probes.

*Future Developments of Electron Energy Loss Spectroscopy:* The investigation of surface phonon dispersions can currently be performed with only two experimental methods: EELS and the scattering of helium atoms of thermal energies. As discussed above, the He-atom technique is limited to low excitation energies. The principal advantage of electron scattering, on the other hand, is the short acquisition time, especially near the zone boundary (typically 20 min compared to 1–3 hours); until now this advantage has not been fully exploited. Two ways of improvement have so far been explored. One way is to use a channel plate instead of a channeltron as a detector. In principle, one can record simultaneously in this way the whole energy loss spectrum, thus gaining at least a factor five to ten in measuring time [1.114]. This improvement is particularly suited for out-of-specular studies where the low dynamic range of the channel plate is no disadvantage since the count rates in the elastic channel and in the different inelastic channels are comparable and small.

Another approach has been opened up recently by the development of spectrometers based on the principle of dispersion compensation. In this way, two orders of magnitude have been gained for the in-specular scattering current, allowing to record energy loss spectra in fractions of a second. With such a time resolution new applications are opened up to EELS in the study of reactive surfaces or, may be, of precursor states of the adsorption process. It remains an open issue whether it will be possible to obtain a sufficiently good energy resolution with these spectrometers so that they can be used also for the study of phonon spectra.

For fast data aquisition and the investigation of dipole active modes EELS has to face the competition of the Fourier-transform infrared spectroscopy (IRS), which has also undergone a rapid development in the last years [1.51,53]. Further improvements are expected from the use of synchrotron radiation. Tunable laser sources, which permit time resolved experiments with a time resolution in the picosecond range, are also under development [1.101]. These spectroscopic techniques are limited to vanishing momentum transfers; the results are therefore complementary to those obtained with off-specular EELS. For this reason, all the characteristic future potential of electron scattering, i.e. not only due to its charge but also due to its momentum and spin respectively, will have to be exploited by EELS investigations in the near future. An example in this context is the observation of the Stoner excitation spectrum on Fe(110) by EELS with spin polarized electrons [1.75].

*Future Development of Atom Scattering:* The techniques for elastic atom scattering experiments are nearing maturity, although there is still potential for improvements. With regard to phenomena and materials to be studied, the range is immense, and hence the future for atom scattering looks very promising [1.89]. The bottleneck of an atomic beam apparatus is the low sensitivity of the detector for He atoms. An increase of the sensitivity would be desirable in order to reduce the measuring time and to increase the angular and energy resolution. The possibility for extended and directed scans in inelastic scattering experiments (e.g., constant $Q$ and constant $\omega$ scans) by the use of a crystal energy analyzer is not fully exploited as yet. With the crystal spectrometer the study of the $Q$ dependency of the cross section could be made more accessible. This would allow the separation of single and multiple scattering at large energy transfers as well as the independent measurement of correlation lengths and correlation times for collective excitations.

Unquestionably the resolution in both elastic and inelastic experiments will be improved in the future and enable new applications of the method. Structural [1.88] and kinetic properties of defects on surfaces may then be studied in more detail as well as space- and time-dependent particle correlations in diffusion processes. Phonon softening accompanied with second-order phase transitions should be directly observable. Atom scattering may also become a useful tool in the study of layer growth phenomena (Chap. 3) and of processes in physical and chemical adsorption [1.60,103].

The application of other atoms than He atoms as probes for the analysis of surfaces by scattering experiments may be favorable in special cases, for example, for H atoms the Armand effect is weaker than for He atoms and Ne atoms have an enhanced sensitivity towards details of the corrugation of metal surfaces [1.88c].

*Future Development of Neutron Scattering:* As already discussed above, neutrons are relatively surface insensitive. In the past they have mainly been applied to the study of adsorbed layers with relatively large scattering cross sections. In these cases the substrate contribution to the scattered intensity were separated by subtraction of the intensity obtained for the bare substrate. Such experiments will profit in the future by improvements of neutron sources and spectrometers. Higher flux and improved energy and momentum resolution will make possible new applications of this method such as the study of   self-correlation functions for surface diffusion in adsorbed layers and the investigation of critical fluctuations on surfaces. In favorable cases neutrons can also be applied to the study of interfaces; for example, by diffraction of polarized neutrons from metal-metal superlattices with one magnetic constituent [1.93]. Both theoretical and experimental exploratory studies have shown that the surface insensitivity of neutrons can be overcome at least in some cases by the application of the grazing incidence method [1.93,106,116]. This method was already successfully applied in x-ray diffraction experiments on single crystals [1.94]. The bulk contribution to the scattering intensity can be reduced by taking advantage of the total reflection which appears at the boundary to an optical thinner medium if the incident angle with respect to the surface is smaller than a critical angle. Under such conditions a layer of only about 5 nm thickness contributes to the scattering. This effect should be useful for surface studies with neutrons also since most substances have an index of refraction for thermal neutrons which is less than unity. Bragg diffraction under grazing incidence conditions has recently been studied on a Si crystal [1.106]. The results confirm theoretical predictions. Thus, in favourable cases the grazing incident method may be applicable to the study of the structure and dynamics of single crystal surfaces by neutron scattering.

*Future Development of X-Ray Scattering:* The number of x-ray diffraction studies of surface structures and related phase transitions are still quite limited. This will change drastically in the near future. Early x-ray diffraction experiments were utilized with rotating anode generators and exfoliated graphite substrates with surface areas of the order of $m^2$. Synchrotron x-ray sources provide high-intensity, high-directional beams which are optimally suited for such surface studies with extreme $Q$ resolution [1.118]. By the application of wiggler and undulator beam lines an intensity increase over the conventional storage rings by several orders of magnitude has already been achieved, and surface science will benefit from this progress. Considerable improvements of surface diffraction experiments using x-rays are enabled by the invention of the grazing incident method which uses the effect of total reflection in order to increase the surface sensitivity. By this method the x-ray study of structures of bare crystal surfaces became possible (Chap. 5) [1.94]. Also the melting and growth of adsorbed metal layers on a metallic substrate were studied [1.119] demonstrating the effectiveness of the grazing incidence x-ray diffraction for such investigations. Altogether, there is a large potential for the application of x-rays in surface science, without the consideration of the application of the surface extended x-ray absorption fine structure (SEXAFS [1.81,120]), the near edge x-ray absorption fine structure (NEXAFS [1.120]) and x-ray standing-wave (XSW [1.121]) techniques to study intramolecular and chemisorption bonds. Finally, the development of x-ray monochromators for ultra-high energy resolution ($\lesssim 10$ meV [1.122]) will make inelastic x-ray scattering experiments on surfaces feasible.

*Future Development of Ion Scattering:* Real-space analysis of surface structures with ions has recently been improved considerably by two innovations: Firstly, the development of an electrostatic energy analyzer with a position sensitive detector for RBS, which allows a depth resolution down to 0.5 nm [1.96b,123]; this means an improvement of about a factor of ten. Secondly, the invention of low energy alkali ion scattering by which the lateral structure of the outermost layer and surface reconstructions can be studied with high resolution

[1.97,111]. These improvements make very detailed studies of the structures on surfaces, on interfaces and of chemical reactions (taking place on surfaces and interfaces) possible.

## 1.3 On the Temperature-Dependence of Surface Properties

As already outlined in [Ref. 1.1, Chap. 1] surface phenomena are strongly influenced by the structure and the dynamics, and the structural and dynamical properties at the surface are obviously much more sensitive to temperature variations than in the bulk of the crystal. In particular, it has been discussed in [Ref. 1.1, Chap. 6] that the *harmonic approximation* is limited to low temperatures. As a rule of thumb, in the case of *noble-gas* crystals to below one sixth of the melting temperature $T_m$ (as compared to $1/3\ T_m$ for the bulk). In the case of krypton ($T_m = 116\,K$) anharmonicities are already effective at 7 K (Chap. 6, Fig. 6.22). It has been shown experimentally [1.18] that the *thermal expansion* (due to the anharmonicities) at a xenon surface is 4–5 times larger than in the bulk of the xenon crystal. Recent experimental data indicate that the *dispersion curves* for an aluminium surface vary strongly with temperature [1.124]. Thus, also in the case of *metals* temperature effects (e.g., anharmonicities) have to be taken into consideration.

### 1.3.1 Pair Potentials at the Surface

*Molecular-dynamics* calculations are important in studying classical many-particle systems with strong anharmonicities, since anharmonicity is treated without approximation. In [Ref. 1.1, Chap. 6] the molecular-dynamics method was introduced and discussed in connection with surface problems. The main problem in such calculations is to find realistic pair potentials. The only group of materials for which the pair potentials are well-known is that of noble-gas solids. In contrast to metals, the potential functions of noble-gas solids do not depend on the temperature, and they are the same at the surface and in the bulk of the crystal. This is not the case for *metals*: At free metal surfaces the local background electron density $n(r)$ is different from its bulk-value, and because the pair potential between the metal ions depends critically on the electron density the pair potential at the surface is different from that in the bulk of the crystal. Because surface properties are sensitive to variations in temperature also the ion-ion potential at metal-surfaces should vary with temperature and, therefore, $n(r)$ at the surface should be sensitive to variations in temperature. The determination of $n(r)$ and other electronic properties of metals is a quantum mechanical problem. Let us consider a system of N electrons of mass m and charge $-e$ which are distributed around positive charges of magnitude $Z_1e$, $Z_2e$, ... at positions $R_1$, $R_2$, .... The Hamiltonian H is given in the second-quantized representation by

$$H = T + U + V \ , \quad \text{where} \tag{1.8}$$

$$T = \frac{\hbar^2}{2m} \int \nabla \psi^+(\mathbf{r}) \nabla \psi(\mathbf{r}) d\mathbf{r} \tag{1.9}$$

is the kinetic energy operator. $\psi(\mathbf{r})$ and $\psi^+(\mathbf{r})$ are annihilation and creation operators. U describes the mutual repulsion, both among the electrons and the ions:

$$U = \frac{1}{2} \iint d\mathbf{r} \, d\mathbf{r}' \frac{e^2}{|\mathbf{r} - \mathbf{r}'|} \psi^+(\mathbf{r}) \psi^+(\mathbf{r}') \psi(\mathbf{r}') \psi(\mathbf{r}) + \frac{1}{2} \sum_{i \neq j} \frac{Z_i Z_j e^2}{|\mathbf{R}_i - \mathbf{R}_j|} \ . \tag{1.10}$$

V in (1.8) is given by

$$V = \int d\mathbf{r} \, v(\mathbf{r}) \psi^+(\mathbf{r}) \psi(\mathbf{r}) \ , \tag{1.11}$$

where $v(\mathbf{r})$ is an external potential arising from the nuclei. The electron density $n(\mathbf{r})$ is given in the ground-state $|\phi\rangle$ by the expection value

$$n(\mathbf{r}) = \langle \phi | \psi^+(\mathbf{r}) \psi(\mathbf{r}) | \phi \rangle \ . \tag{1.12}$$

As already mentioned above, the pair potential between the ions depends critically on $n(\mathbf{r})$. At the surface $n(\mathbf{r})$ is different from its bulk-value and because surface properties are very sensitive to variations in temperatue, $n(\mathbf{r})$ at the surface should be determined *as a function of temperature*.

### 1.3.2 Density-Functional Formalism for Nonzero Temperatures

For *zero temperatures Kohn* and co-workers [1.125,126] showed that for a non-degenerate $|\phi\rangle$, $v(\mathbf{r})$ is (to within a constant) a unique functional of $n(\mathbf{r})$ and the correct $n(\mathbf{r})$ minimizes the ground-state energy

$$E = \langle \phi | H | \phi \rangle \tag{1.13}$$

which is a unique functional of $n(\mathbf{r})$

$$E = E\{n(\mathbf{r})\} \ . \tag{1.14}$$

With (1.12) we have

$$E\{n(\mathbf{r})\} = \int v(\mathbf{r}) n(\mathbf{r}) d\mathbf{r} + F\{n(\mathbf{r})\} \ , \quad \text{where} \tag{1.15}$$

$$F\{n(\mathbf{r})\} = \langle \phi | (T + U) | \phi \rangle \ . \tag{1.16}$$

Equation (1.15) defines a *variational principle* for the ground-state energy of an electron gas in an external potential $v(r)$, in which $n(r)$ is the variable function. However, from this density-functional formalism we can only extract *zero-temperature properties* for the electron gas. It is, however, possible to extend this formalism to *nonzero* temperatures, and this has been done by *Merwin* [1.127].

In the case of *nonzero temperatures* the equilibrium electron density in a grand canonical ensemble is given by

$$n(r) = \mathrm{Tr}\left\{\varrho_0 \psi^+(r)\psi(r)\right\} , \tag{1.17}$$

where $\varrho_0$ is the grand canonical density matrix

$$\varrho_0 = \frac{\exp\left(-\frac{H-\mu\hat{N}}{k_B T}\right)}{\mathrm{Tr}\left\{\exp\left(-\frac{H-\mu\hat{N}}{k_B T}\right)\right\}} , \tag{1.18}$$

$\mu$ is the chemical potential and $\hat{N}$ is the particle number operator. *Merwin* showed [1.127] that in a grand canonical ensemble for a given temperature T, chemical potential $\mu$ and external potential $v(r)$ the quantity

$$\Omega\{n(r)\} = \int dr\, v(r)n(r) + F\{n(r)\} - \mu \int dr\, n(r) \tag{1.19}$$

with

$$F\{n(r)\} = \mathrm{Tr}\left\{\varrho_0(T + U + k_B T \ln \varrho_0)\right\} \tag{1.20}$$

is equal to the *grand potential*

$$\Omega = k_B T \ln \mathrm{Tr}\left\{\exp\left(-\frac{H - \mu\hat{N}}{k_B T}\right)\right\} \tag{1.21}$$

when $n(r)$, expressed by (1.17), is the correct equilibrium electron density. In particular, *Merwin* showed that the correct density *minimizes* (1.19) over all density functions that can be associated with some external potential $v(r)$.

### 1.3.3 Conclusions

No numerical results based on the *Merwin* formalism seem as yet to be available. It could therefore be of great value to do such nonzero-temperature calculations. With the resulting $n(r)$ phenomenological pair potentials for the ions could be pictured, and on the basis of such potentials the *structure* and *dynamics* of metal surfaces could be determined realistically. Such results could be confirmed by molecular dynamics calculations and temperature dependent elastic and inelastic scattering experiments.

# 1.4 Book Outline

The purpose of the preceding discussion was to supplement both Volume I (Topics Curr. Phys., Vol. 41) and Volume II (this volume) of *Structure and Dynamics of Surfaces*. In the first volume, some typical examples of the influence of structure and dynamics on surface phenomena were given and six up-to-date reviews were presented: experimental methods for determining surface structures and surface corrugations (K.H. Rieder); high-resolution electron microscopy of surfaces (L.D. Marks); surface channeling and its application to surface structures and location of adsorbates (C. Varelas); dynamical surface properties in the harmonic approximation (J.E. Black); molecular dynamics and the study of anharmonic surface effects (W. Schommers); surface phonon dispersion of surface and adsorbate layers (M. Rocca, H. Ibach, S. Lehwald, and T.S. Rahman).

In this Topics volume like in the first volume, the chapters include the *background* of the subject, *typical results,* and in most cases also trends of *future developments.*

In Chap. 2, L. Miglio and G. Benedek discuss the standard theory of the Green's function method for solid-state problems and the authors apply it to the case of surface vibrations. The role of electrons is discussed within the framework of the breathing-shell model, which is used to calculate surface phonon dispersion curves and surface projected phonon densities for NaCl-type crystals.

Surface diffusion and layer growth are reviewed in Chap. 3 by P. von Blanckenhagen. Particular emphasis is placed on scattering experiments using x-rays, neutrons, electrons and He atoms as probes. Some typical experimental results on surface diffusion are presented, e.g., the anisotropy and the temperature dependence of surface diffusion on metal surfaces. A second part of Chap. 3 deals with epitaxial growth and the analysis of growth modes.

In Chap. 4, E. Bauer discusses phase transitions in two dimensions with emphasis on systems in which the lateral interactions between the atoms are weak compared to their interaction with the supporting substrate. Experimental techniques as well as the theoretical foundations for the understanding of the phenomena are presented. The results for specific systems obtained to date, both theoretically and experimentally, are reviewed.

With the development of very intense and highly collimated x-ray beams, so-called synchrotron radiation, from electron storage rings operating at electron energies in the GeV region new possibilities, including x-ray surface diffraction, have appeared in x-ray physics. In Chap. 5, J. Als-Nielsen discusses this technique and its rapidly growing application to solid and liquid surfaces.

The structure of liquid surfaces is described in terms of distribution and correlation functions. These functions are introduced in Chap. 6 by W. Schommers on the basis of statistical mechanics. Simple models are discussed for the single-particle distribution function, which is important in connection with the determination of the density variation through the liquid-vapor interface. Models are also given for the two-particle distribution function, which were often used in the determination of thermodynamic functions. A second part of Chap. 6 deals with the effect of premelting.

In Chap. 7, H. van Beijeren and I. Nolden give an introduction to the roughening transition. Besides the roughening transition the relation between a Wulff plot (a polar plot of surface tension vs surface orientation) and the equilibrium crystal shape is discussed, in general, and various possibilities for equilibrium crystal shapes are mentioned. A qualitative explanation is given why vicinal areas bordering to facets should exhibit a universal non-analytic shape. Finally van Beijeren and Nolden discuss aspects of surface melting vs roughening and some possibilities for roughening transitions that are not of Kosterlitz-Thouless type.

Experimental and theoretical aspects of adsorption and desorption are reviewed in Chap. 8 by G. Doyen. The emphasis lies on general ideas and interconnections between various approaches. Some recent promising developments are presented.

In Chap. 9, A. Muramatsu and W. Hanke discuss surface elementary excitations and response functions on a microscopic scale. Many-body effects of random-phase and electron-hole type are included. Various applications of the theory to the special case of an ideal Si(111) surface are reviewed.

*Acknowledgements.* The authors would like to thank H. Ibach and M. Rocca for the contribution "Future developments of electron energy loss spectroscopy".

# References

1.1   W. Schommers, P. von Blanckenhagen (eds.): *Structure and Dynamics of Surfaces I*, Topics Curr. Phys., Vol. 41 (Springer, Berlin, Heidelberg 1986)
1.2   G.A. Samorjai: *Chemistry in Two-Dimensions* (Cornell Univ. Press, Ithaca 1981)
      G. Ertl, J. Küppers: *Low Energy Electrons and Surface Chemistry* (Verlag Chemie, Weinheim 1985)
1.3   M.A. van Hove, G.A. Samorjai: "Adsorbed Monolayers on Solids", in *Structure and Bonding*, Vol. 38 (Springer, Berlin, Heidelberg 1979)
1.4   R. Vanselow, R. Howe (eds.): *Chemistry and Physics of Solid Surfaces IV*, Springer Ser. Chem. Phys., Vol. 20 (Springer, Berlin, Heidelberg 1982)
1.5   R. Vanselow, R. Howe (eds.): *Chemistry and Physics of Solid Surfaces V*, Springer Ser. Chem. Phys., Vol. 35 (Springer, Berlin, Heidelberg 1984)
1.6   P.J. Estrup: "Reconstructions of Metal-Surfaces" in [Ref. 1.5, p. 205]
      J.E. Ingelsfield: Prog. Surf. Sci. **20**, 105 (1985)
      K. Müller: Ber. Bunsenges. Phys. Chem. **90**, 184 (1986)
1.7   H.-J. Gossmann, L.C. Feldmann: "Molecular Beam Epitaxy and Reconstructed Surfaces", Appl. Phys. A**38**, 271–279 (1985)
1.8   A. Kahn: in Surf. Sci. Rpt. **3**, 193–300 (1983)
1.9   D.H. Chadi: Vacuum **33**, 613–619 (1983)
      C. Mailhiot, C.B. Duke, D.J. Chadi: Surf. Sci. **149**, 366 (1985)
1.10  M.A. van Hove, S.Y. Tong: *The Structure of Surfaces*, Springer Ser. Surf. Sci., Vol. 2 (Springer, Berlin, Heidelberg 1985)
1.11  M.L. Cohen: Theory of Surface Reconstruction, in [Ref. 1.10, pp. 4–11]
1.12  U. Landman, R.N. Hill, M. Mostoller: Phys. Rev. B**21**, 448 (1980)

1.13   U. Landman, R.N. Barnett, C.L. Cleveland, R.H. Rast: J. Vac. Sci. Technol. A3, 1574 (1985)
1.14   D.L. Adam, L.E. Petersen, C.S. Sorensen: J. Phys. 18, 1753 (1985)
1.15   Y. Kuck, C.L. Feldmann: Phys. Rev. B30, 5811 (1984)
1.16   H.L. Davis, J.R. Nooman: Phys. Rev. Lett. 54, 566 (1985)
1.17   W. Schommers, P. von Blanckenhagen: Vacuum 33, 733 (1983)
1.18   A. Ignatiev, T.N. Rhodin: Phys. Rev. B8, 893 (1973)
1.19   F. Nizzoli, K.H. Rieder, R.F. Willis: *Dynamical Phenomena at Surfaces, Interfaces and Superlattices*, Springer Ser. Surf. Sci., Vol. 3 (Springer, Berlin, Heidelberg 1985)
1.20   R.F. Willis: Surface Reconstructions Phase Transformations, in [Ref. 1.19, pp. 126–147]
       R.F. Willis: Ber. Bunsenges. Phys. Chemie 90, 190 (1986)
1.21   K. Heinz, G. Schmidt, L. Hammer, K. Müller: Phys. Rev. B32, 6214 (1985)
1.22   S.C. Fain: Low Energy Electron Diffraction Studies of Physically Adsorbed Films, in [Ref. 1.4, pp. 203–216]
1.23   M.B. Webb, E.R. Moog: LEED Studies of Physisorbed Noble Gases on Metals and Interatoms Interaction, in [Ref. 1.10, pp. 397–403]
1.24   R.J. Birgeneau, P.M. Horn, D.E. Moncton: "Phase Transitions in Two Dimensional Systems with Competing Interacctions", in [Ref. 1.10, pp. 404–412]
       R.J. Birgeneau, P.M. Horn: Science 232, 329 (1986)
1.25   B. Bak: Phase Transitions on Surfaces, in [Ref. 1.5, pp. 317–337]
       H.J. Kreuzer, Z.W. Gortel: *Physisorption Kinetics*, Springer Ser. Surf. Sci., Vol. 1 (Springer, Berlin, Heidelberg 1986)
1.26   E.G. McRae, R.A. Malic: Surf. Sci. 148, 551 (1984)
1.27   H.W. Diehl, S. Dietrich: Adv. Solid State Physics 25, 39–52 (1985)
1.28   K. Binder: "Critical Behavior at Surfaces", in *Phase Transitions and Critical Phenomena*, Vol. 8, ed. by C. Domb, J.L. Lebowitz (Academic, London 1983) pp. 1–144
       A. Sadig, K. Binder: J. Stat. Phys. 35, 517–585 (1984)
1.29   L.D. Roelofs: Monte Carlo Simulations of Chemisorbed Overlayers, in [Ref. 1.4, pp. 219–249]
1.30   F.F. Abraham: J. Vac. Sci. Technol. B2, 534 (1984)
1.31   J.M. Bowman (ed.): *Molecular Collision Dynamics*, Topics Curr. Phys., Vol. 33 (Springer, Berlin, Heidelberg 1983)
       B.C. Eu: *Semiclassical Theories of Molecular Scattering*, Springer Ser. Chem. Phys. Vol. 26 (Springer, Berlin, Heidelberg 1984)
1.32   P. Keban: Finite Size Effects, Surface Steps and Phase Transitions, in [Ref. 1.5, pp. 339–363]
1.33   J. Wendeken, G.C. Wang: Phys. Rev. B32, 7542 (1985)
1.34   M.G. Lagally: Structural Defects in Surfaces and Overlayers, in [Ref. 1.4, pp. 281–313]
1.35   M. Henzler: Defects at Surfaces, in [Ref. 1.19, pp. 14–34]
1.36   G. Brusdeylins, R.B. Doak, J.R. Toennies: Phys. Rev. Lett. 46, 437 (1981)
       J.P. Toennies: Phonons Interaction in Atom Scattering from Surfaces, in *Dynamics of Gas-Surface-Interaction*, ed. by G. Benedek, U. Valbusa, Springer Ser. Chem. Phys., Vol. 21 (Springer, Berlin, Heidelberg 1982) p. 208
       J.P. Toennies: J. Vac. Sci. Technol. A2, 1055 (1984)
       U. Harten, J.P. Toennies, Ch. Wöll: In [Ref. 1.19, p. 117]
1.37   B. Feuerbacher: Inelastic Scattering from Metal Surfaces, in *Dynamics of Gas-Surface-Interaction*, ed. by B. Benedek, U. Valbusa, Springer Ser. Chem. Phys., Vol. 21 (Springer, Berlin, Heidelberg 1982)
       B. Feuerbacher: Inelastic Molecular Beam Scattering from Surfaces, in [Ref. 1.49a, pp. 91–110]
1.38   M. Cates, D.R. Miller: Phys. Rev. B28, 3615 (1983)
1.39   K.D. Gibson, S.J. Sibener: Phys. Rev. Lett. 55, 1514 (1985)
1.40   J.E. Black: Dynamical Surface Properties in the Harmonic Approximation, in [Ref. 1.1, Chap. 5]
1.41   M. Rocca, H. Ibach, S. Lehwald, T.S. Rahmann: Surface Phonon Dispersion of Surface and Adsorbate Layers, in [Ref. 1.1, Chap. 7]
1.42   G. Brusdeylins, R. Rechsteiner, J.G. Skofronick, J.P. Toennies: Phys. Rev. Lett. 54, 466 (1985)

1.43  U. Harten, J.P. Toennies, Ch. Wöll, G. Zang: Phys. Rev. Lett. **55**, 2308 (1985)
1.44  C. Calandra, Cattellani: Pseudopotentials and Dynamical Properties of Metallic Surfaces, in [Ref. 1.19, pp. 80–91]
1.45  W. Schommers: Phys. Rev. B**32**, 6845 (1985)
1.46  M.G. Lagally: Surface Vibrations, in *Surface Physics of Materials*, Vol. II, ed. by J.M. Blakeley (Academic, New York 1975) pp. 419–473
      C.S. Jananthi, E. Tosatti, L. Pietronero: Phys. Rev. B**31**, 3456 (1985)
      C.S. Jananthi, E. Tosatti, A. Fasolino: Phys. Rev. B**31**, 470 (1985)
1.47  G. Boato, P. Cantini: Adv. Electronics and Electron Physics, **60**, 95–160 (1983)
1.48  V. Celli, D. Eichenauer, A. Kaufhold, J.P. Toennies: "Pairwise additive semi ab initio potential for the elastic scattering of the atoms from the LiF(001) crystal surface", J. Chem. Phys. **83**, 2504 (1985)
1.49  R.F. Willis (ed.): *Vibrational Spectroscopy of Adsorbates*, Springer Ser. Chem. Phys., Vol. 15 (Springer, Berlin. Heidelberg 1980)
      R. Candano. J.-M. Gilles, A.A Lucas (eds.): *Vibrations at Surfaces* (Plenum, New York 1982)
      A. Yoshimori, M. Tsukuda (eds.): *Dynamical Processes and Ordering on Solid Surfaces*, Springer Ser. Solid-State Sci., Vol. 59 (Springer, Berlin, Heidelberg 1985)
1.50  H. Ibach, D.L. Mills: *Electron Energy Loss Spectroscopy and Surface Vibrations* (Academic, San Diego 1982)
1.51  C.R. Brundle, H. Morawitz: *Vibration at Surfaces* (Elsevier, Amsterdam 1983)
      D.A. King, N.V. Richardson, S. Holloway (eds.): "Vibrations at Surfaces 1985", Surface Sci. **38**, (1986)
1.52  H. Ibach: Electron Energy Loss Spectroscopy of Surfaces and Adsorbates, in [Ref. 1.19, pp. 109–116]
1.53  W.H. Weinberg: Vibration at Overlayers, in [Ref. 1.84a, pp. 23–125]
1.54  N.J. Wu, V. Kumykov, A. Ignatiev: Surf. Sci. **163**, 51 (1985)
1.55  S. Ioannotta, G. Scoles, U. Valbusa: Surf. Sci. **161**, 411 (1985)
1.56  J. Lapujoulade, J. Perrau, A. Kara: Surf. Sci. **129**, 59 (1983) and references quoted therein
      J. Villain, D.R. Grempel, J. Lapujoulade: J. Phys. F**15**, 809 (1985)
1.57  G. Ehrlich: "Diffusion in Surface Layers", CRC Crit. Revs. in Solid State and Materials Sci. **10**, 391 (1982)
      G. Ehrlich: Diffusion and Interactions of Adatoms, in [Ref. 1.10, p. 375]
      Vu Thien Binh (ed.): *Surface Mobilities on Solid Materials – Fundamental Concepts and Applications* (Plenum, New York 1983)
      N. Ernst, G. Ehrlich: "Field Ion Microscopy", in *Microscopic Methods in Metals*, ed. by U. Gonser, Topics Curr. Phys., Vol. 40 (Springer, Berlin, Heidelberg 1986) Chap. 4
1.58  G.F. Mazenko: Statistical Mechanical Models and Surface Diffusion, in [Ref. 1.57c, p. 27]
1.59  A. Cassuto: "Thermal Desorption and Comprehension of Adsorption-Desorption Mechanism", in Proc. of the 9th Int. Vacuum Congress and the 5th Int. Conf. on Solid Surfaces, ed. by J.L. Segovia (ASEVA, Madrid 1983) p. 179
      R. Gomer (ed.): *Interaction at Metal Surfaces*, Topics Appl. Phys., Vol. 4 (Springer, Berlin, Heidelberg 1975)
      D. Menzel: Thermal Desorption, in [Ref. 1.4, pp. 389–406]
      H.J. Kreuzer: *Physisorption Kinetics*, Springer Ser. Surf. Sci., Vol. 1 (Springer, Berlin, Heidelberg 1986)
1.60  G. Comsa: "The Dynamical Parameters of Desorbing Molecules", in *Dynamics of Gas-Surface Interaction*, ed. by Benedek, U. Valbusa, Springer Ser. Chem. Phys., Vol. 21 (Springer, Berlin, Heidelberg 1982) pp. 117–127
      G. Comsa, R. David: Surf. Sci. Rpt. **5**, 145 (1985)
      C.R. Hery, C. Chapon, B. Muttaffschiev: Surf. Sci. **163**, 409 (1985)
1.61  M.P. D'Evelyn, R.J. Madix: Surf. Sci. Rpt. **3**, 413–495 (1984)
1.62  G.A. Samorjai: The Molecular Surface Science of Heterogeneous Catalysis: History and Perspective, in [Ref. 1.5, pp. 1–22]
1.63  T.M. Buck: "Segregation and Ordering at Alloy Surfaces Studied by Low Energy Ion Scattering", in [Ref. 1.4, pp. 435–464]
      H. Viefhaus, M. Rüsenberg: Surf. Sci. **159**, 1 (1985)

1.64  C.L. Briant: The Effects of Internal Surface Chemistry on Metallurgical Properties, in [Ref. 1.4, pp. 465–485]

1.65  R.S. Bauer (ed.): *Surfaces and Interfaces; Physics and Electronics* (North Holland, Amsterdam 1983)
G. Bauer, F. Kucker, H. Heinrich (eds.): *Two-Dimensional Systems: Physics and New Devices*, Springer Ser. Solid-State Sci., Vol. 67 (Springer, Berlin, Heidelberg 1986)

1.66  C.M. Wilmsen: *Physics and Chemistry of III–V Compound Semiconductors Interfaces* (Plenum, New York 1985)

1.67  G.W. Rubloff: Metal-Semiconductor Interfaces and Schottky Barriers, in [Ref. 1.19, pp. 220–243]

1.68  L. Esaki: Advances in Semiconductor Superlattices, Quantum Wells and Heterostructures, in [Ref. 1.19, pp. 48–59]

1.69  Ch.M. Falco: Metal-Metal Superlattices, in [Ref. 1.19, pp. 35–47]

1.70  G. Benedek, V. Velasco: Phonons at Interfaces and Superlattices, in [Ref. 1.19, pp. 66–79]

1.71  K. Binder, D.P.L. Landau: Surf. Sci. **151**, 409 (1985)

1.72  U. Gradmann: J. Magn. Magn. Mater. **6**, 173 (1977)

1.73  M. Campagna: J. Vac. Sci. Technol. A**3**, 1491 (1985)

1.74  H.C. Siegmann: Surface Magnetism by Spin Polarized Electrons, in [Ref. 1.19, pp. 306–315]

1.75  J. Kirschner: Phys. Rev. Lett. **55**, 973 (1985)
J. Kirschner: *Polarized Electrons at Surfaces*, Springer Tracts Mod. Phys., Vol. 106 (Springer, Berlin, Heidelberg 1985)
R. Fender (ed.): *Polarized Electrons in Surface Physics* (World Scientific, Singapore 1985)
J. Kessler: *Polarized Electrons*, 2nd ed., Spriner Ser. Atom. Plasm., Vol. 1 (Springer, Berlin, Heidelberg 1985)

1.76  J.E. Ingelsfield: "Electron at Surfaces", in *Chemical Physics of Solid Surfaces and Heterogeneous Catalysis*, Vol. 1, ed. by D.A. King and D.P. Woodruff, (Elsevier, Amsterdam 1981) pp. 183–363
J.E. Ingelsfield: Rep. Prog. Phys. **45**, 224–284 (1982)

1.77  J.A. Freemann: Electronic Structure of Surfaces, Interfaces and Superlattices, in [Ref. 1.19, pp. 162–175]
M.L. Cohen: Ann. Rev. Phys. Chem. **35**, 537 (1984)

1.78  F.J. Himpsel: Adv. Phys. **32**, 1–57 (1983)
F.J. Himpsel: Appl. Phys. A**38**, 205–212 (1985)

1.79  C.S. Fadley: Prog. Surf. Sci. **16**, pp. 275–388 (1984)
D.J. Dow, R.E. Allen, O.F. Sankey: Intrinsic and Extrinsic Surface Electronic States of Semiconductors, in [Ref. 1.5, pp. 483–500]
B. Feuerbacher, B. Fitton, R.F. Willis (eds.): *Photoemission and the Electronic Properties of Surfaces* (Wiley, New York 1978)
G.K. Wertheim: X-Ray Photoelectron Spectroscopy, in *Microscopic Methods in Metals*, ed. by U. Gonser, Topics Curr. Phys., Vol. 40 (Springer, Berlin, Heidelberg 1986) Chap. 7

1.80  V. Dose: Surf. Sci. Rep. **5**, 337 (1985)

1.81  H.K. Rieder: Experimental Methods for Determining Surface Structures and Surface Corrugations, in [Ref. 1.1, Chap. 2]

1.82  F. Jona: J. Phys. C**11**, 4271–4306 (1978)
F. Jona, J.A. Stroizier, Jr., P.M. Marcus: Determination of Surface Structures by LEED, in [Ref. 1.10, pp. 92–99]
P.M. Marcus, F. Jona (eds.): *Determination of Surface Structure by LEED* (Plenum, New York, 1984)
M.A. van Hove, S.Y. Tong: *Surface Crystallography by LEED*, Springer Ser. Chem. Phys., Vol. 2 (Springer, Berlin, Heidelberg 1979)

1.83  L.J. Clark: *Surface Crystallography – An Introduction to Low Energy Electron Diffraction* (Wiley, Chichester 1985)
M. van Hove, W.H. Weinberg, K.M. Chang: *Low-Energy Electron Diffraction*, Springer Ser. Surf. Sci., Vol. 6 (Springer, Berlin, Heidelberg 1986)

31

1.84   R.L. Park, M.G. Lagally (eds.): *Methods in Experimental Physics*, Vol. 22, "Solid State Physics: Surfaces" (Academic, Orlando 1985)
       R. Vanselow, R. Howe (eds.): *Chemistry and Physics of Solid Surfaces VI*, Springer Ser. Surf. Sci., Vol. 5 (Springer, Berlin, Heidelberg 1986)
1.85   M.G. Lagally: Diffraction Techniques, in [Ref. 1.84a, pp. 237–298]
1.86   P.K. Larsen: RHEED and Photoemission Studies of Semiconductors Grown in-situ by MBE, in [Ref. 1.19, pp. 237–298]
1.87   H. Ibach (ed.): *Electron Spectroscopy for Surface Analysis*, Topics Curr. Phys., Vol. 4 (Springer, Berlin, Heidelberg 1977)
1.88   T. Engel, K.H. Rieder: "Structural Studies of Surfaces", in *Springer Tracts Mod. Phys.*, Vol. 91 (Springer, Berlin, Heidelberg 1982) pp. 55
       W.A. Schlump, K.H. Rieder: Phys. Rev. Lett. **56**, 73 (1986)
       K.H. Rieder, M. Baumberger, W. Stocker: Phys. Rev. Lett. **55**, 390 (1985)
1.89   D.R. Frankl: Prog. Surf. Sci. **13**, 285–356 (1983)
       D.R. Frankl: Crit. Revs. Solid State and Mat. Sci. **10**, 411 (1982)
1.90   J. Lapujoulade, B. Salanon, D. Gorse: Surface Structure Analysis by Atomic Beam Diffraction, in [Ref. 1.19, pp. 176–16]
1.91   J.P. McTague, M. Nielsen, L. Passel: Crit. Revs. Solid State and Materials Sci. **7**, 135–155 (1979)
       M. Nielsen, J.P.T. McTague, L. Passel: Neutron Scattering of Physisorbed Monolayers on Graphite, in *Phase Transitions in Surface Films*, ed. by J.G. Dash, J. Ruvalds (Plenum, New York 1980) pp. 127–163
       R.K. Thomas: Prog. Solid State Chem. **14**, 1–93 (1982)
1.92   C.J. Wright: Neutron Scattering Studies in Vibrational Spectroscopy of Adsorbates, in [Ref. 1.5, pp. 111–124]
       L.J. Wright, C.M. Sayers: Rep. Progr. Phys. **46**, 773 (1983)
1.93   G.P. Felcher: Magnetism of Interfaces and Surfaces as Probed by Neutron Scattering, in [Ref. 1.19, pp. 316–326]
       M.B. Salomon, S. Sinha, J.J. Rhyne, J.E. Cunningham, R.W. Erwin, J. Borchers, C.P. Flynn: Phys. Rev. Lett. **56**, 259 (1986)
1.94   R. Feidenhans'l, J. Bohr, M. Nielsen, M. Toney, R.Z. Johnson, F. Grey: Adv. Solid State Phys. **25**, 545–554 (1985)
       B. Brenan: Surf. Sci. **152/153**, 1 (1985)
       I.K. Robinson: Surface Structure by X-Ray Diffraction, in [Ref. 1.10, p. 60]
       M. Nielsen: Z. Physik B**61**, 415 (1985)
1.95   C. Varelas: Surafce Channelling and its Application to Surface Structures and Location of Adsorbates, in [Ref. 1.1, Chap. 4]
1.96   C. Feldmann, J.W. Mayer, S.T. Picraux: *Materials Analysis by Ion Channelling, Submicron Crystallography* (Academic, New York 1982)
       J.F. van der Veen: Surf. Sci. Rep. **5**, 199 (1985)
1.97   M. Aono: Nucl. Instr. Meth. B**2**, 374 (1984)
1.98   J.A. Panitz: "High-Field Techniques", in [Ref. 1.84a, pp. 349]
1.99   L.D. Marks: High-Resolution Electron Microscopy of Surfaces, in [Ref. 1.1, Chap. 3]
1.100  W. Heiland, E. Taglauer: Ion Scattering and Secondary Ion Mass Spectrometry, in [Ref. 1.84a, pp. 299–348]
       H. Oechsner (ed.): *Thin Film and Depth Profile Analysis*, Topics Curr. Phys., Vol. 37 (Springer, Berlin, Heidelberg 1984)
1.101  R.R. Cavanagh, D.S. King: Laser Studies of Surface Chemical Reactions, in [Ref. 1.5, pp. 141–158]
       F.R. Aussenegg, A. Leitner, M.E. Lippitsch (eds.): *Surface Studies with Lasers*, Springer Ser. Chem. Phys., Vol. 33 (Springer, Berlin, Heidelberg 1983)
1.102  M. Heuer, T.M. Rice: Surf. Sci. **115**, L269 (1985)
       M. Heuer, T.M. Rice: Z. Phys. B**59**, 299 (1985)
1.103  G. Comsa, B. Polsema. Appl. Phys. A**38**. 153–160 (1985)
1.104  G. Brusdeylins, R.B. Doak, J.P. Toennies: Phys. Rev. B**27**, 3662 (1983)
1.105  B.F. Mason, B.R. Williams: Rev. Sci. Instrum. **49**, 897 (1978)
       B.F. Mason, B.R. Williams: J. Chem. Phys. **75**, 2199 (1981)
       B.F. Mason, B.R. Williams: Surf. Sci. **148**, L686 (1984)
1.106  A. Zeilinger, T.J. Beatty: Phys. Rev. B**27**, 7239 (1983)

1.107 R. Beame, J. Suzanne, J.P. Coulomb, A. Glachaut, G. Bomchil: Surf. Sci. **137**, L117 (1984)
1.108 J.P. Coulomb, M. Bienfait, P. Thorel: J. Physique Coll. **38**, C4–31 (1977)
1.109 L. van Hove: Phys. Rev. **95**, 249 (1954); **95**, 1347 (1958)
   P. Egelstaff: *Thermal Neutron Scattering* (Academic, London 1965)
1.110 P.A. Heiney, P.W. Stephens, R.J. Birgeneau, P.M. Horn, D.E. Moncton: Phys. Rev. **B28**, 6413–6434 (1983)
1.111 H. Niehus, G. Comsa: Surf. Sci. **151**, L171 (1985)
1.112 D.R. Cook, T.N. Horsky, P.E. Coleman: Appl. Phys. **A34**, 237 (1984)
   A.P. Mills, Jr., W.S. Crane: Phys. Rev. **B31**, 3988 (1985)
   W.E. Frieze, D.W. Gidley, K.G. Lynn: Phys. Rev. **B31**, 5628 (1985)
1.113 H. Ibach, T.S. Rahman: Surface Phonon Dispersion, in [Ref. 1.5, pp. 455–482]
   H. Ibach: Surface Phonon Spectroscopy, in Proc. 9th Int'l. Vacuum Congress and the 5th Int'l. Conf. on Solid Surfaces, ed. by J.L. de Segovia (ASEVA, Madrid 1983) p. 17
1.114 R. Franchy, H. Ibach: Surf. Sci. **155**, 15 (1985)
1.115 G. Comsa: Surf. Sci. **81**, 57 (1979)
1.116 G.H. Vineyard: Phys. Rev. **B26**, 4146 (1982)
1.117 S.E. Nagler, P.M. Horn, T.F. Rosenbaum, R.J. Birgeneau, M. Sutton, S.G.J. Mochrie, D.E. Moncton, R. Clarke: Phys. Rev. **B32**, 7373 (1985) and references therein
1.118 E.-E. Koch, D.E. Eastman, Y. Farge: Synchrotron Radiation – a Powerful Tool in Science, in *Handbook on Synchrotron Radiation*, Vol. 1A, ed. by E.-E. Koch (North Holland, Amsterdam 1983)
1.119 S. Brennan, P.H. Fuoss, P. Eisenberger: In [Ref. 1.10, p. 421]
1.120 J. Stöhr: "NEXAFS and SEXAFS Studies of Chemisorbed Molecules: Bonding, Structure and Chemical Transformation", in [Ref. 1.10, p. 140]
   P.H. Citrin: "Current Status and New Applications of SEXAFS": in [Ref. 1.10, pp. 149–155]
   J. Haase: Appl. Phys. **A38**, 181 (1985)
1.121 G. Materlik: Z. Phys. **B61**, 405 (1985)
1.122 B. Dorner, T.D. Benda, E. Burkel, J. Peisl: Adv. Solid State Phys. **25**, 685–688 (1985)
1.123 F.W. Saris: Nucl. Instr. Meth. **194**, 625 (1983)
1.124 J.P. Toennies: private communication
1.125 P. Hohenberg, W. Kohn: Phys. Rev. **136**, B864 (1964)
1.126 W. Kohn, L.J. Sham: Phys. Rev. **140**, A1133 (1965)
1.127 N.D. Merwin: Phys. Rev. **137**, A1441 (1965)

# Additional References with Titles

Anderson, S., Wilzen, L.: Resonant sticking at surfaces. Phys. Rev. Lett. **57**, 1603 (1986)
Amirar, A., Cardillo, M.J.: Electron-hole pair creation by atomic scattering at surfaces. Phys. Rev. Lett. **57**, 2299 (1986)
Baumberger, M., Stocker, W., Rieder, K.H.: Investigation of the selective population of hydrogen subsurface sites on Pd(110) using the diffraction and thermal-desorption spectroscopy. Appl. Phys. **A41**, 151 (1986)
Chen, S.P., Voter, A.F., Srolowitz, D.J.: Oscillatory surface relaxation in Ni, Al and their ordered alloys. Phys. Rev. Lett. **57**, 1308 (1986)
Erskine, J.L.: High resolution electron energy loss spectroscopy: Explored regions and the frontier. J. Vac. Sci. Technol. **4**, 1282 (1986)
Ertl, G.: Reactive transformation of surface structure. Ber. Bunsenges. Phys. Chem. **90**, 284 (1986)
Frenken, J.W.M., Hunssen, F., van der Veen, J.F.: Evidence for anomalous thermal expansion at a crystal surface. Phys. Rev. Lett. **56**, 734 (1987)
Hansma, P.K.: Scanning tunneling microscopy. J. Appl. Phys. **61**, R1 (1987)
Heinz, K., Müller, K., Popp, W., Lindner, H.: Measurement of diffuse LEED intensities. Surface Sci. **173**, 66 (1986)
Ibach, H., Lehwald, S.: Elastic diffuse and inelastic electron scattering from surfaces with disordered overlayers. Surface Sci. **176**, 629 (1986)

Kern, K., David, R., Palmer, R.L. Comsa, G.: Adsorbate induced Rayleigh phonon gap of p(2 × 2) of Pt(111). Phys. Rev. Lett. **56**, 2064 (1986)

Lahee, A.M., Manson, J.R., Toennies, J.P., Wöll, Ch.: Observation of interference oscillations in Helium scattering from single surface defects. Phys. Rev. Lett. **57**, 471 (1986)

Lapujoulade, J.: Molecular beam study of surface roughening transition. Surface Sci. **178**, 406 (1986)

Madey, T.E.: Electron stimulated desorption and its relation to molecular structure at surfaces. J. Vac. Sci. Technol. A**4**, 257 (1986)

Mayer, R., Zhang, Chun-Si, Lynn, K.G., Frieze, W.E.: Low-energy electron and positron diffraction measurements and analysis on Cu(100). Phys. Rev. B**35**, 3102 (1987)

Möller, J., Snowdon, K.J., Heiland, W., Niehus, H.: Low energy ion scattering from the Au(110) surface. Surface Sci. **178**, 475 (1986)

Müller, J.E., Wuttig, M., Ibach, H.: Adsorbate-induced surface stress: Phonon anomaly and reconstruction on Ni(001) surfaces. Phys. Rev. Lett. **56**, 1583 (1986)

Neuhaus, D., Joo, F., Feuerbacher, B.: Adsorbate-induced surface-phonon softening on Pt(111). Phys. Rev. Lett. **58**, 694 (1987)

Rau, C., Liu, C., Schmalzbauer, A., Xing, G.: Ferrogmagnetic order at (100) p(1 × 1) surfaces of bulk paramagnetic vanadium. Phys. Rev. **57**, 2311 (1986)

Scheithauer, U., Meyer, G., Henzler, M.: A new LEED instrument for quantitative spot profile analysis. Surface Sci. **178**, 441 (1986)

Smith, D.J., Bursill, L.A., Jefferson, D.A.: Atomic imaging of oxide surfaces. Surface Sci. **175**, 673 (1986)

Tromp, R.M., Hamers, R.J., Demuth, J.E.: Quantum states and atomic structure of silicon surfaces. Science **234**, 304 (1986)

Witt, J., Müller, K.: Evidence for the Ir(100) reconstruction by field-ion microscopy. Phys. Rev. Lett. **57**, 1153 (1986)

# 2. Study of Surface Phonons by Means of the Green's Function Method

L. Miglio and G. Benedek

With 18 Figures

We discuss the standard theory of the Green's function method for solid-state problems and we apply it to the case of surface vibrations. The perturbation in the force-constant matrix, induced by the creation of a surface is then expressed through the invariance conditions in terms of inverse unperturbed Green's functions. The role of electrons is discussed in the framework of the breathing shell model, which is used to calculate surface phonon dispersion curves and surface projected phonon densities for NaCl-type crystals. Green's function calculations of surface vibrations for other materials as well as future developments of this field are finally sketched.

## 2.1 Introductory Remarks

The application of the time-independent Green's function (GF) methods to solid-state problems dates back to the original works of *Lifshitz* [2.1], and *Koster* and *Slater* [2.2]. They have stimulated a great development of the GF methods as a powerful tool of the quantum theory of scattering [2.3–5]. Although the general formulation of the GF method in solid-state physics has become a textbook subject [2.6–8], only in the recent decades has the method been extensively used in the dynamics of real systems, such as solids with defects, disorder, or boundary surfaces [2.9,10].

This contribution is devoted to the GF method in surface lattice dynamics, a subject which nowadays is registering a rapid progress thanks to the recent advance in surface phonon spectroscopy [2.11–13].

Besides the classical Lifshitz GF theory [2.1], which has been used in a variety of models with short-range interactions [2.14–18], there are several variations of the GF calculation for surface phonons, applicable to more specific and complex situations. We mention, among the others, the GF matching [2.19], the continued-fration technique [2.20], the method of generating coefficients [2.21], and the invariant GF method, which we have developed for treating the ionic crystal surfaces [2.22–24].

Since there is substantial equivalence among these formulations [2.23] we shall restrict the present theoretical survey to the invariant method. We want nevertheless to stress a peculiar aspect of the method which makes it applicable to ionic crystals: the reduction of long-range Coulomb interactions, as modified by the surface, to an effective short-range perturbation, which is in turn selfconsistently defined through invariance conditions. The finite range of the perturbation subspace is a prerequisite for the standard GF method.

## 2.2 The Time-Independent Green's Function Method in Summary

Consider the eigenvalue problem for the linear Hermitean operators $L_0$ in the direct-space representation

$$\int L_0(\boldsymbol{r},\boldsymbol{r}')\psi_0(\boldsymbol{r}')\mathrm{d}^3\mathrm{r}' - \lambda\psi_0(\boldsymbol{r}) = 0 \ , \tag{2.1}$$

whose eigenvalues are arranged along the real axis either in a discrete set $\{\lambda_n\}$ or a continuum $\{\lambda_c\}$ (or both). The corresponding inhomogeneous equation

$$\int L_0(\boldsymbol{r},\boldsymbol{r}')\psi(\boldsymbol{r}')\mathrm{d}^3\mathrm{r}' - \lambda\psi(\boldsymbol{r}) = \mathrm{f}(\boldsymbol{r}) \ , \tag{2.2}$$

produced by a perturbing field $\mathrm{f}(\boldsymbol{r})$, admits different types of solutions according to whether $\lambda$ belongs to or is outside the continuous spectrum $\{\lambda_c\}$.

For $\lambda \in \{\lambda_c\}$ the perturbed solution can be written as a sum of any eigenvector $\psi_0(\boldsymbol{r})$ of eigenvalue $\lambda$ and a particular solution of (2.2), i.e.

$$\psi(\boldsymbol{r}) = \alpha\psi_0(\boldsymbol{r}) + \int G_0(\boldsymbol{r},\boldsymbol{r}';\lambda)\mathrm{f}(\boldsymbol{r}')\mathrm{d}^3\mathrm{r}' \ , \tag{2.3}$$

where the Green's function $G_0(\boldsymbol{r},\boldsymbol{r}';\lambda)$ fulfills the equation [2.26]

$$\int [L_0(\boldsymbol{r},\mathrm{r}'') - \lambda\delta(\mathrm{r} - \mathrm{r}'')]G_0(\mathrm{r}'',\boldsymbol{r}';\lambda)\mathrm{d}\mathrm{r}'' = \delta(\boldsymbol{r} - \boldsymbol{r}') \ . \tag{2.4}$$

This is just (2.2) for a perturbing field consisting in a unitary stimulus $\delta(\boldsymbol{r} - \boldsymbol{r}')$ at position $\boldsymbol{r}'$. Thus $G_0(\boldsymbol{r},\boldsymbol{r}';\lambda)$ represents the $\lambda^{\mathrm{th}}$ component of the response of the system at position $\boldsymbol{r}$ to the unitary stimulus at $\boldsymbol{r}'$. The constant $\alpha$ is arbitrary, but has clearly to be zero if $\lambda \notin \{\lambda_c\}$, namely for localized solutions of the perturbed problem.

The important point is that (2.4) is formally solved once we know the complete set of eigenvalues $\lambda_j$ and eigenvectors $\psi_{0j}(\boldsymbol{r})$. One immediately sees that the complex functions

$$G_0^{\pm}(\boldsymbol{r},\boldsymbol{r}';\lambda) = \sum_{j=\{n,c\}} \frac{\psi_{0j}(\boldsymbol{r})\psi_{0j}^*(\boldsymbol{r}')}{\lambda_j - \lambda \pm i0^+} \tag{2.5}$$

are solutions of (2.4) for any $\lambda$, by virtue of completeness of the eigenvector set.

The continuous spectrum of $L_0$ gives rise to a branch cut for $G_0$ along a portion of the real axis.

Here $G_0^+$ and $G_0^-$ are different and complex non-hermitean operators, whereas outside the eigenvalue spectrum $G_0^+$ and $G_0^-$ are equal and hermitean. In general, we can split $G_0^\pm$ into their hermitean and anti-hermitean parts

$$G_0^\pm = \tfrac{1}{2}(G_0^+ + G_0^-) \pm \tfrac{1}{2}(G_0^+ - G_0^-) \; ; \tag{2.6}$$

The anti-hermitean and hermitean parts are, respectively, related to the (hermitean) *spectral* and *dispersive* parts of the GF operator

$$\begin{aligned} Sp\{G_0^\pm\} &= \mp i\tfrac{1}{2}(G_0^+ - G_0^-) \\ &= \pm\pi \sum_{j=\{n,c\}} \psi_{0j}(\boldsymbol{r})\psi_{0j}^*(\boldsymbol{r}')\delta(\lambda_j - \lambda) \end{aligned} \tag{2.7}$$

$$\begin{aligned} Dp\{G_0\} &= \tfrac{1}{2}(G_0^+ + G_0^-) \\ &= P \sum_{j=\{n,c\}} \psi_{0j}(\boldsymbol{r})\psi_{0j}^*(\boldsymbol{r}')(\lambda_j - \lambda)^{-1} \;, \end{aligned} \tag{2.8}$$

where P means that Cauchy principal part has to be taken in the spectral integration.

Since $G_0^\pm$ are eigenfunctions of the Hilbert transforms with eigenvalues $\pm i$, i.e.,

$$\frac{1}{\pi}P \int\limits_{-\infty}^{+\infty} G_0^\pm(\lambda')\frac{d\lambda'}{\lambda' - \lambda} = \pm i G_0^\pm(\lambda) \;. \tag{2.9}$$

The spectral and dispersive parts, as well as the imaginary and real parts fulfill the Kramers-Kronig relations

$$\frac{1}{\pi}P \int\limits_{-\infty}^{+\infty} Sp\{G_0^\pm(\lambda')\}\frac{d\lambda'}{\lambda' - \lambda} = \pm Dp\{G_0^\pm(\lambda)\} \;, \tag{2.10}$$

$$\frac{1}{\pi}P \int\limits_{-\infty}^{+\infty} Dp\{G_0^\pm(\lambda')\}\frac{d\lambda'}{\lambda' - \lambda} = \mp Sp\{G_0^\pm(\lambda)\} \;, \tag{2.10'}$$

$$\frac{1}{\pi}P \int\limits_{-\infty}^{+\infty} Im\{G_0^\pm(\lambda')\}\frac{d\lambda'}{\lambda' - \lambda} = \pm\, Re\{G_0^\pm(\lambda)\} \;, \tag{2.11}$$

$$\frac{1}{\pi}P \int\limits_{-\infty}^{+\infty} Re\{G_0^\pm(\lambda')\}\frac{d\lambda'}{\lambda' - \lambda} = \mp\, Im\{G_0^\pm(\lambda)\} \;. \tag{2.11'}$$

The density of states (DOS) is directly obtained from the spectral part, (2.7),

$$\varrho_0(\lambda) = \pm \frac{1}{\pi} \operatorname{Tr}\{SpG_0(\lambda)\} = \pm \frac{1}{\pi} \operatorname{Tr}\{\operatorname{Im} G_0(\lambda)\} \ , \tag{2.12}$$

Note that outside the continuum $G_0^+ = G_0^-$ have poles at $\lambda = \lambda_n$ so that the localized states of the system are found at the singularities of the GF.

The time-dependent GFs, obtained by Fourier transforming $G_0^+$ and $G_0^-$, are seen to represent the response of the system either following or anticipating the stimulus, respectively. This causality implies that $G_0^+$ is physically meaningful. Since we shall always use causal Green's functions, we shall hereafter drop the apex $+$.

The GF formalism is particularly useful also in the homogeneous problem for the operator L, whenever it admits the decomposition

$$L = L_0 + \Lambda \ , \tag{2.13}$$

$L_0$ being an operator whose eigenvalues and eigenvectors are known, and $\Lambda$ is a well defined localized perturbation. Here we use abstract matrix notations as we do not need to specify the coordinate space where $\Lambda$ exhibits localization.

In the direct space we deal with

$$f(\mathbf{r}) = - \int \Lambda(\mathbf{r}, \mathbf{r}')\psi(\mathbf{r}')\mathrm{d}^3\mathrm{r}' \ , \tag{2.14}$$

and (2.3) transforms into the Lippmann-Schwinger equation for the perturbed eigenvector

$$\psi(\mathbf{r}) = \alpha\psi_0(\mathbf{r}) - \iint G_0(\mathrm{r}, \mathrm{r}''; \lambda)\Lambda(\mathbf{r}'', \mathbf{r}')\psi(\mathbf{r}')\mathrm{d}^3\mathrm{r}'\mathrm{d}^3\mathrm{r}'' \ , \tag{2.15}$$

where $\alpha$ acts now as a normalization constant.

This equation can be solved in practice provided we dispose of the particular representation where $\Lambda$ is localized. Hence we proceed using abstract notations by re-writing (2.15) as

$$|\psi\rangle = \alpha|\psi_0\rangle - G_0\Lambda|\psi\rangle \ , \quad \text{where} \tag{2.16}$$

$$G_0 = (L_0 - z)^{-1} \quad \text{and} \tag{2.17}$$

$$z \equiv \lambda + iO^+ \ . \tag{2.18}$$

We expect from (2.9) three types of solutions: (i) solutions for $\lambda$ inside the continuous spectrum $\{\lambda_c\}$ consisting in the distortion of the unperturbed solutions $\psi_0$; (ii) localized solutions with $\lambda \notin \{\lambda_c\}$, but possibly close to $\lambda_n$, as modifications of the unperturbed localized states (if any); (iii) additional localized solutions with $\lambda \notin \{\lambda_c\}$, peeled off from the edges of the continuous spectrum as an effect of the perturbation.

We consider first the localized solutions, $\lambda \notin \{\lambda_c\}$. In this case (2.16) reduces to

$$|\psi\rangle = -G_0 \Lambda |\psi\rangle \tag{2.19}$$

and the new eigenvalues are solutions of the secular equation

$$\det \{1 + G_0 \Lambda\} = 0 \ . \tag{2.20}$$

By introducing the perturbed GF

$$\begin{aligned} G &= (L - z)^{-1} \\ &= (L_0 + \Lambda - z)^{-1} \\ &= (1 + G_0 \Lambda)^{-1} G_0 \end{aligned} \tag{2.21}$$

we see that the solution of (2.20), i.e., the perturbed localized eigenvalues, are poles of G. Note that the perturbed GF is solution of the Dyson equation

$$G = G_0 - G_0 \Lambda G \ , \tag{2.22}$$

which is equivalent to (2.16) for the wavefunction.

When $\lambda \in \{\lambda_c\}$ we deal with (2.16). This can be formally solved to give

$$|\psi\rangle = \alpha(1 + G_0 \Lambda)^{-1} |\psi_0\rangle \tag{2.23}$$

$$= \alpha|\psi_0\rangle - \alpha G_0 T |\psi_0\rangle \ , \tag{2.23'}$$

where the transition matrix

$$T \equiv \Lambda(1 + G_0 \Lambda)^{-1} \tag{2.24}$$

has non-zero elements only in the perturbation subspace. It is in turn solution of the Dyson equation

$$T = \Lambda - \Lambda G_0 T \ . \tag{2.25}$$

Equation (2.23) expresses the solution of a perturbative problem as a linear superposition of unperturbed wavefunctions, whereas (2.23') represents the solution in the language of scattering theory, i.e., as a sum of incident wave $|\psi_0\rangle$ and diffused wave $- G_0 T |\psi_0\rangle$.

In the first-order Born approximation, i.e., when (2.25) is solved to first order by replacing T with V,

$$|\psi\rangle = \alpha(1 - G_0 \Lambda)|\psi_0\rangle \ . \tag{2.26}$$

Here, however, the peculiar effects of the denominator in (2.24) are lost. Indeed, the factor $1 + G_0\Lambda$ may be responsible for resonant enhancement of the scattered wave. In general, we speak of resonance when the perturbation induces a peak in the spectral part of the perturbed GF, (2.21). This occurs for values of $\lambda = \lambda_R$ fulfilling the resonance condition

$$\mathrm{Re}\left\{\det(1 + G_0\Lambda)\right\} = 0 \ . \tag{2.27}$$

By expanding around $\lambda_R$ one sees that a resonance appears as a Lorentzian peak in the perturbed DOS

$$\varrho(\lambda) = \frac{1}{\pi}\,\mathrm{Tr}\left\{\mathrm{Im}\,G(\lambda)\right\} \ . \tag{2.28}$$

Clearly inside the continuum $\mathrm{Im}\left\{\det(1 + G_0\Lambda)\right\}$ is non-vanishing. If this is a slowly varying function around $\lambda_R$ the peak half-width is approximately given by

$$\tfrac{1}{2}\Gamma = \frac{\mathrm{Im}\left\{\det\left[1 + G_0(\lambda_R)\Lambda\right]\right\}}{\frac{\partial}{\partial\lambda}\,\mathrm{Re}\left\{\det\left[1 + G_0(\lambda_R)\Lambda\right]\right\}} \ . \tag{2.29}$$

In order for the peak to be sharp and to produce a significant feature in the perturbed DOS, $\mathrm{Im}\left\{\det(1 + G_0\Lambda)\right\}$ has to be small. Under these conditions a resonance gives $G(z)$ and $T(z)$ a pole in the complex plane at

$$z \cong \lambda_R - \tfrac{1}{2}i\Gamma \ . \tag{2.30}$$

Thus $G(z)$ takes a physical meaning in the whole complex plane, in that the complex poles indicate resonances as well as real poles identify localized solutions. Since we are actually working with the causal GF, $\Gamma$ has to be positive. Therefore, a value of $\lambda_R$ yielding in (2.29) a negative $\Gamma$ does not contribute a real resonance, but an "anti-resonance", i.e., a depletion region in $\varrho(\lambda)$, with respect to $\varrho_0(\lambda)$, in favour of the intensity which has been concentrated in local and resonant modes.

This can be better understood by considering the change of DOS induced by the perturbation. By remembering that, for any operator $\hat{O}$, $\mathrm{Tr}\left\{\ln\hat{O}\right\} = \ln\left\{\det\hat{O}\right\}$, we can rewrite (2.12 and 28) as

$$\varrho_0(\lambda) = \frac{1}{\pi}\frac{\partial}{\partial\lambda}\,\mathrm{Im}\left\{\ln\det G_0(\lambda)\right\} \ , \tag{2.31}$$

$$\varrho(\lambda) = \frac{1}{\pi}\frac{\partial}{\partial\lambda}\,\mathrm{Im}\left\{\ln\det G(\lambda)\right\} \ . \tag{2.32}$$

Hence $\Delta\varrho \equiv \varrho - \varrho_0$ is given by

$$\Delta\varrho(\lambda) = -\frac{1}{\pi}\frac{\partial}{\partial\lambda}\operatorname{Im}\left\{\ln\det\left[1 + G_0(\lambda)\varLambda\right]\right\} . \tag{2.33}$$

The change of DOS at a resonance can be related with little algebra to the halfwidth;

$$\Delta\varrho(\lambda_R) = 2/\pi\varGamma . \tag{2.34}$$

Thus a positive $\varGamma$ implies a resonance enhancement; a negative $\varGamma$ a depletion. The larger $\Delta\varrho(\lambda_R)$ the smaller $\varGamma$. For a local mode $\Delta\varrho(\lambda_R) = +\infty$ and $\varGamma = 0^+$.

Expression (2.33) offers a straightforward way for calculating $\Delta\varrho(\lambda)$ directly from the resonant denominator $\det(1 + G_0\varLambda)$. Indeed, such determinant is identically equal to the same determinant evaluated in the perturbation subspace. In the latter only the elements of $G_0$ in the $\varLambda$ subspace (in the representation where it exhibits localization), i.e., the elements of the *projected* GF $g_0$ are needed. However, $\det G_0 \neq \det g_0$, so that distinction has to be made between total and projected DOS according to whether whole-space or projected GFs are used in (2.31 and 32). For projected DOS the trace is intended over the perturbation subspace in (2.12 and 28). As a general concept, spectral localization and localization in space of perturbed waves are strictly related, which means that the sharper the resonance the larger is its amplitude in the perturbation subspace. While a resonant state tends to a plane-wave behaviour at large distances from the perturbation region, a localized state cannot propagate at all in the unperturbed region of the system and therefore it must decay exponentially out of the perturbation subspace.

## 2.3 The Green's Function Method in Surface Dynamics

We consider a crystal lattice constituted by N unit cells at positions $r_\ell$ ($\ell = 1, 2, \ldots, N$) with s atoms per unit cell located at $r_\kappa$ ($\kappa = 1, 2, \ldots, s$) with respect to the conventional cell center. Thus the equilibrium atomic positions are

$$r_{\ell\kappa} = r_\ell + r_\kappa . \tag{2.35}$$

Then we write the secular equation for lattice vibrations in the harmonic approximation as

$$\sum_\beta [\phi_{\alpha\beta}(\ell\kappa, \ell'\kappa') - \omega^2 M_\kappa \delta_{\ell\ell'}\delta_{\kappa\kappa'}\delta_{\alpha\beta}]u_\beta(\ell'\kappa') = 0 , \tag{2.36}$$

where $\phi$ is the force constant matrix, $M_\kappa$ the $\kappa^{\text{th}}$ atom mass, $\omega$ the phonon angular frequency and $u(\ell'\kappa')$ the atomic displacement vector from the equilibrium position. The invariance of the crystal potential energy with respect to infinitesimal rigid translations and rotations implies the following conditions for the tensor $\phi(\ell\kappa, \ell'\kappa')$:

$$\text{TI}: \quad \mathcal{T}\phi \equiv \sum_{\ell\kappa} \phi(\ell\kappa, \ell'\kappa') = 0 \; , \quad \forall \ell'\kappa' \quad \text{and} \tag{2.37}$$

$$\text{RI}: \quad \mathcal{R}\phi \equiv \sum_{\ell\kappa} \mathbf{r}_{\ell\kappa} \times \phi(\ell\kappa, \ell'\kappa') = 0 \; , \quad \forall \ell'\kappa' \; , \tag{2.38}$$

respectively.

The linear operators $\mathcal{T}$ and $\mathcal{R}$ in the $(\ell\kappa\alpha)$-space, respectively, determining TI and RI conditions, are given by the $3 \times 3Ns$ rectangular matrices

$$\langle \alpha | \mathcal{T} | \ell\kappa\beta \rangle = \delta_{\alpha\beta} \; , \tag{2.39}$$

$$\langle \alpha | \mathcal{R} | \ell\kappa\beta \rangle = (\underset{\sim}{\delta} \times \mathbf{r}_{\ell\kappa})_{\alpha\beta} \; , \tag{2.40}$$

where $\underset{\sim}{\delta}$ is the unit tensor in the cartesian subspace of components $\delta_{\alpha\beta}$.

For a three-dimensional periodic lattice $\ell$ represents a set of three integer numbers $(\ell_1, \ell_2, \ell_3)$ such that $\mathbf{r}_\ell = \ell_1\mathbf{a}_1 + \ell_2\mathbf{a}_2 + \ell_3\mathbf{a}_3$, $\{\mathbf{a}_j\}$ being any set of direct lattice vector. For a semi-infinite lattice with a single surface (or a slab with two parallel surfaces) one can always choose $\mathbf{a}_1$ and $\mathbf{a}_2$ in the surface plane, so that the pairs $(\ell_1, \ell_2) = L$ label the translations along to the surface, and $\ell_3$ labels the atomic layers parallel to the surface.

We consider an infinitely thick slab originated by perturbing an infinite lattice with three-dimensional cyclic boundary conditions [2.22,23]. The free surfaces are constructed by cutting the infinite lattice along an ideal plane $\Sigma$, as shown in Fig. 2.1; the resultant perturbation of the cyclic lattice force constants $\phi_{0\alpha\beta}(\ell\kappa, \ell'\kappa') = \phi_{0\alpha\beta}(L - L'; \ell_3 - \ell'_3, \kappa\kappa')$ is then described by setting to zero all the interatomic force constants crossing the plane $\Sigma$; more precisely we define the perturbation matrix according to

$$\begin{aligned} \Lambda_{\alpha\beta}&(L - L'; \ell_3\kappa, \ell'_3\kappa') \\ &= \phi_{\alpha\beta}(L - L'; \ell_3\kappa, \ell'_3\kappa') - \phi_{0\alpha\beta}(L - L'; \ell_3\kappa, \ell'_3\kappa') \; , \end{aligned} \tag{2.41}$$

where both $\phi$, the slab force constant matrix, and $\phi_0$ are assumed to satisfy the TI and the RI conditions. Thus, also $\Lambda$ fulfills the TI condition and in the absence of elastic relaxation, also the RI condition [2.29]. In (2.41) it is intended that the slab force constant elements connecting atoms across $\Sigma$ are zero. In practice, $\Lambda$ works as a perturbation only when its nonzero elements are restricted within a small perturbation subspace $\sigma$ (Fig. 2.1); for example, if $\ell_3 = 1$ and $\ell_3 = N_L$ denote the two surface layers, the subspace $\sigma$ includes $\ell_3 = 1, 2, \ldots, N_p$ and $N_L - N_p + 1$, where $N_p$, the number of atomic layers on each side of the slab involved in the cutting procedure, is "small". This

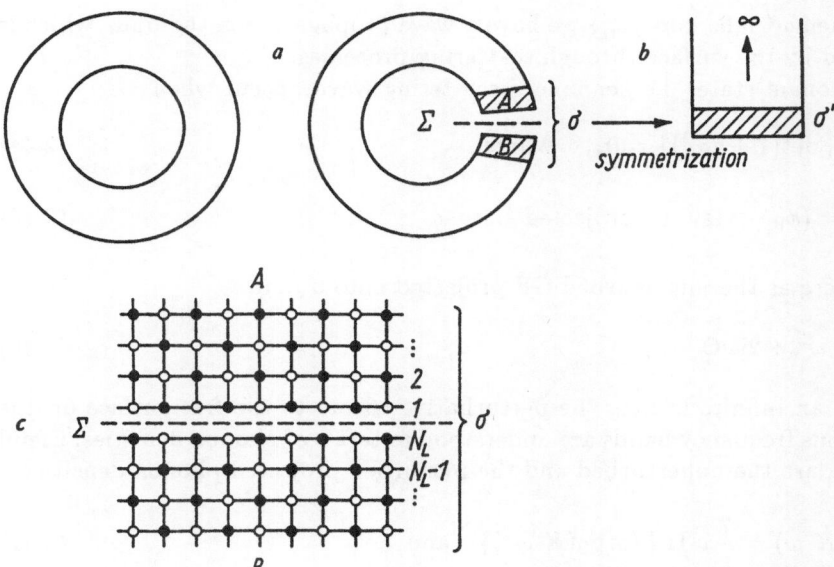

**Fig. 2.1a–c.** Representation of an ideal lattice with cyclic boundary conditions (a), and of the same lattice after two free surfaces are created by a cut along the plane $\Sigma$ (b), $\sigma$ denotes the perturbation subspace. In the representation of symmetrized coordinates we obtain a semi-infinite lattice with a perturbation subspace $\sigma'$, restricted to a single surface (c). For a ionic lattice (NaCl like) the inversion symmetry with respect to $\Sigma$ is recovered after a shear translation of the surface B with respect to A

requirement seems to rule out the applicability of the perturbation method to lattices with long-range forces, like ionic crystals. However, we shall see that in the limit $N_L \to \infty$ it is possible to work with a small perturbation matrix also for ionic crystals.

In abstract notations, (2.36) for stationary vibrational states reads

$$(\phi_0 + \Lambda - M\omega^2)\mathbf{u} = 0 \ , \tag{2.42}$$

where M is the mass matrix. We assume that (2.42) has already been Fourier transformed with respect to the coordinate along which translational symmetry is maintained, i.e., we work in the surface wave-vector representation.

The eigenvalue spectrum of $D_0 = M^{-1/2}\phi_0 M^{-1/2}$, the dynamical matrix of infinite lattice, is represented for each wavevector $\mathbf{K}$ by a set of 3s continuous bands. According to the analysis made in Sect. 2.1 we first consider $\omega^2$ outside the spectrum $\{\omega_0^2\}$ of $D_0$; the values of $\omega^2$ for which

$$\det\{1 + (\phi_0 - M\omega^2)^{-1}\Lambda\} = 0 \tag{2.43}$$

at any $\mathbf{K}$ give the dispersion relations of the surface modes; they are localized at the surface since no frequency outside $\{\omega_0^2\}$ can propagate inside the lattice.

When $\omega^2$ falls into $\{\omega_0^2\}$ we have a wave propagating in the bulk, which is distorted by the surface through scattering processes.

Resonant states, i.e., enhanced scattering waves, occur when

$$\mathrm{Re}\,\{\,\det(1 + \mathrm{g}_0 \Lambda)\} = 0 \;, \quad \text{where} \tag{2.44}$$

$$\mathrm{g}_0 = (\phi_0 - \mathrm{Mz})^{-1} \quad \text{projected onto } \sigma \tag{2.45}$$

works here as the unperturbed GF projected onto $\sigma$, and

$$\mathrm{z} = \omega^2 + 2\mathrm{i}\omega\mathrm{O}^+ \;. \tag{2.46}$$

For an infinite lattice, the perturbative effects to the free surface on the continuous frequency bands are understood in terms of phonon densities: useful concepts are the unperturbed and the perturbed projected phonon densities

$$\varrho_0(\boldsymbol{K}, \omega) = \frac{2}{\pi}\omega\,\mathrm{Tr}\,\{\,\mathrm{Im}\,\mathrm{g}_0(\boldsymbol{K}, \omega^2)\} \quad \text{and} \tag{2.47}$$

$$\varrho(\boldsymbol{K}, \omega) = \frac{2}{\pi}\omega\,\mathrm{Tr}\,\{\,\mathrm{Im}\,\mathrm{g}(\boldsymbol{K}, \omega^2)\} \;, \tag{2.48}$$

respectively, where Tr denotes trace in the subspace $\sigma$ of the perturbation, and

$$\mathrm{g} = (\phi - \mathrm{Mz})^{-1} \quad \text{projected onto } \sigma \tag{2.49}$$

$$= (1 + \mathrm{g}_0 \Lambda)^{-1}\mathrm{g}_0 \tag{2.50}$$

is the perturbed projected Green's function matrix [2.30].

For $\boldsymbol{K}$ along the symmetry directions of the surface Brillouin zone all modes have a polarization which is either in the plane defined by the directions of the $\boldsymbol{K}$ vector and the normal to the surface [sagittal ($\perp$) polarization] or transverse and parallel to the surface [parallel ($\parallel$) polarization]. This yields the factorization

$$\{\omega_0^2\} = \{\omega_0^2\}_\perp \otimes \{\omega_0^2\}_\parallel \tag{2.51}$$

and the block diagonalization of all the above matrices. Having in mind (2.50), the following cases for the values $\omega^2$ fulfilling (2.44) occur:

1. Surface modes: $\omega^2$ is outside $\{\omega_0^2\}$. The imaginary part of $\mathrm{g}_0$ is infinitesimal and $\varrho(\boldsymbol{K}, \omega)$ exhibits a $\delta$-peak.
2. Pseudo-surface modes: $\omega^2$ falls into a band of $\{\omega_0^2\}$, but the displacement vector $\boldsymbol{u}$ is orthogonal to all vectors $\boldsymbol{u}_0$ of that band; again we have a local mode, crossing a transparent bulk band. This occurs, e.g. when $\omega^2$ belongs to $\{\omega_0^2\}_\perp$ but is outside $\{\omega_0^2\}_\parallel$ (or viceversa). Clearly, pseudo-surface modes exist only along symmetry directions.

3.  Surface resonances: $\omega^2$ falls into a band of $\{\omega_0^2\}$ whose $\boldsymbol{u}_0$ are not orthogonal to $\boldsymbol{u}$. In this case we have a resonance: $\varrho(\boldsymbol{K}, \omega)$ displays a Lorentzian-shaped peak, whose width is proportional to $\text{Im}\{g_0\}$, which is finite. When deviating from a symmetry direction any pseudo-surface mode transforms into a resonance. However, surface resonances may exist also along symmetry directions in addition to local modes.

## 2.4 From a Slab to a Semiinfinite Lattice: Selfconsistent Definition of the Perturbation Matrix

For a slab with identical surfaces we can always find a symmetry transformation $S$ producing a simultaneous block diagonalization of the Green's function and perturbation matrices, namely

$$S g_0 S^{-1} = \begin{vmatrix} g_+ & 0 \\ 0 & g_- \end{vmatrix} \;, \quad S \varLambda S^{-1} = \begin{vmatrix} \varLambda_+ & 0 \\ 0 & \varLambda_- \end{vmatrix} \tag{2.52}$$

and therefore

$$S g S^{-1} = \begin{vmatrix} \tilde{g}_+ & 0 \\ 0 & \tilde{g}_- \end{vmatrix} \qquad \text{where} \tag{2.53}$$

$$\tilde{g}_\pm = (1 + g_\pm \varLambda_\pm)^{-1} g_\pm \;. \tag{2.54}$$

For example, in many cases the lattice exhibits a mirror symmetry with respect to the plane $\varSigma$; sometimes, the mirror symmetry is recovered by a shear translation of the surface B with respect to A, as for the (001)-surfaces of a NaCl lattice (Fig. 2.1c). In these cases, by re-labelling the layers of A and B sides by $\ell_3 = 1/2, 3/2 \ldots N_P - 1/2, \ldots$ and $\ell_3 = -1/2, -3/2, \ldots, -N_p + 1/2, \ldots$, respectively, it is possible to replace the set of coordinates $(\ell_3 \kappa \alpha)$ with a new set of symmetrized coordinates $(|\ell_3|, \kappa \alpha, \text{p})$ defined by the transformation $S$

$$S |\ell_3 \kappa \alpha \rangle = \||\ell_3|, \kappa \alpha, \text{p} \rangle = \frac{1}{\sqrt{2}} [|\ell_3, \kappa \alpha \rangle + \text{p sgn}(\alpha)| - \ell_3, \kappa \alpha \rangle] \;, \tag{2.55}$$

where $\text{p} = \pm 1$ is the parity index, and $\text{sgn}(\alpha) = 1$ for $\alpha = \text{z}$ (the surface orientation), and $= -1$ for $\alpha = \text{x}, \text{y}$. This enables us to work in a reduced $3sN_\text{p}$-dimensional subspace $\sigma'$.

When $N_\text{L} \to \infty$ a slab with identical surfaces becomes equivalent to a semiinfinite lattice with a single surface, provided that the displacement correlation between the two slab surfaces tends to zero all over the spectral region. This means that all the modes of the slab become degenerate in pairs. Therefore in the subspace $\sigma'$ we have

$$\tilde{g}_+ = \tilde{g}_- \equiv \tilde{g} \;, \tag{2.56}$$

45

$\tilde{g}$ is interpreted as the perturbed projected Green's function for the single surface of the semiinfinite lattice. Equation (2.54) can be written as

$$\tilde{g} = (1 + \bar{g}\bar{\Lambda})^{-1}\bar{g} \ , \quad \text{where} \tag{2.57}$$

$$\bar{g}^{-1} = \tfrac{1}{2}(g_+^{-1} + g_-^{-1}) \quad \text{and} \tag{2.58}$$

$$\bar{\Lambda} = \tfrac{1}{2}(\Lambda_+ + \Lambda_-) \tag{2.59}$$

are the inverse of the unperturbed projected Green's functions for the semiinfinite lattice, and the pertaining perturbation, respectively. Notice that all the inversions are performed in the subspace $\sigma'$. The resonance condition, (2.44), in the subspace $\sigma'$ reads

$$\text{Re}\{\det(1 + \bar{g}\bar{\Lambda}\} = 0 \ . \tag{2.60}$$

Apart from dimensional reduction, the symmetrization (2.55) has the advantage that the symmetrized perturbation $\bar{\Lambda}$ takes a very simple form. Indeed, for the case of symmetry mentioned above, we have

$$\Lambda(\boldsymbol{K}; \ell_3\kappa, \ell_3'\kappa') = \text{sgn}(\alpha)\text{sgn}(\beta)\Lambda_{\alpha\beta}(\boldsymbol{K}; -\ell_3\kappa, -\ell_3'\kappa') \ , \tag{2.61}$$

since the same property holds for $\phi_0$ and $\phi$ [2.31]. The symmetrized components are readily obtained by means of (2.55)

$$\begin{aligned}
\Lambda(\boldsymbol{K}; |\ell_3|\kappa\text{p}; |\ell_3'|\kappa'\text{p}') \\
= \delta_{\text{pp}'}[\Lambda_{\alpha\beta}(\boldsymbol{K}; |\ell_3|\kappa, |\ell_3'|\kappa') + \text{sgn}(\alpha)\text{p}\Lambda_{\alpha\beta}(\boldsymbol{K}; |\ell_3|\kappa, |-\ell_3'|\kappa')]
\end{aligned} \tag{2.62}$$

and according to (2.59),

$$\bar{\Lambda}_{\alpha\beta}(\boldsymbol{K}; \ell_3\kappa, \ell_3'\kappa') = \Lambda_{\alpha\beta}(\boldsymbol{K}; \ell_3\kappa, \ell_3'\kappa') \ , \quad \ell_3, \ell_3' > 0 \ . \tag{2.63}$$

Thus the semiinfinite lattice perturbation $\bar{\Lambda}$ contains only the elements of $\Lambda$ which connect pairs of atoms on one side with respect to the cuttinig plane $\Sigma$. These elements represent the reaction of the lattice in the region A to the band cutting as required by conservation of total momentum and total angular momentum. The two conservation laws are reflected in the force constant matrix through the TI and RI conditions, respectively. Therefore the matrix elements of $\bar{\Lambda}$ contain exclusively what is implied by TI and RI conditions. More precisely the diagonal part $\Lambda^T$ of $\bar{\Lambda}$ (the Einstein part) comes from TI condition, so that we can write

$$\bar{\Lambda} = \Lambda^T + \Lambda^R \ . \tag{2.64}$$

The procedure for obtaining $\Lambda^{\mathrm{T}}$ and $\Lambda^{\mathrm{R}}$ is based on the identity

$$\Lambda_{\mathrm{p}} = \overline{\Lambda} - \tfrac{1}{2}(\mathbf{g}_{\mathrm{p}}^{-1} - \mathbf{g}_{-\mathrm{p}}^{-1}) \ , \quad (\mathrm{p} = \pm) \tag{2.65}$$

directly obtainable from (2.54–59). By applying to both numbers the blocks $\mathcal{T}_{\mathrm{p}}$ and $\mathcal{R}_{\mathrm{p}}$ of the symmetrized TI and RI operators $\mathcal{T}\,S^{-1}$ and $\mathcal{R}\,S^{-1}$, respectively, and noting that $\mathcal{T}_{\mathrm{p}}\Lambda_{\mathrm{p}} = \mathcal{R}_{\mathrm{p}}\Lambda_{\mathrm{p}} = 0$, we have

$$\mathcal{T}_{\mathrm{p}}\overline{\Lambda} = \tfrac{1}{2}\mathcal{T}_{\mathrm{p}}(\mathbf{g}_{\mathrm{p}}^{-1} - \mathbf{g}_{-\mathrm{p}}^{-1}) \ , \tag{2.66}$$

$$\mathcal{R}_{\mathrm{p}}\overline{\Lambda} = \tfrac{1}{2}\mathcal{R}_{\mathrm{p}}(\mathbf{g}_{\mathrm{p}}^{-1} - \mathbf{g}_{-\mathrm{p}}^{-1}) \ . \tag{2.67}$$

This is a set of linear inhomogeneous equations having the elements of $\overline{\Lambda}$ as unknown. One can prove that, whatever the extension of the *intrinsic* surface perturbation, the set (2.66,67) contains as many non trivial independent equations as many independent elements of $\overline{\Lambda}$ are needed.

We see that the perturbation matrix $\overline{\Lambda}$ turns out to be entirely self-consistently defined in terms of elements of inverse unperturbed GFs, i.e., in terms of the intrinsic bulk dynamics. This is an interesting illustration of the general statement that all what happens at the free surface of a solid, such as the change of force constants, of atomic equilibrium positions, of electronic structure, is due to intrinsic properties of the bulk Hamiltonian manifesting themselves through the symmetry breaking. Thus a full knowledge of the dynamical structure of the bulk, no matter whether obtained ab-initio or in a phenomenological way, should be sufficient to account for all the surface intrinsic dynamical properties. In the GF method applied to surface problems the bulk dynamical structure is the basic ingredient, whereas the perturbation of an intrinsic surface, induced by the symmetry breaking, turns out to be fully described through the TI and RI conditions. This procedure has been shown [2.25] to be equivalent to a more general theory of interface dynamics based on the Green function matching when applied to free surfaces.

The elements of the unperturbed bulk GF matrix contain all the information we need about both bulk dynamics and intrinsic surface perturbation. Thus the GF method, in the form it has developed, is just a formal procedure to transfer to the surface all what we know *a priori*, in the ideal lattice with cyclic boundary conditions.

Another interesting property of $\overline{\Lambda}$ is its sharp localization. The application of $\mathcal{T}_{\mathrm{p}}$ and $\mathcal{R}_{\mathrm{p}}$ on the right-hand member of (2.66 and 67) implies a sum over the atom index $\kappa$. In ionic crystals the sum over positive and negative ions yields a cancellation of the long-range Coulomb contributions and the elements of (2.63), fall off rapidly for increasing $\ell_3$ or/and $\ell_3'$. Thus $\overline{\Lambda}$ can be truncated and restricted to a very small number of layers. Although truncation would prevent a complete separation of the two sides A and B, the free-surface behaviour is always recovered through TI and RI conditions, which ensure the existence and a correct behaviour of the Rayleigh wave in the long-wave length limit.

Let us consider an application of these concepts to the (001) surface of alkali halides with the rocksalt structure. The Einstein part of $\bar{\Lambda}$ has the form

$$\Lambda^{\mathrm{T}}_{\alpha\beta}(\boldsymbol{K}; \ell_3\kappa, \ell_3'\kappa') = \delta_{\ell_3\ell_3'}\delta_{\kappa\kappa'}\delta_{\alpha\beta}\mathrm{f}_{\kappa\alpha}(\ell_3) \qquad (2.68)$$

with $\ell_3 > 0$, and the force constants

$$\mathrm{f}_{\kappa\alpha}(\ell_3) \equiv \sum_{\boldsymbol{L},\kappa',\ell_3'>0} \phi_{0\alpha\alpha}(\boldsymbol{L}; \ell_3\kappa, -\ell_3'\kappa') \qquad (2.69)$$

are obtained from the bulk dynamical model. A calculation performed for NaCl(001) (Fig. 2.2) [2.22] shows for $\mathrm{f}_{\kappa\alpha}(\ell_3)$ a steep exponential decay for increasing $\ell_3$, to such an extent that $\bar{\Lambda}$ can be restricted to the first surface layer ($\ell_3 = 1/2$). After truncation we have to replace the force constants (2.69) with those self consistently obtained from the TI condition

$$\mathrm{f}_{\kappa\alpha} = \tfrac{1}{2}\mathrm{sgn}(\alpha) \sum_{\kappa'} [(\kappa\alpha|\mathrm{g}_-^{-1} - \mathrm{g}_+^{-1}|\kappa'\alpha)]_{\boldsymbol{K}=0} \ . \qquad (2.70)$$

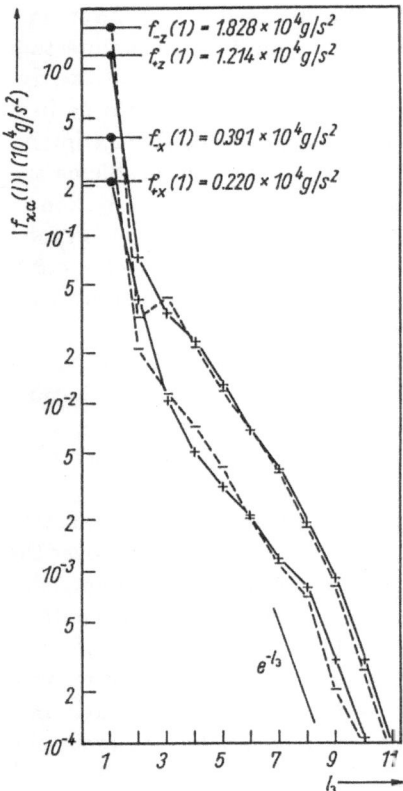

Fig. 2.2. Spatial behaviour along the direction perpendicular to the surface of the double Fourier transformed force constants, calculated using the breathing shell model and T=0 K input data for the infinite lattice

48

Then from the RI condition we obtain, for $\ell_3 = \ell'_3 = 1/2$,

$$\Lambda^R_{\alpha\beta}(\boldsymbol{K}; \kappa\kappa') = i(1 - \delta_{\kappa\kappa'})\varepsilon_{\alpha\beta}(\boldsymbol{K})(f_{\kappa\alpha}\delta_{\beta z} + f_{\kappa'\beta}\delta_{\alpha z}) \ , \tag{2.71}$$

where

$$\begin{aligned}
\varepsilon_{\alpha\beta}(\boldsymbol{K}) &\equiv (\delta_{\alpha x}\delta_{\beta z} - \delta_{\alpha z}\delta_{\beta x}) \sin r_0 K_x \\
&\quad + (\delta_{\alpha y}\delta_{\beta z} - \delta_{\alpha z}\delta_{\beta y}) \sin r_0 K_y
\end{aligned} \tag{2.72}$$

and $r_0$ is the interionic distance.

The elements of the unperturbed GF matrices for $\ell_3 = \ell'_3 = 1/2$ are found to be

$$\langle \kappa\alpha | g_p(\boldsymbol{K}, \omega^2) | \kappa'\beta \rangle = \frac{1}{\pi\sqrt{M_\kappa M_{\kappa'}}} \int_0^\pi \sum_j \frac{e_\alpha(\kappa|\boldsymbol{q}j)e_\beta(\kappa'|\boldsymbol{q}j)}{\omega^2_{\boldsymbol{q}j} - \omega^2 - 2i\omega 0^+}$$

$$[(1 + \text{sgn}\,\alpha\,\text{sgn}\,\beta)(1 + p\,\text{sgn}\,\alpha\,\cos\,\xi) + ip(\text{sgn}\,\alpha - \text{sgn}\,\beta)\sin\,\xi]d\xi \ , \quad (2.73)$$

where $\omega_{\boldsymbol{q}j}$ and $e(\kappa|\boldsymbol{q}j)$ are bulk phonon angular frequencies and polarization vectors of wavevector $\boldsymbol{q} \equiv (\boldsymbol{K}, \xi/r_0)$ and branch index j.

# 2.5 The Electronic Contribution to Surface Dynamics in the Framework of Shell Models

Shell models [2.10,32–34] are perhaps the simplest way to incorporate the effects of electrons dynamics in the lattice-dynamical matrix. Virtual excitations of the electron system modify the interionic force constants for the ground state. This happens through a change in the shapes of the external electron orbitals surrounding ion cores, during their vibrational motion. These charge modifications are monopolar, dipolar, quadrupolar, etc. according to the nature of the corresponding electron virtual excitations. From the point of view of lattice dynamics all electron transitions can be grouped according to their symmetries and their overall conditions. They are parametrized through the definition of shell monopolar (breathing) deformabilities, dipolar deformabilities (polarizabilities), quadrupolar deformabilities, etc.

In this description the dynamical equations of the crystal are written as

$$M\omega^2 \boldsymbol{u} = (R + ZCZ)\boldsymbol{u} + (T + ZCY)\boldsymbol{w} + Q\boldsymbol{v} \ , \tag{2.74a}$$

$$m_d\omega^2 \boldsymbol{w} = (T^+ + YCZ)\boldsymbol{u} + (S + YCY)\boldsymbol{w} + Q\boldsymbol{v} \ , \tag{2.74b}$$

$$m_q\omega^2 \boldsymbol{v} = Q^+(\boldsymbol{u} + \boldsymbol{w}) + H\boldsymbol{v} \ , \tag{2.74c}$$

49

where (2.74a) describes the behaviour of cores, (2.74b,c) that of electronic shells in the adiabatic approximation; $u$ represents the core displacements, $w$ the shell-core displacements and $v$ accounts for the isotropic and quadrupolar deformations of ions. R, T and S are the short-range core-core, core-shell and shell-shell parts of the dynamical matrix, respectively, while the long-range Coulomb contribution is provided by C, through the ion-charge matrix Z and shell-charge matrix Y. Q represents the interactions of isotropic and quadrupolar deformations with core and shell displacements and H the interaction of isotropic and quadrupolar deformations with themselves. M is the core-mass matrix, $m_d$ and $m_q$ represent the effective shell masses for the various types of deformations. By grouping the electronic degrees of freedom into the vector

$$ d = \begin{bmatrix} w \\ V \end{bmatrix} \tag{2.75} $$

the dynamical matrix can be divided into four blocks

$$ \phi_{zz} = [R - ZCZ] \ , \qquad \phi_{zy} = [T + ZCY, \quad Q] \ , $$

$$ \phi_{yz} = \begin{bmatrix} T^+ + YCZ \\ Q^+ \end{bmatrix} \ , \qquad \phi_{yy} = \begin{bmatrix} S + YCY & Q \\ Q^+ & H \end{bmatrix} \ . \tag{2.76} $$

The surface perturbation affects in principle both rigid-ion and electron dynamics, so that we have a general perturbation

$$ \Lambda = \begin{bmatrix} \Lambda_{zz} & \Lambda_{zy} \\ \Lambda_{yz} & \Lambda_{yy} \end{bmatrix} \ . \tag{2.77} $$

Here $\Lambda_{zz}$ accounts by itself for the appearance of certain surface modes, such as Rayleigh waves (RW) and Lucas modes (LM). $\Lambda_{yy}$ accounts for the possible increase of ion polarizabilities at the surface and the associated decrease of excitonic frequencies, which can be related to the appearance of surface states in the gap between valence and conduction bands. The change in surface polarizabilities may have a direct influence on RW and LM dispersion curves, particularly at the zone boundary [2.36].

In calculations of phonon dispersion curves the exciton dynamics is generally ignored by setting $m_d = m_q = 0$ and eliminating the components $d$ of the total displacement vector by reducing the problem to the subspace $\sigma_z$ of ion displacements. Here the secular equation for the unperturbed lattice reads

$$ \det [\phi_{zz} - \phi_{zy}\phi_{yy}^{-1}\phi_{yz} - M\omega^2] = 0 \ . \tag{2.78} $$

In surface problems, the reduction subspace $\sigma_z$ should occur after perturbation and symmetrization have been applied, in order to treat rigorously the perturbation of electron dynamics. Only by working in the extended space of nuclei and electron coordinates does the semiinfinite perturbation $\overline{\Lambda}$, as defined through TI and RI conditions, attain self-consistency. This is particularly important in those crystals where important anomalies driven by electron-phonon interaction appear in the phonon dispersion, e.g., in charge-density-wave (CDW) materials, mixed-valence crystals and ferroelectrics.

Since we can define a symmetry transformation which is block-diagonal with respect to $u$ and $d$ coordinates, namely

$$S = \begin{bmatrix} S_z & 0 \\ 0 & S_y \end{bmatrix} \tag{2.79}$$

the perturbed GF matrix $G_{zz}$ obtained after the final reduction to the subspace $\sigma_z$ can be expressed exclusively in terms of the unperturbed GF and perturbation matrix blocks in the subspaces of nuclear and electron coordinates. We find

$$\tilde{G}_{zz} = (1 + \overline{G}^*_{zz}\overline{\Lambda}_{zz})^{-1}\,\overline{G}^*_{zz} \, , \qquad \text{where} \tag{2.80}$$

$$\overline{G}^*_{zz} = \overline{g}_{zz}(1 - \tilde{g}_{zy}^{-1}\tilde{g}_{yy}\tilde{g}_{yz}^{-1}\overline{g}_{zz})^{-1} \tag{2.81}$$

$$\tilde{g}_{ij} \equiv (1 + \overline{g}_{ij}\overline{\Lambda}_{ij})^{-1}\overline{g}_{ij} \, , \quad ij = yz, zy, yy \tag{2.82}$$

and the inversions are performed in the respective subspaces. As usual

$$\overline{g}_{ij}^{-1} = \tfrac{1}{2}(g_{ij+}^{-1} + g_{ij-}^{-1}) \, , \tag{2.83}$$

$$\overline{\Lambda}_{ij} = \tfrac{1}{2}(\Lambda_{ij+} + \Lambda_{ij-}) \, , \tag{2.84}$$

$$\overline{\Lambda}_{ij} = \Lambda_{ijp} + \tfrac{1}{2}(g_{ijp}^{-1} - g_{ij-p}^{-1}) \, , \quad p = +, - \qquad \text{where} \tag{2.85}$$

$$\Lambda_{ij\pm} = S_i \Lambda_{ij} S_j^{-1} \, , \tag{2.86}$$

$$g_{ij+} = S_i g_{ij} S_j^{-1} \, , \tag{2.87}$$

$$g_{ij} = (\phi_{ij} - \delta_{ij}M\omega^2)^{-1} \, , \tag{2.88}$$

and $i, j = z, y$. Now the matrix elements of the blocks $\overline{\Lambda}_{ij}$ can be obtained by applying TI and RI conditions to (2.85). In particular, the non-diagonal elements of $\Lambda_{yy}$ derived from TI will contain in a natural way the information on the shell equilibrium configuration and the possible static polarization at the surface.

51

The secular determinant yielding surface modes in the shell model framework is then given by

$$\text{Re}\{\det(1 + \overline{G}^*_{zz}\overline{\Lambda}_{zz})\} = 0 \ . \tag{2.89}$$

Note that $\overline{G}^*_{zz}$ plays the role of an effective GF, which is partially perturbed by the electronic contributions $\overline{\Lambda}_{ij}$ in (2.82). Actually, the term $-\tilde{g}^{-1}_{zy}\tilde{g}_{yy}\tilde{g}^{-1}_{yz}$ in (2.81) works as a pseudo-perturbation due to electron excitations. In this case the localization of such a pseudo-perturbation is ensured by the short-range nature of T, S, Q and H force constant matrices. Thus all subspaces where inversions are performed are just perturbation subspaces, and all $\tilde{g}_{ij}$ in (2.82) are actually projected GFs. This makes the procedure practicable. However, the evaluation of so many GFs and perturbation matrices is rather cumbersome. It is hard to say whether this procedure is faster than the direct calculation of $\overline{g}$ and $\overline{\Lambda}$ in the extended perturbation subspace, without eliminating internal shell coordinates.

When the perturbation on electron dynamics is argued to have only a small effect on surface dynamics, as in alkali halides, a strong simplification is achieved by eliminating internal shell coordinate *before* applying surface perturbation and symmetrization. Within this approximation we have

$$\tilde{G}_{zz} = (1 + \overline{G}_{zz}\overline{\Lambda}^*_{zz})^{-1}\overline{G}_{zz} \ , \qquad \text{where} \tag{2.90}$$

$$\overline{G}^{-1}_{zz} = \tfrac{1}{2}(G^{-1}_{zz+} + G^{-1}_{zz-}) \ , \tag{2.91}$$

$$G_{zz+} = S_z G_{zz} S^{-1}_z \ , \tag{2.92}$$

$$G_{zz} \equiv g_{zz}(1 - \phi_{zy}g_{yy}\phi_{yz}g_{zz})^{-1} \ , \tag{2.93}$$

and the effective perturbation $\overline{\Lambda}^*_{zz}$ is defined by

$$\overline{\Lambda}^*_{zz} = \Lambda_{zz,p} + \tfrac{1}{2}(G^{-1}_{zz,p} - G^{-1}_{zz,-p}) \ . \tag{2.94}$$

Thus the TI and RI conditions connect the elements of $\overline{\Lambda}^*_{zz}$ to $G_{zz}$ rather than $g_{zz}$.

Calculations based on this approximate method have been performed in the framework of the breathing shell model (BSM) [2.35] for LiF and the cluster model [2.36] for TiN (Sect. 2.6). In $\overline{\Lambda}^*_{zz}$ any explicit information on the surface perturbation of electronic structure and ion polarizabilities is lost, in the sense that the elements of $\overline{\Lambda}^*_{zz}$ deduced from the TI and RI conditions applied to $G^{-1}_{zz+}$ refer to a surface whose shell deformabilities are the same as in the bulk. In this case the perturbation in ion polarizabilities could be introduced *ad hoc*, e.g., by adding to $\overline{\Lambda}^*_{zz}$ the term

$$\bar{A}_{\text{pol}} = \tfrac{1}{2}\Delta_{\alpha\to\alpha_s}(G_{zz+}^{-1} - G_{zz-}^{-1}) \ , \tag{2.95}$$

where $\Delta_{\alpha\to\alpha_s}$ means the difference between the expression calculated for polarizabilities equal to their surface values $\alpha_s$ and that for bulk polarizabilities $\alpha$. One should take into account that $G_{zz+}$ and $G_{zz-}$ in (2.95) retain their meaning of *projected* GF matrices.

## 2.6 Surface Vibrations in Alkali Halides

In this section we review our GF calculations of surface phonons on the (001) surface of eight alkali halides with the rocksalt structure [2.22,24,37,38]: LiF, NaF, NaCl, NaI, KI, KCl, KBr and KI. All these calculationis are based on the BSM. For LiF, NaF, NaCl, NaI, RbF, RbCl and MgO shell model calculations based on the direct dynamical matrix diagonalization for the 15-layer slab, have been reported by *Chen* et al. [2.39], so that for some of these crystals a comparison between the GF and the slab methods is possible.

With respect to the direct dynamical matrix diagonalization applied to a thin slab, the GF method has the formal advantage of reducing a big problem to a small one: it enables one to work in the peturbation subspace rather than in the large slab space. In practice, however, the GF method has a drawback in the severe computational difficulties originated from the singular nature of the surface-projected Green's function [2.26].

The BSM dynamical matrix is directly related to a set of phenomenological input data such as the elastic constants $c_{ij}$, ionic polarizabilities $\alpha_\pm$, net charge Z, bulk transverse optical frequency $\omega_{\text{TO}}$, static and high-frequency dielectric constants $\varepsilon_s$ and $\varepsilon_0$. In addition to nearest neighbour repulsion, the model allows for a repulsive second neighbour (2n) interaction only between larger ions. Thus 2n repulsion between halogen ions is considered in all crystals, except in KF, where the 2n interaction is between potassium ions.

The GFs are calculated for each surface wavevector $K$ over a mesh of 101 equally-spaced values of the frequency from zero to the maximum value of the crystal. The numerical integration over the wave-vector component normal to the surface from 0 to $\pi/r_0$ has been performed on a mesh of 97 points, which is equivalent to a slab calculation with 192 layers.

In such a large value rests the main difference between the GF method and the slab calculations of Chen et al. Here, unlike slab calculations, there is no interference between deeply penetrating surface modes, pertaining to opposite surfaces, such as long-wavelength Rayleigh modes; moreover, the surface projected bulk bands form a continuum, well separated from local modes. Actually, the main advantage of the GF method is to allow for an accurate determination of the surface-projected phonon densities which are useful quantities in the calculation of surface thermodynamical properties and vibrational response functions.

In order to offer a guideline for understanding the calculated dispersion curves we present a classification of surface phonons according to their character and polarization.

Along symmetry directions $\overline{\Gamma X} : \mathbf{K} = (\xi, \xi)$; $\overline{\Gamma M} : \mathbf{K} = (\xi, 0)$ the bulk bands and related surface modes have either sagittal ($\perp$) or parallel ($\parallel$) polarization. Sagittal modes can be either quasi shear vertical (SV) or quasi longitudinal. Parallel and shear horizontal (SH) polarizations are synonymous. We associate $TA_1$, LA, $TO_1$ and LO bulk modes with sagittal bands; $TA_2$ and $TO_2$ bands with parallel bands. Often, but not always, the band edges correspond to certain dispersion curves of the bulk along symmetry directions.

With regard to the identification and classification of surface modes the simplest way is to associate each surface mode with the corresponding bulk band which the surface mode comes from. Roughly speaking we expect at least one surface mode from each of the six bulk modes.

In doing that we follow, and strongly suggest as a convention, the general principle that labels should correspond to well defined spectroscopic entities, with special regard to polarization and (optical or acoustic) character. Therefore we adopted as long as possible the notations $(S_j)$ of Chen et al., but we were forced to change them wherever they do not fulfill the above requirement. This often occurs for the non-crossing behaviour of quasi-shear horizontal and quasi-sagittal branches in nonsymmetry directions.

In Table 2.1 we display the classification of surface modes according to the above concepts.

**Table 2.1.** Classification and conventional name of surface mode according to their bulk bands of origin and polarization

| Bulk band | Surface mode | Polarization | Conventional name |
|-----------|--------------|--------------|-------------------|
| $TA_1$ | $S_1$ | Quasi SV | Rayleigh wave |
|  | $S_8$ |  | Crossing mode |
| $TA_2$ | $S_7$ | SH |  |
| LA | $S_6$ | Quasi longitudinal |  |
| $TO_1$ | $S_4$ | Quasi longitudinal | Sagittal Lucas mode |
| $TO_2$ | $S_5$ | SH | Parallel Lucas mode |
| LO | $S_2$ | Quasi SV |  |
|  | $S_3$ |  |  |

The sagittal mode $S_1$ corresponds to a Rayleigh wave in the continuum limit ($\mathbf{K} \to 0$) [2.40]. The SH mode $S_7$ is another acoustic surface mode with $\parallel$ polarization existing along the (110) direction in cubic crystals. It was found by G.P. Alldredge [2.41]. $S_6$ is an acoustic mode peeled off from the LA lower edge; normally it is localized at the zone boundary. The sagittal resonance $S_8$, crossing the LA band in all directions, may appear in some crystals as a folded prolongation of the Rayleigh wave. Its intensity is appreciable in crystals with nearly equal ionic masses (NaF, KCl).

$S_4$ and $S_5$ form the pair of Lucas modes (LM) [2.42]. They have optical character and become degenerate at the zone center ($\overline{\Gamma}$ point). For $K \rightarrow 0$ along x, $S_4$ and $S_5$ become linearly polarized along x and y, respectively. $S_2$ is a microscopic optical mode whose polarization is quasi SV everywhere, and exactly normal to the surface at $\overline{\Gamma}$ and $\overline{M}$. At the zone boundary $S_2$ turns out to be associated with $TO_1$ whereas another mode, $S_3$, comes from the edge of LO.

According to *Maradudin* et al. [2.9], and also according to *de Wette* [2.43], $S_2$ identifies with the surface-phonon polariton (the so-called Fuchs-Kliever mode) of the semiinfinite crystal provided the dispersion of the optical branches is so effective to push this mode down in the gap below the LO band [2.44]. Calculations for NaCl and MgO slabs in the rigid ion model ([2.43] and [2.45], respectively) offer an example.

The connection between the polarization of a surface mode and that of the related band modes might appear to be rather complicated at first glance. We give here a simple guideline. Sagittal surface modes are normally elliptically polarized except at $K = (1,0)\,\pi/r_0$ ($\overline{M}$-point), where all modes have linear polarization for symmetry reasons. We note that at $K = 0$ ($\overline{\Gamma}$-point) longitudinal and transverse bulk modes are polarized, respectively, along z and x (or y), whereas at the $\overline{M}$ point the bulk modes at the lower edge are polarized along x if longitudinal, and along z if transverse ($TA_1$ or $TO_1$). Therefore $S_1$ is normally z-polarized at $\overline{M}$ and becomes more and more elliptical as $K \rightarrow 0$. The polarization, however, remains elliptical in the continuum limit due to the macroscopic nature of RW. On the contrary, the microscopic optical modes $S_4$ and $S_2$ become linearly polarized as $K \rightarrow 0$, obviously along x and z, respectively, like the corresponding $TO_1$ and LO bands at $\overline{\Gamma}$.

The behaviour of optical $S_4$ and $S_2$ ($S_3$) modes for $K$ varying across the surface Brillouin zone up to the zone boundary cannot be reduced to the simple scheme shown above due to their mutual hybridization. Labels and polarizations are often interchanged. This is seen, for instance, at $\overline{M}$, where x- and y-polarized surface modes are degenerated in pairs for symmetry reasons, like at $\overline{\Gamma}$ : in some cases (e.g., Na halides) the degenerated pair is ($S_5$ and $S_2$) whereas at $\overline{\Gamma}$ the degenerated pair is ($S_4$, $S_5$).

### 2.6.1 LiF and NaF

We show first the surface phonon dispersion curves of LiF and NaF for which extensive experimental investigation has been performed (Figs. 2.3–8). For sake of clarity, along symmetry directions (Figs. 2.3,4,6,7) we plot sagittal and parallel dispersion curves separately. Along the zone boundary $\overline{MX}$, however, sagittal and parallel components are mixed together and are shown superimposed in Figs. 2.5 and 2.8. Heavy lines are surface modes. Thin lines are band edges of bulk modes. When a dispersion curve enters a band, the surface-localized mode transforms into a resonance. The RW dispersion curve ($S_1$) appearing in Figs. 2.3 and 2.6 is compared to the data obtained from He scattering time-of-flight (TOF) spectra (open circles [2.46,47] and black squares [2.48]).

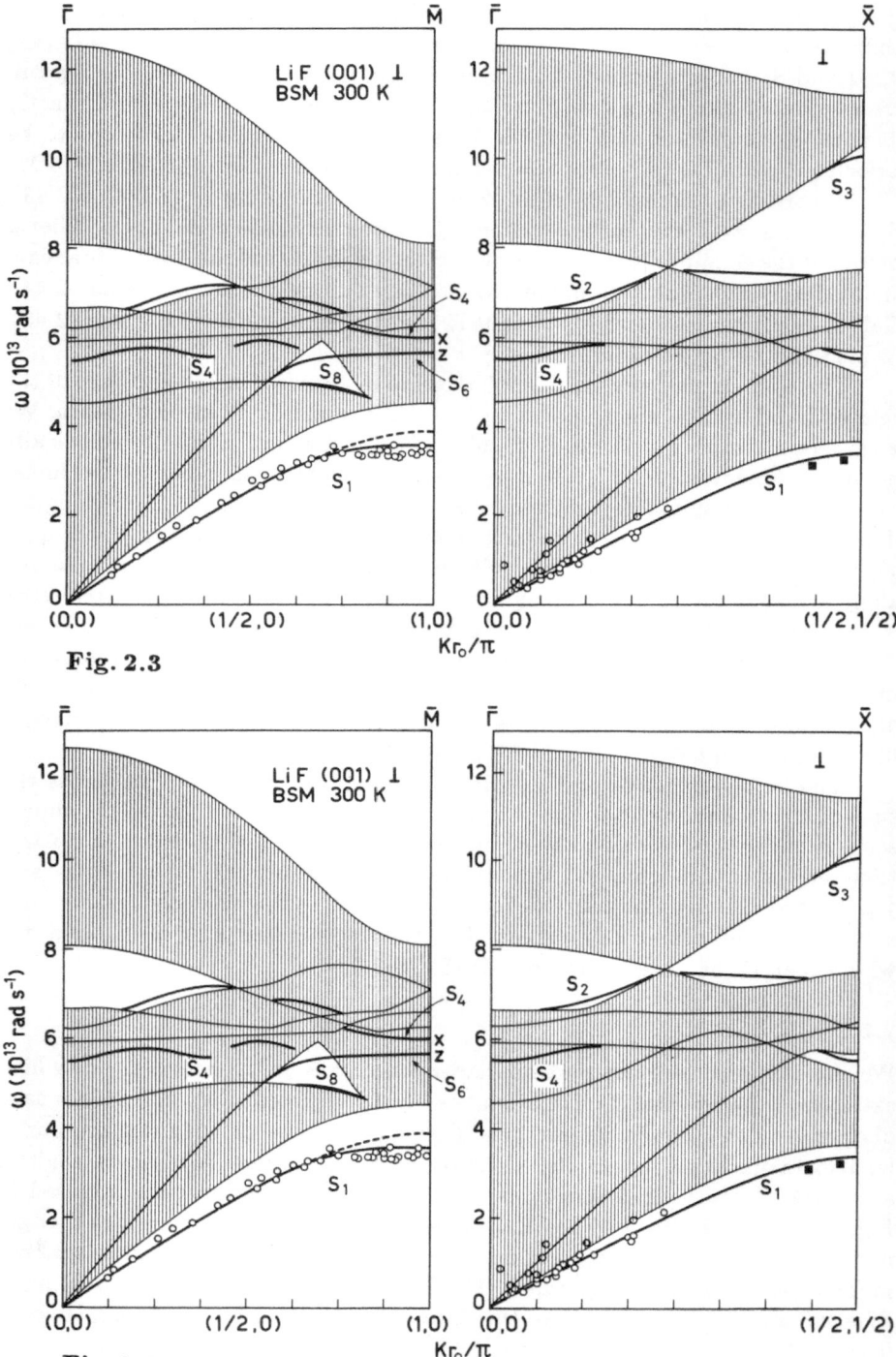

Fig. 2.3

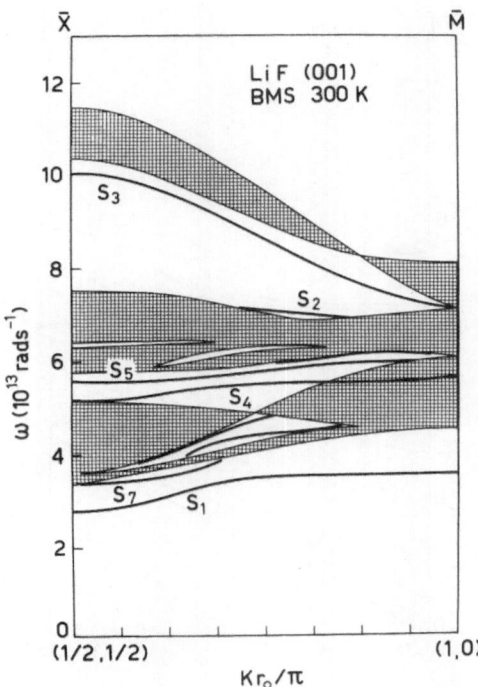

LiF (001)
BMS 300 K

$S_3$

$S_2$

$S_5$

$S_4$

$S_7$   $S_1$

$\omega$ ($10^{13}$ rads$^{-1}$)

(1/2,1/2)            (1,0)

$Kr_o/\pi$

**Fig. 2.5.** Surface phonon dispersion curves of LiF(001), along $\overline{MX}$ direction. Here sagittal and parallel components intermix and we display them altogether

The experimental points follow very well the theoretical dispersion curve along the two directions in both crystals. However, both slab [2.39] and GF calculations [2.37] do not predict correctly the Rayleigh frequency at the $\bar{M}$ point (broken line in Fig. 2.3) of LiF. An adjustment of bulk fluorine ion polarizability gives a better result (heavy line); still a 17 % increase of the surface F$^-$ polarizability is needed to obtain a perfect agreement [2.37]. Among the various mechanisms which can lower the zone-boundary frequency [2.49], we considered the polarizability change because the misfit is found when the anions move in the $\bar{M}$-point Rayleigh mode, as in LiF, and is not found when the anions are at rest, as in NaF (Fig. 2.6, open circles). Moreover, the Rayleigh wave at the $\bar{M}$ point is much more sensitive to the ionic polarizabilities than any other surface mode [2.38], so that no alteration of the good agreement at the $\bar{X}$ point is produced by the change of F$^-$ polarizability.

**Fig. 2.3.** Surface phonon dispersion curves of LiF(001) along symmetry directions for sagittal polarization ($\perp$). Calculations have been made by the GF method with BSM room-temperature data. *Heavy lines* are surface modes; *shaded areas* correspond to bulk bands projected to the surface. Atom scattering data are displayed: *open circles* from [2.46]; *black squares* from [2.48]

**Fig. 2.4.** The same as Fig. 2.3 for parallel polarization ($\parallel$)

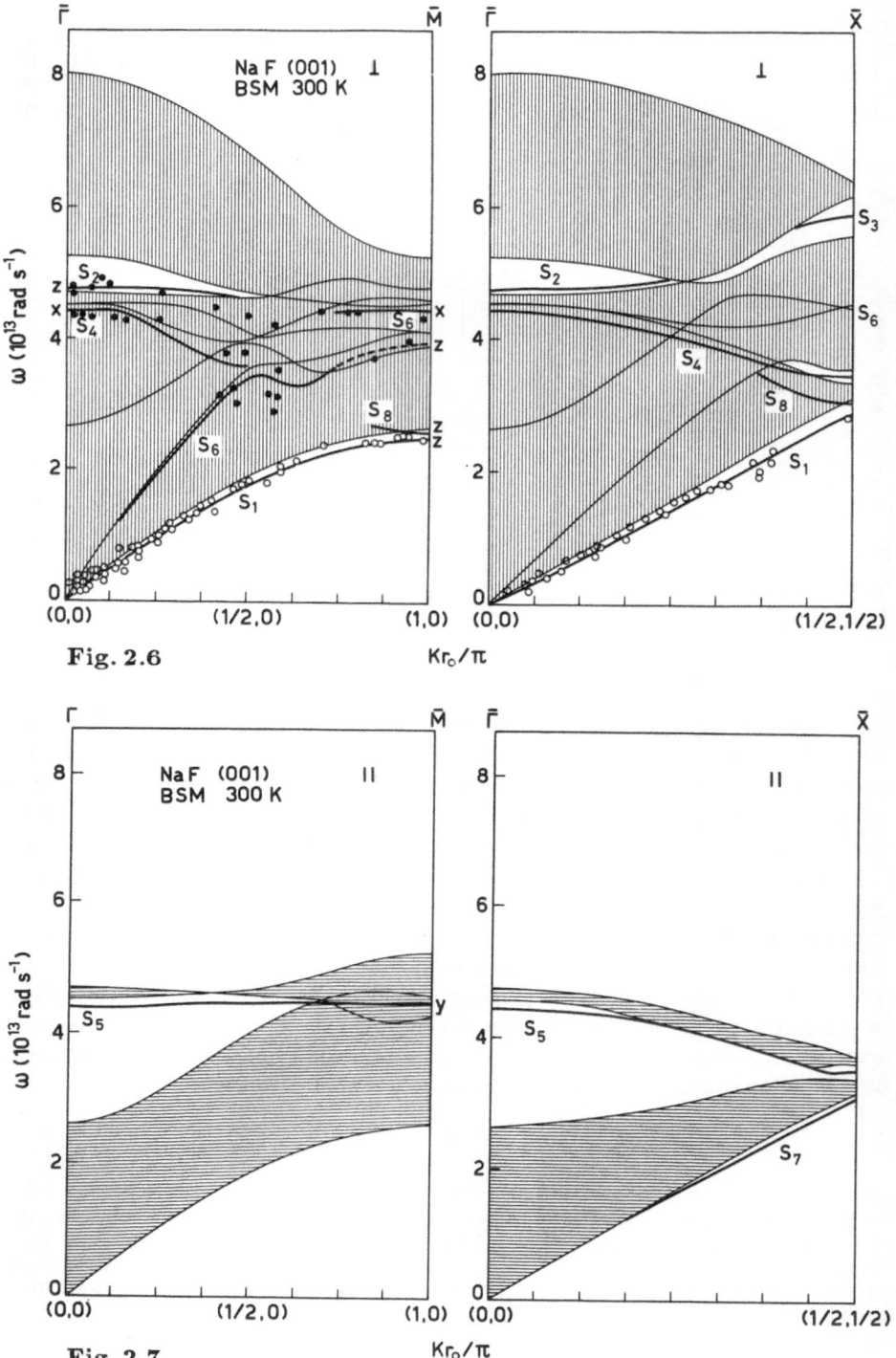

Fig. 2.6

Fig. 2.7

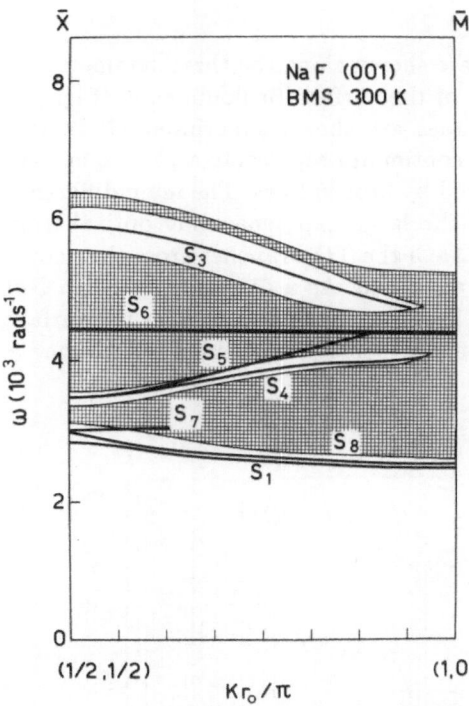

**Fig. 2.8.** The same as Fig. 2.5 for NaF(001)

The problem of relaxation in LiF(001) has recently been reinvestigated by *de Wette* et al. in a shell-model slab calculation, where surface dynamics is consistently based on the surface equilibrium configuration. The change of three-body forces is found to be the most efficient mechanism for shifting the $\overline{M}$ Rayleigh mode [2.50].

In NaF fluorine ions move mainly in the optical modes. Since anions have usually a larger interaction with He atoms than cations, owing to their large size, optical surface phonons have been detected in NaF(001) (Fig. 2.6, black dots) [2.51]. The experimental points are in good agreement with the theoretical $S_2$, $S_4$ and $S_6$ branches. However, for $S_4$ and $S_6$ a simpler hybridization scheme is suggested by experiment: $S_4$ should stay above and continue up to the zone boundary, while $S_6$ bends down connecting with $S_4$.

**Fig. 2.6.** Surface phonon dispersion curves of NaF(001) along symmetry directions for sagittal polarization ($\perp$). Calculations have been made by the GF method with BSM room temperature data. *Heavy lines* are surface modes; *shaded areas* correspond to bulk bands projected to the surface. *Open circles* from [2.47], *black dots* from [2.51]

**Fig. 2.7.** The same as Fig. 2.6 for parallel polarization ($\parallel$)

### 2.6.2 NaCl and NaI

The dispersion curves of these crystals are shown along the three symmetry directions delimiting the irreducible part of the surface Brillouin zone (Figs. 2.9 and 11). Here sagittal and parallel modes are shown superimposed. In this case a thick line embedded in the bulk continuum represents a pseudo-surface mode, whereas resonances are represented by broken lines. The main difference between these two crystals consists of the large gap separating optical from acoustical branches in NaI, whereas in NaCl the TO branches cross the acoustical bands and very small gaps occur at $K = 0$. As a consequence $S_4$ in NaI exists as an almost dispersionless mode localized in the gap all over the surface Brillouin zone, while in NaCl $S_4$ presents a rather complicated dispersion. A similar behaviour is found for $S_5$.

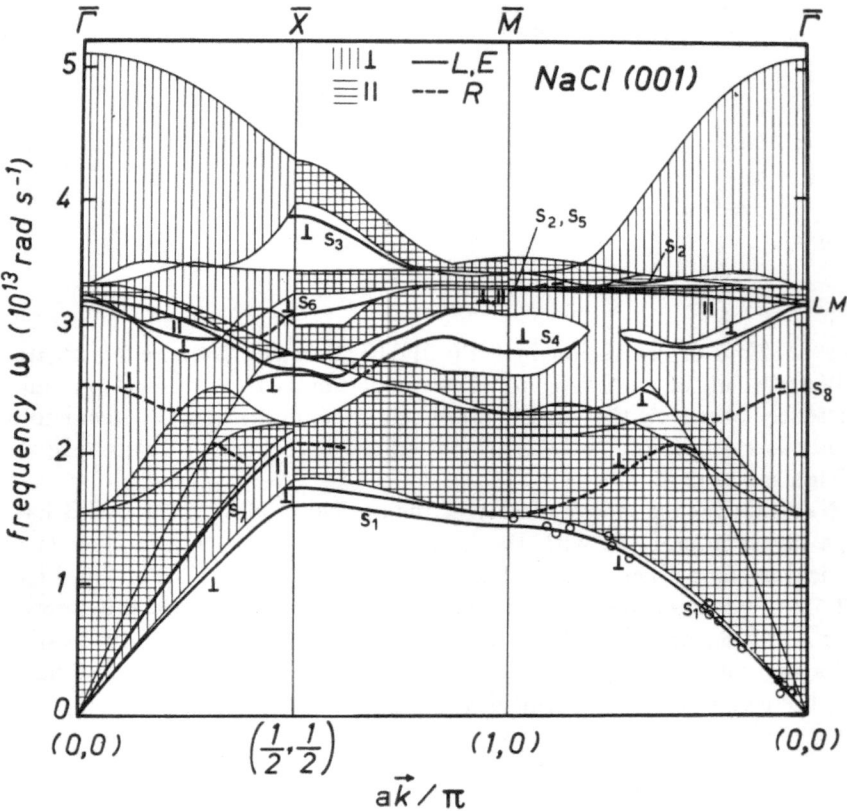

**Fig. 2.9.** Surface phonon dispersion curves of NaCl(001) calculated by BSM (extrapolated zero-temperature data) along the borders of irreducible part of the surface Brillouin zone. Sagittal and parallel modes are superimposed. *Thick lines* are localized modes, *broken lines* display resonances. Experimental *open points* from [2.52]

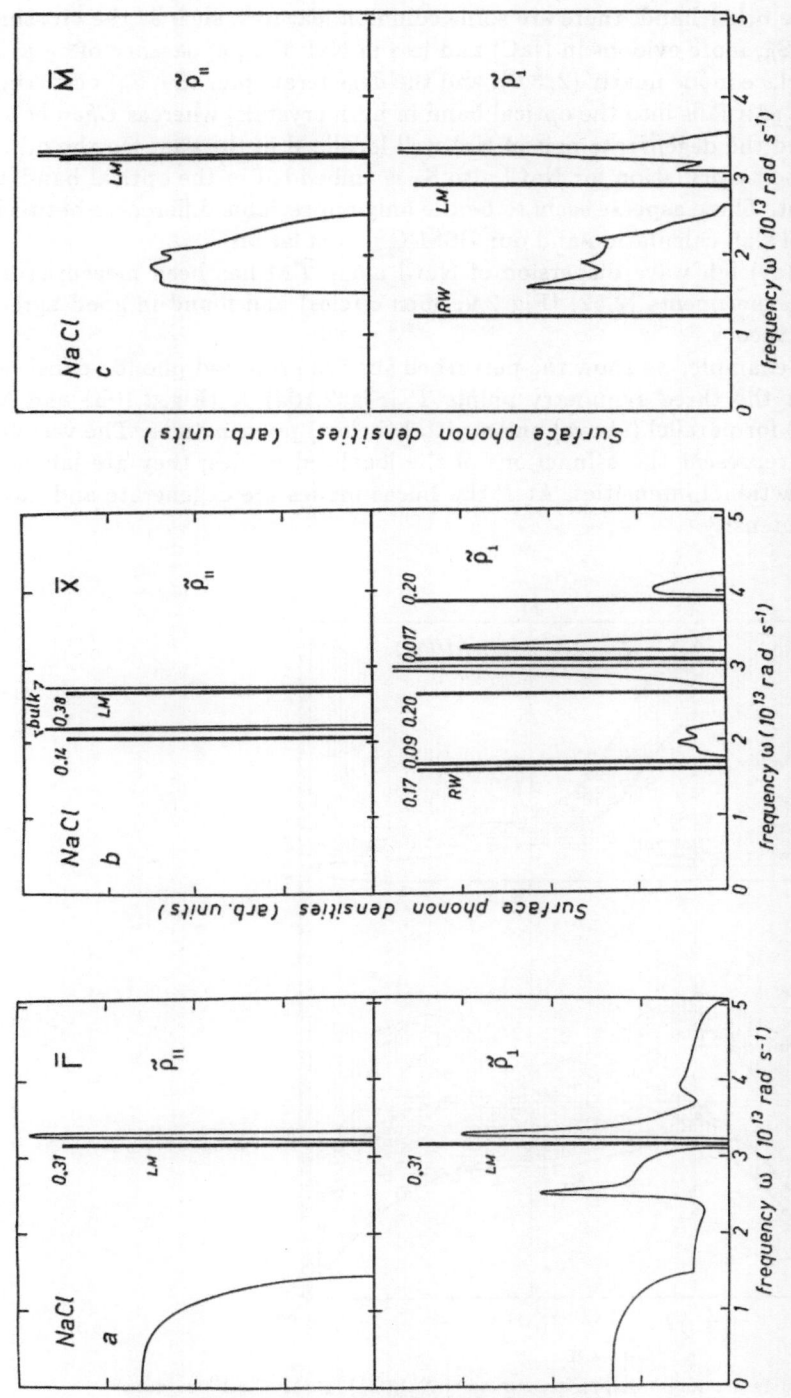

**Fig. 2.10a–c.** Perturbed surface projected phonon densities of NaCl at the three symmetry points $\overline{\Gamma}$ (a), $\overline{X}$ (b), and $\overline{M}$ (c) for parallel (*above*) and sagittal (*below*) polarizations. The vertical bars represent the δ-functions of the localized modes which are labelled by their fractional intensity

On the other hand, there are some common features, such as the crossing resonance $S_8$, more evident in NaCl and less in NaI; the appearance of $S_2$ as a pseudo-surface mode nearly $(2/3,0)$, and the degenerate pair $(S_2, S_5)$ occurring at $\overline{M}$. This pair falls into the optical band in both crystals, whereas *Chen* et al. [2.39] found the degenerate pair of NaI well localized in the gap. On the other hand, in their calculation for NaCl also $S_4$ is embedded in the optical band at the $\overline{M}$ point. These aspects seem to be the only appreciable differences between shell-model slab calculation and our BSM-GF calculation.

The Rayleigh wave dispersion of NaCl along $\overline{\Gamma M}$ has been measured by scattering experiments [2.52] (Fig. 2.9, open circles) and found in good agreement with theory.

As an example, we show the perturbed surface projected phonon densities of NaCl at the three symmetry points $\overline{\Gamma}$ (Fig. 2.10a) $\overline{X}$ (Fig. 2.10b) and $\overline{M}$ (Fig. 2.10c) for parallel (above) and sagittal (below) polarizations. The vertical thick bars represent the $\delta$-functions of the localized modes; they are labelled by their fractional intensities. At $\overline{\Gamma}$ the Lucas modes are degenerate and have the same intensity.

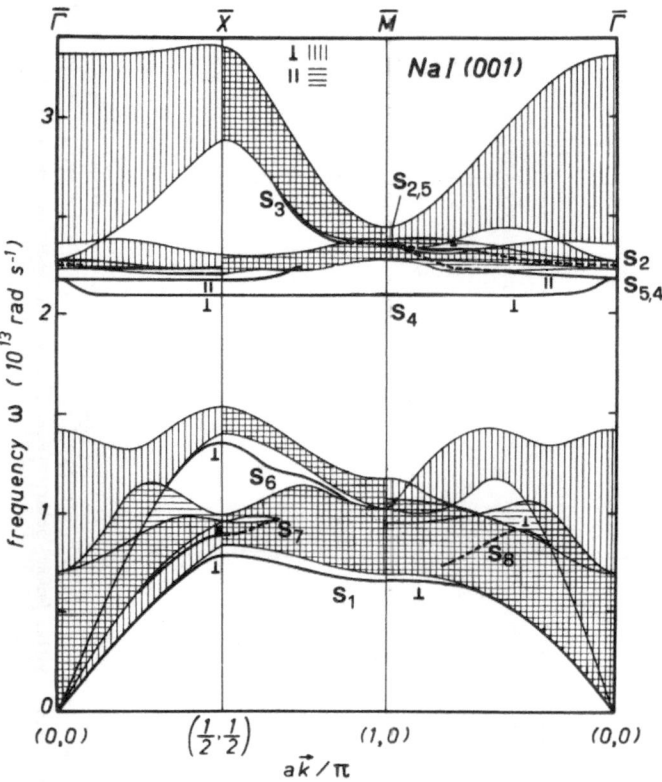

Fig. 2.11. Surface phonon dispersion curves of NaI(001), BSM (T=100 K data)

### 2.6.3 Potassium Halides

The most interesting feature common to all potassium halides (Figs. 2.12–15) is the appearance of the $S_2$ all over the surface Brillouin zone and always below the Lucas mode pair. Even at $K = 0$ its frequency is lower than $\omega_{TO}$. A similar situation was found in RbCl by *Chen* et al. [2.39]. In KBr and KI, $S_2$ is completely localized in the gap. Due to its polarization, which is everwhere approximately normal to the surface, $S_2$ is a good candidate for the experimental observation by means of inelastic atom scattering. At present we have only a few data on KCl (Fig. 2.12, black points) obtained by analyzing the time-of-flight (TOF) spectra of He$^4$ scattering measured in Göttingen [2.11,53]. Only few points, corresponding to rather weak TOF peaks, are found to be

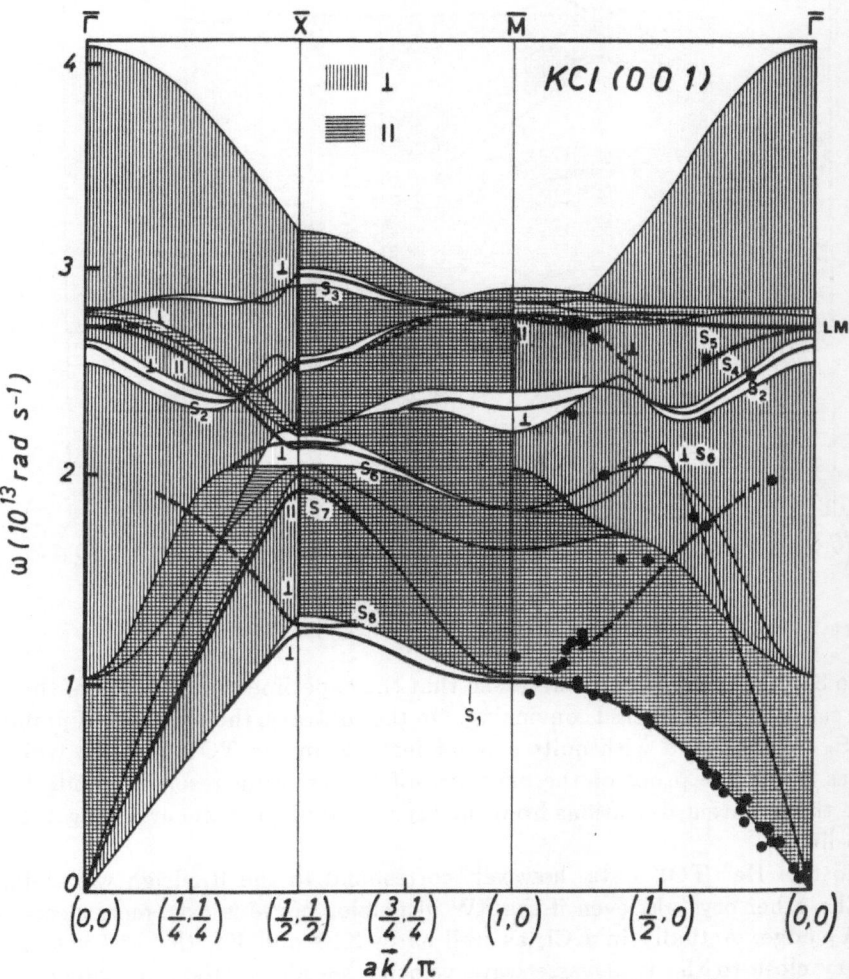

**Fig. 2.12.** Surface phonon dispersion curves of KCl(001), BSM (room-temperature data). Experimental *black points* from [2.11,53]

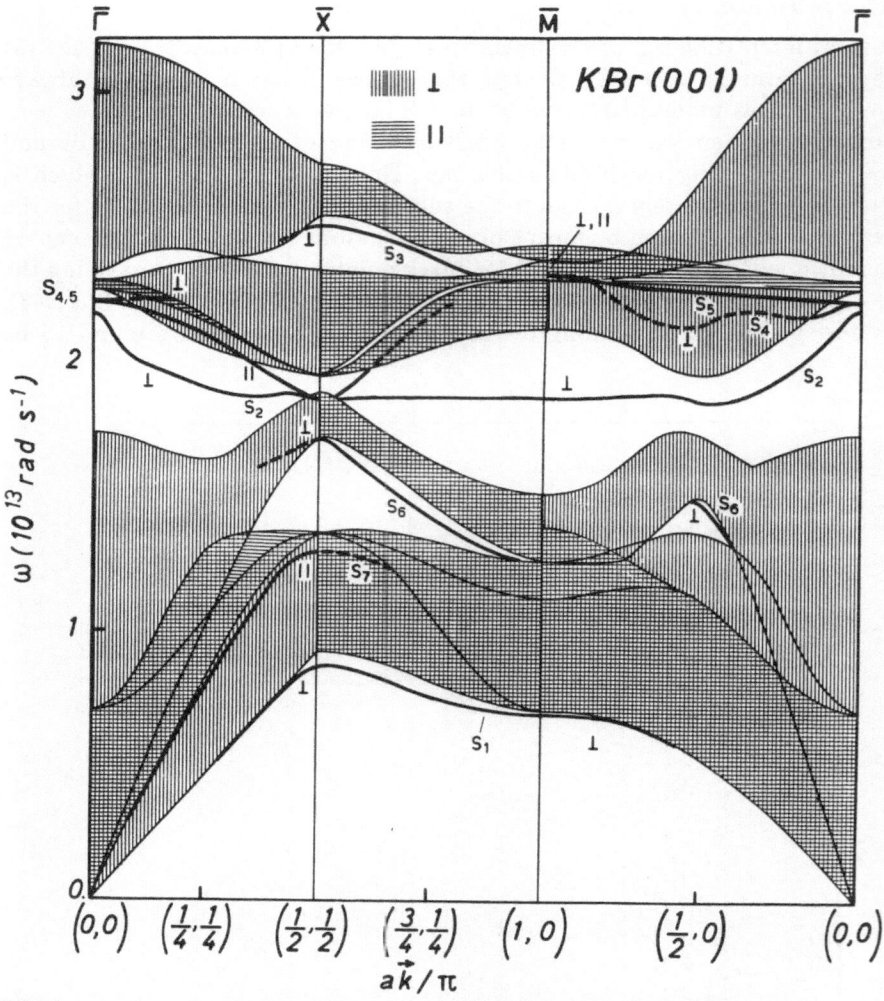

**Fig. 2.13.** Surface phonon dispersion curves of KBr(001), BSM (T=0 K data)

close to $S_2$ and $S_4$ dispersion curves, so that the experimental evidence of these modes cannot be considered convincing. On the contrary, the sequence of points along $S_8$ is associated with quite evident features in the TOF spectra, which provides a reliable proof of the existence of the crossing resonant mode $S_8$, despite the apparent deviations from the calculated dispersion curve (Fig. 2.12, broken line).

Most of He$^4$ TOF data, however, correspond to the Rayleigh wave, like in all the other crystals, even if the RW dispersion curve is extremely close to the TA$_1$ edge. Actually in KCl, as well as in KBr and KI, the RW velocity v is very close to the transverse wave velocity $v_T$ along (100), in agreement with the prediction of continuum elastic theory ($v/v_T$ = 0.988, 0.992, 0.994, respectively).

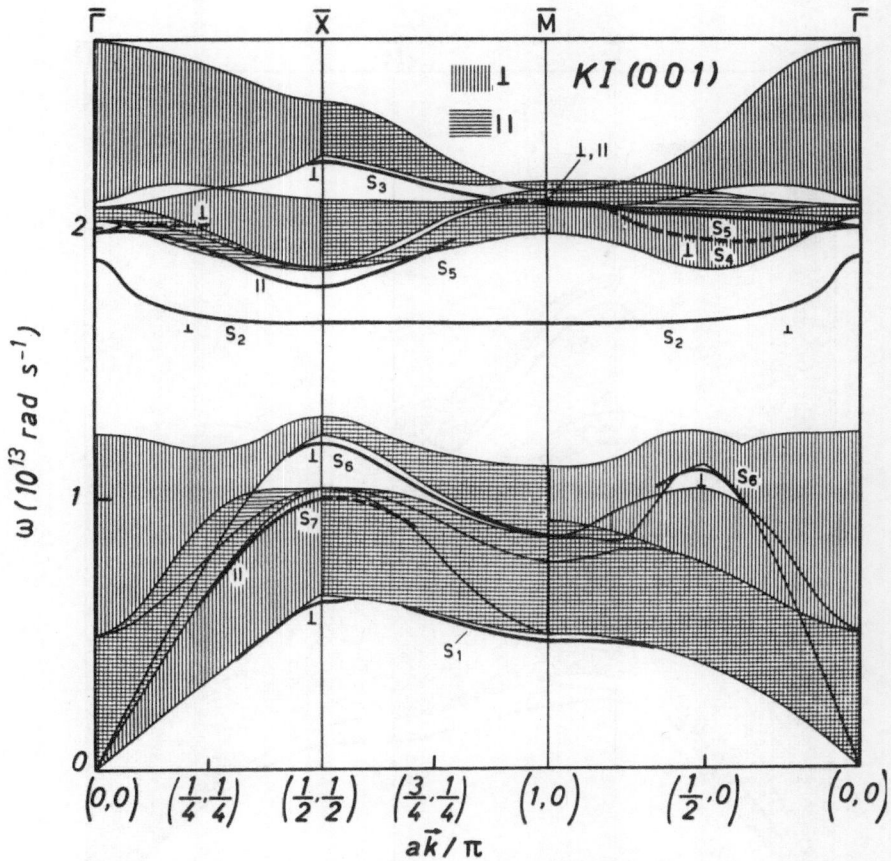

Fig. 2.14. Surface phonon dispersion curves of KI(001), BSM (T=0 K data)

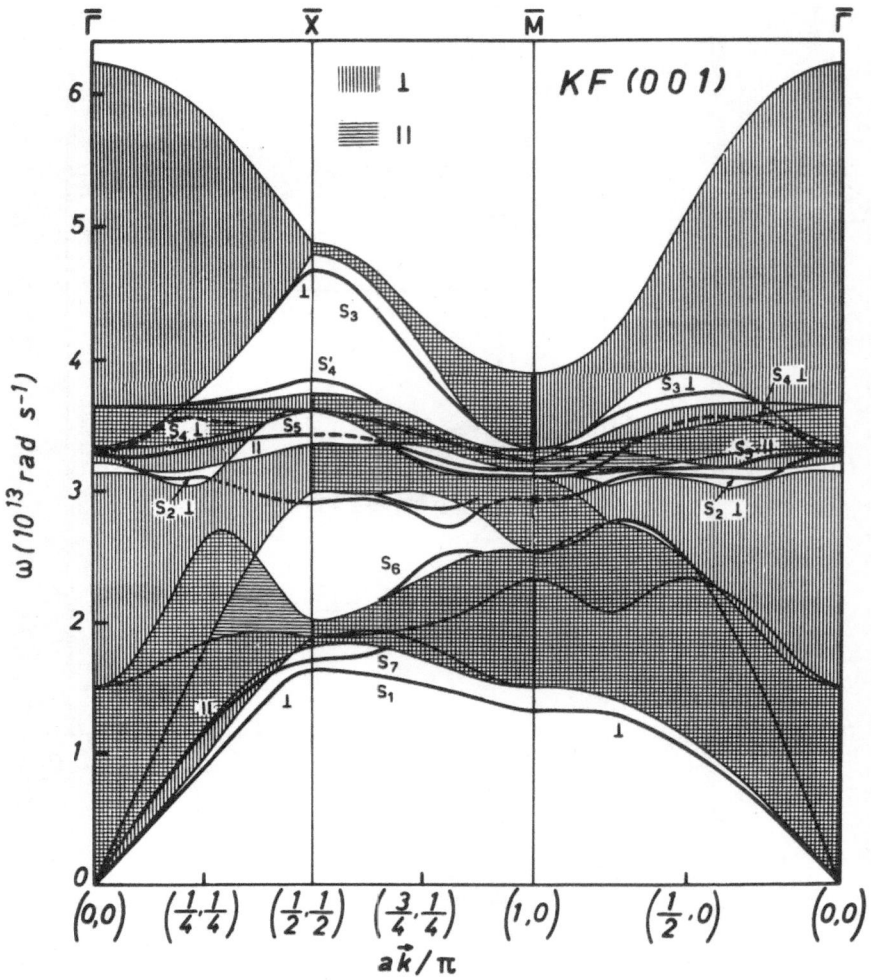

**Fig. 2.15.** Surface phonon dispersion curves of KF(001). BSM (room-temperature data)

## 2.7 Further Developments:
## The Study of Surface Phonon Anomalies

The GF method applies well to more complex perturbations, e.g., a change in surface force constants [2.22,26]. The difficult cases of stepped surfaces and reconstruction have been successfully treated by the method of GF generating coefficients developed by *Armand* and *Masri* [2.54]. More recently, *Goldhammer* et al. have approached the problem of the $(7 \times 7)$ reconstruction of the Si(111) and the $(8 \times 8)$ reconstruction of Ge(111) surface with related phonon instabilities by means of the GF method (Fig. 2.16) [2.17]. Many of the surface dynamical phenomena owe their complexity to the role of electron-phonon interaction. He-scattering measurements are revealing aspects of surface dynamics that were unexpected on the basis of the underlying bulk dynamics. Many microscopic mechanisms which are forbidden or inhibited in the bulk for symmetry reasons, are promoted by a symmetry reduction and manifest themselves at the surface. A typical example is that of surface phonon anomalies induced by electron-phonon coupling, which may eventually occur also in crystals with regular bulk phonon dispersion.

Most of the existing theories of surface lattice dynamics have been conceived for ideal, unrelaxed surfaces. There the change of atomic equilibrium positions at the surface (elastic relaxation) of electron density of states (electronic relaxation), and the consequent change in electron-phonon interaction are not taken into account, or at best, not in a self-consistent way.

The case of LiF(001) gave indication that the change in the electronic polarizabilities at the surface may induce important deviations from the ideal surface behaviour in the surface phonon spectrum. The role of surface elec-

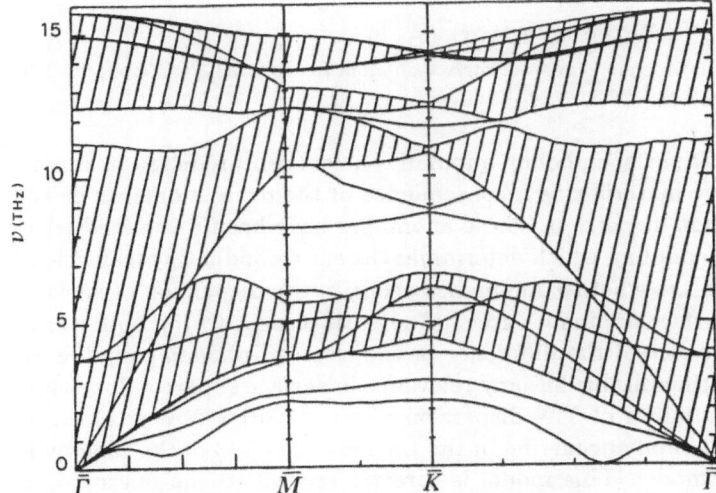

**Fig. 2.16.** GF calculated dispersion curves of surface phonons in Si(111) with a choice of surface force constants which yield phonon anomaly driving $(7 \times 7)$ reconstruction. From [2.17]

tronic states and surface electron-phonon interaction becomes apparently dramatic in noble-metal surfaces. High-resolution He TOF spectra from Ag(111) give evidence, in addition to the Rayleigh mode, for resonance below the LA bulk edge [2.55]. The downward shift of this resonance with respect to the LA edge is not expected in the ideal surface GF calculation peformed by *Armand* (Fig. 2.17) by means of the generating coefficients [2.56]. This behaviour is reproduced in a slab calculation peformed by *Bortolani* et al. [2.57] with a suitable parametrization of bulk dynamics and a softening of in-plane surface force constants. Again the microscopic origin of such softening apparently rests on the surface enhancement of the electronic susceptibility related to phonon-induced d-s hybridization of the electronic states [2.58].

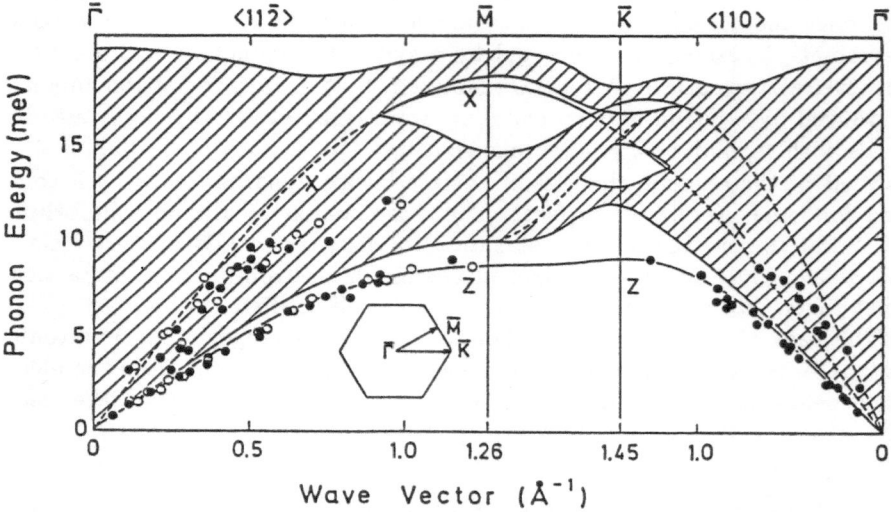

**Fig. 2.17.** GF calculated dispersion curves of surface phonons for Ag(111) are compared with experimental results (*filled* and *open circles*) from [2.55]

More spectacular surface phonon anomalies have been predicted in superconducting refractory materials as a consequence of the deep anomalies existing in the bulk phonon dispersion. These anomalies have been associated with the electron-phonon coupling which determines the superconducting transition. Interesting questions are whether anomalies occur in the dispersion of surface phonons; whether surface anomalies are shifted with respect to the bulk counterpart, as a consequence of the difference in the electron-phonon coupling at the surface; and whether all this has any relevance in surface superconductivity.

In a recent calculation of TiN dispersion curves [2.36] *Miura* et al. have explained the deep anomaly occurring in the LA branch at 2/3 of the zone by a cluster-deformation model. This model is directly related to the microscopic properties and emphasizes the importance of excitations of p-d hybridized states near the Fermi surface for the lattice vibrations in transition metal compounds. The calculated surface dynamics of TiN(001) by the GF method (Fig. 2.18) show that the dispersion curves of the Rayleigh wave ($S_1$) and the resonance $S_6$, and the optical mode $S_2$ are anomalous in the (100) direction [2.59]. The $S_6$ anomaly remains very close to the bulk anomaly, as a consequence of its prevailing bulk character, while the Rayleigh wave anomaly is shifted back to 1/2 of the zone. This means that the effective range of the electron-phonon interaction, determining the position of anomalies, is changed at the surface. The $S_2$ anomaly consists in the large softening towards the zone boundary, in contrast with the optical band edge increase and with the regular behaviour of $S_2$ in large-gap insulators (NaI, Fig. 2.11).

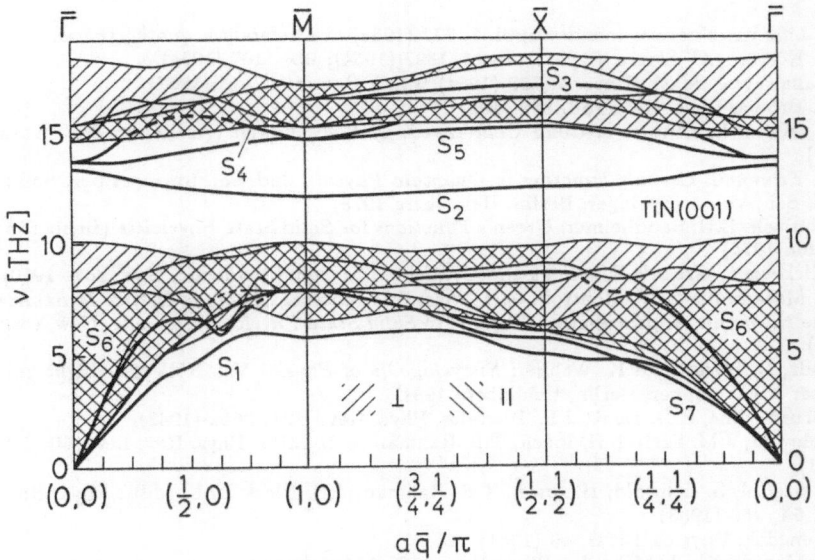

**Fig. 2.18.** GF calculated dispersion curves of surface phonons for TiN(001)

Such an anomaly has been recently found in transition metal carbides by means of high-resolution electron energy loss spectroscopy by *Oshima* et al. [2.60]. Moreover, the $S_1$ mode in $NbC_{0.96}(001)$ appears to be localized below the acoustic edge above 1/2 of the zone and suddenly raises up into the band above 3/4 of the zone, which suggests an anomaly shift with respect to the LA bulk anomaly at 2/3 of the zone, in agreement with the prediction for TiN(001).

Clearly, the interpretation of surface phonon anomalies points directly to the microscopic theory of surface dynamics, and represents a powerful significant test for the present models of electron-phonon interaction. Like the advent of neutron spectroscopy thirty years ago which stimulated the first microscopic models and theories of lattice dynamics, the present development in surface phonon spectroscopy will trigger more extensive work in the first-principle calculation of surface phonons, particularly in those systems where stability and reconstruction problems have a primary importance.

In this case the difficulty of dealing with too many degrees of freedom, as implied by slab calculations, becomes crucial, and the GF method seems to offer more practicable paths for future developments.

*Acknowledgements.* We thank Professor J.G. Skofronick (Florida State University, Tallahassee) for a critical reading of the manuscript and Professor J.P. Toennies (Max-Planck-Institut für Strömungsforschung, Göttingen) for keeping us constantly informed on the progress in He-scattering surface-phonon spectroscopy. This work was partially supported by NATO through Grant No. SA.5-2·05 RG (571/83) 463/83/TT.

# References

2.1    I.M. Lifshitz: Nuovo Cimento Suppl. **3**, 732 (1956) and references quoted therein
2.2    G.F. Koster, I.C. Slater: Phys. Rev. **94**, 1392 (1954); **95**, 1167 (1954)
2.3    J. Callaway: J. Math. Phys. **5**, 783 (1964); Phys. Rev. **154**, 515 (1967)
2.4    R.A. Brown: Phys. Rev. **156**, 889 (1967)
2.5    M.V. Klein: In *Physics of Color Centers*, ed. by W.B. Fowler (Academic, New York 1968)
2.6    E.N. Economu: *Green's Function in Quantum Physics*, 2nd. ed., Springer Ser. Solid-State Sci., Vol. 7 (Springer, Berlin, Heidelberg 1978)
2.7    S. Doniach, E.H. Sandheimer: *Green's Functions for Solid State Physicists* (Benjamin, London 1974)
2.8    G. Rickeyzen: *Green's Functions and Condensed Matter* (Academic, New York 1980)
2.9    A.A. Maradudin, E.W. Montroll, G.H. Weiss, I.P. Ipatova: *Theory of Lattice Dynamics in the Harmonic Approximation,* Supp. 3 to *Solid State Physics* (Academic, New York 1971)
2.10   H. Bilz, D. Strauch, R.K. Wehner: *Encyclopedia of Physics* Vol. XXV/2d (Light and Matter Id) (Springer, Berlin, Heidelberg 1984)
2.11   G. Brusdeylins, R.B. Doak, J.P. Toennies: Phys. Rev. **B27**, 3662 (1983)
2.12   S. Lehwald, J.M. Szeftel, H. Ibach, T.S. Rahman, D.L. Mills: Phys. Rev. Lett. **50**, 518 (1981);
       J.M. Szeftel, S. Lehwald, H. Ibach, T.S. Rahman, J.E. Black, D.L. Mills: Phys. Rev. Lett. **51**, 268 (1983)
2.13   G. Benedek: Physica **127B**, 49 (1984)
2.14   A.A. Maradudin, J. Mengalis: Phys. Rev. **133**, A1188 (1964)
2.15   L. Dobrzynski, G. Leman: J. Phys. (Paris) **30**, 116 (1969)
2.16   S.W. Musser, K.H. Rieder: Phys. Rev. **B2**, 3034 (1970)
2.17   W. Goldhammer, W. Ludwig, W. Zierau, C. Falter: Surf. Sci. **141**, 139 (1984);
       W. Zierau, W. Goldhammer, C. Falter, W. Ludwig: Proc. Int'l. Conf. on Superlattices, Urbana, Ill. (USA) (1984)
2.18   J.E. Black, B. Lacks, D.L. Mills: Phys. Rev. **B22**, 1818 (1980)
2.19   F. Garcia Moliner: Ann. Phys. (Paris) **2**, 179 (1977)
2.20   R.E. Allen: Surf. Sci. **76**, 91 (1978)
2.21   G. Arman: Phys. Rev. **B14**, 2218 (1976)
2.22   G. Benedek: Phys. Stat. Sol. **B58**, 661 (1973)
2.23   G. Benedek: Surf. Sci. **61**, 603 (1976)
2.24   G. Benedek, L. Miglio: in *Ab initio calculation of phonon spectra*, ed. by J.T. Devreese, V.E. van Doren, P.E. van Canp [Plenum, New York 1983)
2.25   G. Platero, V.R. Velasco, F. Garcia Moliner, G. Benedek, L. Miglio: Surf. Sci. **143**, 243 (1984)
2.26   Often the GF is defined with opposite sign. Here we keep the convention used in our previous works and adopted in [2.10]
2.27   A. Messiah: *Mécanique Quantique* (Dunod, Paris 1959)
2.28   J.M. Ziman: *Elements of Advanced Quantum Theory* (Cambridge U. Press, Cambridge 1969)

2.29　We note that the operator $\mathcal{R}$ is a function of the equilibrium positions. Only when they do not change by cutting the bands across $\Sigma$ the operator $\mathcal{R}$ is the same for $\phi$ and $\phi_0$. However, the RI condition is proved to be equivalent to equilibrium condition and therefore it contributes a fictious force field at the surface, required to equilibrate the surface in the unrelaxed configuration. The relationship between rotational invariance and equilibrium, and the effect of elastic relaxation have been discussed in [2.24]

2.30　Since the eigenvalue in (2.45 and 49) is multiplied by the mass matrix $\varrho_0$ and $\varrho$ are not merely frequency densities. They are normalized to $1/s\mu$, $\mu$ being the unit cell reduced mass

2.31　R.E. Allen, G.P. Alldredge, F.W. de Wette: Phys. Rev. B**4**, 1648, 1661, 1682 (1971); B**2**, 2570 (1970)

2.32　B.G. Dick, Jr., A.W. Overhauser: Phys. Rev. **112**, 90 (1958)

2.33　A.D.B. Woods, W. Cochran, B.N. Brockhouse: Phys. Rev. **119**, 980 (1960)

2.34　W. Cochran: CRC Crit. Rev. Solid State Sci. **2**, 1 (1971)

2.35　U. Schröder: Solid State Commun. **4**, 347 (1966);
U. Schröder, V. Nüsslein: Phys. Stat. Sol. **21**, 309 (1967)

2.36　M. Miura, W. Kress, H. Bilz: Z. Physik B**54**, 103 (1984)

2.37　G. Benedek, G.P. Brivio, L. Miglio, V.R. Velasco: Phys. Rev. B**26**, 497 (1982)

2.38　G. Benedek, F. Galimberti: Surf. Sci. **71**, 87 (1978); **118**, 713 (1982)

2.39　T.S. Chen, F.W. de Wette, G.P. Alldredge: Phys. Rev. B**15**, 1167 (1977)

2.40　A.A. Maradudin: *Festkörperprobleme* **21**, 25 (Vieweg, Braunschweig 1981)

2.41　G.P. Alldredge: Phys. Rev. Lett. **41A**, 281 (1972)

2.42　A.A. Lucas: J. Chem. Phys. **48**, 3156 (1968)

2.43　F.W. de Wette: In *Lattice Dynamics*, ed. by M. Balkanski (Flammrion, Paris 1978)

2.44　In the absence of retardation the macroscopic field producing the LO-TO splitting of optical modes works as an external field. When passing from the cyclic to the semiinfinite lattice the peturbation $\tilde{\varLambda}$ should also contain the change of the macroscopic field due to the depolarization effect of a single surface. This contribution to $\tilde{\varLambda}$ is seen to produce an extra pole in $\tilde{g}$ at the (complex) frequency where the dielectric constant is equal to $-1$ (surface-phonon-polariton): a proof has been given by G. Benedek in *Excited State Spectroscopy*, ed. by N. Terzi and U. Grassano (Editrice Compositori, Bologna 1986). In most cases, however, the polariton frequency falls into the LO continuum and the amplitude decays slowly into the solid on a much longer scale than that of microscopic surface modes. Thus its contribution to the surface projected phonon density is vanishingly small.

2.45　G. Lakshmi, F.W. de Wette: Phys. Rev. **22**, 5009 (1980)

2.46　G. Brusdeylins, R.B. Doak, J.P. Toennies: Phys. Rev. Lett. **47**, 1417 (1980); **16**, 437 (1981)

2.47　G. Benedek, J.P. Toennies, R.B. Doak: Phys. Rev. B**28**, 7277 (1983)

2.48　G. Bracco, E. Cavanna, A. Gussoni, C. Salvo, R. Tatarek, S. Terreni, F. Tommasini: Vuoto Sci. Tecn. **16** (1986) and to be published

2.49　E.R. Cowley, J.A. Barker: Phys. Rev. B**28**, 3124 (1983)

2.50　F.W. de Wette, U. Schröder, W. Kress: to be published

2.51　G. Brusdeylins, R. Rechsteiner, J.G. Skofronick, J.P. Toennies; G. Benedek, L. Miglio: Phys. Rev. Lett. **54**, 466 (1985)

2.52　G. Benedek, G. Brusdeylins, R.B. Doak, J.G. Skofronick, J.P. Toennies: Phys. Rev. B**28**, 2104 (1983)

2.53　G. Benedek, G. Brusdeylins, R.B. Doak, J.P. Toennies: J. Phys. (Paris) **42**, C6–793 (1981)

2.54　G. Armand, P. Masri: Surf. Sci. **130**, 89 (1983)

2.55　R.B. Doak, U. Harten, J.P. Toennies: Phys. Rev. Lett. **51**, 578 (1983)

2.56　G. Armand: Solid State Comm. **48**, 261 (1983)

2.57　V. Bortolani, A. Franchini, F. Nizzoli, G. Santoro: Phys. Rev. Lett. **52**, 429 (1984)

2.58　H. Bilz: private communication

2.59　G. Benedek, M. Miura, W. Kress, H. Bilz: Phys. Rev. Lett. **52**, 1907 (1984)

2.60　C. Oshima, R. Souda, M. Aono, S. Otani, Y. Ishizawa: Phys. Rev. Lett. **56**, 2401 (1986)

# 3. Surface Diffusion and Layer Growth

P. von Blanckenhagen
With 20 Figures

This chapter reviews surface diffusion and layer growth in relation to experimental methods for their analysis, where particular emphasis is placed on scattering experiments using x-rays, neutrons, electrons and He atoms as probes. Following a brief introduction to the microscopic description of surface diffusion by correlation functions and scattering laws, some typical experimental results on surface diffusion are presented, e.g., the anisotropy and the temperature dependence of surface diffusion on metal surfaces. A second part of this chapter then deals with epitaxial growth and the analysis of growth modes. Selected results of various experimental methods are discussed, e.g., the results of x-ray diffraction experiments on the growth of physisorbed overlayers. It is shown that scattering experiments are important for the study of surface diffusion and layer growth in order to prove microscopic theories for these phenomena.

## 3.1 Mass Transport and Growth Phenomena at Surfaces

Mass transport by diffusion of particles on surfaces is an important step in many surface phenomena such as layer growth, heterogeneous catalysis, phase transitions, segregation, sintering, etc. In recent years considerable progress has been achieved in the study of surface diffusion and the related phenomena owing mainly to significant improvements of experimental techniques. The growth of solid-surface layers involves a series of atomic steps which can be classified as follows [3.1]: condensation by adsorption, migration of adatoms by surface diffusion, incorporation of adatoms at growth sites, and nucleation of clusters. The events upon which the formation of overlayers depends are the same for the growth of crystals and of layers on substrates (epitaxial growth). The atomic steps as illustrated schematically in Fig. 3.1 take place not only on bare crystal or substrate surfaces, but also within and on each overlayer as it forms. These processes are usually thermally activated, involving, for example, activation energies for surface diffusion, desorption of adatoms and interdiffusion between different layers [3.2]. The activation energy for desorption is typically larger, by a factor of about 5, than the activation energy for surface diffusion [3.3]. This finding shows that diffusion events at surfaces may occur much more frequently than desorption events. Hence, diffusion processes play an important role in growth processes. The adsorbed atoms have to migrate to special lattice

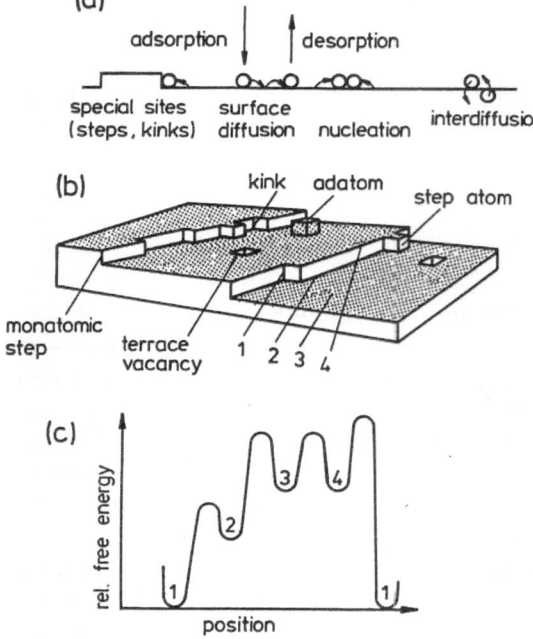

(a)

adsorption | desorption

special sites    surface     nucleation    interdiffusion
(steps, kinks)  diffusion

(b)

kink   adatom    step atom

monatomic
step    terrace    1  2  3  4
        vacancy

(c)

rel. free energy

position

Fig. 3.1a–c. Atomic steps in layer growth: (a) kinetic processes during deposition and growth (as considered in [3.2]); (b) the structure of real single crystalline surfaces with defects and adatoms; (c) relative free energies at different surface positions (*1-4,* as indicated in (b)) as discussed in [3.17]

sites where the surface layer is completed. Due to the reduced bonds of atoms at the surface, one expects a larger diffusion coefficient for the surface than for the bulk. Figure 3.2 shows an enhancement of the diffusion coefficient at the surface by several orders of magnitude. This is analogous to diffusion at grain boundaries, important in metallurgical processes, which is also strongly enhanced in comparison to diffusion in the bulk [3.4–6]. Surprisingly, near the melting point of metals the surface diffusion coefficient is considerably larger than expected from a linear extrapolation of an Arrhenius plot of the data measured at lower temperatures. Furthermore, it exceeds the diffusion coefficient of bulk liquids up to a factor of about 100 [3.7,8]. This extremely fast diffusion is not fully understood as yet, but it may be related to the effect of premelting (Sect. 3.3.5). Another diffusion phenomenon which attracts attention, in particular related with growth of multilayers for practical applications (e.g. for superlattices), is the interdiffusion between different layers [3.9,10]. By this process the properties of superlattices can be strongly modified.

In this chapter a brief review of surface diffusion and layer growth is given, which is intended to be illustrative rather than comprehensive. Only few typical results are presented and emphasis is upon scattering experiments. Experimental methods for the detection of surface diffusion and layer growth are tabulated because here is no space to discuss all of them in detail. For fuller appreciation of the field the reader may consult specialized literature on surface diffusion [3.1,8,11–21,30], on layer growth [3.2,22–34], and on superlattices and interfaces [3.9,35–43].

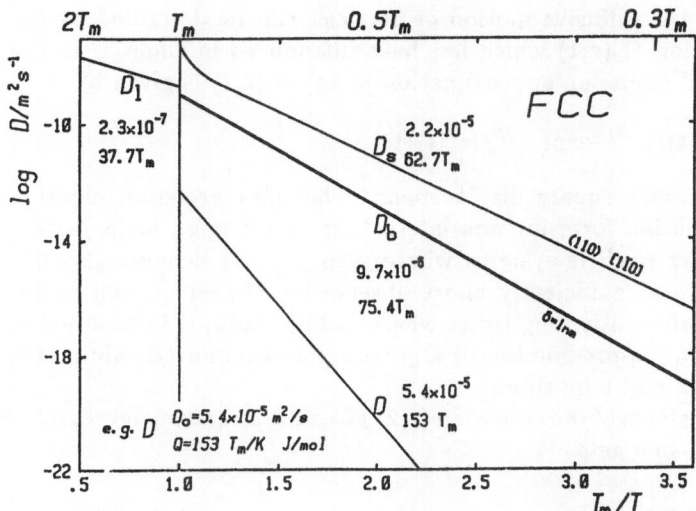

**Fig. 3.2.** Arrhenius diagram for the self-diffusion coefficient of fcc metals [3.4]: D(bulk), $D_b$ (grain boundaries), $D_l$ (liquid), $D_s$ (surface). The numbers in the diagram give the diffusion parameters, activation energies Q and preexponential factors $D_0$ for the diffusion coefficients, $T_m$ is the melting point and $\delta$ is the thickness of the grain boundaries

## 3.2 Description of Surface Diffusion

Due to the reduced bonding at the surface, diffusion processes are much more probable than in the bulk (Fig. 3.2). The surface diffusion depends strongly on the lattice structure of the crystal or the substrate. The lateral structure causes an anisotropy of the probability for diffusion on a perfect surface. An extreme anisotropy may restrict the diffusion at low temperatures to one-dimensional diffusion. On a real surface the diffusion is additionally influenced by steps and local defects.

By observing the migration of diffusing particles during a time t, the diffusion coefficient D can be obtained from the time-dependent ratio of the mean-square displacement $\langle r^2(t)\rangle$ to t

$$D = \frac{\langle r^2(t)\rangle}{ct} , \tag{3.1}$$

where c is a constant which may have the values 2 or 4 for one- or two-dimensional diffusion, respectively. For isotropic diffusion in the bulk c has the value 6. Equation (3.1) holds on the macroscopic scale and also on the microscopic scale for small displacements if the observation time is long in comparison to the lifetime of correlations between the diffusing particles (loss of memory condition).

*Microscopically* the diffusive motion of particles can be described by the self-correlation function $G_s(\boldsymbol{r}, t)$ which has been introduced in Chap. 1. In the frame of the so-called Gaussian approximation [3.44] $G_s(\boldsymbol{r}, t)$ is given by

$$G_s(\boldsymbol{r}, t) = [\pi \langle r^2(t) \rangle]^{-3/2} \exp(-r^2/\langle r^2(t) \rangle) , \qquad (3.2)$$

where $\langle r^2(t) \rangle$ is the mean-square displacement. The self-correlation function has exactly this Gaussian form for non-interacting atoms (e.g., for a perfect gas and for real many particle systems where each particle behaves also like a perfect gas particle at sufficiently short observation times) as well as for diffusing atoms for sufficiently long times where (3.1) is valid. It is assumed in (3.2) that the Gaussian approximation is a good description for all values of t, i.e., not only for short and long times.

The Fourier transform of the Gaussian (3.2) gives the scattering law $S_s(\boldsymbol{Q}, \omega)$ (Chap. 1). With (3.1) one gets

$$S_s(\boldsymbol{Q}, \omega) = \frac{DQ^2}{\pi[(DQ^2)^2 + \omega^2]} . \qquad (3.3)$$

The energy width of this Lorentzian is related to the diffusion coefficient by

$$\Delta E = 2\hbar DQ^2 . \qquad (3.4)$$

Since the surface diffusion is usually anisotropic a single Lorentzian is expected only for a fixed direction on a singly crystal surface. For a powder $S_s(Q, \omega)$ must be averaged over all orientations, i.e. it consists of a sum of Lorentzians of different widths. By scattering experiments the diffusion coefficient can be derived not only from the width of the scattering law at small Q values but also by an absolute evaluation of the scattering law at small energy transfers. From (3.3) it follows

$$S_s(\boldsymbol{Q}, 0) = \frac{1}{\pi DQ^2} . \qquad (3.5)$$

The scattering law for diffusion in adsorbed layers has been studied by inelastic neutron scattering for which the scattering cross-section is proportional to $S_s(\boldsymbol{Q}, \omega)$ (Sect. 3.3.2).

Generally the diffusion at surfaces may be more complicated than described by (3.3). Due to a strong corrugation of the surface, the particle may be trapped and oscillate at a fixed position for a time $\tau_0$ before it jumps during a time $\tau_1$ to a neighboring quasi-equilibrium position. If $\tau_1 \ll \tau_0$ this process is called jump diffusion. The correlation function for such a jump diffusion model yields also a Lorentzian scattering law, however, the width $\Delta E$ is proportional to $Q^2$ only at small values of Q (when $\langle r^2 \rangle Q^2 \ll 1$) and becomes independent of Q if $\langle r^2 \rangle Q^2 \gg 1$. In contrast to the case of continuous diffusion, the scattering law depends now also on the mean-square vibrational amplitudes of the particles in the quasi-equilibrium positions via the Debye-Waller factor [3.45].

The influence of diffusive motion on microscopic surface properties, as reflected in scattering laws, self-correlation functions and frequency distributions, has in few cases already been derived from inelastic neutron scattering data [3.45–50], from molecular-dynamics calculations (Chap. 6) and from model calculations [3.51–55]. Some of these results will be discussed in the following Sects. 3.3 and 7. The relation between the diffusion coefficient and the velocity autocorrelation function is discussed in Chap. 6.

The diffusion in a concentration gradient can be described by Fick's laws. In this case the diffusion coefficient is defined as the proportionality constant relating the mass flux to a concentration gradient. Applying the second Fick's law to the space and time dependent particle density $n(\mathbf{r}, t)$, the diffusion coefficient can be evaluated from experimental data for the position and time dependence of particle density gradients [3.7]. In the one-dimensional case $[n(\mathbf{r}, t) = n(x, t)]$ we have

$$\frac{dn(x, t)}{dt} = D \frac{d^2 n(x, t)}{dx^2} \; . \tag{3.6}$$

The development of $n(x, t)$ as a function of x and t in a density gradient depends also on the ratio of attractive and repulsive forces in the lateral interaction between the diffusing particles [3.51]. Equation (3.6) defines the so-called chemical diffusion coefficient which is, in general, not identical to the coefficient defined

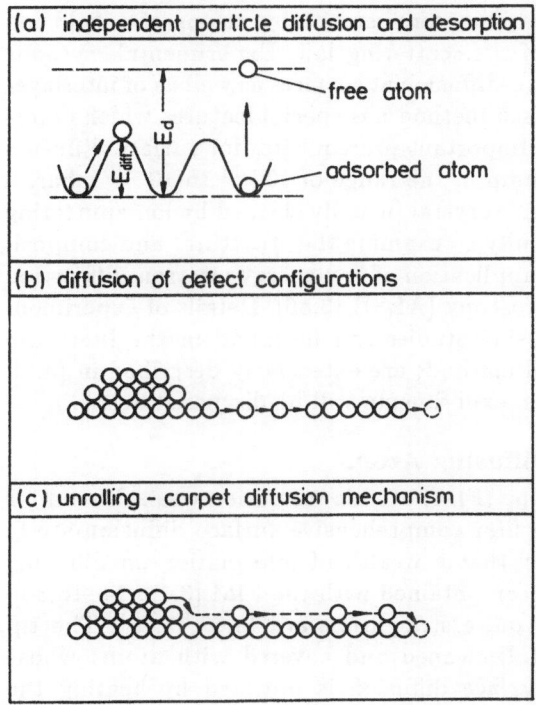

Fig. 3.3a–c. Diffusion mechanisms at surfaces: (a) independent particle diffusion and desorption. Energy barriers: $E_{diff}$ is the activation energy for diffusion and $E_d$ the activation energy for desorption; (b) diffusion by migration of defects; (c) migration of adatoms by the unrolling carpet mechanism

by (3.1), the so-called tracer diffusion coefficient. It was shown in [3.15b] that both equations yield the same values for D only if the interactions between the diffusing particles can be neglected.

If the diffusion occurs by thermally activated jumps over a barrier (Fig. 3.3), then the temperature dependence of the diffusion coefficient can be expressed by

$$D = D_0 \exp(-E_{diff}/k_B T) \ , \tag{3.7}$$

where $E_{diff}$ represents the activation energy and the preexponential factor $D_0$ is related to the mean-square jump length and to the mean vibrational frequency of the particles in the quasi-equilibrium positions [3.7]. Examples for the measurement of surface diffusion parameters are given in the following (Sect. 3.3). Interdiffusion between different layers of a superlattice is discussed in Sect. 3.6.5, and results of model calculations and simulations in Sect. 3.7.

## 3.3 Experimental Methods for the Study of Surface Diffusion and Typical Results

For a characterization of thermally activated diffusion processes at surfaces, the activation energy $E_{diff}$ and the preexponential factor $D_0$ according to (3.7) have to be measured for specified crystallographic directions. More detailed studies are needed to prove the self-correlation of the diffusing particles as can be achieved by direct observations, for instance by means of a field ion microscope, or by measurement of the scattering law. Experimental methods useful for analysis of self- and hetero-diffusion at surfaces as well as of interlayer diffusion are listed in Table 3.1. Each method has special features which determine the range of its application. Important prerequisites for surface diffusion studies are: a base ultra-high vacuum in the range of $10^{-11}$ to $10^{-10}$ mbar, a properly oriented and cleaned sample crystal (usually cleaned by ion-sputtering and/or annealing), and the possibility to examine the structural and compositional surface quality (usually by application of low energy electron diffraction (LEED) and Auger electron spectroscopy (AES)) [3.56]. Details of experimental arrangements for surface diffusion studies can be found in the literature quoted in Table 3.1. The standard methods are extensively described in [3.7]. A few examples of surface diffusion experiments will be discussed below.

### 3.3.1 Direct Observation of Diffusing Atoms

By means of the field ion microscope (FIM) atomic resolution was achieved for the first time in 1951 [3.67]. The first comprehensive surface diffusion study was reported in 1966 [3.71]. After that a wealth of information on diffusing atoms at crystalline surfaces has been obtained with the FIM [3.1,3,14–18,30]. The experimental procedures are quite straightforward [3.68,77,78]. The tip is made from the substrate crystal, cleaned and covered with atoms whose diffusion shall be studied. The surface diffusion is initiated by heating the

**Table 3.1.** Methods for the study of surface and interlayer diffusion

| Methods | Special features and applications | References |
|---|---|---|
| *Direct observation of diffusing atoms:* | | |
| Field ion microscopy | Single atoms and clusters on a small metal surface area, strongly bound adsorbates (low coverage up to high temperatures) | [3.3,67,68,78] |
| Electron microscopy | Heavy atoms | [3.66] |
| Scanning tunnel microscopy | Migrating atoms and detailed surface structures (simultaneously) | [3.65] |
| *Concentration gradient methods:* | | |
| Work function measurement | Hetero-diffusion, calibration necessary | [3.19] |
| Scanning AES analysis | Hetero-diffusion, material specific | [3.19] |
| Laser light diffraction | Self-diffusion on metal surfaces up to high temperatures | [3.8] |
| Scanning photoelectron analysis | Hetero-diffusion, material specific | [3.57] |
| Scanning secondary electron microscopy | Combined with collimated molecular beams for adsorbate deposition | [3.11,58] |
| Radioactive tracer detection | If appropriate isotopes available | [3.20,59] |
| Tip profile evoluton | Self-diffusion at high temperatures | [3.60] |
| Pulsed light or electron stimulated desorption | Low coverages, large temperature range | [3.20,61] |
| Surface ionization ion microscopy | Adsorbed alkali atoms, large temperature range | [3.69] |
| He-atom scattering | Elastic scattering from adsorbed particles | [3.62] |
| *Equilibrium methods:* | | |
| Field emission fluctuation method | Adsorbed gases up to high coverages in presence of an electrical field | [3.15] |
| Neutron scattering | Diffusion of adsorbates, inelastic scattering | [3.45–50, Chap. 1] |
| He-atom scattering | Diffusion of adsorbates, inelastic scattering (proposed) | [3.55, Chap. 1] |
| *Interlayer diffusion:* | | |
| X-ray diffraction | Superlattices (nondestructive) | [3.9,63,119] |
| Electron microscopy | Heterostructures (destructive) | [3.9,64] |
| Ion scattering | Mass dispersive thickness, concentration and position analysis of heterostructures (nondestructive) | [3.114a] |
| AES depth profiling | High resolution concentration analysis (destructive) | [3.103] |

tip to a certain temperature (typically 200–500 K) for a fixed interval of time. Generally, the electrical field is removed during this diffusion interval. Following this time interval the tip is quickly cooled. Then a field-ion picture of the tip can be taken of the new adatom positions on the surface. From a series of such pictures the mean-square displacement of a migrating atom can be measured as a function of time and temperature. A measurement accuracy down to 0.01 nm is possible. With (3.1) the diffusion constant can be obtained directly from the FIM data. According to (3.7) one gets $E_{diff}$ from the slope of the graph $\log D$ vs $1/T$.

Pictures of diffusing Re atoms on a W(211) surface taken at different times are shown in Fig. 3.4 as an example for FIM results together with pictures from corresponding hard sphere models. Two Re atoms move, in this case, as a cluster. They are separated by an empty channel. The dimer migrates by jumps of one of the atoms at a time, alternating between straight and staggered configurations. It migrates about five times as fast as single atoms. The activation energy is about 10 % lower for cluster diffusion than for single atom diffusion [3.3b]. The formation of clusters seem to be typical of diffusion of metal atoms on metal surfaces [3.1,3,17,18]. Clusters of 2, 3, 4, 5 or more atoms have been observed [3.3d]. From these studies it can be concluded that mobile clusters take part in layer growth processes on metal surfaces.

Most of the FIM data have been obtained so far for W surfaces. This is not due to limitations inherent in the experimental technique; it rather reflects the fact that such studies are just beginning [3.1].

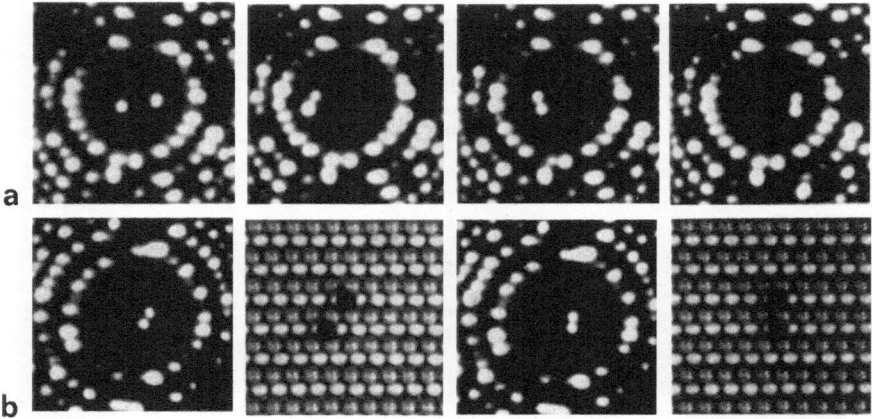

**Fig. 3.4a,b.** Diffusion of Re-atoms on a W(211) surface visualized by a field ion microscope [3.3b]: **(a)** Two individual atoms are initially in adjoining channels *(left image)*. Once they approach each other they coalesce into a pair and the resulting dimer moves as stable unit. The dimer diffusion shown in the following images occurs during time intervals of three seconds at 375 K. **(b)** Straigth and staggered Re-dimers (hard sphere models and FIM pictures)

### 3.3.2 The Study of Particle Correlations in Equilibrium Systems

As pointed out in Sect. 3.2, the surface diffusion may be described by the space and time dependent self-correlation function $G_s(\mathbf{r}, t)$. This correlation function is accessible by the following experimental methods: (1) direct observation of individual particles, (2) measurement of the field emission current fluctuations, and (3) measurement of the scattering law $S_s(\mathbf{Q}, \omega)$ by inelastic scattering experiments.

As described above the motion of diffusing atoms can, for example, be studied by a FIM. However, since the observed surface area is relatively small (it has a diameter of about 20 times the lattice constant) the results are influenced by boundary effects. Nevertheless, interatomic forces, or pair energy as a function of distance, can be directly derived by measuring the pair distribution [3.18b].

By method (2) thermal adsorbate concentration fluctuations about the equilibrium coverage are measured. As described in [3.15,16], the time auto-correlation function is obtained corresponding to an integration of $G_s(\mathbf{r}, t)$ over both initial and final locations of atoms extended over the entire emitting area. The application of this method is limited by the fact that it can only be applied in the presence of an electrical field. For adsorbates with high dipole moments or polarizabilities this causes a serious perturbation of the zero-field behaviour.

The diffusion of light adsorbed particles, which are not visible in the FIM, have been studied on W surfaces using the *current fluctuation method* [3.15c,72,73]. As an example the temperature dependence of the surface diffusion coefficient of $^1$H and $^2$H, respectively, at various coverages should be mentioned [3.72,73]. These results are suggestive of tunneling effects in surface diffusion. For diffusion by tunneling, a much weaker temperature dependence is expected than for the "barrier climbing" diffusion, and the diffusion coefficient of $^1$H should be much larger than that for $^2$H. Both effects are observed, in addition to thermally activated diffusion. However, there are discrepancies compared with results from other experiments on the $^1$H/W(110) system, where such a strong enhanced low temperature mobility of $^1$H was not detected [3.20]. In field emission experiments as described above, an enhancement of the hydrogen mobility by the emission current cannot be completely excluded. Hence, further experiments are needed to elucidate surface tunneling diffusion. Inelastic scattering experiments on equilibrium systems are desirable, because the self-correlation functions and their temperature dependence could be tested in the absence of an electric field by such experiments.

Recently, the quantum mechanical contribution to the diffusion of $^1$H, $^2$H and $^3$H, respectively, on a Cu(100) surface as a function of temperature has been investigated by model calculations using an anharmonic, temperature dependent potential for the interaction between the adatoms and a static Cu surface [3.76a]. These results show that at 100 K the quantum mechanical contribution to the diffusion coefficient may be several orders of magnitude larger than the contributions from thermally activated diffusion. Similar results have also been obtained by a completely different method [3.76b].

*Inelastic scattering experiments* performed within the framework of diffusion studies on equilibrium systems have hitherto been performed with neutrons only. For systems with large surface areas, like condensed gases on exfoliated graphite, the scattering law due to diffusive motions at the surface can be measured by quasi-elastic neutron scattering. By an analysis of the width of the quasi-elastic line for a $CH_4$ layer on graphite, the diffusion coefficient was obtained for different temperatures indicating that the layer is in the liquid state at T = 80 K and a coverage of $\theta = 0.9$ (Chap. 1). At T<76.6 K the $CH_4$ layer solidifies and the center of mass of the $CH_4$ molecules becomes fixed [3.47,48]. However, the molecules may still be free to rotate and this has been proved also by neutron scattering experiments. In Fig. 3.5a–c the shape of the quasi-elastic neutron scattering line is shown schematically for different phases existing in adsorbed layers at different temperatures. The different contributions to quasi-elastic scattering can be separated by analysis of the line shape if the resolution of the spectrometer is sufficient. The quasi-elastic scattering line for a solid $CH_4$ layer at 55 K measured with a high energy resolution is shown in Fig. 3.5d. This result provides the evidence that the molecules are still free to rotate in the solid layer.

In contrast to the case of translational diffusion, where $\Delta E$ is proportional to $Q^2$, for fixed rotators the width of the quasi-elastic line is independent of the momentum transfer $Q$ [3.45,48]. This property is used to separate the different contributions to quasi-elastic scattering. For several other adsorbed layers the surface diffusion has been studied by similar neutron scattering experiments [3.45–50]. In all cases samples with very large surface areas had to be used and, due to the large penetration depth of neutrons, the substrate contribution to the scattering intensity had to be substracted. More extended studies including measurements at larger momentum transfers would be desirable in

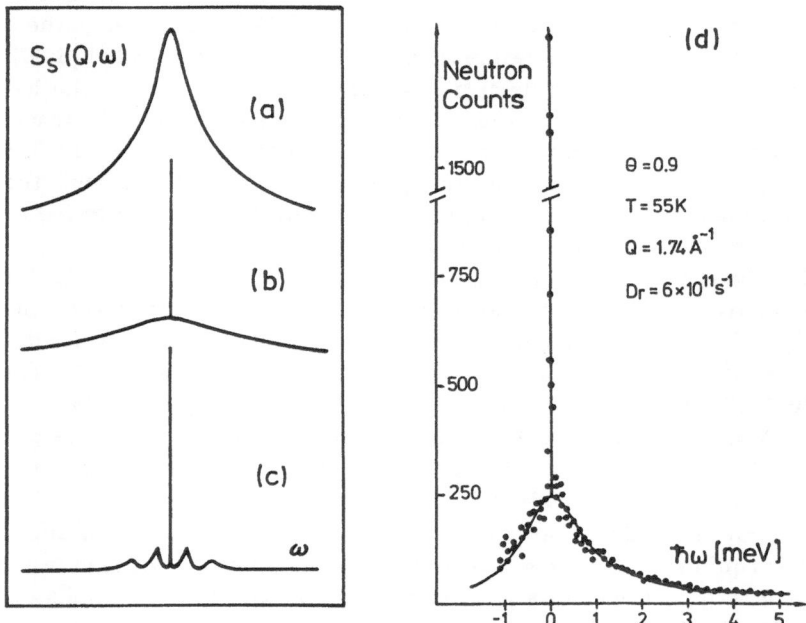

**Fig. 3.5.** Scattering law $S_s(Q, \omega)$ for diffusing particles: schematic for translational *(a)* and rotational *(b)* diffusion, and for a surface layer without diffusion *(c)*. A scattering law of type *(b)* measured by neutron scattering is shown in *(d)* for a solid $CH_4$-layer at 55 K and a coverage of 0.9. $D_r$ is the rotational diffusion coefficient and Q the momentum transfer [3.48]

order to develop detailed models for surface diffusion. Besides neutrons, He atoms can be also used for surface diffusion studies. He-atom beams are now widely applied in the analysis of surface structures [3.77,78] and lattice dynamics at surfaces (Chap. 1). Due to their high surface sensitivity and the high energy resolution which can be achieved in inelastic scattering experiments, He atoms are appropriate probes also for the study of the scattering law caused by diffusing particles. Inelastic He-atom scattering from diffusing adatoms has already been treated by model calculations [3.55]. However, so far only elastic He-atom scattering has been applied to diffusion studies (next Section).

### 3.3.3 Diffusion Studies in a Concentration Gradient

Concentration gradients of metallic adsorbates are usually created by applying an appropriate mask to shape the beam of evaporating atoms. This procedure was developed for many metals at room temperature, but for alkali and alkaline-earth elements cooling of the sample was necessary to get a sharp gradient. Closely fitting masks were used for the deposition of concentration gradients of condensed gases. Concentration profiles of different adsorbates have also been established with the aid of light or electron stimulated desorption [3.20]. For the measurement of self-diffusion coefficients of metals by the light-diffraction method periodic surface profiles are prepared by the photo-resist technique [3.8]. Surface defects may induce gradients in the adsorbate coverage on the atomic scale which can be detected by elastic scattering experiments (see below).

The change of macroscopic concentration gradients due to surface diffusion can be measured by a series of methods (Table 3.1). The work function method and the scanning Auger electron spectroscopy were applied to a broad range of adsorbate-substrate systems including gas-metal and metal-metal systems [3.19,20]. On the other hand, the light-diffraction methods were used especially to study self-diffusion on metal surfaces at high temperatures (up to the melting point) [3.8].

As an example of concentration gradient measurements let us consider the results for the system Li on W(110) obtained by *work function measurements* [3.20,80]. The concentration gradient is detected by the measurement of the change of the work function as a function of the position of the probe. A resolution of 20 to $50\,\mu$m has been achieved by different methods [1.19]. The adsorption of elements with low ionization energies, such as Li or Ba on refractory metals, are characterized by a variety of two-dimensional lattices being formed in the course of layer growth. This behaviour is due to the strongly polar, repulsive lateral interactions. The diffusion coefficient of Li on W(110) varies strongly with coverage $\theta$, for example, at 225 K by more than 3 orders of magnitude. Distinct maxima are observed at $\theta=0.13$, 0.22, 0.32, and 0.5. At all temperatures D decreases as the coverage approaches a monolayer. Using (3.7) and the diffusion data for different temperatures the dependence of $E_{diff}$

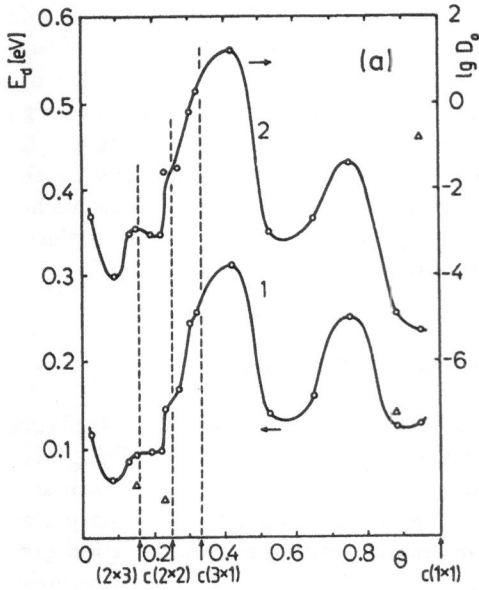

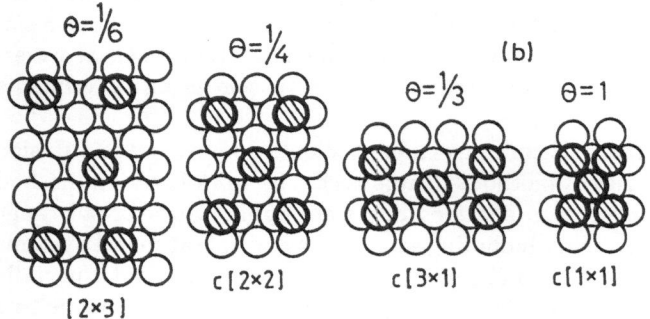

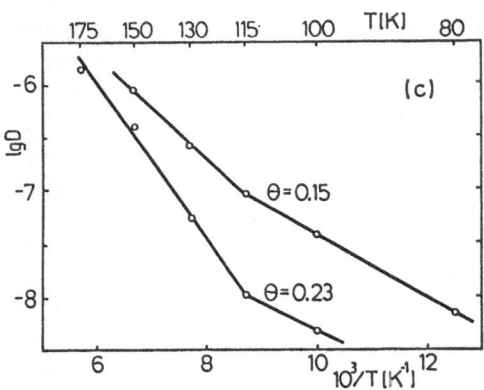

Fig. 3.6a–c. Diffusion parameters and structures for Li on W(011) [3.80]: (a) activation energy $E_d$ and preexponential factor $D_0$ as a function of coverage $\theta$ (the Li-layer structures are marked). (b) Elementary cells of two-dimensional Li lattices at different $\theta$. (c) Arrhenius plot of the diffusion coefficient $D$ for two different $\theta$ indicating phase transitions by changes of the slopes

and $D_0$ on $\theta$ was calculated (Fig. 3.6a). Four different two-dimensional lattices are formed by the Li adatoms as concluded from LEED studies and shown in Fig. 3.6b. The breaks in the Arrhenius plots of D data are observed for well-defined coverages (Fig. 3.6c) indicating order-disorder transitions. $E_{diff}$ values are lower in the disordered lattice. These results show clearly that there are strong relations between surface diffusion, layer growth and structural phase transitions. For the same system a relatively fast spreading has been observed for a monolayer or a double layer from an initial multilayer step on a precovered surface [3.80]. The diffusion coefficient estimated for this spreading process is of the order of $10^{-4} \, cm^2 \, s^{-1}$, this value is at least by one order greater than the D values in the submonolayer range. It is remarkable that this fast diffusion occurs through the completed monolayer. It might be due to a high mobility of defect configurations appearing in the commensurate layer by a local change in adsorbate density or due to diffusion of weakly bound adatoms of the second layer (unrolling carpet mechanism) (Fig. 3.3). Such a phenomenon, which has also been observed for the system Ba/Mo(110), could be described in another way by the migration of misfit dislocations or domain walls corresponding to the formation of solitons [3.20]. The soliton represent a linear region of destroyed commensurability, and a theoretical analysis predicts that the soliton diffusion can provide a much faster adsorbate transport than the migration of defect configurations. This proposed collective character of the extremely fast adsorbate diffusion has to be proved by further work including detailed scattering experiments.

Mass transfer studies of surface self-diffusion up to high temperatures became possible with the introduction of the *laser-light diffraction method* [3.8]. In these experiments the amplitude of a surface profile is measured at a constant temperature as a function of annealing time by recording the intensity distribution of laser-light diffraction patterns generated by the periodic profile. A comparison of recorded with calculated intensity distributions provides values for the amplitude A. The surface self-diffusion coefficient is then determined from the time dependence of A. This method was in particular applied to self-diffusion on metal surfaces at high temperatures up to the melting point (Sect. 3.3.5).

The sensitivity of thermal *He-atom scattering* to surface diffusion is demonstrated by the experimental result shown in Fig. 3.7. A steep increase of the specular intensity with increasing temperature appears which is caused by the onset of CO diffusion on the Pt(111) surface. At low temperatures the molecules are adsorbed at terraces and step sites more or less randomly. If the thermal energy is sufficiently high the ad-molecules on the terraces become mobile and start to migrate driven by local gradients in the binding energy. They can be trapped at step sites, where they are more strongly bound. The scattered He-atom intensity then increases because the bare terraces have a higher reflectivity than the covered terraces. The activation energy for the surface diffusion has been estimated from the characteristic temperature at which the scattered

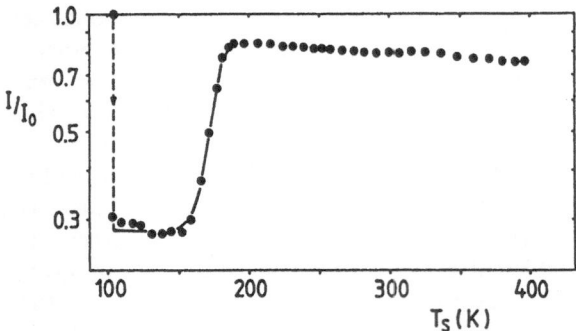

**Fig. 3.7.** Observation of the onset of surface diffusion by He-atom scattering for CO on Pt(111): The relative specular intensity I as a function of the surface temperature $T_s$. I is normalized to 1 for the bare Pt(111) surface at low temperatures. After deposition of CO at the initial temperature the intensity decreases strongly and remains nearly constant with increasing temperature until the CO molecules start to migrate [3.62]

intensity increases. The large cross section of adsorbed particles for He-atom scattering makes the study of surface diffusion over a large range of coverages possible.

### 3.3.4 Anisotropy of Surface Diffusion

Anisotropy effects may have different origins: (1) The activation energy and preexponential factor (according to (3.7)) for diffusing atoms or clusters may depend on the crystallographic direction. (2) The presence of steps with preferential orientations related to a special crystallographic direction may cause an anisotropy of the diffusion [3.7]. Both effects have been established by experiments. *Directional dependence* in the single atom and dimer migration on the W(110) surface have been studied using the FIM [3.18b]. The rate of diffusion is much faster in the [110] direction parallel to the closed packed rows of atoms than across these rows in the [100] direction. Such anisotropy is expected on the basis of simple arguments about the height of the energy barriers. Mass transfer studies by the profile decay technique have also shown such an anisotropy effect [3.8,70]. Below a certain temperature diffusion on the Ni(110) surface is considerable faster in the [110] direction than in the [100] direction (Fig. 3.8). However, near the melting point, where the surface diffusion coefficient becomes anomalously large only a relatively small anisotropy remains. A strong anisotropy on the Pt(110) surface was recently also observed by the same method [3.70b]. The isotropy observed earlier at low temperatures for surface diffusion on Pt(110) seems to be a result of the (2 × 1) surface reconstruction, which is obviously absent at higher temperatures.

For heterogeneous diffusion, anisotropy was observed for many systems in FIM experiments [3.17]. Studies by means of the surface ionization ion microscope [3.69] revealed directional anisotropy for the diffusion of K on a W(112) surface. Here the diffusion along the channels ([$\bar{1}\bar{1}1$] direction) is considerably faster than perpendicular to the channels ([$1\bar{1}0$] direction).

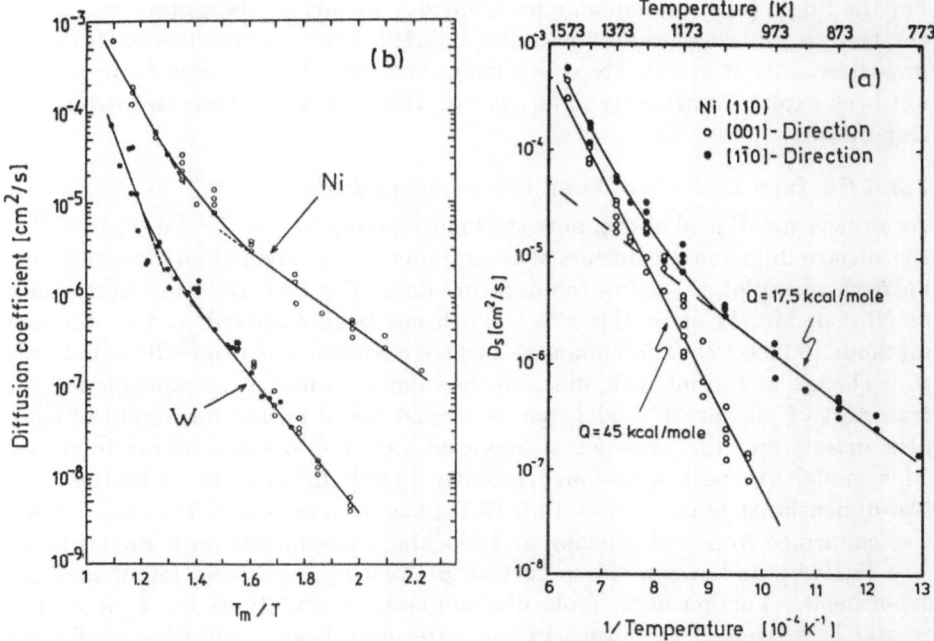

**Fig. 3.8a,b.** Temperature and direction dependence of the surface self-diffusion coefficient on W and Ni(110): (a) shows the anisotropy of the self-diffusion coefficient $D_s$ and of the activation energy Q on Ni(110) [3.8b]. (b) shows the enhancement of the surface diffusion near the melting point $T_m$ for W and Ni(110) [3.82]

The *influence of surface steps* on surface diffusion was studied for Ni on W(110) and for other systems by the concentration gradient method. Diffusion proceeds more rapidly along the steps in the Ni/W(110) system [3.20], and a pronounced difference of the diffusion coefficients upstairs and downstairs was observed (Fig. 3.9). Diffusion enhancement along the steps was detected for a Ni monolayer, whereas at $\theta < 1$ the anisotropy due to the steps was negligible.

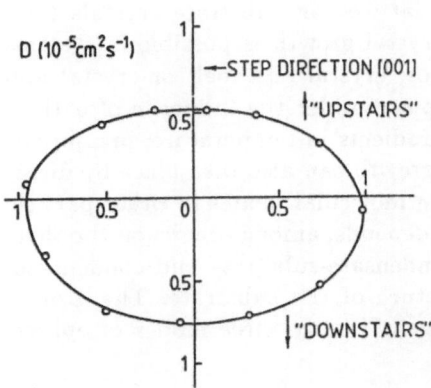

**Fig. 3.9.** Polardiagram for the diffusion coefficient D (in units of $10^{-5}\,\mathrm{cm^2s^{-1}}$) showing the anisotropy of the surface diffusion of Ni atoms on a stepped W(110) surface at $T = 1170\,\mathrm{K}$ [3.74]

For the Pd/W(650) system an anisotropy due to surface steps appeared only if the terrace was wider than about 400 Å [3.19]. These results demonstrate the important role of atomic steps in surface kinetics, though some findings have not been explained properly so far, as e.g., the relation between anisotropy and step density.

### 3.3.5 Surface Diffusion Near the Melting Point

As already mentioned above, near the melting temperature $T_m$ (at $T \gtrsim 0.75\,T_m$) the surface diffusion coefficients of several metals are larger than expected from an Arrhenius plot of the low temperature data (Fig. 3.8). Data for metals such as Ni, Cu, Mo, W show this effect which has been observed by two different methods [3.8,80,82]. The enhanced surface diffusion was originally attributed to a change in the intrinsic diffusion mechanism which is responsible for the transport of matter. A model was developed based on the formation of complex defects and the presence of localized and non-localized diffusion states. This model has been questioned recently [3.82]. Instead the formation of a two-dimensional dense fluid at $T > 0.75\,T_m$ has been assumed. This assumption was confirmed by ion shadowing and blocking experiments on a Pb(110) surface [3.100b]. Indications for a surface premelting have been found in these experiments. Furthermore, molecular-dynamics calculations for a noble gas crystal have yielded for atoms in the outermost layer a diffusion coefficient ($\sim 10^{-5}\,cm^2/s$) similar to that of a liquid even considerably below the melting temperature of the bulk $T_m$ (Chap. 6 and [3.84]). Moreover, this simulation showed that the structure factor for the outermost layer was similar to that of a liquid at a temperature below $T_m$, where the large diffusion coefficient was found [3.84b].

To elucidate the phenomenon of enhanced surface diffusion near the melting point, measurements of the structure factor and of the quasi-elastic scattering law as a function of temperature are required.

## 3.4 Growth Modes

In this section the growth of crystalline lattices on substrate crystals (epitaxial growth) is considered. Generally, crystal growth is possible in systems with two phases such as crystal and vapor, crystal and melt or crystal and solution. In those systems growth takes place under the influence of a thermodynamic driving force determined by gradients in temperature, pressure or concentration [3.31]. On the other hand, growth can also take place by direct deposition of particles on a substrate using molecular beams or other particle sources [3.31,36–40]. The mode of growth depends, among others, on the thermodynamic growth conditions, on the condensate-substrate and condensate-condensate interactions and on the structure of the substrate. The growth processes start with the formation of nuclei. Usually three modes of epitaxial growth are distinguished [3.2,24,86]:

- layer-by-layer growth (Frank-van der Merwe "FM" mode),
- island growth (Volmer-Weber "VW" mode), and
- layer growth followed by island growth (Stranski-Krastanov "SK" mode).

These growth modes are illustrated in Fig. 3.10. The energetically different situations can be described by the difference of specific free surface energies

$$\Delta f = f_{cv} + f_{sc} - f_{sv} \ . \tag{3.8}$$

The *layer-by-layer growth mode* is favored if the sum of the condensate-vacuum and the substrate-condensate free interface energies ($f_{cv} + f_{sc}$) is smaller than the substrate-vacuum free surface energy ($f_{sv}$) [3.31]. In this mode nucleation and lateral growth occur in every layer, leading to alternating nucleation and growth with a period of one monolayer. The observation of such oscillations by diffraction experiments is reported in Sect. 3.6.2. The FM mode was, for example, observed for some noble gases adsorbed on graphite, metals on metals (Au/Ag(100)) and for homoepitaxial growth of Si [3.2,24,29]. In heteroepitaxial layer-by-layer growth not only the condition (3.8) has to be fulfilled but also the lattice misfit between substrate and condensate has to be sufficiently small.

The *VW growth mode* generally appears if the adatom-adatom interaction exceeds the adatom-substrate interaction leading to $\Delta f > 0$. The nucleation is followed by the growth of three-dimensional crystalline islands (Fig. 3.10). The spatial distribution of the islands is typically not random, because nucleation tends to be less likely in the immediate neighborhood of existing islands [3.105].

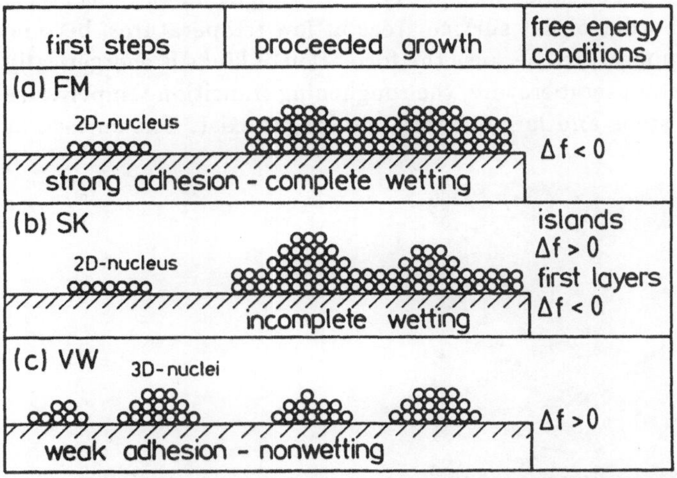

| first steps | proceeded growth | free energy conditions |
|---|---|---|
| (a) FM <br> 2D-nucleus <br> strong adhesion – complete wetting | | $\Delta f < 0$ |
| (b) SK <br> 2D-nucleus <br> incomplete wetting | | islands <br> $\Delta f > 0$ <br> first layers <br> $\Delta f < 0$ |
| (c) VW   3D-nuclei <br> weak adhesion – nonwetting | | $\Delta f > 0$ |

**Fig. 3.10.** Growth modes (schematic cross sections through growing layers and islands at two different stages, the assumed flat substrate is hatched, $\Delta f$=difference in specific free energy). *(a)* layer-by-layer growth (Frank-van der Merwe "FM" mode), *(b)* layer growth followed by island growth (Stranski-Krastanov "SK" mode), *(c)* island growth (Volmer-Weber "VW" mode). The first steps shown are the formation of two (2D)- and three (3D)-dimensional nuclei, respectively

This phenomenon was confirmed by experimentally determined radial distribution functions of islands [3.24]. The VW mode occurs typically for metals with large cohesive energy (e.g., Au, Ag, Cu, Ni ...) deposited on substrates with lower cohesive energy such as ionic crystals. The islands may coalesce by migration of the islands or with increasing coverage by the formation of a homogeneous bulk crystal due to single particle diffusion if the temperature is sufficiently high [3.2,24].

The *SK mode* is an intermediate case between the FM and the VW mode and occurs for instance in systems for which layer growth would be anticipated from the surface free energy condition (3.8). Initially growth leads to the formation of one or more strained layers, whose structure is strongly modified by the influence of the underlaying substrate. Further deposition produces an inhomogeneous adatom layer from which three-dimensional islands may nucleate and grow (Figs. 3.10,11). The number of completed layers and the number and shape of the islands depend on deposition conditions and on the properties of the particular adsorbate-substrate system. The SK growth mode has been identified for many metal-metal (e.g., Pb/W(110)) and metal-semiconductor (e.g., Ag/Si(111)) systems [3.2,24,40,43].

On real surfaces the free energy depends not only on the adatom-substrate interaction as expected for an ideal substrate surface structure but also on the surface *roughness* determined by the presence of steps, kinks, defects and impurities (Fig. 3.1) by which the local free-surface energy may be reduced. Due to the surface roughness the probability for processes like adsorption, diffusion, nucleation, and desorption is changed in comparison to ideally flat surfaces. The roughness increases with increasing temperature; for example, the kink density increases usually in an exponential way [3.31]. The straight monatomic steps, which appear at the border of flat surface areas at low temperatures, become irregular at higher temperatures because the formation of kinks is energetically favorable. Above a certain temperature, the roughening transition temperature $T_R$, long monatomic steps and large terraces no longer exist. The surface at

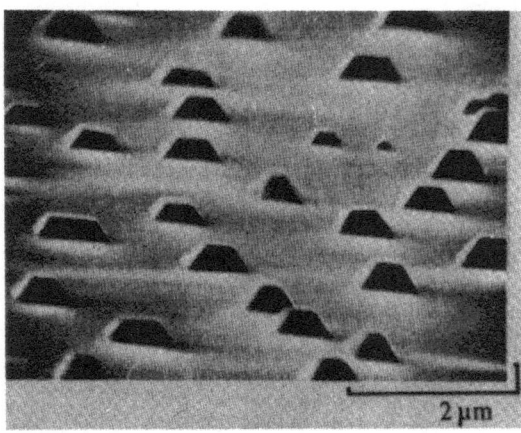

Fig. 3.11. A scanning electron microscope picture of Ag islands and layers grown by the SK mode on Mo(100) (at 10° glancing incidence) [3.105]

2 μm

$T > T_R$ consists of irregular atomic clusters with fluctuating height and lateral dimensions ([3.31] and Chap. 7). Obviously the layer-by-layer growth is limited to temperatures below $T_R$.

For the formation of clusters of adsorbed particles, i.e., for the *nucleation* on an ideal surface a potential barrier has to be overcome as long as the clusters is smaller than a critical size. The barrier is reduced on real surfaces at singular lattice positions such as kinks, defects or impurities. This effect is used for the decoration technique by which monatomic steps, dislocations or other structural defects at surfaces can be visualized. For example, Au atoms deposited on the surface of an ionic crystal aggregate preferentially near defect positions, and, hence, the trace of extended defects can be observed, e.g., by means of a transmission electron microscope [3.85]. The thermally activated and diffusion-controlled migration of monatomic steps is shown in Fig. 3.12. Not only the migration of steps and other defects at surfaces depends on *surface diffusion*. More generally, it plays a leading role in growth from vapor and molecular beams. The density of adsorbed particles will be constant if the desorption flux of particles is equal to the incident flux. In such a situation the frequency of arrival of particles at each surface lattice position from neighboring positions in the adsorbed layer is usually much larger than that of arrival directly from the gas phase. This is mainly due to the fact that the activation energy for desorption is several times larger than the activation energy for surface diffusion $E_{diff}$. In heteroepitaxial growth the *lattice misfit* between substrate and condensate influences the growth mode. Presupposition for epitaxial growth is a lattice misfit not larger than 10–15 % as has been found from empirical studies and model calculations [3.89]. The misfit accommodation by misfit dislocations and misfit strain is illustrated in Fig. 3.13 for an interface

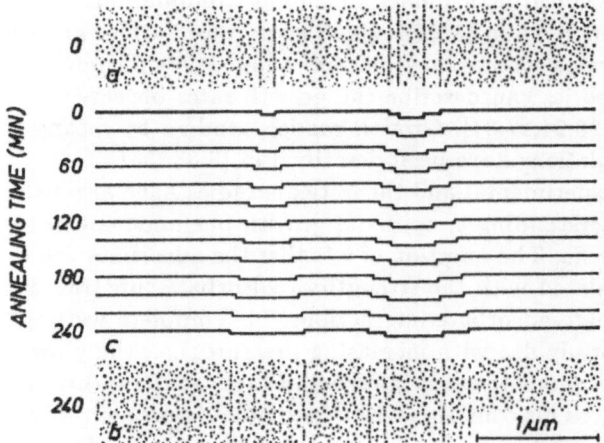

**Fig. 3.12.** Thermally activated migration of monatomic surface steps on cleaved NaCl(100) visualized by decoration with Au islands (a transmission electron microscope picture) [3.85]: *(a)* initial step configuration, *(b)* final step configuration; *(c)* successive stages of step propagation during annealing (simulated)

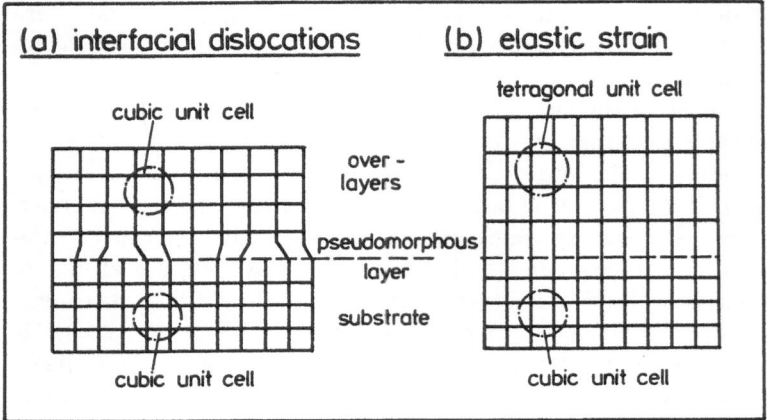

**Fig. 3.13.** Interface between a cubic substrate and epitaxial multilayers of a cubic condensate (schematic): *(a)* misfit accommodation by interfacial dislocations in a so-called pseudomorphous layer, *(b)* misfit accommodation by elastic strain

between a cubic substrate and a cubic condensate. Misfit strained overlayers may match the substrate completely without defects. The coherency of the misfit is lost, however, beyond a critical thickness. Only for condensates that have exactly the same lattice structure as the substrate can one add layers ad infinitum. Superlattices and other heterostructures where the lattice misfit between different layers is accommodated totally by strain have promising electronic properties (Sect. 3.6.5). At a critical value of the misfit the pseudomorphic layer between substrate and the overlayers becomes unstable. In the case of an adsorbed monolayer a commensurate-incommensurate phase transition then occurs (Chap. 4).

Comprehensive reviews on growth modes, including compilations of experimental results, have been given in [3.2,24–29]. Theories for surface growth rates which consider the atomic processes have been developed on a phenomenological basis. These rate theories can describe the growth rates observed as a function of temperature [3.23,24,27,31]. Growth models studied by means of Monte-Carlo computer simulations have been described in [3.33–35].

During the last years experimental and theoretical studies have also been aimed at a microscopic understanding of epitaxial growth phenomena considered now as *wetting phenomena*. The question of whether the adsorbate will or will not wet is directly correlated with the strength of adsorbate-substrate interactions relative to adsorbate-adsorbate interactions. In "complete wetting" the adsorbate remains uniformly flat with increasing pressure, i.e., the growth proceeds in the layer-by-layer mode towards infinite thickness at the saturation pressure of the bulk phase $p_0$. This contrasts with "incomplete wetting", where uniform overlayers grow only up to a finite condensate thickness and the bulk phase is formed abruptly at $p_0$. In extreme cases, termed "nonwetting", there is no formation of uniform overlayers at pressures $p < p_0$, but only bulk condensation at the saturation pressure $p_0$ [3.87,112]. If a system shows incomplete wetting at low temperatures, there may be a transition to complete wetting as the temperature is increased. Such wetting transitions can be discontinuous or continuous. A comprehensive review on wetting phenomena is given in [3.32], it turns out that the understanding of these phenomena is far from being complete. Among others, detailed scattering experiments and model calculations which consider the realistic form and strength of the adsorbate-substrate and adsorbate-adsorbate interactions (including their temperature dependence) are needed. For comparison of experimental and theoretical results, surface defects and lattice misfit have to be treated, too.

## 3.5 Layer-Growth Techniques

A broad spectrum of layer-growth techniques has been developed [3.31,36–39,42,43]. The choice of a specific growing method depends on the material to be grown. For crystal growth from the vapor-phase supersaturation and substrate temperature must be chosen properly. If the temperature is too low or the supersaturation too high, polycrystalline or amorphous films will grow [3.31]. The following methods for condensate vapor supply to the substrate are commonly used: molecular beam epitaxy (MBE), cathode sputtering, chemical-vapor deposition (CVD), vapor-phase growth in a closed system and layer growth in a gas flow. Typically, cathode sputtering is used for the growth of metal crystals and MBE for the growth of semiconductor crystals.

An MBE system is basically an ultra-high vacuum chamber equipped with several evaporation cells (Knudsen cells) and with different analytical instruments to examine the composition and the structure of the substrate and the overlayers [3.42,43,107]. The pressure in the growth chamber is maintained below $10^{-9}$ mbar. The substrate crystal is finally cleaned by ion sputtering and/or annealing. The cleanliness of the substrate is examined by AES and the structure by LEED or reflection high-energy electron diffraction (RHEED).

## 3.6 Methods for the Examination of Growth Processes and Typical Results

Owing to the development of sophisticated methods for preparation and analysis of surfaces, layer-growth processes can be studied in great detail. For a short survey we have compiled different methods in Table 3.2. They can be classified into four groups: thermodynamic methods, scattering experiments, spectroscopic methods, and microscopy. Results obtained with some of these methods are discussed in the following sections. The combination of scanning AES and electron microscopy has been successfully applied in the study of VW and SK modes of growth, for example, and scattering experiments in the study of the FM mode of growth. By diffraction experiments not only the growth mode can be analyzed but also the detailed structure of overlayers and

**Table 3.2.** Methods for the examination of layer growth processes

| Methods | Typical applications | References |
|---|---|---|
| *Thermodynamic methods:* | | |
| Volumetry | Adsorption isotherms of condensed gases on large substrate surfaces | [3.26] |
| Microbalance method | Adsorption isotherm for single crystal surfaces | [3.90] |
| Specific heat measurements | Mulitlayer growth on large substrate surfaces | [3.25,91] |
| Thermal desorption | Layerwise desorption | [3.24,92] |
| Work function measurements | Initial stage of growth | [3.93] |
| *Scattering methods:* | | |
| Reflection high-energy electron diffraction | Layer-by-layer and island growth | [3.75,94] |
| Low energy electron diffraction | Substrate spot attenuation, layer growth, island analysis | [3.75,108] |
| X-ray diffraction | Wetting phenomena, phase transitions, superlattice structure and composition | [3.75,97,119] |
| Neutron scattering | Structural phase transitions, excitations in multilayers | [3.44,75,98] |
| He-atom scattering | Layer-by-layer growth | [3.99] |
| Ion scattering | Composition and depth analysis, interface structures | [3.100] |
| Ellipsometry | Layer-by-layer growth, transparent adsorbates | [3.24,101] |
| Soft x-ray reflection | Multilayer growth | [3.106] |
| *Spectroscopic methods:* | | |
| Auger electron spectroscopy (AES) | Detection of growth modes | [3.2,24,102] |
| AES depth profiling | Heterostructures | [3.103] |
| Photoelectron spectroscopy | Layer dependent binding energy | [3.104] |
| Electron energy loss spectroscopy | Layer dependent binding energy | [3.81] |
| Nuclear magnetic resonance | Phase transitions in overlayers | [3.25,79] |
| *Microscopy, topography:* | | |
| Electron microscopy | Island growth, interface formation | [3.64,104,115] |
| Scanning tunnel microscopy | Surface topography with atomic resolution | [3.74,77] |
| Field ion microscopy | Cluster formation and migration | [3.30] |

**Fig. 3.14a–c.** Layer-by-layer growth (typical examples): (a) adsorption isotherm at 77.3 K of krypton on exfoliated graphite obtained by volumetry ($p_\infty$ is the vapor pressure of the bulk). The high steps are due to layer-by-layer condensation and the sub-steps correspond to two-dimensional (2D) phase transitions in the layers as indicated for the first layer [3.24], (b) variation of the (10) graphite LEED spot intensity as a function of pressure during the condensation of $CF_4$ on graphite(0001) at 55.7 K showing the formation of 4 layers [3.96], (c) xenon coverage on a Ag(111) surface as a function of temperature for an effective pressure of $2 \times 10^{-8}$ Torr. The fractional area covered has been determined from the attenuation of the Ag(10) LEED spot after correction for Debye-Waller factor and change of the Xe-Xe spacing with temperature. The plateau between monolayer and bilayer is flat and the two steps are equal [3.88a]

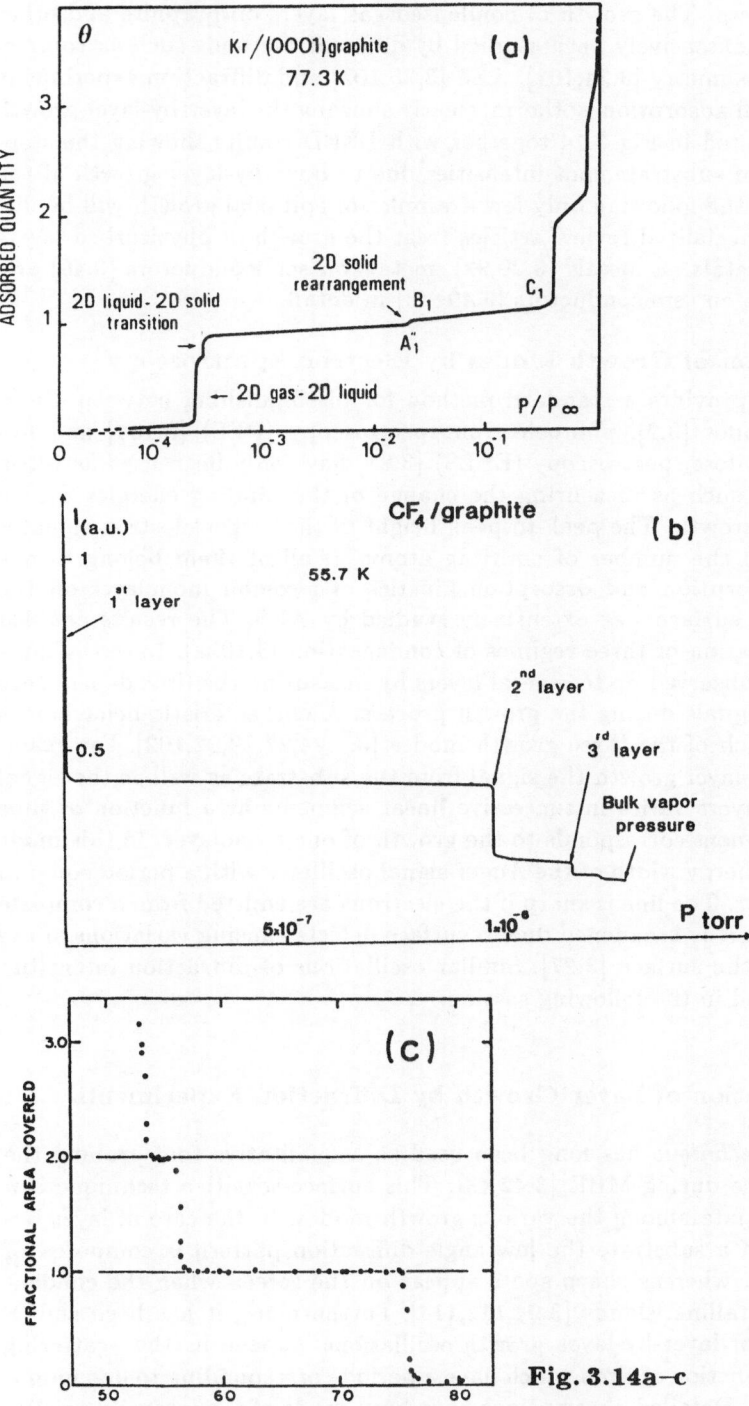

Fig. 3.14a–c

structural defects. The growth of condensed gas layers on graphite and other substrates has extensively been studied by different methods such as volumetry [3.26], ellipsometry [3.24,101], AES [3.25,102] and diffraction experiments [3.88]. A typical adsorption isotherm, clearly showing the layer-by-layer growth mode, is presented in Fig. 3.14 together with LEED results showing the stepwise decrease of substrate spot intensities due to layer-by-layer growth of the condensate. In the following only few examples of epitaxial growth will be discussed, since specialized review articles treat the growth of physisorbed layers [3.25,32,88], metals on metals [3.29,92], metals on semiconductors [3.40] and semiconductors on semiconductors [3.40–43] in detail.

### 3.6.1 Detection of Growth Modes by Electron Spectroscopy

Whereas AES provides a standard method for distinguishing between different growth modes [3.2], photoelectron spectroscopy (PES) [3.104] and low energy electron loss spectroscopy (EELS) [3.81] have only been used as yet in special studies such as measuring the change of the binding energies during layer-by-layer growth. The peak-to-peak height of the Auger electron signal is proportional to the number of emitting atoms, if all of them belong to one layer. The adsorption and desorption kinetics of a xenon monolayer on the graphite(1000) surface was extensively studied by AES. The results enabled the characterization of three regimes of condensation [3.102a]. Layer-by-layer growth can be observed up to several layers by measuring the time dependence of the Auger signals during the growth process. A characteristic behaviour is observed for each of the three growth modes [3.2,24,27,29,92,102]. For example, in layer-by-layer growth the signal from the substrate as well as the signal from the overlayers varies in successive linear segments as a function of time where each segment corresponds to the growth of one monolayer. In this mode of growth the energy width of the Auger signal oscillates with a period equal to one atomic layer. The line is sharp if the electrons are emitted from a complete layer, otherwise it is broadened due to surface defects causing variations of the free energy at the surface [3.27]. Similar oscillations of diffraction intensities will be discussed in the following section.

### 3.6.2 Observation of Layer Growth by Diffraction Experiments

The *RHEED technique* has long been used as a qualitative tool probing the crystal structure during MBE [3.42,43]. This surface-sensitive technique can clearly differentiate among the various growth modes. In the case of layer-by-layer growth on a substrate the low angle diffraction pattern is composed of parallel streaks, whereas sharp spots appear on the screen when the condensate forms crystalline islands [3.94,112,114]. Furthermore, it has been shown recently that for layer-by-layer growth oscillations appear in the scattering intensity as a function of time which have a period corresponding to one monolayer [3.95,114]. Detailed observations have been made of the intensity oscillations in the specularly reflected and various diffracted beams in the RHEED pattern occurring immediately after initiation of growth on a GaAs(001) surface (Fig. 3.15a–c). The oscillations are associated with a layer-by-layer growth process, i.e., with two-dimensional nucleations and subsequent completion of individual layers. A real space representation of the formation of layers (Fig. 3.15d) illustrates how the oscillations in the intensity occur. For the initial and the final smooth surfaces there is a maximum in reflectivity and a minimum for the intermediate stage when the growing layer is approximately half complete. The oscillations are gradually damped. This effect can be explained by the fact that new layers start to grow before the preceding one has been completed. The oscillation period of the RHEED beam intensity provides an absolute growth rate monitor with atomic layer precision.

The formation of overlayers has also been observed by the *LEED technique* [3.88,108]. The shape of the diffraction spots provides information on atomic steps and islands, in contrast to RHEED experiments where the resolution is generally worse. The height of islands or steps can be derived from the spot profile as a function of the electron energy [3.108a]. An oscillation of the spot intensity due to layer-by-layer growth was observed by this method during the vapor deposition of Si on Si(111) [3.108b]. The growth of individual adsorbate overlayers has been observed by the stepwise attenuation of LEED spots from the substrate (Fig. 3.14b,c and [3.88a,96]).

Nucleation and layer growth has recently been studied by *He-atom scattering* [3.99]. In contrast to scattering of high- and low-energy electrons (RHEED and LEED), which have a finite penetration depth into the bulk, the He atoms are sensitive only to the outermost layer (to charge corrugation and structure) and are therefore especially suited to give the average terrace size or step density at the surface of a growing crystal. For weakly bound adsorbate layers these probes have, moreover, the advantage that they are absolute nondestructive. The specular He-atom scattering from a Cu(100) surface at different temperatures was measured as a function of Cu coverage during Cu vapor deposition. An oscillatory intensity is detected for a given temperature. The amplitudes of the oscillations decrease with increasing coverage. In the temperature range studied, a layer-by-layer growth was expected. However, the observations indicate that this growth mode is not perfect. This means that the successive adlayers are not completed when the next layer starts to grow. From an analysis of the normalized intensity as a function of the angle of incidence for a fixed temperature $(T = 280\,K)$ and coverage $(\theta = 1/2)$ the height of the islands was estimated. The obtained height is in agreement with the bulk interlayer spacing giving evidence that only two-dimensional islands are formed in the process of crystal growth on Cu(100).

The adsorption of ethylene on exfoliated graphite has extensively been studied by *x-ray diffraction* [3.97]. At temperatures well below the critical temperature $T_c$ of bulk ethylene adsorption only occurs up to a finite number of layers. Any further condensation results in the formation of crystalline islands. At a coverage of 0.9 and at a temperature of 83 K a broad asymmetric diffrac-

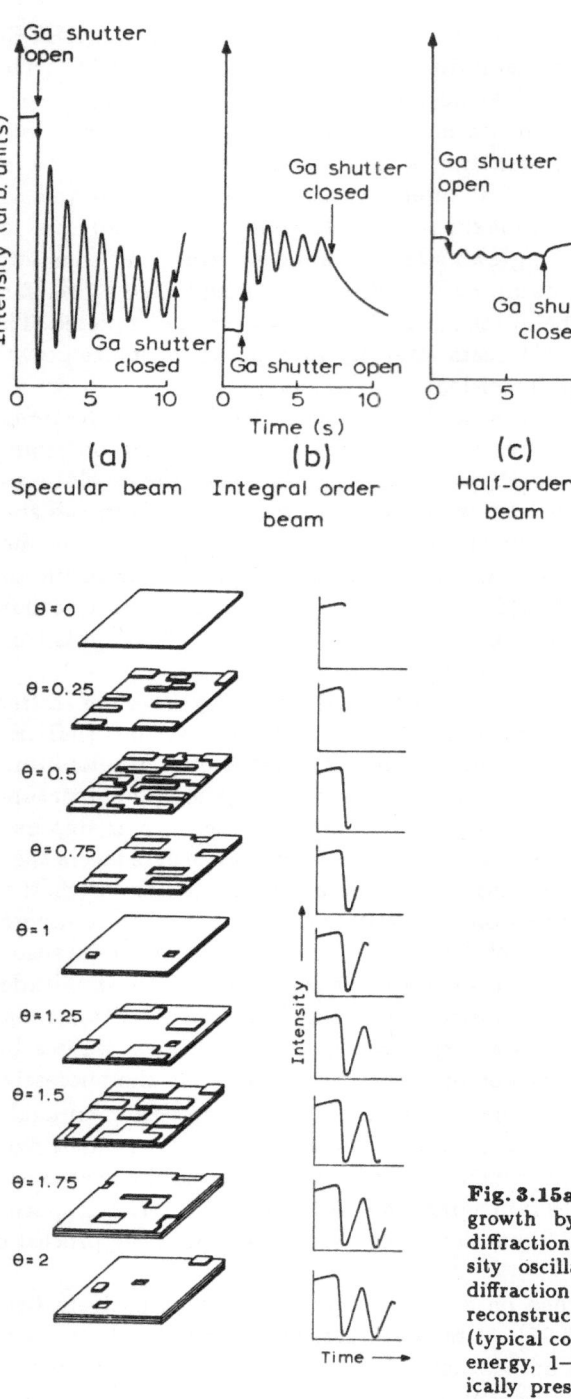

Fig. 3.15a–d. Detection of layer-by-layer growth by reflection high-energy electron diffraction (RHEED) [3.114b]: (a–c) Intensity oscillations of various beams in the diffraction pattern from a GaAs(001)-(1 × 4) reconstructed surface in the [111] azimuth (typical conditions: 10–20 keV primary beam energy, 1–4° incident angle); (d) schematically presentation of the formation of two monolayers in relation to the RHEED intensity behavior for different coverages θ

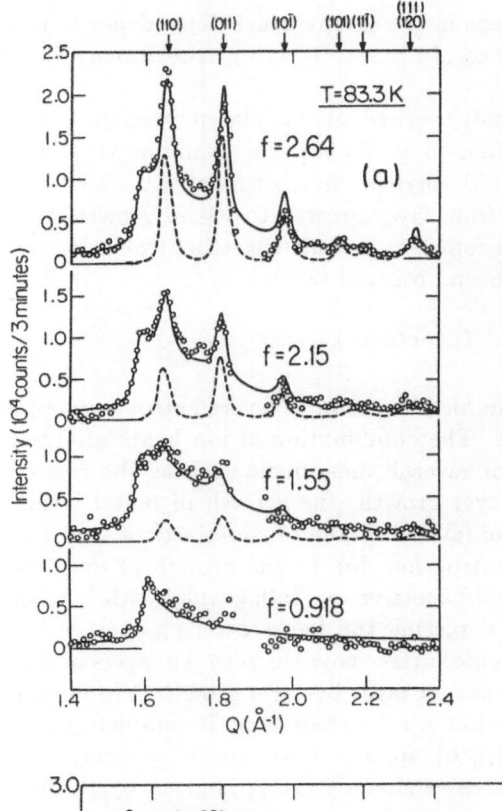

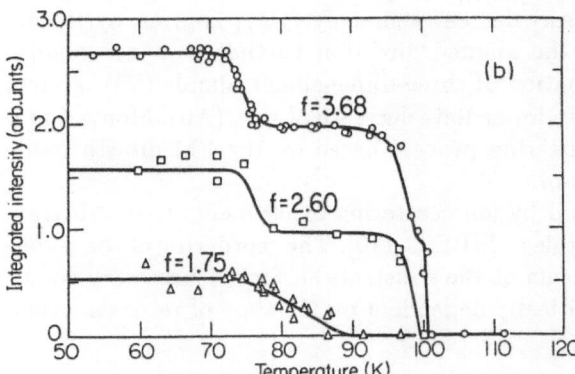

**Fig. 3.16a,b.** Detection of layer growth and wetting by x-ray diffraction [3.97]: (a) the evolution of the x-ray diffraction profile of ethylene adsorbed on exfoliated graphite with increasing coverage f at constant temperature. The *solid lines* are the results of fits to scattering intensities due to two- and three-dimensional structures. The arrows indicate the bulk peaks whose contribution is shown by the *dashed line*. The background due to the graphite substrate have been subtracted. The appearance of the narrow peaks at f > 1 is typical for the SK growth mode. (b) integrated intensity of the bulk ethylene peaks for several coverages f as a function of temperature (intensity is normalized). The size and position of the steps are evident for discrete layering transition temperatures. The decrease of the intensity with increasing temperature indicates the transition to complete wetting

tion profile is observed which is characteristic for two-dimensional structures (Fig. 3.16a). In addition, narrow symmetric peaks appear at coverages beyond one monolayer. The position of these peaks corresponds to the structure of bulk ethylene. This result demonstrates conclusively the coexistence of one monolayer of ethylene with bulk islands in this temperature range (corresponding to incomplete wetting and SK growth mode). The bulk peaks of the diffraction pattern disappear at temperatures near $T_c$ indicating the transition to complete wetting (Fig. 3.16b). The steps in the temperature dependence of the integrated peak intensities are evidence of discrete layering transitions which still have to be studied in more detail.

Noble gases and other species physisorbed on exfoliated graphite have also been studied by *neutron diffraction* [3.45–50,98]. For example, the layer growth has been analyzed for deuterated ethylene. In agreement with the x-ray results discussed above, a transition from layer growth to island growth at a coverage $\theta \gtrsim 1$ and indications for an incomplete-complete wetting transition by an increase of the temperature have been observed [3.98a].

### 3.6.3 Examination of Surface and Interface Layers by High-Energy Ion Scattering

High-energy ions are well suited for the investigation of layered structures and interfaces as well as of growth modes. The combination of ion beam analysis with MBE made possible the study of several phenomena such as the role of lattice mismatch in heteroepitaxial layer growth, the growth of metal layers on different substrates, the structure of interfaces and of superlattices, and the change of the substrate-surface reconstruction due to the growth of overlayers [3.100]. The sensitivity of the mass-dispersive crystallography with ions as probes is demonstrated in Fig. 3.17a depicting the backscattering spectra for the bare Ag(111) surface and for the same surface covered with Au layers of different thicknesses [3.109]. Direct evidence for layer-by-layer growth (FM mode) has been obtained by the observation that Au-Au shadowing is completely absent (i.e., $\chi_{min}(Au) = 1.0$, see Fig. 3.17b) up to a coverage of precisely one monolayer, where upon $\chi_{min}(Au)$ decreases sharply ($\chi_{min}$ is given by the ratio of the scattering yield in the aligned direction to that in a non-aligned (random) direction). The formation of three-dimensional islands (VW growth mode) would have resulted in an immediate decrease of $\chi_{min}(Au)$. Monte-Carlo simulations results for the scattering process based on the FM growth mode indeed fit the ion scattering data.

Recently, it has been found by ion scattering experiments that substrate reconstruction plays a critical role in MBE [3.110]. The reordering of the reconstruction – induced displacements at the substrate surface, a necessary condition for epitaxial growth, is critically dependent on the type of reconstruction:

---

**Fig. 3.17a–c.** Analysis of layer growth by high-energy ion scattering: (a) backscattering spectra for 1 MeV He$^+$ ions incident in [110] direction: *(1)* bare Ag(111) surface; *(2–4)* Ag(111) covered with 0.7–3.8 monolayers (ml) of Au [3.109]. (b) Ag(111) surface peak area and $\chi_{min}(Au)$ as a function of coverage. The *solid* and the *dashed curves* are derived from computer simulations assuming layer-by-layer growth [3.109]. (c) surface peak intensity (SP) of backscattered 0.5 MeV He$^+$ ions in off-normal incidence for Ge deposition at room-temperature on Si(100)-(2 × 1) as a function of Ge coverage. The observed LEED pattern is indicated at the bottom. The *solid line* describes the expected dependence of the SP if the overlayer is highly disordered. The displacements present in the bare Si(100) surface due to (2 × 1) reconstruction vanish upon Ge deposition [3.110]. The insets schematically show the scattering geometry

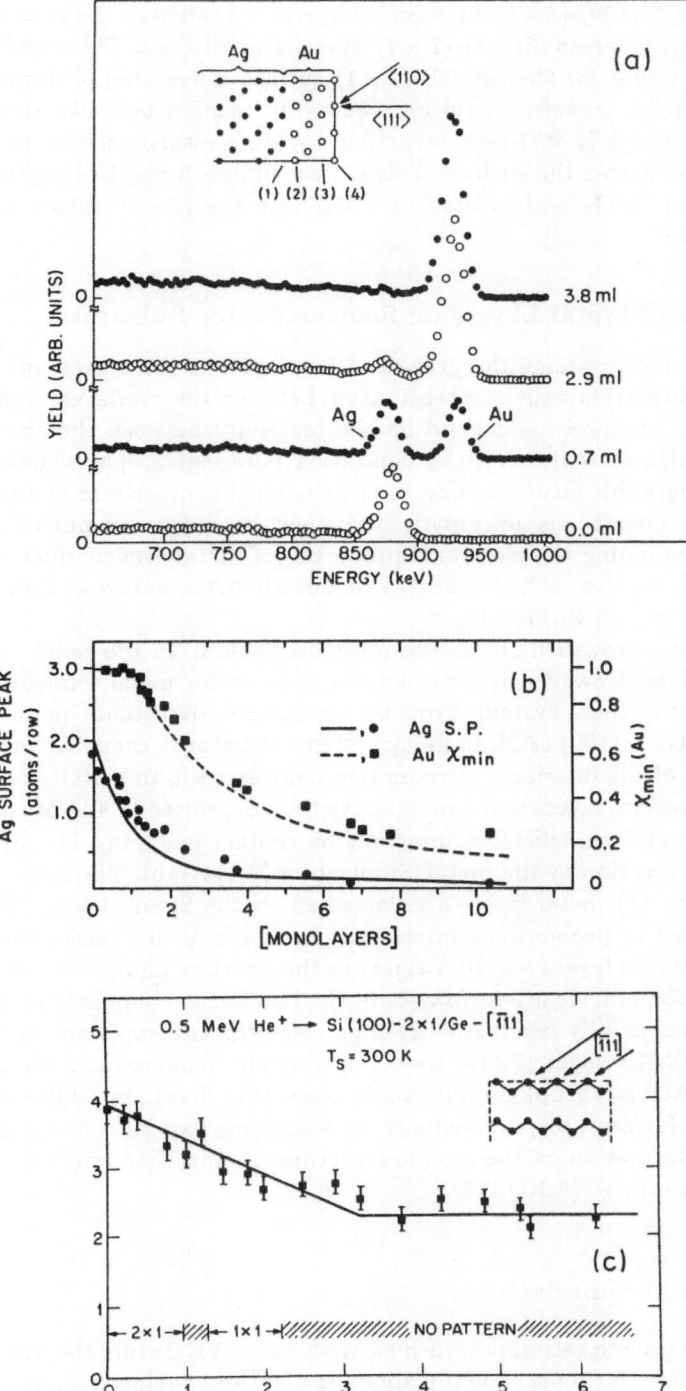

**Fig. 3.17a–c**

Deposition of Ge on Si(100)-(2 × 1) at room temperature removes the reconstruction (Fig. 3.17c), whereas Si(111)-(7 × 7) appears unaffected. This result is related to the fact that for the Si(100)-(2 × 1) surface a breaking of dimer bonds without any mass transfer would be required to reach a bulk-like configuration, while the rough (7 × 7)-reconstruction would necessarily involve the migration of Si atoms across the surface. This surface diffusion requires higher temperatures. Indeed, if Ge is deposited at 570 K also the (7 × 7) substrate reconstruction reorders.

### 3.6.4 The Growth of Metal Layers on Semiconductor Substrates

For metal-semiconductor systems the growth of overlayers is often accompanied by compound formation and/or interdiffusion between the overlayers and the substrate. These mechanisms depend on the reactivity between the constituent atomic species as well as on the deposition parameters. The knowledge of the crystallographic structure and of the chemical composition at the metal-semiconductor interface is not only a fundamental requirement but also important for understanding the electronic properties of metal-semiconductor interfaces in practical devices such as Schottky diodes where the nature of these interfaces play a key role [3.40,113,115].

The layer-by-layer growth mode is the most desirable from the point of view of the applications. However, it is not known to occur for metal-semiconductor systems. Most of these systems grow by the Stranski-Krastanov mode, e.g. Au/Si(111), Al/GaAs(100) [3.2]. In many systems structural, chemical and electronic properties of the interfaces at room temperature seem to be fully established after only several layers of the metal have been deposited [3.40]. Most of the transition metals form silicide compounds by contact reactions, i.e., by a thermally induced reaction at the metal/Si interface [3.115,116]. The stages of these reactions are: (1) metal layers are deposited on the Si substrate; (2) the sample is annealed to promote the interfacial reaction and this causes the formation of a thin silicide layer (~3–10 Å thick) at the interface and (3) by further annealing a thicker silicide layer grows until the reaction is completed and all the deposited metal is fully reacted. Ultra-high resoluton transmission electron microscope (TEM) images [3.116] have impressively demonstrated that the silicide/Si interface is abrupt on an atomic scale (Fig. 3.18). In addition to the TEM studies, for example, high-energy ion scattering experiments have provided valuable information on the atomic structure, composition and reactivity of silicide-Si interfaces [3.100,115].

### 3.6.5 Examination of Superlattices

The improvements in growth techniques such as MBE and CVD during the last decade made it possible to prepare high-quality superlattices having designed profiles of concentration and, related to it, of electronic band structure. Three

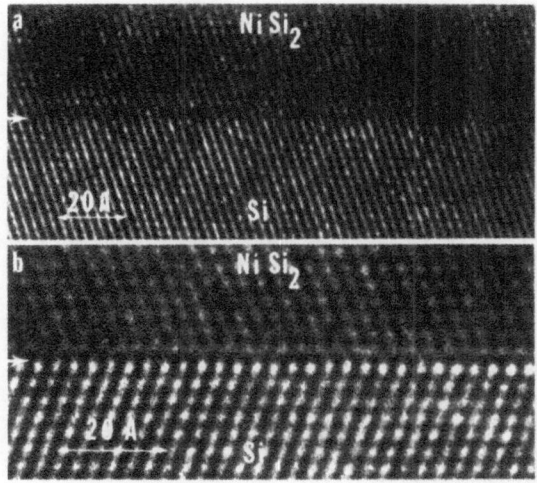

**Fig. 3.18a,b.** High resolution electron microscope images of $NiSi_2/Si$ interfaces in cross-section. On Si(111) the Ni-silicide grow in two orientations. In (a) the cubic silicide layer is aligned exactly with the substrate, in (b) it is rotated about the [111] direction by 180° [3.116]

types of superlattices composed of semiconductors or metals have meanwhile been studied both experimentally and theoretically [3.37]:

— *compositional semiconductor superlattices,* with a periodic sequence of different semiconductor layers (e.g., GaAs-$Ga_{1-x}Al_xAs$ (Fig. 3.19a) [3.117];
— *doping semiconductor superlattices* with a sequence of layers of the same material, however, with opposite sign of doping (n- and p-doped layers, possibly with intrinsic layers in between, as in so-called "nipi crystals") (Fig. 3.19b) [3.118]; and
— *metallic multilayers,* consisting of alternating layers of two different constituents with sharp interfaces between the different layers (e.g., NbCu [3.122]).

Nearly perfect semiconductor superlattices with a few monolayers or thicker deposition periods can now be produced. The electronic, the optical and the transport properties of superlattices and non-periodic heterostructures can be tailored by the selection of the constituents and of the layer thicknesses of these artificial structures. Several novel electronic devices such as solid-state lasers with programmed operating wavelengths have already been developed and many other new applications of semiconductor superlattices and other heterostructures are under development. Prerequisite of this material engineering is an excellent epitaxial layer growth. Very sensitive techniques have been developed for the examination of superlattices by various methods. For example, the interlayer diffusion and the structure in $(GaAs)_n(AlAs)_m$ multilayers has been studied by x-ray diffraction [3.119]. Diffraction pattern from a material in which the lattice parameter and/or the scattering amplitude is modulated in one dimension are characterized by satellites around each Bragg peak of the average lattice (Fig. 3.20a). The intensities of these satellite peaks are propor-

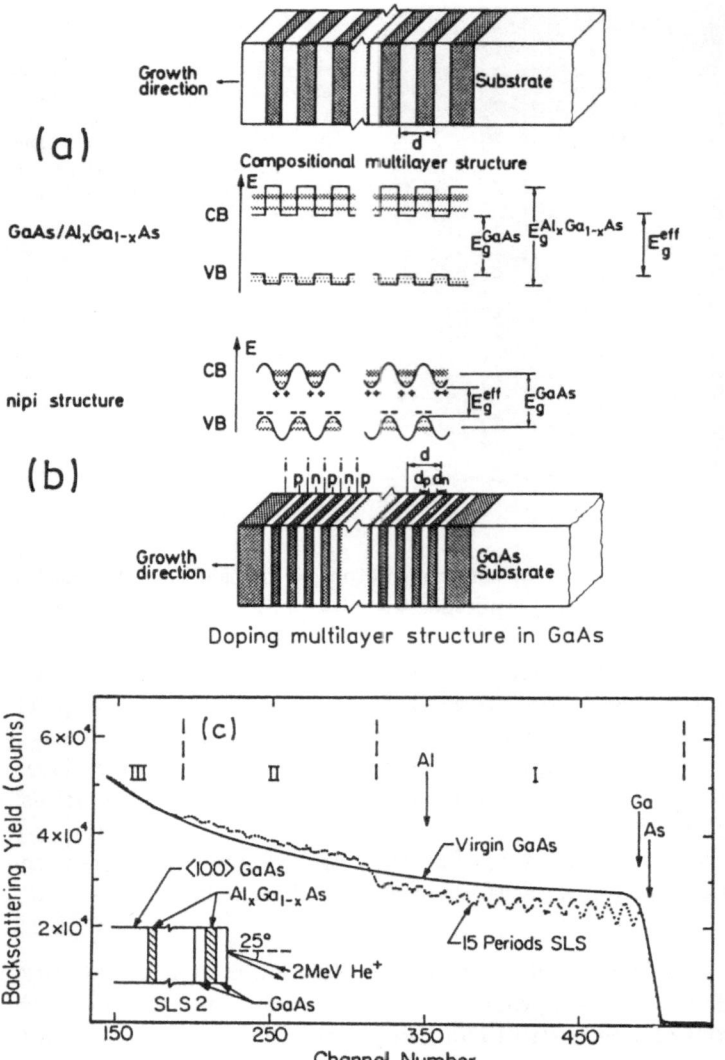

**Fig. 3.19a–c.** Semiconductor superlattices: **(a)** compositional superlattice consisting of alternating layers of GaAs and $Ga_{1-x}Al_xAs$ ($d=50$–$500$ Å), and schematic electronic band structures (CB: conduction band, VB: valence band, $E_g$: band gap), **(b)** superlattice in the otherwise homogeneous semiconductor GaAs and related band structures: n=doping by donors, p=doping by acceptors, i=intrinsic layers [3.107], **(c)** He$^+$ backscattering spectra from a 15 period GaAs/$Ga_{1-x}Al_xAs$ superlattice. The intensity oscillations correspond to the individual layers (SLS: strained-layer-superlattice). These data allow an accurate nondestructive measurement of the compositional period [3.129]

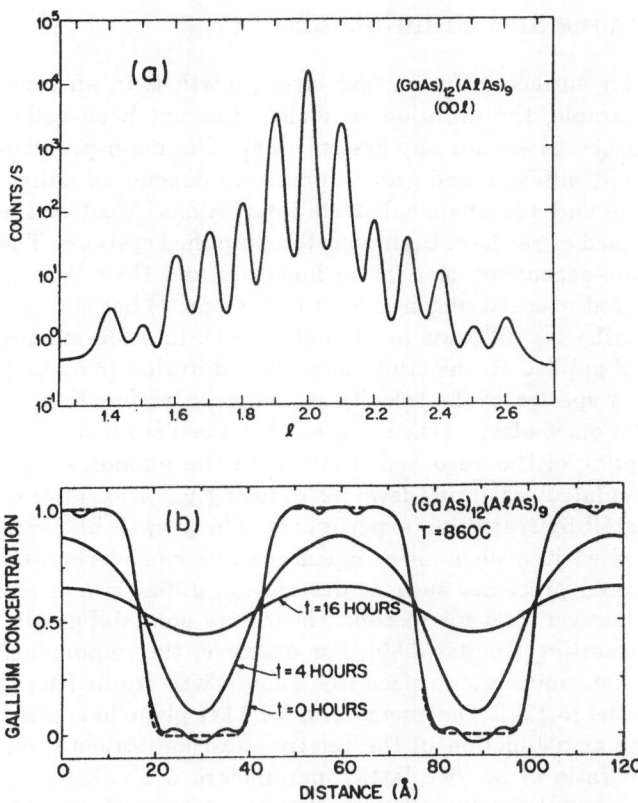

**Fig. 3.20a,b.** X-ray analysis of superlattices [3.119]: **(a)** longitudinal scan along the [00ℓ] direction of a (GaAs)$_{12}$(AlAs)$_9$ sample showing six harmonics on each side of the (002) reflex which are related to the period of the superlattice, **(b)** composition wave obtained from the Fourier transformation of the x-ray data which have been measured before and after 4 and 16 h annealing of the sample. The *dashed curve* is the initial concentration wave used in the diffusion calculation

tional to the Fourier components of composition modulation in the direct space; therefore, the structure and the composition as well as the interdiffusion of the constituents can be studied by measuring the satellite intensity as a function of the annealing time at a constant temperature (Fig. 3.20b). In addition, measurements of the widths of the satellites yield information about the overall quality of the superlattice (see also [3.120]). In addition to diffraction experiments, the structure of superlattices can be examined also by high-resolution electron microscopy and by ion scattering experiments. An example for an examination by high-energy ion scattering is shown in Fig. 3.19c [3.121]. Superlattices and non-periodic heterostructures can be proven by this method directly in real space and in a nondestructive way.

## 3.7 Model Calculations and Simulations

The microscopic theory for surface diffusion and layer growth is in an early state [3.7,83,123]. For example, the diffusion coefficient has not been calculated with sufficient accuracy as yet for any system [3.8]. The main problem of *model calculations* is that diffusion and growth processes depend on anharmonic adsorbate-adsorbate and adsorbate-substrate interactions. Most of the model calculations performed as yet have been done for simplified systems. The concept of space- and time-dependent correlation functions and their Fourier transforms, which are related to scattering cross sections (Chap. 1) has successfully been applied to describe the diffusion in the bulk [3.44]. In a few studies this concept has also been applied to the study of surface diffusion [3.52–55]. For example, the frequency spectra of the velocity autocorrelation function for adsorbed particles and the quasi-elastic scattering law has been studied under consideration of the coupling of the adsorbed particles to the phonon modes [3.54]. However, such calculated scattering laws for diffusing particles have so far not been proven in detail by scattering experiments. The growth phenomena have mostly been treated in a phenomenological manner considering, for example, thermally activated processes such as desorption, diffusion and nucleation [3.2,24,33–35]. However, first microscopic theories of epitaxial growth have already revealed interesting results [3.89]. For example, the importance of different terms which determine the surface free energy was studied for a rigid lattice, atomistic model [3.125]. The energy of a fcc(111) plane in contact with a bcc(110) substrate as a function of the relative position, orientation, vertical displacement and ratio of bcc/fcc lattice parameters was calculated. A comparison with experimental results indicates that the growth of metallic superlattices is controlled not only by interface energies but also by the stacking of various constituents. A full theory would require, however, to take into account also the effect of lattice vibrations.

Surface diffusion and epitaxial growth have also been studied by *computer simulations* (Monte-Carlo "MC" and molecular-dynamics "MD" calculations). These methods can be applied if realistic potentials describing the interatomic interactions at surfaces are available. In the classical approximation the anharmonicity of the interactions is fully taken into account in such calculations. The MC and the MD method and their application to surface studies is discussed in Chap. 6 and in [3.33–35,111,123,126–129]. For the surface of a krypton crystal near the melting temperature a liquid-like diffusion coefficient was obtained from a MD calculation using a realistic pair-potential (Chap. 6). The velocity autocorrelation function and the diffusion coefficient for a liquid $CH_4$ layer was obtained by the MD method using a Lennard-Jones potential [3.127]. This result agrees well with the D value measured by quasi-elastic neutron scattering (Sect. 3.3.2). MD surface diffusion studies for metal surfaces are just at the beginning [3.123b]. This is mainly due to the problem of finding realistic potentials valid on surfaces and applicable also at high temperatures. A rough estimation of lateral pair interaction energies of adsorbed metal atoms can be made by analyzing the pair correlation as observed by means of an FIM [3.124]. Layer growth has been studied at solid-liquid as well as at solid-vapor interfaces [3.126,128,129]. For example, the epitaxial growth of a system with Lennard-Jones interactions has been analyzed as a function of the substrate temperature $T_s$ and the deposition rate [3.128]. For all substrate temperatures the growth leads to well-ordered layers, which are fully completed at intermediate $T_s$. At very low $T_s$ the layers contain defects and voids. These results confirm the layer growth mode found experimentally. Recently, the MD method has also been applied to more complex covalent bond systems. It has been found that the Si(111) crystal-melt interface propagates in the layer-by-layer growth mode [3.129a], whereas the Si(100) crystal-melt interface exhibits a faceted structure [3.129b].

## 3.8 Conclusion and Outlook

Only few results on surface diffusion and layer-growth studies have been presented in this chapter. An attempt has been made to outline typical results and to show trends in recent developments. Research on surface diffusion and layer growth has profited from several break-throughs in experimental techniques during the last decades, for example, by the development of:

—   the field ion microscope by which the observation of individual atoms has become possible,
—   the high-resolution electron microscope for the imaging of solid surfaces and interfaces with atomic resolution, and
—   the molecular beam epitaxy as a tool for growing of nearly perfect multilayer structures.

As shown in the preceding sections very detailed results have been obtained by using these advanced techniques on surface diffusion as well as on layer growth. However, more systematic studies on these surface phenomena will be necessary in order to provide data for the development of related microscopic theories. It would be very useful to study surface diffusion in more detail by inelastic scattering experiments and to analyze the scattering law in relation to models for the space- and time-dependent correlation functions describing surface diffusion on the atomic scale. Such scattering experiments have been performed as yet only with neutrons in the study of diffusion on exfoliated graphite. Similar experiments on single crystal surfaces may be possible in the near future also with He atoms as probes since a sufficient energy resolution for high-intensity beams can now be achieved with these probes. In the future scanning tunnel microscopy (STM) will also become a tool for surface diffusion and layer growth studies. First studies of adsorbed particles by STM [3.65] have demonstrated that STM is capable of detecting simultaneously the migration of individual particles and the structure of the substrate with all defects like steps and grain boundaries. STM may also allow to detect the decay of concentration gradients or – as has been recently proposed – to apply the field emission fluctuation method [3.130].

Considerable progress in growth studies is due to the application of improved scattering techniques using various probes. By x-ray diffraction the study of layer structures, layer growth, and interdiffusion in superlattices will become more important in the future. This is due to the facts that more intense synchrotron x-ray sources will be available for surface research and that new experimental methods have been developed (Chaps. 1 and 5). The study of adsorbed layers on single crystalline substrates is now possible. For example, the temperature dependence of the structure of Pb layers deposited on Cu(110) was studied using the x-ray grazing incident method [3.131].

Surface diffusion and layer growth are studied in order to understand the fundamental processes; however, the technical importance of surfaces and interfaces (e.g., in electronic devices) make this knowledge useful for practical applications. For example, by increasing the quality and reducing the layer thickness in superlattices it may become possible to produce ultra-high speed devices with ballistic electron transport [3.133].

In summary, it was shown that experimental studies of surface diffusion and layer growth yield detailed informations on the related atomic processes taking place at surfaces and interfaces. The mode of growth depends not only on the ratio of the condensate-condensate to the condensate-substrate interaction but also on the substrate temperature as indicated by the wetting transition. The relation between the temperature dependence of the diffusion coefficient and structural phase transitions in overlayers has been illustrated only by one example in this chapter. However, there is already a large amount of information on structural phase transitions at surfaces (see also Chaps. 5–7) even if not all transitions have been studied in such a detail. New experimental techniques promise a great variety of interesting results on surface diffusion and layer growth in the near future. Clearly, however, much work remains to be done, for example, in the development of realistic microscopic models for these phenomena. Some results of the work as briefly described in this chapter are useful for technical developments which are now going on in the world of two- to three-dimensional submicron structures.

*Acknowledgements.* The author wishes to thank Dr. W. Schommers and Prof. J.R. Thompson for helpful discussions and a critical reading of the manuscript.

# References

3.1    G. Ehrlich: Layer Growth on Atomic Picture, in Proc. 9. Int'l. Vac. Congress and 5. Int'l. Conf. Solid Surf., ed. by J.L. de Segovia (ASEVA, Madrid 1983) p. 26

3.2    J.A. Venables, G.D.T. Spiller, M. Hanbücken: Rpt. Progr. Phys. **47**, 399–459 (1984)

3.3    G. Ehrlich: CRC Crit. Rev. Colid State Mat. Sci. **10**, 391 (1982)
       G. Ehrlich, K. Stolt: Ann. Rev. Chem. Phys. **31**, 603 (1980)
       H.W. Fink, G. Ehrlich: Surf. Sci. **143**, 125 (1984)
       H.W. Fink, G. Ehrlich: Surf. Sci. **150**, 419 (1985)

3.4    G. Gust, S. Mayer, A. Bögel, B. Preudel: J. Physique **46**, C4-537 (1985)

3.5    R.W. Balluffi: Grain Boundary Diffusion Mechanisms in Metals, in *Diffusion in Crystalline Solids*, ed. by G.E. Murch, A.S. Nowick (Academic, London 1984)

3.6    N.L. Peterson: Int. Metals Rev. **28**, 65 (1983)

3.7    Vu Thien Binh (ed.): *Surface Mobilities on Solid Materials, Fundamental Concepts and Applications* (Plenum, New York 1983)

3.8    H.P. Bonzel: Mass Transport by Surface Self-Diffusion, in [Ref. 3.7, p. 195]

3.9    A.C. Gossard: Treat. Mat. Sci. Technol. **24**, 13 (1982)

3.10   M. McLean, E.K. Hondros: Influence of Interfacial Diffusion on Material Microstructure and Behaviour, in [Ref. 3.7, p. 459]

3.11   D.A. King: J. Vac. Sci. Technol. **17**, 241 (1980)

3.12   G.E. Rhead: Surf. Sci. **47**, 207 (1975)

3.13   V. Levi: Diffusion at Crystal Surfaces, in *Handbook of Surfaces and Interfaces*, Vol. 1, ed. by L. Dobrzynski (Garland, New York 1978), p. 253

3.14   T.T Tsong: Prog. Surf. Sci. **10**, 165 (1981)

3.15   R. Gomer: Surface Diffusion – Some General Remarks in [Ref. 3.7, p. 1]
       R. Gomer: Relations Among Various Diffusion Coefficients – Diffusion and Density Autocorrelation Function, in [Ref. 3.7, p. 7]
       R. Gomer: Field Emission Studies of Surface Diffusion of Adsorbates, in [Ref. 3.7, p. 127]

3.16   M. Tringides, R. Gomer: Surface Sci. **166**, 419 (1986)

3.17   D.W. Basset: Observing Surface Diffusion at the Atomic Level, in [Ref. 3.7, p. 63]

3.18   T.T. Tsong: FIM Studies of Surface Migration of Single Adatoms and Diatomic Clusters With and Without a Driving Force, in [Ref. 3.7, p. 109]
       T.T. Tsong: Adatom Pair Interactions and Adlayer Superstructure Formation, in [Ref. 3.7, p. 247]
       T.T. Tsong, R. Casanova: Phys. Rev. B**22**, 4632 (1980)

3.19   H. Wagner: Transport of Adsorbed Species: Correlations with Concentration and Step Structure, in [Ref. 3.7, p. 161]

3.20   A.G. Naumovets, Y.S. Vadula: Surf. Sci. Rep. **4**, 365 (1985)
       A.G. Naumovets: Surface Diffusion of Adsorbates and Phase Transitions, in Proc. 9. Int'l. Vac. Congress and 5. Int'l. Conf. Solid Surf., ed. by J.L. de Segovia (ASEVA, Madrid 1983) p. 90

3.21   M.A. van Hove, S.Y. Tong (eds.): *"The Structure of Surfaces"*, Springer Ser. Surf. Sci., Vol. 2 (Springer, Berlin, Heidelberg 1985)

3.22   J.G. Dash: *Films on Solid Surfaces* (Academic, New York 1975)
       J.G. Dash, J. Ruvalds: *Phase Transitions in Surface Films* (Plenum, New York 1980)

3.23   B. Lewis, J.C. Anderson: *Nucleation and Growth of Thin Films* (Academic, New York 1978)

3.24   R. Kern, G. Le Lay, J.J. Metois: Basis Mechanisms in the early Stages of Epitaxy, in Current Topics in Mat. Sci. **3**, 131 (North Holland, Amsterdam 1979)

3.25   M. Bienfait: Two-Dimensional Phase Transitions of Simple Molecules Adsorbed on Graphite, in Current Topics in Mat. Sci. **4**, 361 (North Holland, Amsterdam 1980)

3.26   A. Thomy, X. Duval, J. Regnier: Surf. Sci. Rep. **1**, 1 (1981)

3.27   R.W. Vook: Int. Metals Rev. **27**, 209 (1982)

3.28   J.A. Venables: Surface Processes in the Nucleation and Growth of Thin Films, in Proc. 9. Int'l. Vac. Congress and 5. Int'l. Conf. Solid Surf., ed. by J.L. de Segovia (ASEVA, Madrid 1983)

3.29   E. Bauer: Metals on Metals, in *Chemical Physics of Solid Surfaces and Heterogeneous Catalysis*, Vol. 3 Pt. B, ed. by D.A. King, D.P. Woodruff (Elsevier, Amsterdam 1981)

3.30   G. Ehrlich: An Atomic View of Crystal Growth, in *Chemistry and Physics of Solid Surfaces V*, ed. by R. Vanselow, R. Howe, Springer Ser. Chem. Phys., Vol. 35 (Springer, Berlin, Heidelberg 1984) p. 283
       G. Ehrlich, K. Stolt: *"Surface Diffusion of Metal Clusters on Metals"*, in *Growth and Properties of Metal Clusters*, ed. by J. Bourdon (Elsevier, Amsterdam 1980) p. 1
       G. Ehrlich: Diffusion and Interaction of Adatoms, in [Ref. 3.21, p. 357]

3.31   A.A. Chernov: *Modern Crystallography III – Crystal Growth*, Springer Ser. Solid-State Sci., Vol. 36 (Springer, Berlin, Heidelberg 1984)

3.32   M. Bienfait: Surf. Sci. **162**, 411 (1985)

3.33   H. Müller-Krumbhaar: Kinetics of Crystal Growth, in Current Topics Mat. Sci., **1**, 1 (North Holland, Amsterdam 1970)

3.34 P. Bennema: Science of Crystal Growth: Rough or Flat Surfaces; Formation and Movement of Steps in Relation to Surface Mobilities and Defects, in [Ref. 3.7, p. 275]

3.35 G.H. Gilmer: Ising Model Simulations of Crystal Growth, in *Chemistry and Physics of Solid Surfaces V*, ed. by R. Vanselow, R. Howe, Springer Ser. Chem. Phys., Vol. 35 (Springer, Berlin, Heidelberg 1984) p. 297

3.36 J.W. Mattews (ed.): *Epitaxial Growth*, Pts. A and B (Academic, New York 1975)

3.37 L.L. Chang, B.C. Giessen (eds.): *Synthetic Modulated Structures* (Academic, Orlando 1985)

3.38 B. Matuaftschiev (ed.): *Interfacial Aspects of Phase Transitions* (Reidel, Doderecht 1982)

3.39 L. Aleksandrov: *Growth of Crystalline Semiconductor Materials on Crystal Surfaces* (Elsevier, Amsterdam 1984)

3.40 C.W. Wilmsen (ed.): *Physics and Chemistry of III–V Compound Semiconductor Interfaces* (Plenum, New York 1985)
R.S. Bauer (ed.): *Surface and Interfaces: Physics and Electronics* (North Holland, Amsterdam 1983)

3.41 F. Nizzoli, K.-H. Rieder, R.F. Willis (eds.): *Dynamical Phenomena at Surfaces, Interfaces and Superlattices*, Springer Ser. Surf. Sci., Vol. 3 (Springer, Berlin, Heidelberg 1985)

3.42 L.L. Chang, K. Ploog (eds.): *Molecular Beam Epitaxy and Heterostructures* (Nijhoff, Dordrecht 1985)

3.43 E.H.C. Parker: *The Technology and Physics of Molecular Beam Epitaxy* (Plenum, New York 1985)

3.44 W. Marshall, S.W. Lovesey: *Theory of Thermal Neutron Scattering* (Clarendon, Oxford 1971)
J.P. Boon, S. Yip: *Molecular Hydrodynamics* (McGraw Hill, New York 1980)

3.45 R.K. Thomas; Prog. Solid State Chem. **14**, 1 (1982)

3.46 J.P. McTague, M. Nielsen, L. Passel: Neutron Scattering by Adsorbed Monolayers. CRC Crit. Rev. Solid State and Mat. Sci. **7**, 135 (1979)

3.47 J.P. Coulomb, M. Bienfait, P. Thorel: J. Physique **38**, C4-31 (1977)

3.48 M. Bienfait, P. Thorel, J.P. Coulomb: Molecular Mobility and Two-Dimensional Transitions in Adsorbed Layers, in [Ref. 3.7, p. 257]

3.49 C.J. Wright, C.M. Sayers: Rep. Prog. Phys. **46**, 773 (1985)

3.50 J.W. White: The Dynamics and Structure of Water at Interfaces, in [Ref. 3.7, p. 527]

3.51 M.W. Cole, F. Toigo: Kinetics of Elementary Processes at Surfaces, in [Ref. 3.38, p. 223]

3.52 G.F. Mazenko: Statistical Mechanical Models and Surface Diffusion, in [Ref. 3.7, p. 27]

3.53 D.K. Charturvedi, M.P. Tosi: Z. Phys. B**58**, 49 (1984)

3.54 G. Wahnström: Surf. Sci. **159**, 311 (1985)
G. Wahnström: Surf. Sci. **164**, 449 (1985)
G. Wahnström: Phys. Rev. B**33**, 1020 (1986)

3.55 M. Heuer, T.M. Rice: Surf. Sci. **155**, L269 (1985)

3.56 G. Ertl, J. Küppers: *Low Energy Electrons and Surface Chemistry* (VCH, Weinheim 1985)

3.57 R. Butz, H. Wagner: Surf. Sci. **63**, 448 (1977)

3.58 M.E. Wells, D.A. King: J. Phys. C**7**, 4052 (1974)

3.59 M. Renard, D. Deloche: Surf. Sci. **35**, 487 (1973)

3.60 Vu Thien Binh, P. Heyde: In [Ref. 3.7, p. 558]

3.61 S.M. George, A.M. de Santolo, R.B. Hall: Surf. Sci. **159**, L425 (1985)

3.62 G. Comsa, P. Poelsema: Appl. Phys. A**38**, 153 (1985)

3.63 R.M. Feming, D.B. McWahn, A.C. Gossard, M. Wiegmann, R.A. Logon: J. Appl. Phys. **51**, 357 (1980)

3.64 P.M. Petroff: J. Vac. Sci. Technol. **14**, 973 (1977)

3.65 G. Binning, H. Fuchs, E. Stoll: Surf. Sci. **169**, L295 (1986)

3.66 G.W. Jones, J.A. Venables: Ultramicroscopy **18**, 439 (1985)
H. Hanbücken, T. Doust, O. Osasona, G. Lelay, J.A. Venables: Surf. Sci. **168**, 133 (1986)

3.67 E.W. Müller, T.T. Tsong: *Field Ion Microscopy* (Elsevier, New York 1969)

3.68 J.A. Panitz: J. Phys. E**15**, 1281 (1982)

3.69    B. Bayat, H.-W. Wassmuth: Surf. Sci. **140**, 511 (1984) and references quoted therein
3.70    H.P. Bonzel, E.E. Latta: Surf. Sci. **76**, 275 (1975)
        N. Freyer, H.P. Bonzel: Surf. Sci. **160**, L501 (1985)
3.71    G. Ehrlich, F.G. Hudda: J. Chem. Phys. **44**, 1039 (1966)
3.72    R. Gomer: Comments Solid State Phys. **10**, 253 (1983)
3.73    R. DiFoggio, R. Gomer: Phys. Rev. B**25**, 3490 (1982)
        C. Dharmadhikari, R. Gomer: Surf. Sci. **143**, 223 (1984)
3.74    W. Hösler, E. Ritter, R.J. Behmißer: Ber. Bunsenges. Phys. Chem. **90**, 205 (1986)
3.75    R. Vanselow, R. Hove (eds.): Chemistry and Physics of Solid Surfaces VI, Springer Ser. Surf. Sci., Vol. 5 (Springer, Berlin, Heidelberg 1986)
3.76    S.M. Valone, A.F. Voter, D. Doll: Surf. Sci. **155**, 687 (1985)
        J.G. Lauderdale, D.G. Truhlar: Surf. Sci. **164**, 558 (1985)
3.77    K.-H. Rieder: In *Structure and Dynamics of Surfaces I*, ed. by W. Schommers, P. von Blanckenhagen, Topics Curr. Phys., Vol. 41 (Springer, Berlin, Heidelberg 1986) Chap. 2
3.78    N. Ernst, G. Ehrlich: In *Microscopic Methods in Metals*, ed. by U. Gonser, Topics Curr. Phys., Vol. 40 (Springer, Berlin, Heidelberg 1986) Chap. 4
3.79    J.Z. Larese, R.J. Rollefson: Surf. Sci. **127**, L172 (1983)
3.80    A.T. Loburets, A.G. Naumovets, Y.S. Vedula: Surf. Sci. **120**, 347 (1982)
3.81    J. Vähäkangas, H. Iwasaki, E.D. Williams, R.L. Park: Surf. Sci. **148**, 453 (1984)
        H. Lüth: Surf. Sci. **168**, 773 (1986)
3.82    Vu Thien Binh, P. Melinon: Surf. Sci. **161**, 234 (1985)
3.83    V.P. Zhdanov: Surf. Sci. **149**, L13 (1985)
        V.P. Zhdanov: Surf. Sci. **161**, L614 (1985)
3.84    W. Schommers, P. v. Blanckenhagen: Surf. Sci. **162**, 144 (1985)
        W. Schommers: Phys. Rev. B**32**, 6845 (1985)
3.85    H. Höchste, H. Bethge: J. Cryst. Growth **52**, 27 (1981)
3.86    E. Bauer: Z. Kristallogr. **110**, 372 (1958)
3.87    R.J. Muirhed, J.G. Dash, J. Kirm: Phys. Rev. B**29**, 5074 (1984)
3.88    M.B. Webb, E.R. Moog: LEED Studies of Physisorbed Noble Gases on Metals and Interatomic Interactions, in [Ref. 3.21, p. 397]
        R.J. Birgeneau, P.M. Horn, D.E. Moncton: Phases and Phase Transitions in Two-Dimensional Systems with Competing Interactions, in [Ref. 3.21, p. 405]
        S.C. Fain, Jr.: "Low-Energy Electron Diffraction Studies of Physically Adsorbed Films", in *Chemistry and Physics of Solid Surfaces, IV*, ed. by R. Vanselow, R. Howe, Springer Ser. Chem. Phys., Vol. 20 (Springer, Berlin, Heidelberg 1982) p. 203
3.89    J.H. van der Merwe: "Recent Developments in the Theory of Epitaxy", in *Chemistry and Physics of Solid Surfaces V*, ed. by R. Vanselow, R. Howe, Springer Ser. Chem. Phys., Vol. 35 (Springer, Berlin, Heidelberg 1984) p. 365
3.90    J. Krim, J.G. Dash, J. Suzanne: Phys. Rev. B**31**, 7643 (1985)
3.91    O.M. Zang, H.K. Kim, M.H.W. Chan: Phys. Rev. B**38**, 413 (1986)
3.92    G.E. Rhead: Contem. Phys. **24**, 535 (1983)
3.93    J. Hölzl, F.K. Schulte: *Work Functions of Metals*, Springer Tracts Mod. Phys., Vol. 85 (Springer, Berlin, Heidelberg 1979)
3.94    S.L. Segin, J. Suzanne, M. Bienfait, J.G. Dash, J.A. Venables: Phys. Rev. Lett. **51**, 122 (1983)
        M. Bienfait, J.L. Seguin, J. Suzanne, E. Lerner, J. Krim, J.G. Dash: Phys. Rev. **29**, 983 (1984)
3.95    J.H. Neare, B.A. Joyce, P.S. Dobson, N. Norton: Appl. Phys. A**31**, 1 (1983)
3.96    J.M. Gay, M. Bienfait, J. Suzanne: J. Physique **45**, 1497 (1984)
3.97    S.G.J. Mochrie, M. Sutton, R.J. Birgeneau, D.E. Moncton, P.M. Horn: Phys. Rev. B**30**, 263 (1984)
3.98    K. Satija, L. Passel, J. Wicksted: Physica **136B**, 7 (1986)
        H.J. Lauter, H. Godfrin, C. Tiby, H. Wichert, P.E. Obermayer: Surf. Sci. **125**, 265 (1983)
3.99    L.J. Gémez, S. Bourgeal, J. Ibánez, M. Salmeron: Phys. Rev. B**31**, 2551 (1985)
3.100  L.C. Feldmann, J.W. Mayer, S.T. Picraux: *Materials Analysis by Ion Channeling* (Academic, New York 1982)
        J.F. van der Veen: Surf. Sci. Rpt. **5**, 199 (1985)
        J.F. van der Veen, E.J. van Loenen: Surf. Sci. **168**, 701 (1986)

3.101 G.A. Bootsma, L.J. Hanekamp, O.L.J. Gigzemann: "Chemisorption Investigated by Ellipsometry", in *Chemistry and Physics of Solid Surfaces IV*, ed. by R. Vanselow, R. Howe, Springer Ser. Chem. Phys., Vol. 20 (Springer, Berlin, Heidelberg 1982)

3.102 M. Bienfait, J.A. Venables: Surf. Sci. **64**, 425 (1977)
G.E. Rhead, M.-G. Barthés, G. Argile: Thin Solid Films **82**, 201 (1981)

3.103 L.P. Erickson, B.F. Phillips: J. Vac. Sci. Technol. B**1**, 158 (1983)

3.104 T.C. Chiang, G. Kaindl, T. Mandel: Phys. Rev. B**33**, 695 (1986)
G. Pirug, A. Winkler, H.P. Bonzel: Surf. Sci. **163**, 153 (1985)

3.105 J.A. Venables: "Analytical Electron Microscopy in Surface Science", in *Chemistry and Physics of Solid Surfaces IV*, ed. by R. Vanselow, R. Howe, Springer Ser. Chem. Phys., Vol. 20 (Springer, Berlin, Heidelberg 1982)

3.106 M.P. Bruijn, H. Müller, J. Verhoeven, M.M. van der Wiel: Surf. Sci. **154**, 601 (1985)

3.107 K. Ploog: Ann. Rev. Mat. Sci. **12**, 123 (1982)
K. Ploog: Retrospect and Prospect of MBE, in [Ref. 3.43, p. 647]

3.108 M. Henzler: Surf. Sci. **168**, 744 (1986)
K.D. Gronwald, M. Henzler: Surf. Sci. **117**, 180 (1982)
D. Saloner, M.G. Lagally,: Domain Size Determination in Heteroepitaxial Systems from LEED Angular Profiles, in [Ref. 3.21, p. 366]
D. Saloner, P.K. Wu, M.G. Lagally: J. Vac. Sci. Technol. A**3**, 1531 (1985)

3.109 L.C. Feldmann, J.M. Poate: Ann. Rev. Mater. Sci. **12**, 149 (1982)

3.110 H.-J. Gossmann, L.C. Feldmann: Appl. Phys. A**38**, 171 (1985)

3.111 W. Schommers, P. von Blanckenhagen (eds.): *Structure and Dynamics of Surfaces I*, Topics Curr. Phys., Vol. 41 (Springer, Berlin, Heidelberg 1986) Chap. 6

3.112 M. Bienfait, J.L. Seguin, J. Suzanne, E. Lerner, J. Krim, J.G. Dash: Phys. Rev. B**29**, 983 (1984)

3.113 L.J. Brillson: Surf. Sci. Rpt. **2**, 123 (1983)

3.114 P.K. Larsen, B.A. Joyce, P.J. Dobson: RHEED and Photoemission Studies of Semiconductors Grown in-situ by MBE, in [Ref. 3.41, p. 196]
B.A. Joyce, P.J. Dobson, J.H. Neave, K. Woodbridge, J. Zhang: Surf. Sci. **168**, 423 (1986)

3.115 G.W. Rubloff: Metal-Semiconductor Interfaces and Schottky Barriers, in [Ref. 3.41, p. 220]
M. Schluter: Surf. Sci. **168**, 285 (1986)
A. Hiraki: Surf. Sci. **168**, 74 (1986)

3.116 J.M. Gibson, R.T. Tung, J.M. Phillips, R. Hull: J. Physique **46**, C4-369 (1985)

3.117 L. Esaki: Advances in Semiconductor Superlattices, Quantum Wells and Heterostructurs, in [Ref. 3.41, p. 48]
L. Esaki: Compositional Superlattices, in [Ref. 3.43, p. 143]

3.118 G.H. Döhler: Physica Scripta **24**, 430 (1981)
G.H. Döhler: Doping Superlattices, in [Ref. 3.42, p. 233]

3.119 D.W. Laidig, C.K. Peng, Y.E. Lin: J. Vac. Sci. Technol. B**2**, 181 (1984)

3.120 L. Tapfer, K. Ploog: Phys. Rev. B**33**, 5565 (1986)

3.121 A.H. Hamdi, V.S. Speriosu, J.L. Tandon, M.-A. Nicolet: Phys. Rev. B**31**, 2343 (1985)

3.122 C.M. Falco, W.R. Bennett, A. Boufelfel: Metal-Metal Superlattices, in [Ref. 3.41, p. 35]

3.123 U. Landmann, R.N. Barnett, C.L. Cleveland, R.H. Rast: J. Vac. Sci. Technol. A**3**, 1574 (1985)
G. de Lorenzi, G. Jaccucci: Surf. Sci. **164**, 526 (1985)

3.124 T.T. Tsong: Physica Scripta T**4**, 17 (1983)
T.T. Tsong: M. Ahmad: In [Ref. 3.21, p. 389]

3.125 R. Ramirez, A. Rahman, I. Schuller: Phys. Rev. B**30**, 6208 (1984)

3.126 F. Abraham: J. Vac. Sci. Technol. B**2**, 534 (1984)
G.H. Gilmer: "Ising Model Simulations of Crystal Growth", in *Chemistry and Physics of Solid Surfaces V*, ed. by R. Vanselow, R. Howe, Springer Ser. Chem. Phys., Vol. 35 (Springer, Berlin, Heidelberg 1984) p. 297

3.127 S. Toxvaerd: Phys. Rev. Lett. **43**, 529 (1979)

3.128 M. Schneider, A. Rahman, I. Schuller: Phys. Rev. Lett. **55**, 604 (1985)

3.129 F. Abraham, J. Broughton: Phys. Rev. Lett. **56**, 734 (1986)
W. Landmann, W. Luedtke, R. Barnett, C. Cleveland, M. Ribarsky, E. Arnold, S. Ramesh, H. Baumgart, A. Martinez, B. Khan: Phys. Rev. Lett. **56**, 155 (1986)
3.130 R. Gomer: Appl. Phys. A**39**, 1 (1986)
3.131 S. Brennan, P.H. Fuoss, D. Eisenberger: In [Ref. 3.21, p. 421]
3.132 L.F. Eastman: "High-Speed Gallium Arsenide Transistors for Logic Applications", in *The Physics of VLSI*, ed. by J.C. Knights (American Institute of Physics, New York 1984)

# Additional References with Titles

Bauer, E., van der Merwe, Jan H.: Structure and growth of crystalline superlattices: From monolayer to superlattice. Phys. Rev. B**33**, 3657 (1986)

Cohen, P.I., Pukite, P.R., van Hove, J.M., Lent, C.S.: Reflection high energy electron diffraction studies of epitaxial growth on semiconductor surfaces. J. Vac. Sci. Technol. A**4**, 1251 (1986)

Heiblum, M., Anderson, I.M., Knoedler, C.M.: DC performance of ballistic tunneling hot-electron transfer amplifiers. Appl. Phys. Lett. **49**, 207 (1986)

Kelly, M.J., Weisbuch, C. (eds.): *The Physics and Fabrication of Microstructures and Microdevices* (Springer, Berlin, Heidelberg 1986)

Kern, K., David, R., Palmer, R.L., Comsa: Complete wetting on "strong" substrates: Xe/Pt (111). Phys. Rev. Lett. **56**, 2823 (1986)

Kobayashi, A., Sun-Mok Paik, K.E. Khor, S.: Das Sarma: A molecular statics and molecular dynamics study of epitaxial growth fronts. Surface Sci. **174**, 48 (1986)

Lo, Tai-Chin, Ehrlich, G.: Activated chemisorption of methane and tunneling. Surface Sci. **179**, L19 (1987)

Mon, K.K.: Monte Carlo simulation of the growth of wetting layers. Phys. Rev. B**35**, 3683 (1987)

Morris, M.C., Barnes, C.J., King, D.A.: Monolayer and multilayer surface diffusion, growth mode and thermal stability of indium on W{100}. Surface Sci. **173**, 619 (1986)

Nham, H.S., Drir, M., Hess, G.B.: Multilayer phase diagram of $CF_4$ adsorbed on graphite. Phys. Rev. B**35**, 3675 (1987)

Preuss, E., Freyer, N.; Bonzel, H.P.: Surface self-diffusion on Pt(110): directional dependence and influence of surface-energy anisotropy. Appl. Phys. A**41**, 137 (1986)

Rys, F.S.: On the roughening of solid surfaces. Surface Science **178**, 419 (1986)

Sanchez, A., Ibanez, J., Miranda, R., Ferrer, S.: The first stages of expitaxial growth of Pb atoms on Cu(100) studied by scattering of thermal helium. Surface Sci. **178**, 917 (1986)

Wilson, R.J., Chiang, S.: Structure of the Ag/Si (111) Surface by Scanning Tunneling Microscopy. Phys. Rev. Lett. **58**, 369 (1987)

Zhdanov, V.P.: Effect of lateral interactions on tunnel diffusion of adsorbed particles. Surface Sci. **177**, L896 (1986)

# 4. Phase Transitions on Single-Crystal Surfaces and in Chemisorbed Layers

E. Bauer

With 24 Figures

Phase transitions in two dimensions are discussed with emphasis on systems in which the lateral interactions between the atoms are weak compared to their interaction with the supporting substrate. In these systems some phases and the associated transitions which can occur in purely two-dimensional systems, i.e. systems on a structureless substrate, are suppressed but are replaced by new phases, resulting in an equally wide or even wider variety of phases and phase transitions. The theoretical foundations for the understanding are presented in a very condensed manner which should give the nonspecialist, in particular the experimentalist, access to the vast theoretical literature for more detailed understanding.

For the theoretician who is not only interested in the beauty of theoretical models but also in physical reality it is important to be aware of the possibilities of the experimentalist and of the limitations set by nature. A section on experimental techniques tries to convey this aspect of two-dimensional phase transitions. The results for specific systems obtained to date, both theoretically and experimentally, are summarized in the last section. It is evident, in particular from the last two sections that the subject of this review is still in its infancy. For example, precise and new methods are still in development and the relative importance of point and line defects in phase transitions of the type discussed here is far from being clear.

An extensive list of references completes the review. It must be kept in mind, however, that the literature on two-dimensional phase transitions is so extensive, that many of the references refer to reviews or to the most recent original publications in which references to older work may be found.

## 4.1 Background

This chapter is written for the experimentalist who wants to find access to this rapidly expanding field without having to struggle through the immense amount of highly sophisticated theoretical work which has been published on this subject (general references [4.1–16]). In distinction from many excellent and/or sophisticated reviews written by theoreticians, which emphasize the critical phenomena aspect of phase transitions, this approach looks at the subject from the point of view of surface physics.

A phase transition is a change of order. Order is determined by the competition between energy U and entropy S, the former wanting to produce order, the latter wanting to destroy it. Thus free energy $F = U - TS$ is the thermodynamic quantity which controls the state of order. The state of order can be characterized by an order parameter $\psi$ which may be a real or complex number or a set of numbers ("multi-component" order parameter) depending upon the complexity of the system. $\psi$ is usually defined in such a manner that $\psi = 1$ when the system is fully ordered and $\psi = 0$ when it is fully disordered. F and $\psi$ depend upon the temperature T, the number N of atoms and – in two dimensions – on the area A.

In the case of single crystal surfaces and chemisorbed layers the two-dimensional density $\varrho = N/A$ of the atoms is constant during a phase transition due to a temperature change. The surface system is, therefore, a closed system in contrast to most physisorption systems which are usually studied in (quasi-) equilibrium with the three-dimensional gas phase. In phase transition studies of physisorbed layers the chemical potential $\mu$, which is the independent variable – in addition to T – in most of the theoretical work, is controlled via the vapour pressure p. In phase transitions on clean surfaces and in chemisorbed layers the vapour pressure is usually negligible and the coverage $\theta = N/N_0 = (N/A) \cdot (A/N_0) = \varrho a_0$ is controlled, where $N_0$ is either the number of atoms per unit area of the substrate or the maximum number of atoms per unit area which can be accommodated at monolayer saturation and $a_0$ the "co-area" of an adsorbed molecule. This has to be kept in mind when the results discussed here are compared with theory and physisorption experiments.

Phase transitions in physisorption systems have been studied extensively [4.17–19] and also for the subject of this chapter a number of reviews [4.22–26] are available. Most of them focus on order-disorder transitions, mainly stimulated by the theoretical interest of such transitions in two dimensions (2d). The interest is due to the unique role which the dimension $d = 2$ plays in the struggle between S and U, for the following reasons. The connectivity of a system of particles increases with increasing dimensionality. Thus in $d = 1$ each atom is directly connected to two neighbours while in $d = 3$ as many as 12 direct connections may exist. The better the connectivity the larger are the possibilities for the ordered phase to remain connected around a local fluctuation and hence the fluctuation has less effect in reducing the degree of order. Thus in $d = 1$, S is so dominating at non-zero temperatures that no long-range order induced by U is possible at all; in $d = 3$ the many interconnections stabilize long-range order to such a degree that most phase changes occur abruptly (first order). In $d = 2$ there is no true long-range order as in the case $d = 3$ – unless some lattice is imposed on the system – and the energy of various imperfections ("topological excitations") is so small that they can be excited thermally. As a consequence, a wide range of phase transitions is possible which makes the case $d = 2$ an ideal testing ground of theoretical models.

The present review tries to give an overview of all *structural* 2d phase transitions which have received attention in recent years. These include not only order-disorder transitions in systems with repulsive nearest neighbour interactions but also 2d sublimation and melting in systems with attractive interactions, displacive order-order transitions, wetting and roughening. Obviously, this cannot be done in depth because of lack of space, it cannot be very coherent because of lack of a common theoretical framework, and rigor will frequently have to be substituted by qualitative and sweeping arguments.

## 4.2 Theoretical Foundations

Phase transitions are classified according to the behaviour of the free energy $F(T, p)$ at the phase transition. When the first derivatives of $F (\partial F / \partial T)_p$ and $(\partial F / \partial p_T$ change discontinuously the transition is called first order, when the second- and higher-order derivatives change discontinuously, it is called second and higher order, respectively, or continuous. When the independent variable is $\theta$ instead of p as is the case in this chapter the distinction is not so simple because $(\partial F / \partial T)_\theta$ and $(\partial F / \partial \theta)_T$ change continuously though with a break during the transition between single phase and two-phase coexistence region. In the presence of attractive interactions the system may exhibit two pure phases, one of low density (gas) and the other of high density (condensate), and between them a coexistence region of the two pure phases. The coexistence region is generally terminated at some temperature $T_c$ by a critical point. If the interactions are repulsive or oscillatory there may be only one phase at given T, $\theta$ and the transition between order ("crystal") and disorder (liquid, vapour) may be either first order or continuous.

There is a vast amount of theoretical work on continuous order-disorder transitions starting with the work of Landau which will be sketched briefly in Sect. 4.2.4. Little work has been done up to now on 2d displacive order-order transitions (Sect. 4.2.5). A particularly frequent transition on surfaces is the commensurate-incommensurate transition in which the surface layer or adsorbate develops a non-rational periodicity with respect to the bulk upon a change of T or $\theta$ (Sect. 4.2.6). Melting which is first order in three dimensions may be a continuous and two-step transition in two dimensions, which will be discussed in Sect. 4.2.7. Interface delocalization transitions have recently received much attention (Sect. 4.2.8). This is not true for two experimentally important subjects, systems with long range repulsive interactions (Sect. 4.2.9) and systems with purely attractive interactions (Sect. 4.2.10).

All these transitions occur also in physisorbed layers and most of the experimental work and detailed calculations were done up to now for such systems. Therefore, some of the general conclusions from this work, which are also valid in the present context, are included here. Before the various phase transitions can be discussed intelligibly it is necessary to define and classify two-dimensional phases (Sect. 4.2.1), to understand their origin and their relation to magnetic phases (Sect. 4.2.2) and to introduce the various models and the nomenclature used to describe them (Sect. 4.2.3).

### 4.2.1 Phases in Two Dimensions

A phase is characterized by its state of order which is described by the order parameter $\psi$. At low temperatures the periodic substrate potential induces long-range positional order in the adsorption layer. With this periodic lattice in real space a reciprocal lattice with reciprocal lattice vectors $\boldsymbol{G}$ is connected. The order parameters are then chosen as the Fourier coefficients $\varrho_G(\boldsymbol{r})$ of the atomic density (occupation probability) distribution over the lattice cells labelled by $\boldsymbol{r}$. The density-density correlation function $\phi_G(\boldsymbol{r}) = \langle \varrho_G(\boldsymbol{r})\varrho_G(0)\rangle$ which is a measure for the probability of finding the same atomic density in the lattice cells $0$ and $\boldsymbol{r}$ is constant in the case of long-range order so that also

$$\lim_{r\to\infty} \phi_G(\boldsymbol{r}) = \text{const} \ . \tag{4.1}$$

This phase is called a locked solid (LSo). There are two other possibilities for the asymptotic behaviour of $\phi_G(\boldsymbol{r})$ : algebraic and exponential decay with increasing distance r:

$$\lim_{r\to\infty} \phi_G(\boldsymbol{r}) = (\text{r}/\text{a})^{-\eta(G)} \ , \tag{4.2}$$

$$\lim_{r\to\infty} \phi_G(\boldsymbol{r}) = \text{const}\, e^{-\text{r}/\xi} \ . \tag{4.3}$$

The corresponding phases have only quasi-long-range order and short-range order and are called floating solid (FSo) and fluid (liquid (Li) or gas), respectively. The fluid phase is always possible – provided that the adsorbate does not desorb or dissolve in the bulk before disordering – the floating solid only above a critical ratio $p_{min}$ of the lattice constants of adsorbate (a) and substrate (b). For a square substrate $p_{min} = 4$, for a triangular substrate $p_{min} = 2\sqrt{3}$ [4.27].

Many substrates are anisotropic and many adsorbates have anisotropic interactions. As a consequence the degree of order may be different in different directions, say x and y, so that combinations of the cases (4.1–3) may occur. The resulting phases are called smectic phases in analogy to 3d smectic liquid crystals which have broken translational order in only one direction. The three possible cases $(x-1, y-2)$, $(x-1, y-3)$, $(x-2, y-3)$ are called partially locked solid (PLS), locked smectic (LSm) and floating smectic (FSm). The possible phases depend upon whether $p_x = a_x/b_x$ and $p_y = a_y/b_y$ are $>$ or $<p_{min}$ and upon the degree of anisotropy, as illustrated in Fig. 4.1 [4.27].

In addition to the 6 phases discussed up to now, there are many intermediate phases with partially broken translational symmetry. One famous 2d phase probably does not occur in the surface systems discussed here: the hexatic phase. This phase has exponential decay of the positional correlations but only algebraic decay of the bond angle correlations. This phase exists only on substrates with rotational invariance between the floating solid and the fluid and is, therefore, not expected on substrates with pronounced lateral periodicity.

The transitions LSo-Li (1.1–3.3), LSo-FSo (1.1–2.2) and FSo-Li (2.2–3.3) will be discussed in Sects. 4.2.4, 6 and 7, respectively. Transitions involving anisotropic phases will be touched upon in Sect. 4.2.4c.

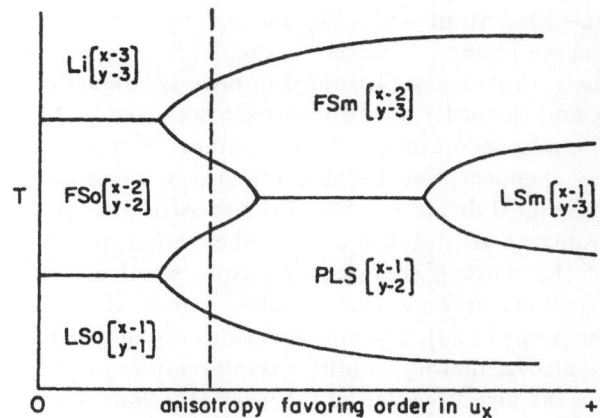

Fig. 4.1. Possible 2d phase transitions in the presence of anisotropy for the case $p_x, p_y > p_{min}$ [4.27]. For explanation see text. $u_x$ describes the x-coordinates of the atomic positions

In the figure, from top to bottom on the left axis T:

$Li \begin{bmatrix} x-3 \\ y-3 \end{bmatrix}$

$FSm \begin{bmatrix} x-2 \\ y-3 \end{bmatrix}$

$FSo \begin{bmatrix} x-2 \\ y-2 \end{bmatrix}$

$LSm \begin{bmatrix} x-1 \\ y-3 \end{bmatrix}$

$PLS \begin{bmatrix} x-1 \\ y-2 \end{bmatrix}$

$LSo \begin{bmatrix} x-1 \\ y-1 \end{bmatrix}$

anisotropy favoring order in $u_x$

## 4.2.2 Origin of Surface Phases and Relation to Magnetic Models

There are two fundamentally different mechanisms which produce surface phases with lateral ordering different from the substrate: in one case the atoms in the topmost layer(s) of the substrate are displaced significantly parallel (and possibly normal) to the surface thus producing the surface phase. This displacement may occur already on the clean surface, e.g. on W(100) or may be assisted by adsorbed atoms, e.g. by hydrogen on W(100). In both cases the surface phase is a consequence of the instability of the surface against displacements of the surface atoms caused by electronic or elastic effects (displacive transition).

In the first case the periodicity of the surface phase whose reciprocal lattice vector $q$ differs from *all* reciprocal lattice vectors $g$ of the substrate creates a band gap in the electronic surface state band structure. This is connected with a reduction of the electronic energy which overcompensates the elastic energy due to the surface distortion and thus minimizes the total surface energy. This "Peierls instability" was originally believed to be the main cause for surface phases with a lateral periodicity different from that of the bulk ("surface reconstruction") [4.28]. More recently [4.29,30] there is mounting evidence that purely elastic models can explain the reconstruction. These models have "soft phonons" with wave vectors $q$ corresponding to the periodicity of the surface phase in their phonon spectrum. Soft phonons have energies $\hbar\omega \approx 0$ and cause, therefore, a spontaneous lattice deformation ("frozen in" or "static" phonon). The driving force is still electronic in nature, but involves states throughout the d-band, rather than just those at the Fermi level with the "spanning" wave vector. The consequences of this more "chemical" driving force include an effectively shorter range and a lack of significant electronic entropy so that an elastic model is appropriate. (For a review, see [4.30a].)

Alternatively the surface atoms of a clean surface may be considered as self-adsorbed atoms or dimers which form an interacting 2d lattice gas. This lattice gas may be transcribed into a spin model (see below) and treated in the same manner as foreign adsorbed atoms [4.31,32]. For a recent review of reconstruction phase transitions see [4.33].

In the second case the substrate atoms are assumed to have the same lateral distribution as in the bulk and the order in the adsorbate is caused by the lateral interactions between the adsorbed atoms. (This assumption is probably never completely correct in chemisorption because the strong adsorbate-substrate interaction causes some local displacements of the substrate atoms.) The substrate has solely the function to stabilize a 2d configuration, to provide a periodic modulation of the binding energy (adsorption sites) and to contribute to the lateral interactions between the adsorbed atoms via indirect electronic and elastic interactions [4.34]. These interactions are in general anisotropic and in the first case also oscillatory. Additional interactions are the direct electronic interactions via the direct overlap of the wave functions of the adsorbed atoms and the dipole-dipole interactions which are important when the adsorbate-substrate complex has a large dipole moment.

The complexity of the interactions forces considerable simplifications when modeling phases and phase transitions. The most drastic approximation is to neglect all interactions except the hard-core repulsion which is described by an infinitely large repulsion energy $\varphi_1 = +\infty$ for first-nearest neighbours (sometimes also for second-nearest neighbours, $\varphi_2 = +\infty$). The first- (and second-) nearest neighbour exclusion leads to the hard square and hard hexagon model for the square and triangular lattice, respectively. The other extreme is to take only long-range interactions between atoms i and j of the form $\varphi_{ij} = \mathrm{const}\,/r_{ij}^3$ into account which is appropriate for dipole-dipole interactions at low coverage.

The most usual ways to model lateral interactions are to assume elastic forces between neighbouring atoms which is appropriate at high coverages and/or strong attractive interactions, or to assume pairwise interactions $\varphi_{ij}$ and trio-interactions $\varphi_{ijk} \equiv \varphi_t$ between the atoms in a lattice gas model. (For a review of trio-interactions see [4.34].) In order to simulate the observed phases pair-interactions up to $7^{\mathrm{th}}$ nearest neighbours have been taken into account but the trio-interactions are always limited to triangles including neighbour atoms and are usually assumed to be independent of the shape of the triangle in order to limit the parameter space. They are necessary in order to take into account the nonadditivity of the interactions. In this lattice gas model the total energy (Hamiltonian) is then given by

$$\mathcal{H} = -\varepsilon \sum_i c_i - \sum_{i \neq j} \varphi_{ij} c_i c_j - \sum_{i \neq j \neq k} \varphi_t c_i c_j c_k \ , \tag{4.4}$$

where $\varepsilon(>0)$ is the binding energy of the isolated adsorbate atom to the substrate, $c_i$ is the occupation number which is either 1 (site occupied) or 0 (site unoccupied) and $\varphi_{ij}, \varphi_t >$ or $<0$ for attractive or repulsive interactions. If N is the number of available adsorption sites then the coverage $\theta$ is given by $\theta = \sum_i \langle c_i \rangle / N$ where $\langle \rangle$ denotes the thermal average.

Calculations are usually not done at fixed $\theta$ – which makes the chemical potential $\mu$ variable with temperature T – (canonical ensemble) but at fixed $\mu$

(grand canonical ensemble) which makes the number $N_a = \sum_{i=1}^{N} c_i$ of adsorbed atoms, i.e. $\theta$ a function of T. Equation (4.4) is then replaced by

$$\mathcal{H} - \mu N_a = -(\varepsilon + \mu) \sum_{i=1}^{N} c_i - \sum_{i \neq j} \varphi_{ij} c_i c_j - \sum_{i \neq j \neq k} \varphi_t c_i c_j c_k \tag{4.5}$$

and $\theta$ is obtained from the thermodynamic relationship

$$\theta = \frac{1}{N} \left( \frac{\partial F}{\partial \mu} \right)_T , \tag{4.6}$$

where F is the free energy of the system which is obtained from $\mathcal{H}$ via the partition function Z

$$F = -k_B T \ln Z = -k_B T \sum_{\{c_i\}} e^{-\mathcal{H}/k_B T} ; \tag{4.7}$$

the summation being over all configurations of the system.

Most models in the theory of phase transitions are not lattice gas models but spin models and most calculations are done for the latter. Spin models in which the spin on each site i may only have two values, say $s_i = \pm 1$ ("Ising pseudo-spin variable") may, however, be transcribed back into a lattice gas model using the relationship $c_i = (1 - s_i)/2$, and vice versa. The Hamiltonian (4.5) leads then to a generalized Ising Hamiltonian

$$\mathcal{H}_{I\sin g} = -H \sum_i s_i - \sum_{i \neq j} J_{ij} s_i s_j - \sum_{i \neq j \neq k} J_t s_i s_j s_k \tag{4.8}$$

with an effective magnetic field

$$-H = \frac{\mu + \varepsilon}{2} + \frac{1}{4} \sum_{i \neq j} \varphi_{ij} + \frac{1}{8} \sum_{i \neq j \neq k} \varphi_t \tag{4.9}$$

and effective 2-spin and 3-spin exchange constants

$$J_{ij} = \frac{1}{4} \varphi_{ij} + \frac{1}{8} \sum_{k(\neq i,j)} \varphi_t , \quad J_t = -\frac{1}{8} \varphi_t . \tag{4.10a, b}$$

Equation (4.9) and (4.10a,b) imply

$$\frac{\varepsilon + \mu}{2} = -H - \sum_{i \neq j} J_{ij} - \sum_{i \neq j \neq k} J_t \tag{4.11}$$

which yields $\mu$ while $\theta$ is obtained from the magnetization $m = \sum_{i=1}^{N} \langle s_i \rangle$ via $c_i = (1 - s_i)/2$ as

$$\theta = (1 - m)/2 . \tag{4.12}$$

The transcription from the spin model to the lattice gas model thus is not trivial and the T-H, T-$\mu$ and T-$\theta$ phase diagrams look quite different from each other. The Ising Hamiltonian (4.8) is invariant under the transformation H, $J_t$, $\{s_i\} \rightarrow -$ H, $-J_t$, $\{-s_i\}$. In the absence of trio-interactions ($J_t = 0$) the T-$\theta$ phase diagram is, therefore, symmetric around $\theta = 1/2$ because $\{s_i\} \rightarrow \{-s_i\}$ means m$\rightarrow -$ m and $\theta \rightarrow 1 - \theta$ ("particle-hole symmetry"). Most experimental phase diagrams are asymmetrical, which demonstrates the importance of non-pairwise interactions.

Another important connection between spin and lattice gas models is that which relates spin models with n possible spin orientations on each lattice site to adsorbates whose unit meshes can have n different but equivalent positions relative to the substrate lattice (q-state Potts models, p-state clock models, etc., see next section).

### 4.2.3 Models and Nomenclature

Many statistical mechanical models have been devised to describe phase transitions [4.12,35,36]. They can be classified in symmetric and asymmetric models which may have discrete or continuous symmetry. Models with discrete symmetry have long-range order at low temperature, models with continuous symmetry have no long-range order at any T>0 in two dimensions.

In the context of this review the most important symmetric models with discrete symmetry are:

1) The q-state Potts model [4.35]. To each lattice site a quantity is assigned which can take q different values $s_i = 1, \ldots, q$. It is usual to represent this quantity by a "spin" $s_i$ with unit length which can point in q different high symmetry directions in $(q-1)$-dimensional space. The Hamiltonian of the model is $\mathcal{H} = -J \sum_{i \neq j} \hat{s}_i \cdot \hat{s}_j$ (Fig. 4.2a).

2) The p-state clock model [4.35] or planar Potts model or $Z(p)$ model is very similar to the Potts model, but the possible directions of the spins $\hat{s}_i$ are limited to a plane. Their orientation can then be characterized by p angles $\theta_i = (2\pi/p)m_i$ ($m_i = 1, \ldots, p$) which leads to the Hamiltonian $\mathcal{H}_N = -J \sum_{i \neq j} \hat{s}_i \cdot \hat{s}_j = -J \sum_{i \neq j} \cos [2\pi(m_i - m_j)/p]$, (Fig. 4.2b).

The comparison of Figs. 4.2a and b shows that the 3-state Potts and the 3-state clock model are identical, as are the corresponding 2-state models which are nothing else but the Ising model with its two possible spin orientations. The 4-state model, however, differs clearly in symmetry. The 4-state clock model $Z(4)$ is a special case of the

3) Ashkin-Teller model [4.12] in which to each lattice site two Ising spins $s_i$, $t_i$ are assigned and whose Hamiltonian is given by $\mathcal{H}_{AT} = -\sum_{i \neq j}(Js_is_j + J't_it_j + J_4s_is_jt_it_j)$. If the 4-spin exchange constant $J_4 = 0$ and $J' = J$ the $Z(4)$

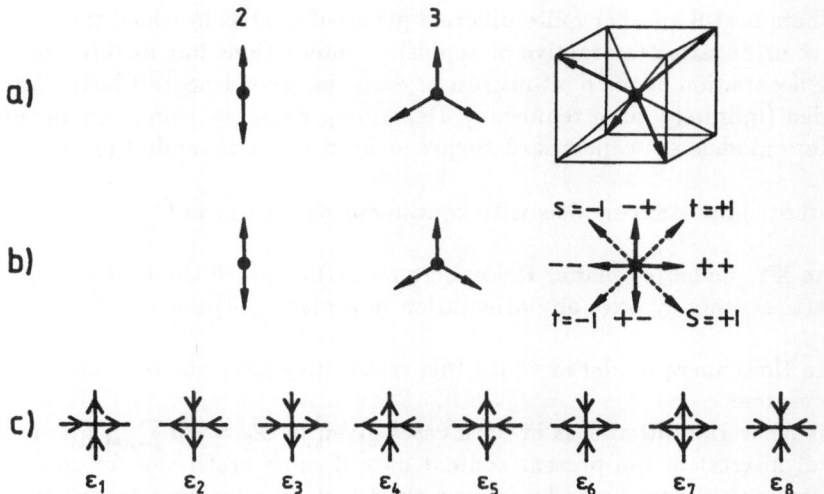

**Fig. 4.2a–c.** Spin representation of some basic models. (a) q-state Potts model. (b) p-state clock (planar Potts, Z(p)) model; for p = 4 the relationship to the Ashkin-Teller spins (– – –) is indicated. (c) Symmetric 8-vertex model

model is obtained from the AT model as indicated in Fig. 4.2b. The special case $J = J' = J_4$ corresponds to the 4-state Potts model.

The 3-state Potts model may also be considered as a special case of a more general model, namely

4) the Blume-Emery-Griffith-model [4.37] which has also 3 possible values $(0, \pm 1)$ of $s_i$ assigned to each lattice site but a more complicated Hamiltonian $\mathcal{H}_{BEG} = -J \sum_{i \neq j} s_i s_j - K \sum_{i,j} s_i^2 s_j^2 + \Delta \sum_i s_i^2$. It contains, in addition to the usual bilinear exchange term, a biquadratic term and a term which depends only upon the occupation ("crystal field interaction"). The model has been extended further by adding a magnet field term $H \sum_i s_i + L \sum_{i \neq j}(s_i s_j^2 + s_i^2 s_j)$ [4.38]. The 3-state Potts model is obtained for $H = L = 0$, $K = 3J$ and $\Delta = 8J$.

Finally, all the models discussed up to now may be viewed as special cases of the

5) Baxter (symmetric 8-Vertex) model [4.12] in which to each lattice site 8 possible 4-arrow combinations ("vertices") are assigned, as indicated in Fig. 4.2c for a rectangular lattice. To each vertex an energy $\varepsilon_i$ is assigned and the Hamiltonian is given by $\mathcal{H}_{8V} = \sum_i n_i \varepsilon_i$, if there are $n_i$ vertices i. The model is restricted to $\varepsilon_1 = \varepsilon_2$, $\varepsilon_3 = \varepsilon_4$, $\varepsilon_5 = \varepsilon_6$, $\varepsilon_7 = \varepsilon_8$, i.e. it is invariant upon reversal of all arrows ("zero field"). It may be transcribed into 2 Ising models which are linked by 4-spin interactions: $\mathcal{H}_{8VI} = J_2' \sum_{i \neq j} s_i s_j + J_4 \sum_{i,j,k,\ell} s_i s_j s_k s_\ell$ where $s_i = \pm 1$ and the i, j sum is over all next-nearest-neighbour pairs of sites and the i, j, k, $\ell$ sum over all quartets of sites forming elementary squares. There are many interrelations between the various models which are useful in the calculation of phase diagrams and of critical exponents (Sect. 4.2.5).

6) There is still another quite different group of models in which there are no positive or negative (attractive or repulsive) interactions but merely exclusions: the occupation of the next-nearest or even the second-nearest lattice site is forbidden (infinitely large repulsion). Depending upon the symmetry of the lattice these models are called hard-square or hard-hexagon models [4.12].

The most important models with continuous symmetry are

7) the XY model (or planar Heisenberg model) in which the unit vector $s_i$ of each lattice site may have any orientation in a plane (2d) and

8) the Heisenberg model in which this vector may have any orientation in three dimensions.

Their basic Hamiltonian is in both cases given by $\mathcal{H}_c = -J \sum_{i \neq j} s_i \cdot s_j$ but they are of interest in the present context only if some preference for certain spin orientations is introduced by adding quadratic or cubic anisotropy terms to $\mathcal{H}_c$.

When the number of equivalent positions of the adsorbate or surface phase mesh with respect to the substrate mesh is larger than 2 then inequivalent walls can be formed between regions with different positions which destroy the up-down or clockwise-counterclockwise symmetry of the models discussed up to now (Fig. 4.3). Uniaxial, "chiral" or "helical" models result in this way from the symmetric models. The most important ones discussed up to now are the

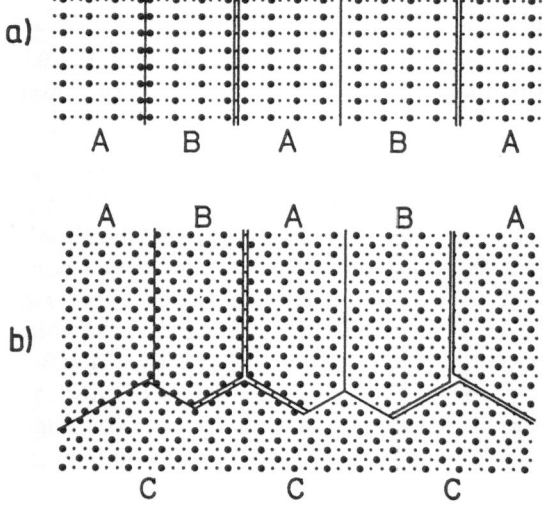

Fig. 4.3. Wall configurations in uniaxial (a) and triaxial (b) chiral (helical) clock models [4.44]. (a) is a p(3 × 1) structure on a rectangular lattice. Note the difference between AB and BA walls. The left side shows "superheavy" (AB) and "heavy" BA) walls, the right side "light" (AB) and "extralight" (BA) walls. (b) is a $(\sqrt{3} \times \sqrt{3})R\ 30°$ structure on a hexagonal lattice. *Left side:* superheavy AB, BC, CA and heavy BA, AC, CB walls, *right side:* corresponding light walls. Heavy, light, ... describes the packing density in the walls

9) asymmetric clock models [4.39], also called chiral clock or Potts models [4.40,41] and the closely related helical Potts model [4.42,43]. The 3-state asymmetric clock model which has been studied in most detail [4.44–46] results from model 2 by adding a "chirality" term to the Hamiltonian: $\mathcal{H} = -J\sum_{i\neq j}\cos\left[2\pi(m_i - m_j - \Delta\cdot\hat{R}_{ij})/p\right]$ with p = 3. Here $\hat{R}_{ij}$ is a unit vector between sites i and j and the "chiral field" $\Delta = \Delta\hat{x}$ favors a continuous rotation of the phase angle $2\pi m_i/p$ with position along the $\hat{x}$-direction ("chirality"). This competes with the restriction of the phase angle to discrete values (see model 2) and lowers the overall symmetry of the model. Chirality is achieved by introduction of a proper sequence of inequivalent domain walls (Fig. 4.3). A variation of $\Delta$ corresponds to a variation of the chemical potential in the adsorbate system, $\Delta \sim \mu - \mu_0$, and, therefore, to a change of $\theta$ which is realized by a change of the number and kind of walls.

10) ANNNI model (axial or anisotropic next-nearest-neighbour Ising) [4.47–50], additional references in [4.51,52]. The anisotropy is introduced by competing nearest-neighbour ($J_i \gtrless 0$) and next-nearest-neighbour interactions ($J_2 < 0$) along only one axis ($\hat{=}i$) : $\mathcal{H} = -\sum_{i\neq j}(J_1 s_{ij}s_{i+1,j} + J_2 s_{ij}s_{i+2,j} + J_0 s_{ij}s_{i,j+1} + Hs_{ij})$, ($J_0 > 0$, i.e. ferromagnetic or attractive).

11) ATNNI model (anisotropic triangular nearest-neighbour Ising) [4.53]. This appears to be the simplest model which gives a realistic phase diagram. It is for a centered rectangular lattice with repulsive nearest-neighbour ($J_1$) and next-nearest-neighbour interactions ($J_2$) : $\mathcal{H} = \sum_i\{J_1(s_i s_{i+1} + s_i s_{i+2}) + J_2 s_i s_{i+3} - Hs_i\}$ with $J_2 = \alpha J_1$ ($0 < \alpha < 1$).

The asymmetric and anisotropic models are necessary whenever the adsorbate mesh can have 3 or more equivalent positions with respect to the substrate in order to describe phase diagrams and transitions realistically.

When the magnetic field h = 0 (or the chemical potential $\mu = \mu_c$) and there is no other independent variable but the temperature T only a critical point can occur, e.g. the Curie point. When there are two or more variables critical lines, surfaces or hypersurfaces may exist which may contact other lines, surfaces or hypersurfaces in so-called "multicritical" points, lines or surfaces. Here only two variables (T,$\mu$ or T,h) will be considered in which case the following types of multicritical points may occur [4.14]: (i) a tricritical point (Fig. 4.4a) where a continuous transition line changes into a first-order line, (ii) a Lifshitz point (Fig. 4.4c) where a first-order line between a simple ordered phase and a modulated phase meets two continuous transition lines and (iii) a bicritical point (Fig. 4.4d) where a first-order line separating two phases with different symmetry, e.g. Ising and XY-symmetry (n = 1,2) meets two continuous transition lines. (ii) can occur in the presence of competing interactions (see above), (iii) in the presence of symmetry-breaking interactions such as in uniaxial anisotropy $g/2\sum_i[s_{iz}^2 - (s_{ix}^2 + s_{iy}^2)/2]$ which makes $\langle\hat{s}\rangle$ preferred paral-

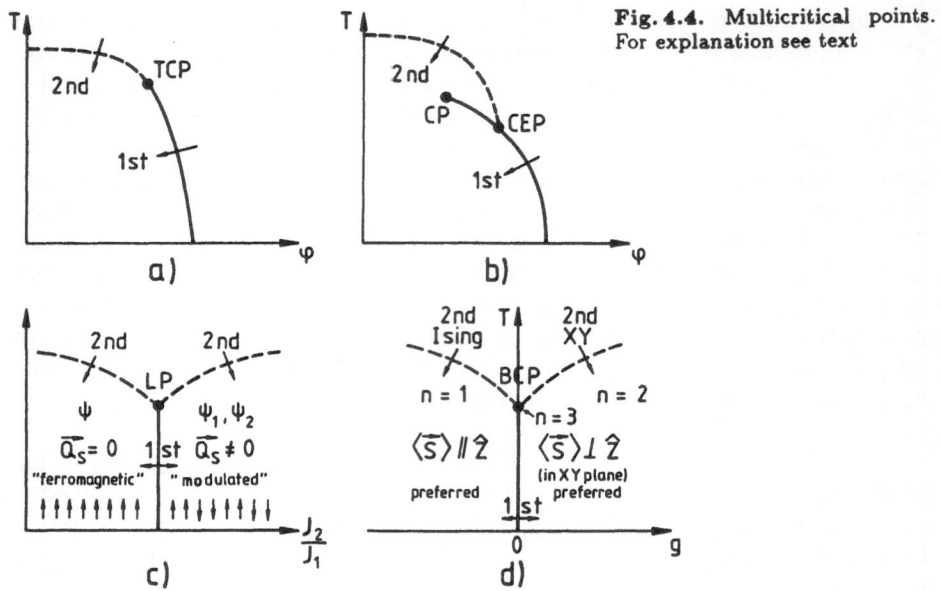

**Fig. 4.4.** Multicritical points. For explanation see text

lel $\hat{z}$ when $g<0$, otherwise preferred in the xy-plane. Other special points are the triple point where three first-order lines meet and the critical end point (Fig. 4.4b) where a continuous transition line ends on a first order line.

### 4.2.4 Continuous Order-Disorder Transitions

#### a) Critical Behaviour and Exponents, Universality and Scaling Laws

In continuous transitions the order changes over a finite temperature or chemical potential $\mu$ ("ordering field h") range until the order parameter vanishes at the critical point $T_c, \mu_c$. Very close to $T_c$ thermodynamic quantities $L(T)$ usually have the asymptotic behaviour

$$L(T) = A_\lambda |t|^\lambda \left(1 + \sum_i a_{\lambda,i}^\pm |t|^{\theta_i}\right) , \tag{4.13}$$

where $A_\lambda$ and $a_{\lambda,i}^\pm$ are constants and $t = (T - T_c)/T_c$ is the reduced temperature. The $+ (-)$ is valid for the approach of $T_c$ from above (below). The expression in parentheses of (4.13) is called the correction to the scaling factor to the asymptotic power law with exponent $\lambda$. It has been set equal to one in all 2d experiments analyzed up to now. It must be kept in mind, however, that even experiments done with the highest precision do not give the theoretical $\lambda$ but a $\lambda_{\text{eff}} = \partial \ln L / \partial \ln |t| \approx \lambda + \sum_i a_{\lambda,i}^\pm \theta_i |\bar{t}|^{\theta_i}$ where $\bar{t}$ is the average of the t range analyzed with (4.13) [4.54].

In the following only the "critical exponents" $\lambda$ and the "critical amplitudes" $A_\lambda$ will be discussed. The importance of the critical exponents rests on the fact that they are "universal", i.e. identical for all systems which have the same symmetry ("order parameter dimensionality" $d_\psi$) and spatial dimensionality (d, here d = 2) irrespective of the specific nature and values of the short-range interactions which cause the order at $T < T_c$. The reason for this universal behaviour is that near the critical point the physical properties are not determined any longer by the short-range interactions but by the interactions between fluctuating domains whose extension $\xi$ is very large compared to the interatomic distances. (These fluctuating domains cause the well-known critical scattering or critical opalescence of liquids just above $T_c$.) The fluctuations are determined by the entropy which, in turn, is determined by the number of possible configurations and the number of degrees of freedom (translation, vibration, rotation). These numbers depend upon d and $d_\psi$. Thus systems with the same d and $d_\psi$ have the same critical exponents. Of course, if long-range interactions are present these have influence, too.

From the foregoing considerations it is obvious that the extension $\xi$ of the fluctuating domains, the so-called correlation length which appeared already in (4.3), is the quantity decisive for the critical behaviour. Its asymptotic form is

$$\xi = \xi_0 t^{-\nu} . \tag{4.14}$$

Another critical exponent characteristic for the correlations is $\eta$ which describes the behaviour of the correlation function $\phi(r)$ at $T = T_c$

$$\phi(r) = \phi_0 r^{-\eta} . \tag{4.15}$$

Three additional critical exponents characterize the behaviour of the order parameter $\psi(h, T)$ and its field derivative

$$\psi(T, 0) = \psi_0 |t|^\beta , \tag{4.16}$$

$$\psi(T_c, h) = \psi_h |h|^{1/\delta} , \tag{4.17}$$

$$\frac{\partial \psi}{\partial h}(T, 0) = \psi_0 |t|^{-\gamma} . \tag{4.18}$$

This derivative is the susceptibility $\chi_T = (\partial M / \partial H)_T$ in spin systems and in particle systems the compressibility

$$K_T = \frac{1}{\varrho}(\partial \varrho / \partial p)_T \quad \text{or} \quad K_T = \frac{a_0}{\theta^2}\left(\frac{\partial \theta}{\partial \mu}\right)_T . \tag{4.19}$$

Finally one exponent characterizes the singular part of the free energy $f_{sing}$ and the specific heat $c_\psi$ near $T_c$

$$c_\psi(T, 0) = c_0 |t|^{-\alpha} , \tag{4.20}$$

$$f_{sing}(T, 0) = f_0 |t|^{2-\alpha} . \tag{4.21}$$

The critical behaviour of $f_{sing}$ is frequently also described by two other exponents

$$f_{sing}(T,0) \sim |t|^{2/y_t} \ , \tag{4.22}$$

$$f_{sing}(T_c,h) \sim |h|^{2/y_h} \ . \tag{4.23}$$

The "thermal exponent" $y_t$ is related to $\alpha$ via $2 - \alpha = 2/y_t$ – compare (4.21 and 22) – and is equal to $1/\nu$, see (4.27), the "magnetic exponent" $y_h$ to $\delta$ via $1 + 1/\delta = 2/y_h$. It should be noted that all relations except (4.16) and (4.23) are for zero ordering field h, i.e. $\mu = \mu_c$, which corresponds to the ideal coverage of the specific adsorbate structure (e.g., p(2 × 1) or c(2 × 2)) considered.

The six critical exponents $\nu$, $\eta$, $\alpha$, $\beta$, $\gamma$, $\delta$ are not independent but interrelated by the "scaling laws"

$$\alpha + 2\beta + \gamma = 2 \ , \tag{4.24}$$

$$\gamma = \beta(\delta - 1) \ , \tag{4.25}$$

$$\gamma = \nu(2 - \eta) \ , \tag{4.26}$$

$$d\nu = 2 - \alpha \quad \text{(here } d = 2) \ . \tag{4.27}$$

They are called scaling laws because they follow from the scaling postulate: the thermodynamic relation $L = L(T,h)$ becomes at small $|t|$ a function W of only one scaled variable $h_s$ if properly scaled by $|t|^\lambda$ : $L_s = W(h_s)$ where W is the scaling function of L.

While t is the experimentally controllable parameter the fundamental quantity determining the critical behaviour is the correlation length $\xi$, as mentioned earlier. At the critical point $1/\xi$ approaches zero. Therefore a theoretically more appropriate expansion of the thermodynamical quantities L is in terms of $1/\xi$. Because of (4.14) this amounts to dividing the critical exponents by $\nu$. The resulting critical exponents $(2 - \alpha)/\nu$, $\beta/\nu$, $\gamma/\nu$ are the same for a much larger group of models ("weak universality" [4.55]) than the original ones. A notable exception is the hard-hexagon model (see below and Fig. 4.5) which is thought to be in the same universality class as the 3-state Potts model and has $\hat{\beta} = \beta/\nu = 2/15$ instead of the weak universality value $\hat{\beta} = 1/8$.

The critical properties of the models briefly sketched in Sect. 4.2.4 have been calculated in part exactly (nearly exclusively at h = 0), in part approximately with a variety of approximation methods, such as high or low temperature or low density series expansions, Monte-Carlo calculations [4.56–58], renormalization-group methods [4.3,59–62], which cannot be discussed here. (See also the general references [4.1–16]). Here only the major results are compiled.

1) The q-state Potts model. It is now well established [4.63,64] that the thermal and magnetic exponents are given by the relationships $y_t = 3(1 - u)/(2-u)$ and $y_h = 0.25(15-8u+u^2)/(2-u)$ where u is defined by $\cos(\pi u/2) = \sqrt{q}/2$. The critical exponents obtained from these relations are listed in Table 4.1. For $q>4$ the phase transition is first order. The critical properties of the Potts models have been studied particularly well. Even the first terms of the correction to scaling, see (4.13), are known for the singular part $f_{sing}$, (4.21), of the free energy [4.65]: $\theta_1 = 4u/3(1 - u)$, and for the correlation function $\phi(r)$, (4.15), [4.64]: $\theta_1 = 4/(2 - u)$. The critical point for $J>0$ (attractive, ferromagnetic interactions) is given by the conditions $[\exp(J/k_BT_c) - 1]^2 = q$ for the square lattice, $[\exp(J/k_BT_c) - 1]^3 + 3[\exp(J/k_BT_c) - 1]^2 = q$ for the triangular lattice and $[\exp(J/k_BT_c) - 1]^3 - 3[\exp(J/k_BT_c) - 1]q = q^2$ for the honeycomb lattice [4.12,35].

**Table 4.1.** Critical exponents of various models

| Quantity | exp. | Eq. | 2-state (Ising) | 3-state Potts | 4-state Potts | Heisenberg Potts with corner cubic anisotropy | XY-model with cubic anisotropy |
|---|---|---|---|---|---|---|---|
| Order parameter $\psi(T,0)$ | $\beta$ | (4.16) | 1/8 | 1/9 | 1/12 | 1/8 | |
| Susceptibility $\partial\psi/\partial h(T,0)$ (compressibility) | $\gamma$ | (4.18) | 7/4 | 13/9 | 7/6 | 7/4 | |
| Correlation length $\xi(T)$ | $\nu$ | (4.14) | 1 | 5/6 | 2/3 | 1 | Non-universal |
| Correlation function $\phi(T_c)$ | $\eta$ | (4.15) | 1/4 | 4/15 | 1/4 | 1/4 | |
| Specific heat $c_\psi(T,0)$ | $\alpha$ | (4.20) | 0 | 1/3 | 2/3 | 0 | |

The $p = 2, 3$ state clock models are identical to the $p = 2, 3$ state Potts model, while the 4-state clock model is the special case for $q = 4$, i.e. $u = 0$ of

2) the Ashkin-Teller model whose exponents $y_t$ and $y_h$ are $y_t = (3 - 2u)/(2 - u)$ and $y_h = 15/8$. In the general case u is changing continuously with the 4-spin exchange constant $J_4$ (Sect. 4.2.3): $\cos(\pi u/2) = 0.5\exp(4J_4/k_BT) - 1$. ($q = 4$ is equivalent to $J_4 = 0$.) This model, thus, has non-universal exponents, i.e. exponents which vary with the interactions. The same is true for

3) the Baxter (8 vertex) model which has the exponents $y_t = 2 - u$ and $y_h = 15/8$ with u changing continuously with the 4-spin exchange constant

$J_4$ in the Ising spin transcription of the model (see Sect. 4.2.3): $\cos(\pi u/2) = \tanh(2J_4/k_BT)$.

4) The hard-cube and hard-hexagon models have the same critical exponents as the corresponding Potts models with the same universality class which can easily be seen by drawing the saturation coverage lattices (Fig. 4.5). The three cases shown correspond to $q = 2, 3$ and 4. The activity of the "lattice gas" $y = \exp(\mu/k_BT)$ plays the role of the temperature in this case so that the critical exponents are defined in terms of fugacities $z$ by $L = A_\lambda |z - z_c|^\lambda$. The critical fugacities $z_c$ are $\approx 3.7966$ [4.66], $0.5(11 + 5\sqrt{5}) \approx 11.0902$ [4.67] and $\approx 5.81$ [4.68], respectively, the critical coverages $\theta_c$ (in units of the saturation coverages of $1/2$, $1/3$ and $1/4$, respectively) are $\approx 0.7355$ [4.66], $3(5 - \sqrt{5})/10 = 0.8292$ [4.67] and $\approx 0.7485$ [4.68].

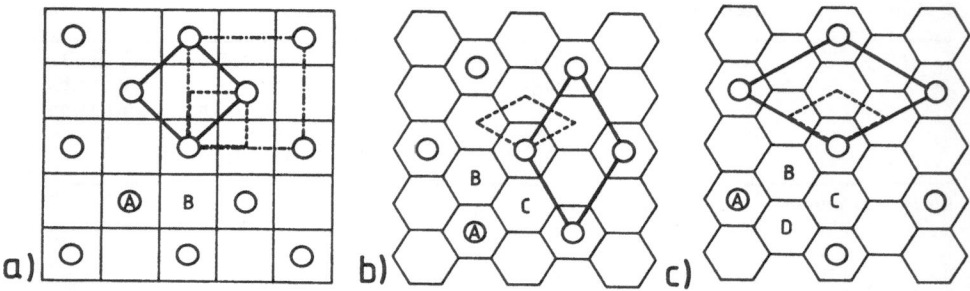

**Fig. 4.5.** Hard square (a) and hard hexagon with nearest-neighbour exclusion (b) and next-nearest-neighbour exclusion (c). Sites of the various sublattices are indicated by A, B, C, D. The resulting structures are ($\sqrt{2} \times \sqrt{2}$)R 45° or c(2 × 2), ($\sqrt{3} \times \sqrt{3}$)R 30° and (2 × 2). The unit meshes are indicated

5) The XY model with cubic anisotropy has non-universal exponents, the Heisenberg model with corner cubic anisotropy the same exponents as the 2-state Plotts (Ising) model [4.69].

### b) Symmetry Analysis and Symmetric Lattice Models

For the theoretical prediction of the critical behaviour of lattice models the free energy of the system is expanded in power(s) of the order parameter(s) $\psi_i$ with T-dependent coefficients. The terms appearing in the expansion must satisfy the symmetry of the lattice model. This "Landau free energy" $f(T, \psi_i)$ is compared with the statistical models with known critical behaviour (Sect. 4.2.3,4). Universality tells then that the critical behaviour of the unknown lattice model is the same as that of the statistical model with the same symmetry (and same type of interaction). There is an excellent description of the procedures used in this endeavour, in particular in the determination of the order parameter(s) [4.69] (for a complete listing, see [4.70]) so that only the results for the most frequent lattice models [4.69] will be shown here (Table 4.2). More complex commensurate (n × m) structures than those shown in Table 4.2 can be reached from the disordered state only via first-order transitions (in symmetric models) but incommensurate structures may be reached by continuous transitions. One can distinguish between commensurate and incommensurate structures by the continuous variation of the wave vector displayed by the latter near a transition.

**Table 4.2.** Relation between commensurate lattice models on common substrates and statistical models [4.69]

| Statistical model / Substrate symmetry | 2-state Potts (Ising) | 3-state Potts | 4-state Potts | Heisenberg with corner cubic anisotropy | XY with cubic anisotropy | Surface examples |
|---|---|---|---|---|---|---|
| Rectangular (p2mm) | (1 × 2) (2 × 1) c(2 × 2) | | | | | fcc(110), bcc(211) |
| Centered rectangular (c2mm) or square (p4mm) | c(2 × 2) | | | | (1 × 2) (2 × 1) (2 × 2) | bcc(110) fcc(100), bcc(100) |
| Triangular (p6mm) | | ($\sqrt{3}$ × $\sqrt{3}$) | (2 × 2) | | | |
| Honeycomb (p6mm) | (1 × 1) | | (2 × 2) | p(2 × 2) | | fcc(111), hcp(0001) |
| Honeycomb in a crystal field (p3ml) | | ($\sqrt{3}$ × $\sqrt{3}$) | (2 × 2) | | | |

## c) Asymmetric and Anisotropic Models

When three or more occupation possibilities of the substrate sites exist, inequivalent walls appear and the surface lattice may become asymmetric or anisotropic in spite of the symmetry or isotropy of the substrate. Various statistical models have been invented to model such surface lattices (Sect. 4.2.3). The phase diagrams resulting from them are shown schematically in Fig. 4.6 [4.39–46]. For all p except $p = 2$ the direct transition from the ordered (commensurate "C") phase into the disordered (fluid "F") phase is replaced by a double transition via a floating (incommensurate "I") phase. At $p \geq 5$ this is true for all values of $\mu$ (or the chiral field $\Delta$) and at $p = 4$ for all $\Delta$ except $\Delta = 0$. The case $p = 3$ is controversial: the IF transition line merges with the CI transition line either at $\Delta = 0$ or in a Lifshitz point L at $\Delta > 0$ (dashed line in Fig. 4.6a). First-order transitions may, of course, pre-empt any of these transitions. Combining the various cases shown in Fig. 4.6a (assuming that a Lifshitz point exists) gives the general schematic phase diagram for an adsorbate forming (p × 1) structures shown in Fig. 4.6b [4.39]. The two transitions involved, CI and IF, will be discussed in Sect. 4.2.6 and 7.

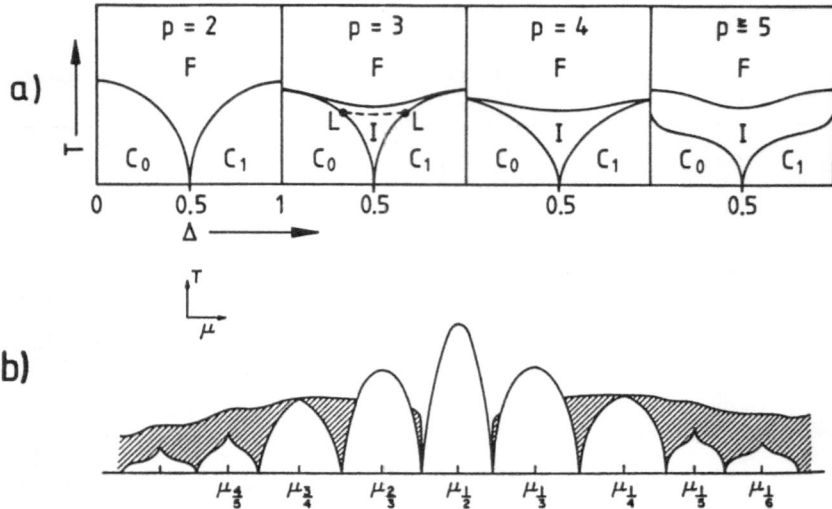

**Fig. 4.6.** Phase diagrams of uniaxial and chiral models (schematically) **(a)** for various p-states, **(b)** as a function of chemical potential $\mu_\theta$ assuming $p = 1/\theta$ and approximate particle-hole symmetry (symmetry with respect to $\theta = 1/2$). $\Delta$ is the chirality (Sect. 4.2.3). For further explanation, see text [4.39]

### 4.2.5 Order-Order Transitions

Continuous transitions may not only take place between ordered and disordered phases but also between two ordered phases via a displacive transition. 3d transitions of this type have been studied extensively and their relation to order-disorder transitions are described well in terms of the potential energy of the system in [Ref. 4.71, Sect. 2.2]. The allowed continuous 2d order-order transitions have been determined by symmetry analysis [4.72]. A selection of them is listed in Table 4.3. The only transition studied to date in some detail is the p4mm-p2mg transition (2d) on the W(100) surface [4.33,73] (Sect. 4.4). It should be noted that these transitions may be continuous but need not be. There is always some strain associated with the transition from one ordered structure to another one and the order parameter is coupled to this strain. Depending upon the boundary conditions for the atomic displacements which occur in the transition, the transition will be first order (discontinuous) or continuous [4.74]. An example of a first-order transition of this type is possibly the (1 × 1) ↔ (7 × 7) transition on the clean Si(111) surface (Sect. 4.4).

**Table 4.3.** Symmetry-allowed continuous order-order transitions between common structures (from [4.72])

| Parent group / Sub-group | p2mm | p2mg | c2mm | pe | p4mm | p3 | p3m1 | p3lm | p6 | p6mm |
|---|---|---|---|---|---|---|---|---|---|---|
| p2mm | 2a,b | 2a | 2c | | | | | | | |
| c2mm | 2c | 2c | 2c | 4 | | | | | | |
| p4mm | 1,2a | 2a,d | 1,2d | 1 | 2d | | | | | |
| p3m1 | | | | | | | 1,3 | 4 | 3 | |
| p6mm | | | | | | 3 | 1,3,4 | 1,3,4 | 1,4 | 3,4 |

The symbols in the table denote the following structures: 1: $(1 \times 1)$, 2a: $p(2 \times 1)$, 2b: $p(1 \times 2)$, 2c: $c(2 \times 2)$, 2d: $p(\sqrt{2} \times \sqrt{2})$ R 45°, 3: $(\sqrt{3} \times \sqrt{3})$ R 30°, 4: $p(2 \times 2)$.

## 4.2.6 Commensurate-Incommensurate Transitions

Commensurate (C) structures have (average) periodicities $a_i$ ($i = 1, 2$) which are simple rational fractions of the substrate periodicities $b_i$, $M_i a_i = N_i b_i$ (M,N integers) while in incommensurate (I) structures the $N_i/M_i$ are irrational fractions. C structures are locked solids (LSo), I structures floating solids (FSo) as defined in Sect. 4.2.1. When the ratios $N_i/M_i$ do not deviate much from simple rational fractions $(N_i/M_i)_r$ then the I structure may be considered as consisting of C regions with $a_i/b_i = (N_i/M_i)_r = a_{ic}/b_i$ separated by linear regions with larger or smaller interatomic spacings ("light or heavy walls", respectively, "solitons", "discommensurations") depending upon whether $a_i \lessgtr a_{ic}$. For example, in the transition from a $(1 \times 3)$ structure at $\theta = 1/3$ to a $(1 \times 2)$ structure at $\theta = 1/2$ commensurate $(1 \times 3)$ domains are separated by heavy walls ($(1 \times 2)$ regions) at $\theta \gtrsim 1/3$ while at $\theta \lesssim 1/2$ commensurate $(1 \times 2)$ domains are separated by light walls ($(1 \times 3)$ regions). Such CI transitions occur frequently when $a_i$ decreases with increasing coverage $\theta$. They may proceed smoothly, as indicated in Fig. 4.7 [4.75] without preference for C phases (a), with locking in at a finite number of commensurate q values (b), or without I structures at all (c, d).

There are two major models used for the description of CI transitions: (a) the Frank-van der Merwe (FvdM) model and its extensions to two dimensions and to T≠0 and (b) the ANNNI model already mentioned in Sect. 4.2.3. In the FvdM model the walls ("misfit dislocations", "solitons") are a result of the competition between the elastic lateral interactions between the surface layer atoms – trying to establish the periodicity a – and the normal interaction with the substrate potential with periodicity b. In the ANNNI model the walls are a result of the competition between nearest and next-nearest neighbour interactions. More recently the asymmetric (chiral) clock model (Sect. 4.2.3) has also been used to analyze CI transitions [4.39–46,76,77]. The various theoretical treatments (for reviews, see [4.46,75,70], for an introduction the beautiful tutorial [4.79]) lead to the following specific predictions:

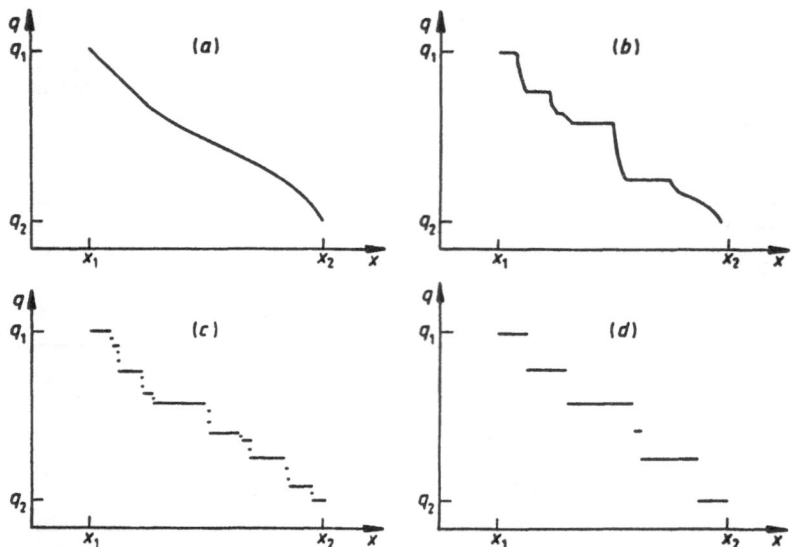

**Fig. 4.7a–d.** Variation of reciprocal lattice vector $q$ of surface structure with independent variable x (e.g., T, $1/\theta$). (a) smooth variation, (b) incomplete and (c) complete devil's staircase and (d) harmless staircases [4.75]

1) The incommensurate phase (I) is a floating phase (Sect. 4.2.1). The exponent $\eta$ characterizing the decay of the density-density correlation function $\phi_G(\mathbf{r})$ with distance, (4.2), has near a C phase with p equivalent positions with respect to the substrate the value $\eta = 2/p^2$. Because dislocations are important when $\eta > 1/4$, by destroying the quasi-long-range order and converting the layer into a fluid (exponential decay of $\phi_G(\mathbf{r})$, (4.3), Sect. 4.2.7) a direct IC transition without intervening fluid phase is possible only for $p > \sqrt{8}$. The width of the fluid region for $p > \sqrt{8}$ may, however, be so small that it is not observable [4.80].

2) The CI transition as a function of coverage is at $T = 0$ in the continuum limit of the FvdM model or in mean field theory at non-zero temperatures (for the simplest case, $N = M$ and unidirectional misfit) characterized by [4.75] $\bar{\delta} \sim - \ln(\delta_c - \delta)$, where $\bar{\delta} = 2\pi(a - b)/b$ is the average misfit which is related to the wall spacing $\ell$ by $\bar{\delta} = 2\pi/p\ell$, $\delta = 2\pi(a_0 - b)/b$ is the natural misfit, $a_0$ being the equilibrium spacing for zero substrate potential amplitude V and $\delta_c$ is the critical misfit, which is given by $4\sqrt{V}/\pi$. The wall width is $\ell_0 = 1/p\sqrt{V}$. The relation between $\bar{\delta}$, $\delta_c$ and $\delta$ is also valid for higher order CI transitions $(Ma = Nb, M{\neq}N)$, with $\delta = 2\pi(a_0 - Nb/M)/b$, $\delta_c \sim V^{M/2}$ and $M = p$. With increasing coverage $\theta$ a series of CI transitions may occur [4.75]. Its details depend especially on the influence of the discreteness of the lattice (Peierls potential). For example, the system may lock into a commensurate structure at every rational value of $a_0(\theta)/b$ within a coverage range $\Delta\theta \sim \Delta a_0 = a_{0c} - a_0$ determined by the critical misfit $\delta_c$. This would produce an incomplete or com-

plete devil's staircase (Figs. 4.7b,c). Some of the more refined details of these staircases may be expected to be washed out at higher temperatures. Another interesting effect may be encountered by taking into account the Peierls potential as one "quenches" a system to low temperatures. Then a metastable chaotic distribution of walls results [4.81].

3) There are two types of wall structures in the I phase (Fig. 4.8): the parallel or "striped" structure and the crossed (rectangular or hexagonal) structure. The parallel structure can occur on surfaces with arbitrary symmetry, the hexagonal structure is typical for surfaces with hexagonal symmetry. The energy of the wall system consists of wall energy, wall repulsion energy and wall crossing energy which add up to the energy density $E = \alpha/\ell + (\beta/\ell)\exp(-\kappa\ell) + \gamma/\ell^2$. When $\gamma > 0$ the striped structure is favoured, when $\gamma < 0$ the hexagonal structure. In the striped structure the walls can have a large number of configurations (Figure 4.8) which is limited by neighbouring walls. The resulting entropy is $\sim 1/\ell^2$, the entropy density $\sim 1/\ell^3$. Minimization of the free energy with respects to $1/\ell$, neglecting the wall repulsion, gives the average misfit at constant T

$$\bar{\delta} \sim \frac{1}{\ell} \sim [\delta_c(T) - \delta]^{1/2} \ . \tag{4.28a}$$

Similarly one obtains at constant $\mu$ the temperature dependence of $\bar{\delta}$

$$\bar{\delta} \sim \frac{1}{\ell} \sim [T - T_c(\mu)]^{1/2} \ . \tag{4.28b}$$

The transition is thus second order. It should be noted the $\ell$ has to be large for this result to be valid, a requirement which is not fulfilled in most experiments (Sect. 4.3.2). Other possible causes for the $1/\ell \sim (\mu - \mu_c)^{1/2}$ relationship (4.28) are dipole-dipole and substrate-mediated elastic interactions [4.34] between the walls which may be important on metal surfaces [4.82]. Furthermore, if the substrate is deformable (4.28a) is valid even in the absence of the entropy term, i.e. at $T = 0$ [4.83]. The critical temperature $T_p$ for a commensurate phase of order p is proportional to $p^{-2}$ which is illustrated qualitatively in Fig. 4.6b.

In the hexagonal phase the domains can change size and shape without increasing the total length of the walls and the number of their intersections, i.e. without significant energy change when the wall repulsion is small. The entropy connected with this "breathing" of the honeycomb network (Fig. 4.8) is very large and its negative contribution to the free energy dominates so that the transition is first order at low temperatures. At higher temperatures an intermediate striped or fluid phase is possible. The experimentally most studied system of this type, Kr on graphite, however, shows no direct first-order CI transition but a transition via a fluid phase. This has also been explained in terms of an asymmetric Potts lattice gas model (Sect. 4.2.3 [4.43]) with trio interactions, with heavy and superheavy walls (Fig. 4.3), wall crossings and dislocations (wall endings, Fig. 4.8). The inclusion of different types of walls and of dislocations leads to more complicated phase diagrams and transitions [4.88] and may have been the cause of some of the controversies regarding the 3-state models (for reviews, see [4.46,78]). The existence of several types of walls is, however, not generally accepted [4.88, 4.91, 4.93].

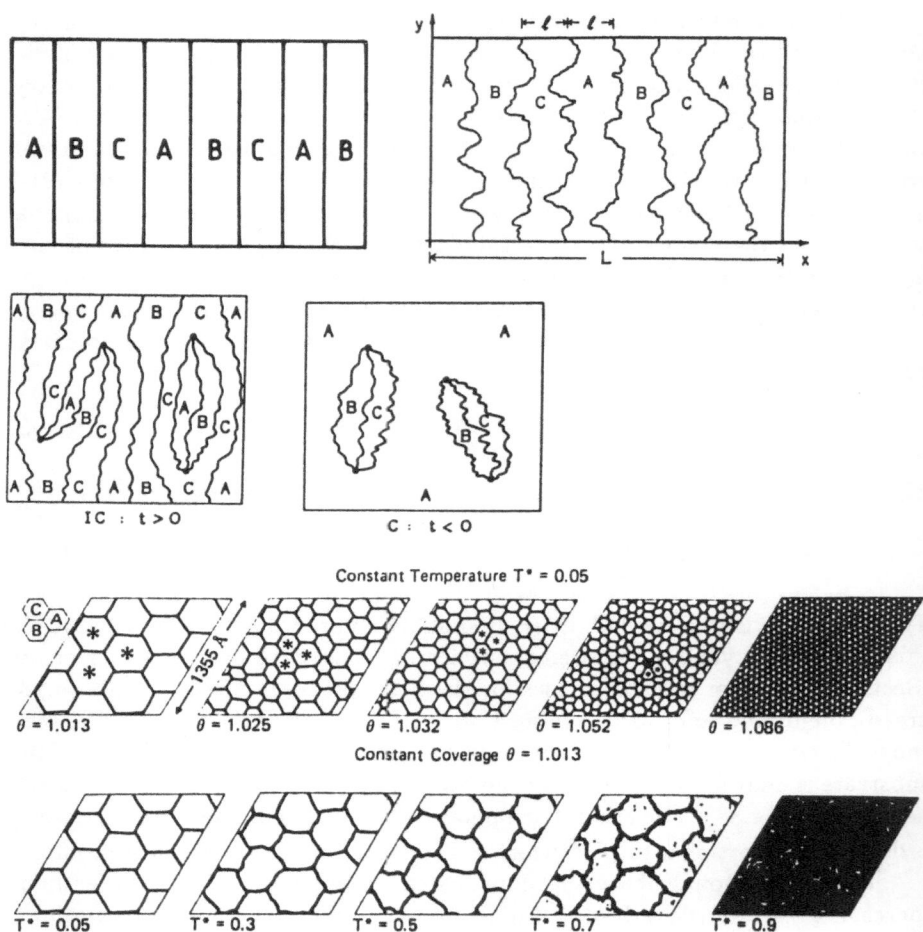

**Fig. 4.8.** Wall configurations near CI transitions. *Top row:* "parallel" walls in (3 × 1) phase (p = 3); from *left* to *right* T = 0, finite T, finite T with dislocations, C phase at finite T with dislocation pairs [4.79]. *Center* and *bottom rows:* "crossed" walls for hexagonal symmetry as a function of $\theta$ and T. The parameters are for Kr on graphite. Molecular-dynamics simulation with Lennard-Jones potential parameters $\varepsilon = 14.6$ meV, $\sigma = 0.36$ nm for Kr-Kr and $\varepsilon = 5.9$ meV, $\sigma = 0.322$ nm for Kr-C interaction, substrate potential amplitude $V = 0.08\varepsilon_{KrKr}$. There are three equivalent domains A, B, C. $T^* = T/170$ K. At constant temperature the thickness of the walls is essentially constant and only their number increases with increasing deviation $\delta\theta = \theta - 1$ from the commensurate coverage, causing a linear decrease of the percentage of commensurate atoms (in open areas) from 90 to 37 % at $T^* = 0.05$. At constant $\theta$ the number of walls is constant but their thickness increases with $T^*$ reducing the percentage of commensurate atoms from 90 to 26 %. 103 041 Kr atoms [4.34]

4) At high coverages uniaxial or isotropic compression of the layer in order to accommodate more atoms becomes more and more difficult. It is then energetically more favourable to introduce transverse strains instead of increasing only the longitudinal strains, e.g. by shifting neighbouring atom rows relative to each other. This shearing process causes a distortion of the adsorbate lattice, in the simplest case – e.g. when adsorbate and substrate have the same symmetry – a pure rotation ("rotational epitaxy"). For large misfit ($\delta \gtrsim 8$–10 %) all theoretical models [4.89–93] predict a linear variation of rotation angle with coverage $\theta$ (or misfit $\delta$) but at small misfit they differ. In the simplest model which describes the rotation by a sequence of discrete coincidence lattices (high order 2d commensurate structures) with increasing rotation angle $\alpha$ is proportional to $\theta$ for all $\theta$ [4.92]. The most sophisticated treatment, a continuum model which takes into account the domain wall energy predicts a strongly nonlinear $\alpha$-$\theta$ relationship and a critical misfit $\delta_c$ (or coverage $\theta_c$) at which rotation sets in [4.91,93]. $\delta_c$ depends upon the elastic constants of the adsorbate and upon the substrate potential amplitude and is for Kr on graphite of the order of a few percent. The exact shape of the $\alpha(\theta)$ curve and the order of the transition depends upon the ratio of these parameters as expressed by the ratio of the longitudinal and transverse sound velocities $C_L/C_T$ [4.91,93]. The most precise measurement to date for Kr on graphite [4.94] gives $\delta_c = 0.035$, a continuous transition except for a possible first-order jump of $\Delta\alpha_1 \approx 0.2°$ and a somewhat better fit of the data by $\alpha \sim (\delta - \delta_c)^{1/2}$ than by the theoretical predictions.

### 4.2.7 Melting in Two Dimensions

In discussing this subject one has to distinguish between 2d systems which are freely suspended or adsorbed on a structureless surface (potential amplitude $V = 0$) and 2d systems on periodic substrates ($V \neq 0$). In the first case (a) order is determined only by the forces between the atoms of the 2d system, in the second case (b) the substrate impresses to a considerable extent its order on the layer. In case (a) which can be described by the XY model (Sect. 4.2.3) thermal fluctuations do not allow long-range order so that the locked solid defined by (4.1) (Sect. 4.2.1) cannot exist. In addition to the (free) floating (quasi) solid (4.2) and the fluid (4.3), there is, however, still a phase possible which has exponential decay of the translational order but algebraic decay of the correlation function of the orientational order, the hexatic fluid. The transitions between the three phases are continuous and occur via the dissociation ("unbinding") of tightly bound dislocation pairs (floating solid $\rightarrow$ hexatic fluid) and of disclination pairs (hexatic fluid $\rightarrow$ normal fluid). The temperature $T_{KT}$ at which the first transition ("Kosterlitz-Thouless 'KT' melting") occurs is determined by the competition between the energy $\Delta U = 0.5K \ln(R/a)$ and the entropy $\Delta S = k_B \ln(R/a)^2$ associated with the creation of a free dislocation. a and R are the dislocation core and the system radius, respectively, and K is determined by the lattice constant and the elastic constants. Melting occurs when with increasing T $\Delta F = \Delta U - T \Delta S$ becomes zero, i.e. at

$T_{KT} = K/4k_B$. Solidification is preceded by a divergence of the correlation length $\xi$: $\xi(T - T_{KT}) \sim \exp[b/(T - T_{KT})^\nu]$ with $\nu = 0.36963\ldots$. The range of validity of this expression depends upon the core energy $E_c(K)$ of the dislocation and is also limited by the size R of the system [4.95]. For sufficiently small $E_c$ pair unbinding occurs discontinuously [4.96] and causes spontaneous generation of grain boundaries, that is arrays of dislocations, which is a first-order transition [4.97,98]. Elastically anisotropic layers contain inequivalent dislocations so that not only one KT temperature exists and different partially disordered phases can occur (see also Sect. 4.2.1) [4.99].

In a 2d system on a periodic substrate potential (case b) energy and entropy of the system are strongly modified by film-substrate interactions. First of all, the hexatic phase whose existence has been questioned even in case (a) [4.97], is suppressed by the substrate potential. Instead, the fluid (disordered) phase (4.3) splits into two disordered phases characterized by $\phi_G(r) \sim \exp(-r/\xi) \cdot \cos(G \cdot r)$, one with $G$ commensurate, i.e. locked to the substrate ("commensurate disordered phase" CD) and the other with $G$ incommensurate, (unlocked, "incommensurate disordered phase" ID) which are separated by a disorder line.

Secondly, the restrictions imposed on the 2d system by the substrate periodicity may suppress the continuous KT transition and lead to a first order transition. This can be best seen by limiting the continuous spin orientations in the XY model with $\mathcal{H}_{XY} = \sum_{i \neq j} V(\theta_i - \theta_j) = 2J \sum_{i \neq j} [1 - \cos^2(\theta_i - \theta_j)/2] \equiv \mathcal{H}_{XY}^1$ ($\theta_i - \theta_j = \theta$ angle between spins of unit length on nearest neighbour sites) to an increasingly smaller angular range be replacing $\mathcal{H}_{XY}^1$ by $\mathcal{H}_{XY}^n = 2J \sum_{i \neq j} [1 - (\cos^2 \theta/2)^n]$ [4.100]. The melting transition becomes first order for $n \lesssim 3$. This is due to the fact that with increasing n the dissociation of dislocation pairs is suppressed at low T until at some T suddenly a large number of them appears causing a first-order transition [4.101]. This mechanism is, however, limited to finite size systems – a typical situation encountered in experiment – while in large systems the KT mechanism prevails [4.102,103].

Thirdly, the point defect(dislocation)-mediated KT melting can be preempted by – also continuous – line defect(wall)-mediated melting ("Pokrovsky-Talapov 'PT' melting") depending on the model parameters and on the model. There are numerous theoretical studies of the melting of uniaxial (p × 1) systems such as the symmetric and asymmetric Potts and clock models (Sect. 4.2.3) [4.27,44–46,51–53,77,79,88,104–106]. In other models the discrete symmetry induced by the substrate is forced on the continuous symmetry of the XY model by introducing fixed correlations between the angles of neighbouring spins (uniformly frustrated XY model) [4.107–109] or by imposing random uniaxial fields [4.110,111]. A bewildering multitude of results is obtained which may be briefly summarized as follows. Melting (disordering) still can occur via the continuous KT transition but other continuous transitions (Ising, Potts, PT, "chiral" (Sect. 4.2.8), …) are frequent, depending upon symmetry, number of states, interaction parameters, field, chirality parameter or coverage.

Two examples illustrate the situation. Figure 4.9 shows the phase diagram of the 2d ANNNI model (Sect. 4.2.3) in a field h (a), its lattice gas analogon (b) and a number of typical equilibrium Monte Carlo configurations close to the

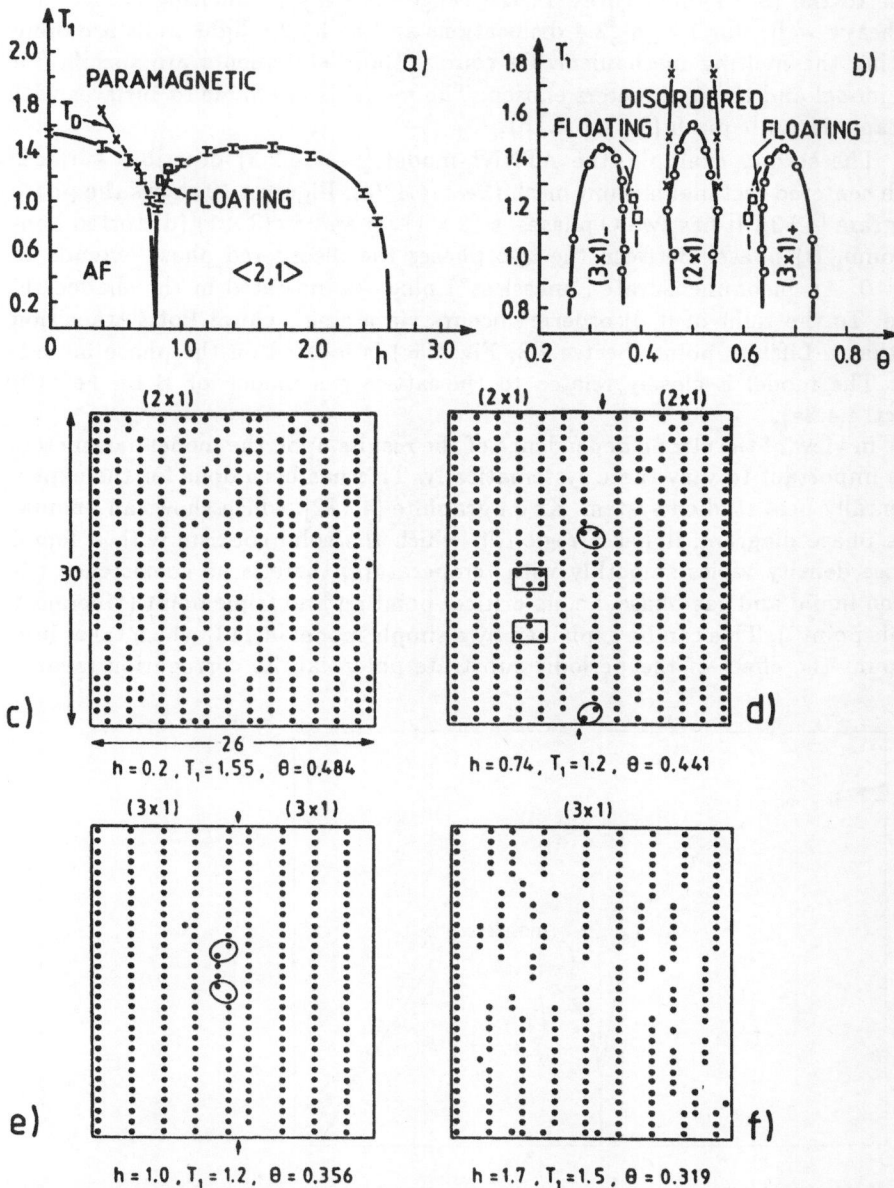

**Fig. 4.9.** Phase diagram of the 2d ANNNI model (**a**) and of the corresponding lattice gas model (**b**) for $J_2/J_1 = \varphi_{02}/\varphi_{01} = 0.3$ and $J_1/J_0 = \varphi_{10}/\varphi_{01} = -1$. $T_1 = k_B T/|J_1|$. (**c–f**) shows typical adatom configurations for various $T_1$ and h ($\theta$) values. Walls are marked by arrows, kinks by ellipses and dislocations by rectangles [4.52]

melting of the commensurate $(2 \times 1)$ $(\theta = 1/2)$ and $(3 \times 1)$ $(\theta = 1/3)$ phases [4.52]. The $(2 \times 1)$ structure melts into a CD structure and this structure into a ID structure at the disorder line marked by crosses. A floating (I) structure is observed only in narrow regions between the $(3 \times 1)$ and the $(2 \times 1)$ structures close to the $(3 \times 1)$ structures. In the range $0.8 \lesssim h \lesssim 1.2$ melting is mediated by heavy walls, for $1.4 \lesssim h \lesssim 2.4$ dislocations and at $h \gtrsim 2.4$ light walls are dominating the melting mechanism. Of course, these statements are specific for the model and the parameters chosen. The model is appliable to surfaces with rectangular unit mesh (Sect. 4.4.1b).

The second example, the ATNNI model (Sect. 4.2.3) describes surfaces with centered rectangular unit mesh (Sect. 4.4.2a). Figure 4.10 shows the phase diagram [4.53]. It has two C phases: a $(2 \times 1)$ (I) and a $c(3 \times 1)$ (distorted honeycomb, II) phase. Between the two phases the disordered phase extends to $T = 0$. An incommensurate ("massless") phase is indicated in the shaded region. To the right of it disordering occurs via a single chiral Potts transition so that a Lifshitz point (Sect. 4.2.3, Fig. 4.5c) is located on the phase boundary. The model is closely related to the lattice gas model for H on Fe(110) (Sect. 4.4.2a).

In view of the strong dependence of the results upon the model parameters it is important to vary these systematically. This has been done for the experimentally best studied system, Kr on graphite [4.1 $\overline{1}$2], which shows an anomalous phase diagram: it has a region in which the solid coexists with a liquid whose density varies smoothly with temperature; there is no coexistence between liquid and gas phase, i.e. no critical point and no triple point ("incipient triple point"). This can be explained by a simple model [4.113] which takes into account the effect of the periodic substrate potential: its corrugation favours

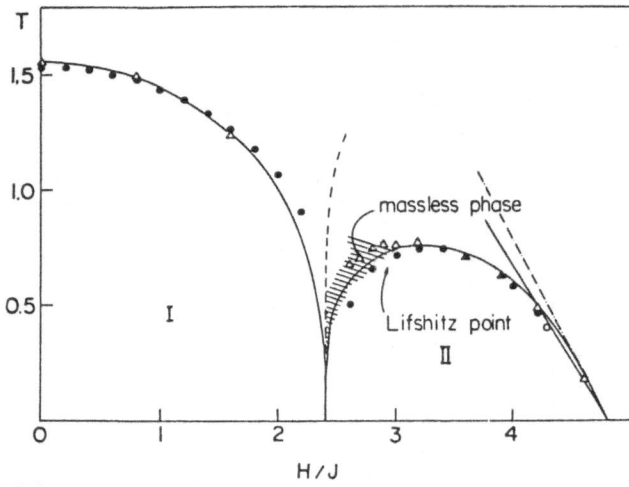

**Fig. 4.10.** Phase diagram for the 2d ATNNI model in a field H for $\alpha = J_2/J_1 = 0.4$. $J_1, J_2 > 0$. The phase boundaries were determined with various methods [4.53]

the commensurate solid phase over the liquid. In the incommensurate phase this stabilization is missing and liquid-gas phase separation occurs as in Xe on graphite. Melting of Xe on graphite has also been a subject of considerable discussion. Experiment [4.114] shows first-order melting at very small coverage $\theta$ and continuous melting at intermediate and large $\theta$. It was suggested that this density dependence of the melting mechanism is connected with the magnitude of the dislocation core energy (see above, [4.96]). Molecular-dynamics simulations (see also [Ref. 4.115, Chap. 6]), however, suggest that the apparent order of melting depends upon whether the experiment is done at constant spreading pressure or at constant area [4.116a]. The crossover from first- to second-order melting with increasing density as suggested by experiment is, therefore, questionable. The simulations appear reliable because they reproduce the different behaviour of Xe, Kr – with its incipient triple point – and Ar – which melt continuously – on graphite with realistic interaction parameters [4.116b]. 2d melting is, thus, still controversial, even in the best studied systems (see also [4.117] (Xe), [4.118] (Ar), [4.119] (He), [4.120]).

### 4.2.8 Interface Delocalization Transitions in Two Dimensions

Interface delocalization is a process in which 2d walls – such as the wall between physically distinct phases or between physically distinct but equivalent domains of a phase – lose their localized approximately straight positions which they have at low temperature when the tempeature is raised. Examples are the edges of 2d islands (walls between $c_i = 1$ and $c_i = 0$ regions or $S_i = +1$ and $S_i = -1$ regions in Sect. 4.2.2) or the energetically unfavourable domain walls in multi-state models with 3 or more states (Sect. 4.2.3). In the first case the process is also called depinning, in the second case it is called interfacial adsorption, interfacial wetting or domain-wall wetting. Whenever these processes can occur they have strong influence on melting or disordering transitions. They have been studied for a variety of models. The depinning transition problem has been solved exactly for the Ising model [4.121] (see also [4.122,79]) and approximately for the q-state Potts model [4.123]. Depinning occurs when the bond $\varphi$ (or spin coupling J) of the wall atoms (spins) to the interior atoms (spin) is weaker than that between interior atoms (spins). This is probably realized in adsorption layers with strong attractive interactions which are partially compensated by dipole-dipole repulsion. This compensation is particularly strong for the island-edge atoms (Sect. 4.4.3).

The domain-wall wetting problem has been studied for many models [4.44, 79,124–132,132b]. The basic mechanism in this process is "wall dissociation". In a p-state system there are p − 1 unequivalent walls (See Fig. 4.3 for p = 3). At the ideal coverages $\theta_i$ of commensurate structures such as $(n_i \times 1)$ structures ($\Delta\mu = \mu - \mu_i$ or $\Delta = 0$, Sect. 4.2.3) the wall energies (interfacial energies) of all walls are equal. With increasing deviation from $\theta_i$ (increasing $\Delta\mu$ or $\Delta$) differences in the wall energies develop. Energetically unfavourable walls decompose, therefore, with increasing temperature into energetically more favourable walls which enclose a new domain as indicated in Fig. 4.11 for p = 3. This new domain is said to wet the original domain. A good illustration for the name wetting or interfacial adsorption is Fig. 4.11c which shows results of Monte-Carlo calculations for a 3-spin state model (BEG model, Sect. 4.2.3) close to the tricritical point (Fig. 4.4a) of this model [4.127]. The dark area which is absent at T = 0 (for $\Delta/J<1$) or only one monolayer thick (for $\Delta/J>1$) is the spin 0 domain which is interfacially adsorbed between the − 1 and +1 domain and replaces the − 1/ + 1 wall by a − 1/0 and a 0/ + 1 wall. It is obvious from Fig. 4.11a,b that the two domain configurations are quite different and that wall-mediated melting should be different in the two cases. (In Fig. 4.11b "chiral" melting occurs.) In the phase diagram a "wetting line" separates the $T, \mu(T, h)$ region with and without wetting. Its location depends on the energy difference between the walls which is determined by the model parameters. Fig-

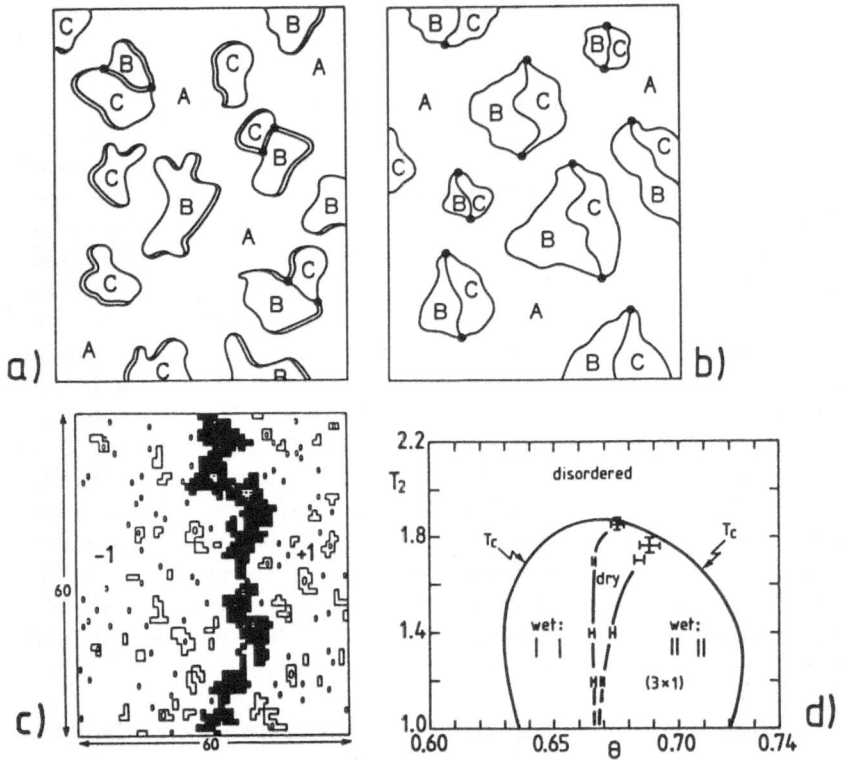

Fig. 4.11a–d. Domain-wall configurations near melting of the commensurate phase of a uniaxial 3-state phase. (a) For $\Delta = 0$, i.e. the ideal coverage, (b) for $\Delta \neq 0$. AB, BC and CA are favourable walls, BA, CB and AC are unfavourable walls between the three domains A, B, C (compare with Fig. 4.3) [4.44]. (c) Wetting layer *(dark region)* near melting as obtained by Monte Carlo calculations for the Blume-Emery-Griffiths model [4.126]. (d) Phase diagram near $\theta = 2/3$ of H on Fe(110) displaying the regions of wetting by light and heavy walls. $T_2 = k_B T/|J_2|$ [4.129]

ure 4.11d shows the region around the $(3 \times 1)$ phase for parameters simulating H on Fe(110) [4.130]. There is only a narrow region in which no wetting occurs. Close to the wetting and transition lines wetting may show critical behaviour. Thus the thickness of the wetting layer diverges at a first-order transition line as $(T - T_c)^{-\omega}$ with $\omega \approx 1/3$ in the BEG model [4.126] and $\omega \approx 1$ in a 20-state Potts model [4.127,128]. Along the continuous transition line of the BEG model, however, the thickness approaches a finite value [4.126]. For additional work on 2d critical wetting, see [4.131,132,132a].

### 4.2.9 Systems with Long-Range Repulsive Interactions

There are many adsorption systems of this type but they have received little theoretical attention. Their physical realization are electropositive adsorbates on electronegative substrates or – more general – adsorbates with small electronegativity on surfaces with larger electronegativity. Examples are alkali, alkaline earth and rare-earth adsorbates on refractory metal surfaces whose phase transitions have been studied experimentally rather thoroughly [4.133]. The dominating lateral interaction is the long-range dipole-dipole interaction $\varphi_{ij} = p^2/|r_i - r_j|$ (p: dipole moment). Nearly all detailed model calculations and most of the general theoretical considerations are not valid for such interactions. The Monte-Carlo (MC) method can, however, be easily extended to long-range interactions if the summation $U = \sum_{i \neq j} \varphi_{ij}/2$ is taken only over the N nearest neighbours and is replaced beyond those by an integral which gives a contribution $E_N = 2(\pi n)^{3/2} \mu^2/N^{1/2}$ which depends only upon concentration n but not upon the structure [4.134]. Experience shows that a summation up to $11^{th}$ nearest neighbours is sufficient [4.135]. T-dependent calculations have been made for the $(2 \times 2)$ structure $(\theta = 1/4)$ [4.134,135] and at $\theta = 1/3$ and $\theta = 1/6$ [4.135] for adsorbates on a centered rectangular lattice, specifically for Na on W(110).

In general, however, only the ground-state configurations have been calculated as a function of coverage for square [4.136], triangular [4.136,137] and centered rectangular lattices [4.135,137]. Comparison with experiments has shown that not all ground-state structures predicted by assuming pure dipole-dipole interactions are observed as a function of coverage, e.g. those at $\theta = 1/5$ and $\theta = 1/7$ for Na on W(110) which requires $\varphi_6 < \varphi_7$. Thus, in addition to the dipole-dipole interaction, other interactions must be taken into account. The appearance of long-periodicity superstructures implies that these additional interactions are also long range, a condition which is fulfilled by indirect electronic interactions [4.34]. Although this oscillatory interaction decays, in general, with distance r asymptotically as $r^{-5}$, it decreases only as $r^{-1}$ or at least not faster than $r^{-2}$ when electronic surface states are present [4.138]. The indirect interactions can be so strong that they overwhelm the dipole-dipole repulsion and cause formation of atomic chains along the $[1\bar{1}0]$ direction on the (112) and (110) surfaces of bcc metals. At low coverages these chains may be quite far apart – up to 9 atoms – so that they interact only very weakly via dipole-dipole and indirect electronic interactions.

As a consequence these systems are characterized by two energy scales, the strong intra-chain and the weak inter-chain interaction. This leads to a new type of phase transition which is not seen in the system discussed up to now, with nonmonotonic decrease of order over a wide temperature range [4.139], in addition to those already discussed. It is interesting to note that due to depolarization with increasing coverage, phase separation into a dilute 2d vapour and a 2d condensate can occur even in the presence of pure dipole-dipole repulsion [4.140,141].

### 4.2.10 Systems with Purely Attractive Interactions

There is a wide range of chemisorption systems in which the lateral interactions are predominantly attractive. This is the case in metal atom adsorbates on metal surfaces with small electronegativity differences between adsorbate and substrate. Although dipole-dipole interactions are present too in these systems, the dipole moments are usually small and the attractive direct electronic interaction dominates. As a consequence, phase separation into a dilute 2d vapour and a dense 2d condensate occurs with a coexistence line separating the vapour-condensate coexistence region from the single-phase vapour and condensate regions. The order parameter for such a system is the (number) density n or more precisely $n - n_c$ where $n_c$ the density at the critical point on the coexistence line. Assuming for the moment that there is only one condensed phase (liquid $\equiv$ solid) then there are only two states and one would expect 2d-Ising-like behaviour. This has indeed been observed in physisorption of $CH_4$ on graphite where the coexistence curve can be fitted over a wide temperature range (up to $|t| = 10^{-1}$) very well with the critical exponent of the order parameter of the 2d-Ising model, $\beta = 1/8$ (Table 4.1, Sect. 4.2.4a). The coexistence curve is quite symmetric though with $n_c(\theta_c) = 0.4$ instead of the theoretical value $1/2$ [4.142]. Also the specific heat exponent $\alpha \approx 0$ agrees well with the 2d Ising value 0 (Table 4.1). For a review of the critical point in physisorption, see [4.143].

In chemisorption of metal atoms on metal surfaces the situation is quite different. Except for the coverage range close to $\theta_c$ which could not be measured with sufficient accuracy (or not at all) the coexistence curve data cannot be fitted by the 2d-Ising model but quite well with the 2d van der Waals equation or with the quasi-chemical approximation. The van der Waals fit for Au on W(110) for example is very good from $|t| = 0.4$ to $10^{-2}$ indicating mean field behaviour much closer to $T_c$ than in the physisorption case. The coexistence curve is very asymmetric with $\theta_c$ varying from 0.2 to 0.35 with the adsorbate on W(110) [4.144]. There are several possible causes for this difference between physisorption and chemisorption of which only two will be discussed.

One is the possibility of a 2d metal-nonmetal transition during vaporization. It has been observed only recently that the coexistence curve of metals in 3d is much more asymmetric [4.145] than that of noble gases and other gases in which no significant change of bonding occurs upon condensation [4.146], although the $\beta$-value is not very different and close to the 3d Ising value 0.325.

It has been suggested that the difference between the two types of systems is connected with the change in bonding during the metal-nonmetal transition [4.146]. Such a bonding change could occur also in 2d in spite of the coupling to the conducting substrate. A second possibility is the difference in the ratio of the amplitude of the substrate potential to the lateral bonding. For Kr on graphite, for example, a ratio of $V_1/\varepsilon \approx 0.05$ gives best agreement with experiment ($V_1$: dominating Fourier component of the substrate potential, $\varepsilon$: Lennard-Jones energy parameter) [4.111]. For metal atoms such as Cu, Ag and Au on W(110) the corresponding ratio is about $1/3$ [4.147]. Thus adsorption is much more localized in metal chemisorption systems than in noble gas physisorption systems.

This suggests a theoretical treatment in terms of lattice-gas models but – as already pointed out – the 2d-Ising model gives a bad description. The fact that the data can be fitted quite well with the quasi-chemical approximation – except for the wrong $\theta_c$ [4.144] – must therefore be considered coincidental. Generalizations of this model to take into account the nonadditivity of the lateral interactions [4.148] can produce the experimental $\theta_c$, but the nonadditivity required for that is too large to be compatible with the presently available most reliable calculations of the interactions [4.147]. The same is true for Monte-Carlo calculations with nonadditive interactions [4.149], which disagree with experiment [4.144] at low coverages even more than MC calculations with only additive interactions. The addition of trio interactions to pairwise interactions, however, could reproduce the experimental results [4.150] with interaction parameters compatible with experiment [4.144]. It remains to be seen if recent improvements and generalizations of mean field theories [4.151–153] or of the quasi-chemical approximation [4.154] will fit the experimental data approximately.

It was assumed above that there are only two phases, the condensed phase and the vapour, frequently also called fluid to encompass the low-density and high-density disordered phase. This allowed a description in terms of the 2d-Ising model. There are, however, systems which have more than one condensed phase, e.g. Ni on W(110) which forms at low coverages a $(1 \times 1)$ phase and at high coverages a C coincidence phase [4.155]. In this case 3-state models such as the BEG model are necessary to describe the phase diagram and phase transitions [4.156]. Neither experimental nor theoretical data for this case of chemisorption are available at present.

The success of the van der Waals equation in fitting experiment suggests – in spite of the strong potential modulation of the substate – that one should use 2d liquid models beyond the van der Waals equation. These are, listed in the sequence of increasing sophistication and computational effort: (i) analytic hard-disk models [4.157–159] with an attractive van der Waals term added [4.160], (ii) numerical hard-disk models to which the attraction is added in terms of a perturbation-theory expansion in powers of the Lennard-Jones interaction energy ("Lennard-Jones fluid") or other numerical approximations [4.161–163], and (iii) Monte-Carlo calculations for Lennard-Jones interactions without

[4.164] or with periodic substrate potential as already discussed (Sects. 4.2.6 and 7) in connection with noble gas layers on graphite (Sect. 4.2.7) [4.84–86,116a,116b]. The influence of the substrate potential modulation has also been taken into account in a nonadditive hard-disk model [4.165]. Approach (i) with the simple equation of state $p/nk_BT = 1/(1 - y)^2 - an/k_BT$ with $y = \pi nd^2/4$ (n: number density, d: atomic diameter) already gives a significant improvement over the van der Waals equation: it reduces $\theta_c$ from $1/3$ to the observed values which depend upon the ratio $d/d_s$ ($d_s$: substrate atomic diameter) because $\theta_c$ is determined by $n_c = 4y_c/\pi d^2 = 4(\sqrt{7} - 2)/3\pi d^2$ and $n_s(d_s)$ [4.160]. More detailed comparisons, in particular with the more sophisticated models are premature.

## 4.3 Experimental Techniques

An experimental technique for the study of structural phase transitions has to respond to the distribution of atoms or to the changes of the physical properties of the system associated with the change of the distribution. In contrast to physisorption where large-area samples may be used and the adsorption layer can be in (quasi)equilibrium with the 3d gas phase, only a small surface area is available in chemisorption and the surface or the adsorbate is in general a closed system. Therefore, most of the techniques used in physisorption systems such as calorimetry (heat capacity measurements), adsorption isotherm measurements, neutron diffraction and standard x-ray diffraction [4.18,19,21] are not applicable to chemisorption. Only low energy electron diffraction (LEED) [4.20], grazing incidence x-ray diffraction [4.166] and possibly He atomic beam scattering [4.167] can be used, too. New techniques (Sect. 4.3.2–4.3.5) are necessary, most of which could be useful in physisorption, too. Many other possibilities have not been explored such as the changes in electronic energy levels connected with changes in the environment of an atom (for valence electrons via UV photoelectron spectroscopy (UPS) [4.168], for core electrons via XPS [4.168a]. Another possibility is the sensitivity of the electron spin alignment to the environment [4.169] which can be studied by spin-polarized electron diffraction or spectroscopy. At present, however, only the techniques described in Sect. 4.3.1–5 have been applied successfully to the study of phase transitions in chemisorption systems and on clean surfaces.

### 4.3.1 Diffraction Methods

These methods measure the average structure function of the system. The average is a twofold one: an amplitude average over the coherence range of the incident wave which is at most a few 100 nm in diameter and an intensity average over the illuminated area which in general is at least 10 $\mu$m in diameter. The structure function has been calculated for a large number of atomic distributions but only the order-disorder transition of a commensurate structure will be discussed as an example. Grazing incidence x-ray diffraction and He atomic beam diffraction have hardly been used for the study of phase transitions except for the roughening transition [4.170–172], which is outside the scope of this review. Therefore, the discussion will furtheron be limited to LEED. There have been several reviews on the use of LEED in the study of disorder and imperfections [4.173–175] so that only the essential points will be stressed.

All studies up to now are based on the assumption that the experimental data may be analyzed in terms of the kinematic theory of diffraction of plane waves because disorder destroys the phase relationships in the multiple scattering processes which are required for dynamical effects to build up. The validity of this assumption for the determination of critical exponents has been questioned recently [4.176], a conclusion which cannot be maintained, however, to the extent to which it was expressed in [4.176,177]. Thus kinematic analysis appears to be permissible but it should be realized that multiple scattering and other small-length-scale phenomena do influence the correction to scaling, (4.13). These have not been adequately studied for experimental systems.

Whether an array of atoms is in the ordered or disordered state it always experiences thermal vibrations. In the case of an order-disorder transition the influence of these vibrations on the scattered intensity distribution has to be eliminated before the state of order can be extracted from the data. This is done by correcting the peak intensities of the diffracted beams with the Debye-Waller (DW) factor $\exp[-W(\boldsymbol{K})]$ with $W(\boldsymbol{K}) \approx \langle u^2 \rangle K^2/2$ where $\langle u^2 \rangle$ is the time average of the square of the displacements of the atoms from their average position ("mean-square displacement") and $\boldsymbol{K} = \boldsymbol{k} - \boldsymbol{k}_0$ the scattering vector (momentum transfer). It is usually assumed implicitly that $\langle u^2 \rangle$ is uninfluenced by the phase transition. This is certainly not true in a first-order phase transition where the atoms change their environment suddenly – and with it the restoring forces – but it is also not quite correct in a second-order transition although W changes here continuously [4.178–180]. Thus care is advisable in the DW factor correction in the determination of critical exponents. Very little quantitative work has been done up to now in the elimination of the thermal (phonon) scattering in the diffuse background which involves one- and multiphonon processes [4.181]. The same is true for the low energy electronic excitations [4.182] whose energy is below the cut-off of the energy filter in front of the detector. Before these problems are solved it will be difficult to extract information on disorder from the background [4.183] unless the background caused by them is sufficiently small.

Finite-size effects are also very important for the data analysis. No surface is perfect but instead has a distribution of steps with varying distances even if the crystal surface is oriented within a few hundredths of a degree of an orientation with low surface energy. Figure 4.12 shows a typical example. Regions with high and low step densities are always present simultaneously which has to be taken into account. In general, the atomic distributions on both sides of a step are not correlated, so that steps determine the size of the system and introduce a new length scale L, in addition to the correlation length. It is evident that this has a strong influence on second-order transitions because $\xi$ which

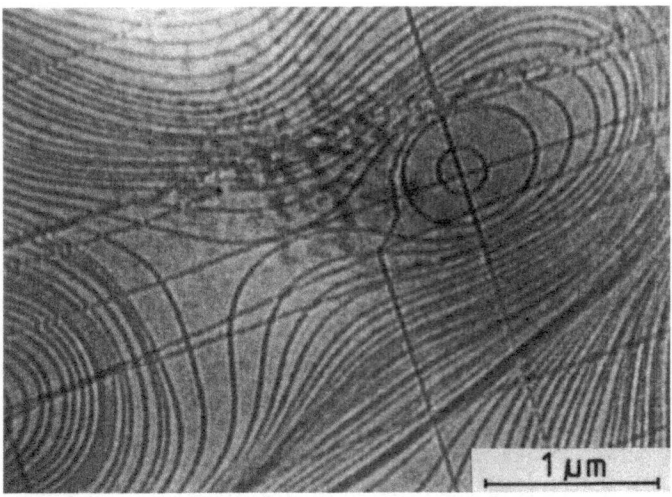

**Fig. 4.12.** Typical distribution of mono-atomic steps on a Mo(110) surface, LEEM micrograph. [4.184]

diverges in the infinite system can now not exceed L. This causes a rounding of the critical region and a shift of $T_c$ which can be taken into account in the data analysis by a Gaussian distribution of critical temperatures [4.185]. Similar but less pronounced rounding effects occur in first-order transition, even to such an extent that they appear as second-order transitions. Finite-size phenomena have been reviewed thoroughly [4.186]. For more recent work, see [4.187,188]. Another important deviation from the ideal crystal are point imperfections. Little is known experimentally about their influence on phase transitions but considerable theoretical work has been done on this subject [4.23,24]. When the defects couple to the order parameter even locally they can have drastic effects, even to the extent of preventing long range order from developing at all. The work delineating these effects in the random field Ising model has been reviewed in [4.188a].

The most important limitation of LEED in the study of critical phenomena has been the inadequacy of the instruments. No instrument produces a monochromatic parallel electron wave and no detector selects in infinitely narrow $K$ range. All these instrumental shortcomings can be described by an instrument response function $T(K)$ with which the structure function $S(K)$ as produced by an ideal instrument is folded. The result is the experimentally measured intensity distribution $J(K) = S(K)*T(K) = \mathcal{F}\{\phi(r)\}*\mathcal{F}\{t(r)\} = \mathcal{F}\{\phi(r)\cdot t(r)\}$, $\mathcal{F}\{\ \}$ denoting the Fourier transform. $\phi(r)$ is the autocorrelation function of the atomic distribution on the surface. The Fourier transform of $T(K)$, the autocorrelation function $t(r)$ may be approximated by a window with a Gaussian transmission distribution with half width $t_w$, the so-called transfer width. In standard LEED instruments it limits coherent scattering to regions of about 10 nm diameter and, thus, introduces an instrumental finite-size effect dimension $L_w = 1/t_w$. Of course, it is possible to determine $T(\boldsymbol{K})$ which is usually approximated by a Gaussian using an "ideal" scatterer, a crystal surface with large step distance (several $10^2$ nm). Then $J_{id}(\boldsymbol{K}) = \delta(\boldsymbol{K})*T(\boldsymbol{K}) = T(\boldsymbol{K})$ for the diffracted beams ($\boldsymbol{K} = 2\pi\boldsymbol{h}$, $\boldsymbol{h} = $ 2d reciprocal lattice vector). The reciprocal of the half width $T_w$ of these beams is then directly a measure for $t_w : t_w = 2\pi/T_w$.

Instead of unfolding $J(\boldsymbol{K})$ with $T(\boldsymbol{K})$ in order to obtain the (average) structure function $S(\boldsymbol{K})$ $t_w$ can be made so large by proper instrument design that the unfolding becomes unnecessary. Such instrumental developments have taken place during the past several years [4.189–192] resulting in transfer widths up to 300 nm. Thus LEED has reached from the instrumental point of view the same coherence widths as has been customary in x-ray and neutron diffraction for years. Figure 4.13 illustrates the improvements achieved. The dashed curve was obtained with a display-type system and optical detection from a Ni(111) surface [4.193], the dotted curve with a retarding field Faraday cup detector from a Si(111) surface [4.194] and the solid curve with a 127° electrostatic analyzer and channeltron detector from a Mo(110) surface [4.195]. The transfer widths measured with the specular beam (00) are 8 nm at 25 eV, 11 nm at 50 eV and 90 nm at 60 eV, respectively. Figure 4.13 shows at the same time another important factor, the instrumental background. The high background of the dotted curve is largely caused by the crystal itself, but that of the dashed curve is caused predominantly by the detection system: in order to minimize the electron optical effects of the retarding grids in front of the fluorescent screen the retarding voltage is chosen so low that a significant number of inelastically scattered electrons are detected. In addition, there is background light which can be eliminated, however [4.189]. Inasmuch as background subtraction is a major problem in the determination of critical pa-

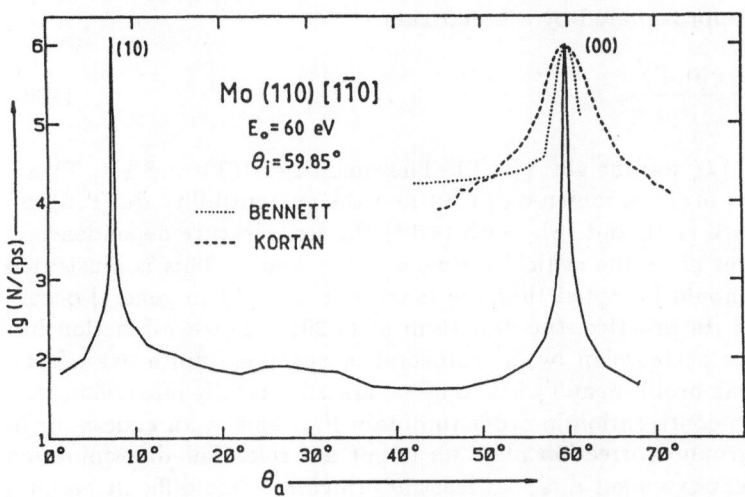

**Fig. 4.13.** Intensity distributions $J(\boldsymbol{K})$ of typical LEED instruments [4.195]. Note that the scale is logarithmic. For explanation see text

rameters display-type systems are not very useful for this purpose. This is not true for applications in which spot positions and semiquantitative spot profiles are sufficient, such as in CI transitions.

LEED allows in principle to determine the critical exponents $\alpha$, $\beta$, $\gamma$ and $\nu$ – defined by (4.14,16,18,20) – and the corresponding critical amplitudes. In principle $\beta$, $\gamma$ and $\nu$ would be sufficient because the other exponents can be obtained from them via the scaling laws (4.24–27). However, in chemisorption all measurements are done at constant coverage $\theta$ while the critical exponents are defined at constant chemical potential $\mu$. They are, therefore, "Fisher-renormalized", that is divided by $1 - \alpha$ [4.196]. Consequently, $\alpha$ must be determined if the coverage deviates from the ideal coverage of the adsorbate structure studied, e.g. of a p($2 \times 1$) or c($2 \times 2$) structure, at which $\mu = \mu_0$ (or h = 0) while T is varied. At all other coverages $\mu$ changes with T.

The critical exponents are determined from the T dependence of the superstructure reflections $h = (h_1, h_2)$ which is given the structure function

$$S(\boldsymbol{K}, T) = I(T)\delta(\boldsymbol{K} - \boldsymbol{K}_0) + \chi(\boldsymbol{K} - \boldsymbol{K}_0, T) , \qquad (4.29)$$

where $\boldsymbol{K}_0 = 2\pi\boldsymbol{h}$ and the $\delta$-function is caused by the long-range order while the "susceptibility" $\chi$ describes the long-range fluctuations at and near the critical temperature. $\chi$ has been calculated for various statistical models. One useful approximate form is $\chi(\boldsymbol{q}, T) = \mathrm{const}\,[(1/\xi)^2 + \phi_1 q^2]^{\eta/2}/[(1/\xi)^2 + \phi_2 q^2]$, where $\boldsymbol{q} \equiv \boldsymbol{K} - \boldsymbol{K}_0$. $\xi$ is the effective range of correlation (correlation length), $\phi_1 \approx 0.03$ and $\phi_2 \approx 1$ close to $T_c$ are slowly varying functions of T. Very close to $T_c$ (($T - T_c)/T = t < 0.005$) $\xi \rightarrow \infty$ so that $\chi(\boldsymbol{q}, T) = \mathrm{const}\,q^{\eta-2}$. (Note the limitation by the coherence length L as determined by surface perfection and the instrument, $\xi \leq L$!).

For small but not too small t and not too small q is $(1/\xi) \ll \phi_1 q^2$ and $\chi(\boldsymbol{q}, T)$ may be approximated by a Lorentzian

$$\chi(\boldsymbol{q}, T) = \frac{\chi(0, T)}{(1/\xi)^2 + q^2} \qquad (4.30)$$

with half width $1/\xi$ and height $\chi(0, T)$. The quantities I(T) and $\chi(0, T)$ are the square of the order parameter $\psi(T, 0)$ and the susceptibility $\partial\psi(T, 0)/\partial h$ of Sect. 4.2.4.a, see (4.16 and 18). With (4.14) the temperature dependence of I, $\chi(0)$ and $\xi$ thus gives the critical exponents $\beta$, $\gamma$ and $\nu$. This is illustrated in Fig. 4.14. It should be noted that the maximum of $\chi(T)$ in general occurs at $T > T_c$ [4.196]. In practice, the first term in (4.29) is not a $\delta$-function but is broadened into a Gaussian by instrumental or specimen finite-size effects. Therefore the peak profile near $T_c$ has to be separated carefully into a Gaussian and a Lorentzian contribution in order to obtain $\beta$, $\gamma$ and $\nu$, an endeavour in which the background correction plays an important role. The determination of the specific heat exponent $\alpha$ is – at least in principle – less difficult because it makes only use of the integrated intensities of the superstructure reflections.

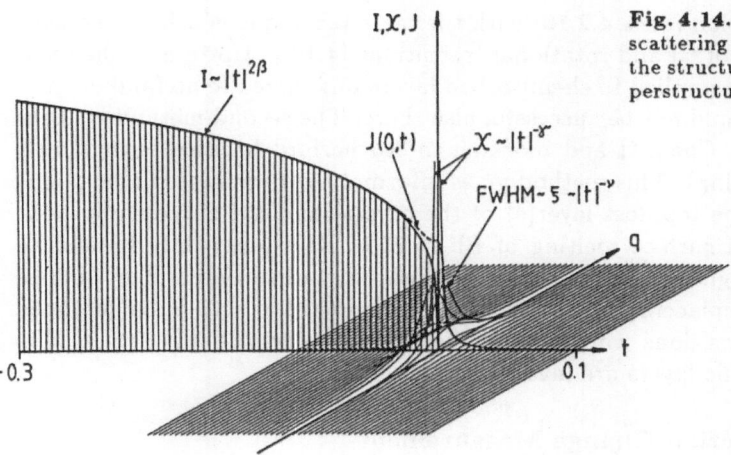

**Fig. 4.14.** Temperature and scattering vector dependence of the structure function near a superstructure reflection [4.195]

When $t_w < \xi \ll L$ where L is the specimen-determined finite-size length, then the temperature dependence of the integrated intensity $I(T) = \int S(q, T)dq$ is given by $I(T) = A \mp B_\pm \cdot |t|^{1-\alpha} - A_1 T + \ldots$ where the $\pm$ signs refer to the approach of $T_c$ from below or above. This is so because the finite-range correlations which are responsible for $I(T)$ are also part of the internal energy u of the system and should, therefore, have the same singular behaviour near $T_c$ as u : $u(T) = \int c(T)dT \sim |t|^{1-\alpha}$ according to (4.20) [4.197]. Model calculations confirm these expectations but no experiments have been analyzed yet in order to obtain $\alpha$.

The determination of the critical exponents $\lambda$ by LEED is limited close to $T_c(t \to 0)$ by instrument- or specimen-determined finite- size effects which suppress the growth of $\xi$ to infinity. On the other end of the T range, it is limited by the breakdown of the asymptotic behaviour of the thermodynamic quantities which require corrections to the asymptotic behaviour, see (4.13). These corrections introduce additional fit parameters and make a reliable $\lambda$ determination impossible. They can be extracted from measurements at large t only after $\lambda$ and $A_\lambda$ have been reliably determined at small t. This requires that finite-size effects must be reduced as much as possible so that the $10^{-4} < t < 10^{-2}$ range may be used in order to obtain $\lambda$ and $A_\lambda$, a requirement which puts high demands on electron optical design, specimen preparation, temperature stabilization and vacuum in the system. For a more detailed discussion, see [4.23–25,33,174].

### 4.3.2 Scattering Methods

Two completely different phenomena are used here. One is the fact that in He atomic beam scattering the differential scattering cross section of an individual adsorbed atom or molecule may differ strongly from that of the same particle incorporated in a 2d crystal [4.167,198]. This allows to study the 2d vapour-condensate transition (Sect. 4.2.10) with the same technique which can be used also for the study of CI and rotational transitions [4.199]. Up to now the technique has not been applied to chemisorbed layers but there are no fundamental reasons why it should not be successful also there. The second method is channeling [Ref. 4.115, Chap. 4] and blocking in Rutherford backscattering (high energy ion scattering). This method gives information on the atomic positions of the atoms in the topmost layer(s) of the crystal and has successfully been used to study the surface melting of Pb [4.200]. It is practically limited to heavy atoms on light surfaces; light atoms can be studied only if they induce substrate atom displacements. The method is competitive with other methods only in phase transitions which cannot be studied very well with them, e.g. when several atomic layers are involved.

### 4.3.3 Work Function Change Measurement

This is the simplest and experimentally least demanding method for the study of 2d phase transitions and is particularly useful in 2d vapour-condensate transitions (Sect. 4.2.10). Its basis is the dependence of the dipole moment of an atom on its neighbourhood. An isolated adsorbed atom has a large dipole moment p while the same atoms incorporated into a 2d crystal have a much smaller p value. A good example is Au on W(110): condensed 2d Au causes a slight work function *increase*, 2d Au vapour a strong work function *decrease* [4.144]. Another example is Ni on W(110). Here the dipole moment in the loosely packed $(1 \times 1)$ ordered phase differs significantly from that in the densely packed coincidence structure [4.155]. The work function change $\Delta\phi(T) = -4\pi\Delta n(T)\Delta p$ due to the dipole moment difference and the change of the atoms from one to the other phase allows, therefore, to determine this number density and fractional coverage of the two phases. It has also been used to study a first-order structural transition in a metal double layer [4.201] and is the only useful method (except Sect. 4.3.4) when the 2d condensate forms a $(1 \times 1)$ structure.

When the dipole moments are large such as in the systems discussed in Sect. 4.2.9 the dipole-dipole repulsion ensures that the average atomic distances do not change much during the phase transition so that $\Delta p$ is small $(\Delta p/p \approx 10^{-3})$ and it is difficult to apply the method. When p is small, however, such as in the low coverage range of oxygen on W(110) and Mo(110) then even small p's can be measured with sufficient accuracy to study order-disorder transitions such as the $p(2 \times 1)\leftrightarrow(1 \times 1)$ and the $p(2 \times 2)\leftrightarrow(1 \times 1)$ transition at $\theta = 1/2$ and $\theta = 1/4$ on W(110) and Mo(110), respectively. Before critical parameters can be extracted from such measurements their accuracy must be improved so that 0.1 meV changes can be measured reliably. The theoretical analysis of [4.197] should cover this case.

### 4.3.4 Thermal Desorption Methods

These methods make use of the fact that the desorption energy and entropy depend upon the coordination of the desorbing species: for an isolated adsorbed atom both are small, for the same atom in a 2d crystal they are large. This is reflected in the desorption rate as a function of coverage which can be analyzed by plotting the logarithm of the desorption rate at constant coverage as a function of 1/T. These "Arrhenius plots" are straight lines with breaks at which the dominating desorption process changes from desorption from the 2d condensed phase to desorption from the 2d vapour. In this manner the coexistence curve or at least the critical temperature can be determined in the T range in which method, Sect. 4.3.3, breaks down because of changing coverage [4.202,203]. Thus, methods, Sects. 4.3.3 and 4, are complementary. Thermal desorption is presently still relatively inaccurate but with some experimental improvements it should be possible to determine the critical exponent $\beta$ of the vapour-condensate transition (Sect. 4.2.10).

### 4.3.5 Electron Microscopy

All methods discussed up to now average over areas from several $10^{-5}$ to several $10^{-3}$ m diameter, depending upon method. As Fig. 4.12 shows surfaces are notoriously heterogeneous. As a consequence not only does the size of the coherent regions vary considerably across the crystal but also the details of the phase transitions which are size-dependent and frequently influenced by steps, e.g. due to preferred nucleation in first-order phase transitions. For a detailed understanding of a 2d phase transition it is therefore indispensable to look at it also with high lateral resolution in order to untangle the information obtained from the laterally averaging techniques. In principle, scanning tunneling microscopy (STM) with its atomic resolution would be the ideal tool. However, the lateral dimensions accessible to STM are too small and the duration of the measurement too long as to make it a practical technique for phase transition studies. Such a technique must allow the observation of phase transitions in areas with diameters from $10^2$ to $10^4$ nm in real time.

Electron microscopy [Ref. 4.115, Chaps. 2 and 3] fulfills this condition. There are two fundamental imaging modes: scanning electron microscopy (SEM) and true image microscopy. In SEM a finely focussed beam is scanned across the surface and the secondary or diffracted electrons produced by the incident beam are used to produce an image via work function and geometry contrast or diffraction contrast, respectively. A resolution of 10 nm or better is obtainable. In the second mode a true image of the surface is formed by a standard optical imaging system. Diffracted electrons or electrons emitted by thermonic, photo- or secondary-electron emission may be used for imaging. The contrast is again due to diffraction or work-function differences. 10–100 nm resolution can be obtained depending upon imaging mode, quality of surface and instrument.

Although all modes, SEM and true imaging emission modes are suited for the study of phase transitions only the true reflection imaging mode has been used up to now, either with high-energy electrons (50–100 keV) at grazing incidence (RHEEM) [4.204] or with low-energy electrons (0–100 eV) at normal incidence (LEEM) [4.184,205]. By switching on an intermediate lens it is pos-

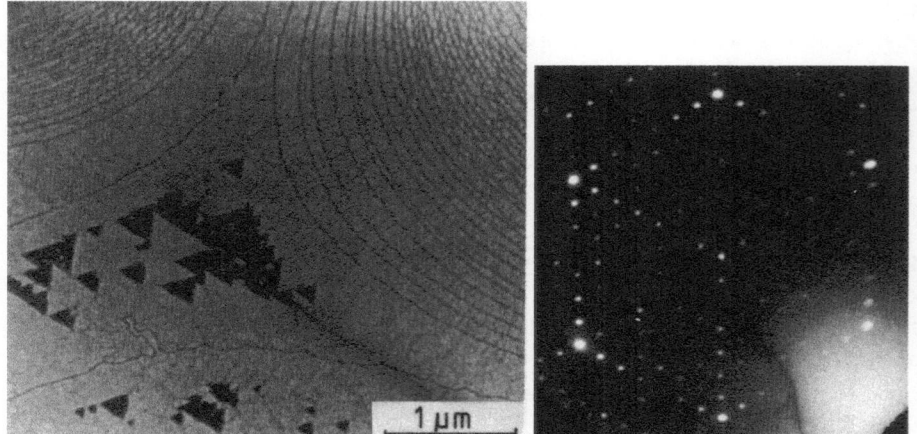

**Fig. 4.15.** LEEM image and LEED pattern of the imaged area during the $(7 \times 7) \leftrightarrow (1 \times 1)$ phase transition of the Si(111) surface. The marker indicates $1\,\mu\mathrm{m}$. Dark areas $(1 \times 1)$ structure, bright areas $(7 \times 7)$ structure, dark bands: location of steps [4.205]

sible to obtain the diffraction pattern of the imaged area so that image and diffraction pattern can be compared immediately. Figure 4.15 shows an example of such a comparison. The system has a large transfer width for LEED which is limited by the resolution of the image detection system consisting of channel plate multiplier and fluorescence screen. This sensitive detection system allows the phase change to be followed in real time via video monitor and thus makes the study of the kinetics of phase transitions possible.

### 4.3.6 Experimental Limitations

Several experimental limitations, mainly of instrumental nature have already been discussed in Sect. 4.3.1. Here they will be amended briefly by the limitations due to the specimen consisting of crystal and adsorbate. First of all, the crystal must be UHV compatible, i.e. it must have a negligible vapour pressure at $T \gtrsim T_c$, e.g. $1.2\,T_c$. It must be possible to clean it in situ and to prepare and maintain a surface as perfect as possible. It may not have a bulk phase transition in the $T$ range which has to be studied in order to obtain a reliable value for the Debye-Waller factor. These three conditions exclude already many elements as substrates. The adsorbate must be reasonably stable against electron-stimulated desorption (ESD) in methods, Sects. 4.3.2 and 5. This is not a very stringent limitation because in most systems of interest ESD is no problem.

With all these conditions fulfilled there is still an important other one which sometimes has not been taken sufficiently serious in the past: the coverage may not change in the course of the experiment unless the change is used to study the phase transition such as in Sect. 4.3.4. That desorption into the vacuum must be avoided is immediately evident but that a coverage change also occurs due to dissolution of the adsorbate in the bulk or segregation from the bulk during the experiment has sometimes only been noticed during the experiment [4.206]. This problem or even worse ones such as the irreversible dissolution of oxygen and oxidation of electropositive metals occur in many adsorbate-substrate systems. Many a theoretician's dream can, therefore, not be realized by the experimentalist.

## 4.4 Results

In this section the experimental and theoretical results obtained to date will be reviewed briefly, emphasizing some selected examples. Unfortunately, the systems studied in experiment and theory frequently do not match so that only few comparisons are possible. From the point of view of the theory it would be desirable to classify the examples according to the type of transition (Sects. 4.2.7–4.2.10) and according to the applicable model. The experimentalist, however, studies a given surface or adsorbate-substrate system and finds several types of transitions in it. The classification below follows the experimental situation with the exception that it separates chemisorbed layers with repulsive interactions from those with attractive interactions. Another classification is that into two-dimensionally close-packed surfaces, one-dimensionally close-packed surfaces and open surfaces, i.e. surfaces with no close-packed directions. This separation is advisable because in the first group the activation barrier for surface diffusion is low and in the last group it is (nearly) insurmountable before disordering sets in. In the middle group it is highly anisotropic and so is the bonding so that it represents good systems for two-state and asymmetric multistate models.

### 4.4.1 Clean Surfaces

#### a) The Si (111)-(7 × 7)  ↔  (1 × 1) Transition

The (7 × 7)  ↔  (1 × 1) transition on this surface is the most thoroughly studied phase transition on a clean surface. In spite (or because?) of several detailed LEED and electron microscopic studies the transition is far from being understood. The first RHEEM study [4.204] led to the conclusion that the transition is of first order while a simultaneous LEED study [4.194,207] suggested that it was second order, a suggestion which was accepted also by the electron microscopists on the basis of additional RHEEM work [4.208]. More recently, refined LEED [4.195,209] and LEEM investigations [4.184,205] revived the interpretation in terms of a first-order transition, though a rather complicated one. The recent studies confirmed to a large extent the previous observations but revealed at the same time that diffraction and microscopic studies were done on surfaces which differed strongly due to their different heat treatment. In the microscopy work well ordered and clean surfaces with large terraces between monoatomic steps were used (Fig. 4.15), in the diffraction work microscopically disordered surfaces which probably contained subsurface carbon. Nevertheless they produced good LEED patterns though with high background. The DW-factor corrected $I(T)$ curves of the superstructure reflections (Fig. 4.16a) show typical second-order behaviour. Because of $I(T) \sim |t|^{2\beta}$, $I(T)^{1/2\beta}$ has to be proportional to $|t|$ for the correct $\beta$, as illustrated in Fig. 4.16b. The correct $\beta$ depends of course on the choice of $T_c$ which is optimized in Fig. 4.16c. The result is $\beta = 0.066$ and $T_c = 1110\,\mathrm{K}$ [4.195,209]. The background increases considerably during the transition so that it appears to be a second-order order-disorder

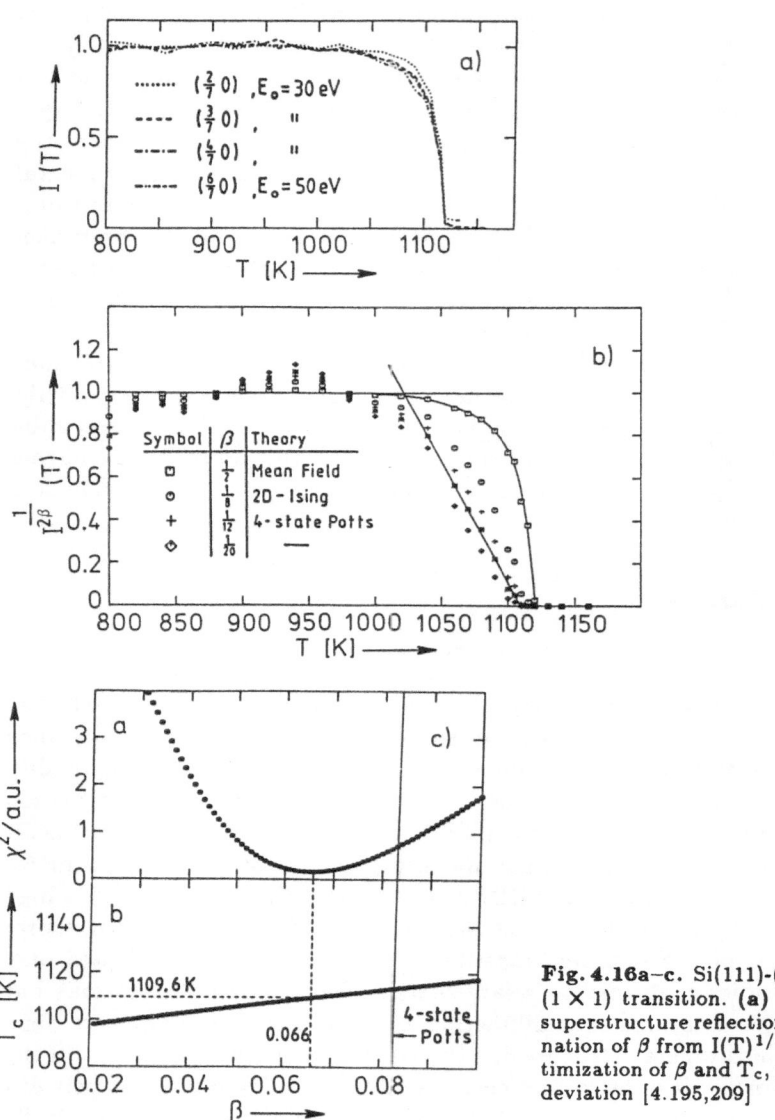

Fig. 4.16a–c. Si(111)-(7 × 7) ↔ (1 × 1) transition. (a) $I(T)$ of several superstructure reflections, (b) determination of $\beta$ from $I(T)^{1/2\beta}$ plot, (c) optimization of $\beta$ and $T_c$, $\chi^2$ squared rms deviation [4.195,209]

transition. Critical scattering is absent, however, and the detailed temperature dependence of the integral order reflections and of the background are not compatible with this interpretation but can be explained by the coexistence of the $(1 \times 1)$ and $(7 \times 7)$ phase over the temperature range in which the apparent second order transition takes place ( $\sim 800\,\mathrm{K}$ to $\sim 1110\,\mathrm{K}$). In contrast to this wide range, the transition temperature range observed in the microscopy studies is only $\sim 20\,\mathrm{K}$ [4.204] or less than $10\,\mathrm{K}$ [4.205].

The present explanation for this discrepancy is as follows. On the relatively perfect and clean surface used in the microscopy work nucleation of the $(7 \times 7)$ phase in the $(1 \times 1)$ high temperature phase occurs at the steps whose strain field assists the transition. After nucleation the $(7 \times 7)$ phase spreads rapidly across the terraces between the steps to cover the whole surface. The surface used in the diffraction work, on the other hand, has a very high density of irregular steps and contains probably subsurface carbon [4.182] with particularly high concentration in the strain field of the steps. The steps are, therefore, poisoned as nucleation sites. The $(7 \times 7)$ phase nucleates then close to the center of the small terraces and can spread rapidly only across the relatively carbon-free region. In the carbon-contaminated region the transition temperature is reduced by a carbon concentration-dependent amount due to the $(1 \times 1)$-stabilizing effect of the carbon so that the transition occurs over a wide T range [4.210]. The transition is believed to be driven by phonon softening [4.30] so that it may be considered to be a displacive transition which is strongly influenced by impurities and strains [4.209]. More work is certainly needed to confirm this picture. For past theoretical work, see [4.28].

## b) The Au(110)-$(1 \times 2)$ $\leftrightarrow$ $(1 \times 1)$ Transition

The Au(110) surface has a $(1 \times 2)$ structure which is due to the fact that every second densely packed row of atoms is missing [4.211]. There are two equivalent row configurations on the rectangular lattice so that the order-disorder transition of this structure belongs to the 2-state Potts (Ising) class (Table 4.2). This transition has been the subject of only one LEED study [4.212] which has, however, produced the largest amount of information on critical properties of a 2d system available to date. Not only were the critical exponents $\beta$, $\gamma$ and $\nu$ determined and the exponent $\eta$ estimated but also the corresponding critical amplitudes, in spite of strong finite-size effects caused by the small average size of the $(1 \times 2)$ domains of about $15\,\mathrm{nm}$. Only one superstructure spot was analyzed at one energy. The quality of the fit to the experimental data I(T), $\chi(0, T)$ and $\xi(T)$ in (4.29,30) (Sect. 4.3.1) is illustrated in Fig. 4.17a, the agreement with theoretical expectations in Fig. 4.17b. The $\beta$, $\gamma$ and $\nu$ values agree within the limits of error with the Ising values, $\eta$ is compatible with the Ising value. For a more detailed discussion, see [4.33,278].

## c) W(100)-c$(2 \times 2)$ $\leftrightarrow$ $(1 \times 1)$ Transition

This transition has been studied extensively both theoretically [4.29,30a,213–221,221a–b] and experimentally [4.222] (for older work see the reviews [4.28,33,

157

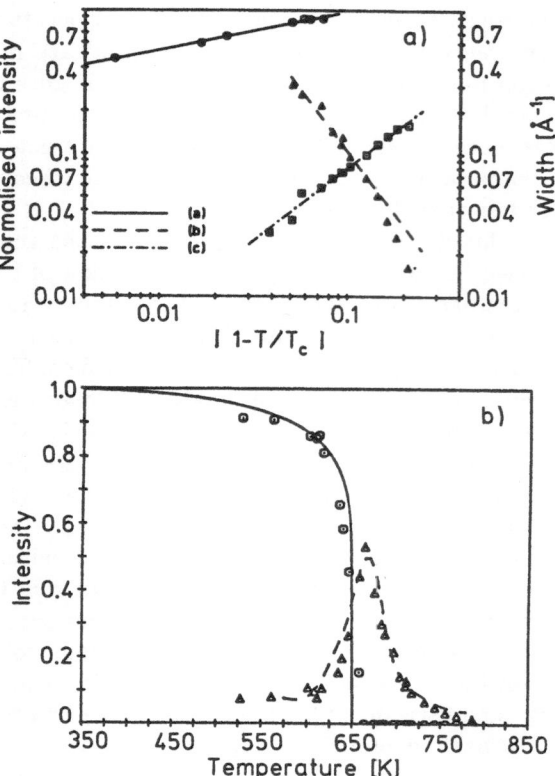

**Fig. 4.17a,b.** Au(110)-(1 × 2) ↔ (1 × 1) transition. (a) Fits of I(T) (Gaussian height, circles), $\chi(0, T)$ (Lorentzian height, triangles) and $1/\xi(T)$ (Lorentzian width, squares). (b) Separation of diffracted intensity into contributions from long range order *(solid line, circles)* and critical fluctuations *(dashed line, triangles)* [4.212]

73]) but has been elusive to detailed understanding. The c(2 × 2) structure consists of zig-zag chains of atoms parallel to the ⟨011⟩ directions (2d⟨11⟩) which results from alternating displacements of the topmost atoms in the directions normal to the chains (Fig. 4.18a). The space group of this structure is p2 mg, that of the substrate p4 mm so that according to Table 4.3 a continuous order-order transition into a (1 × 1) phase is symmetry-allowed. The appropriate statistical model is the XY model with cubic anisotropy (Table 4.2). The theoretical work [4.213–217] aimed at an understanding why the c(2 × 2) structure forms and why it is so unstable – the transition occurs around 200 K – has led to the following picture [4.216]. The surface energy band structure of W(100) is such that a surface band crosses the Fermi level midway between the $\Gamma$ and the M point. The creation of a band gap at $2\pi h = (1/2, 1/2)\, 2\pi/a$ will, therefore, lower the electronic energy of the system. If the elastic energy increase due to the associated lattice distortion is less than the electronic energy decrease the reconstructed state will have a lower total energy $E_T$. Figure 4.18b shows that this is indeed the case [4.217]. $\delta$ is the displacement of the atoms in the topmost layer from the centered position in Fig. 4.18a in ⟨11⟩ direction. The location $\delta_0$ of the $E_T$ minimum is in excellent agreement with experiment. The variation of $E_T$ with $\delta$ is quite small and the potential well for the c(2 × 2) structure

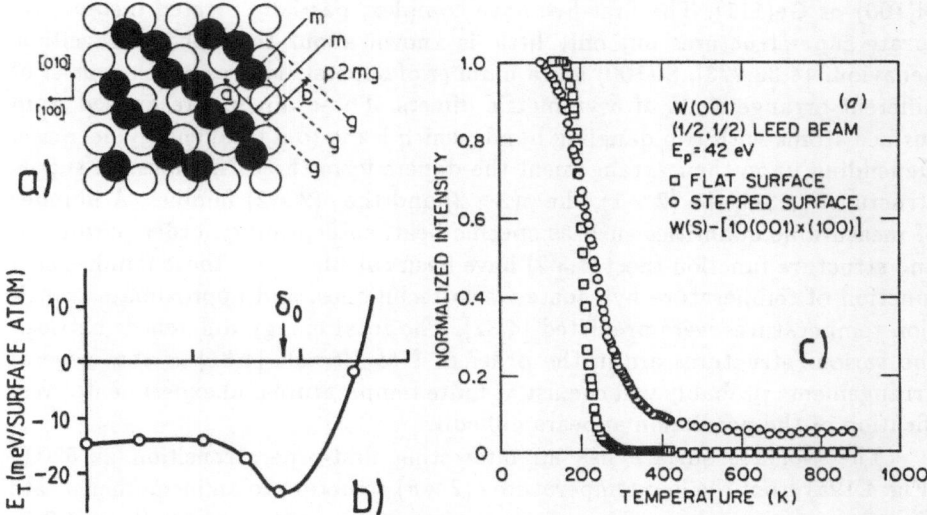

**Fig. 4.18a–c.** W(100)-c(2 × 2) ↔ (1 × 1) transition. **(a)** Atomic arrangement of c(2 × 2) structure. Only one orientation is shown, the other is rotated 90° [4.73]. **(b)** Total energy $E_T$ referred to surface with bulk structure as a function of displacement $\delta$ from bulk position in ⟨011⟩ direction [4.217]. **(c)** Debye-Waller factor corrected I(T) of superstructure reflection for well and poorly oriented surface [4.222]

is quite shallow ( $\approx 0.01$ eV/atom). Therefore, the surface atoms can perform large-amplitude ( $\lesssim 2\delta_0$) oscillations along ⟨011⟩ with increasing temperature with the center of gravity at $\delta = 0$ (soft phonon) which represents the phase transitions c(2 × 2) ↔ (1 × 1). However, also an order-disorder process has been proposed [4.279].

The theoretical work aimed at describing the phase transition in the language of the phase transition theory [4.28,30a,33,218–221] has produced a number of effective Hamiltonians but the only numerical calculations [4.219,221b–c] have little experimental data for comparison. This is not surprising in view of the scarce quantitative experimental material available. The experiments are difficult because of residual gas adsorption at the low temperatures required and because of the strong finite-size influence on the transition which demands very good specimen preparation. Figure 4.18c shows how the transition differs between surfaces with $\lesssim 0.1°$ and $\approx 3.2°$ deviation from the (100) plane [4.222]. The data for the "perfect" crystal can be fitted by $T_c = 211$ K and $\beta = 0.14$ but the fit is not very satisfactory. For a more detailed discussion of this transition, see [4.33].

### d) Other Transitions on Clean Surfaces

There are a number of other clean reconstructed surfaces with phase transitions to a (1 × 1) configuration at sufficiently high temperatures: Au, Pt, Ir(100), Au(111), Mo(100) and many elemental semiconductor surfaces such as

Si(100) or Ge(111). The first five have complex, partially rotated incommensurate superstructures but only little is known about their phase transition behaviour [4.33,223]. Si(100) has a number of superstructures which consist of different arrangements of asymmetric dimers. These dimers are formed from surface atoms with two dangling bonds which leads to a total energy decrease. Depending upon their arrangement the dimers form three "families" of superstructures [4.31], the $(2 \times 1)$, the p$(2 \times 2)$ and the c$(2 \times 2)$ families. A number of measurable quantities such as specific heat, susceptibility, order parameter and structure function (Sect. 4.3.2) have been calculated for those families as a function of temperature by Monte-Carlo techniques, and approximate transition temperatures were predicted [4.32]. The total energy differences between the various structures are in the order of 1 to 10 meV [4.31] so that several arrangements probably will coexist at finite temperature and experimental verification of the prediction appears difficult.

The Ge(111) surface has an interesting first-order transition at 570 K (Fig. 4.19a) from the low-temperature c$(2 \times 8)$ structure to an incommensurate high-temperature phase. This phase is either (i) a striped phase (Sect. 4.2.6) with a perodicity close to $(2 \times 1)$ consisting of $(2 \times 1)$ regions and walls primarily perpendicular to the undoubled period of the $(2 \times 1)$ unit mesh or (ii) a hexagonal phase consisting of $(2 \times 2)$ regions separated by antiphase walls which form an irregular honeycomb structure. This honeycomb structure is similar to that shown in Fig. 4.8 for Kr on graphite but differs in the basic periodicity ($(2 \times 2)$ instead of $(\sqrt{3} \times \sqrt{3})$) and in the atomic density difference between walls and the interior. Furthermore the walls do not simply grow in thickness with increasing T as seen in Fig. 4.8 but also their mean distance decreases from about 3.7 nm at 570 K to about 1.7 nm at 870 K (Fig. 4.19b)

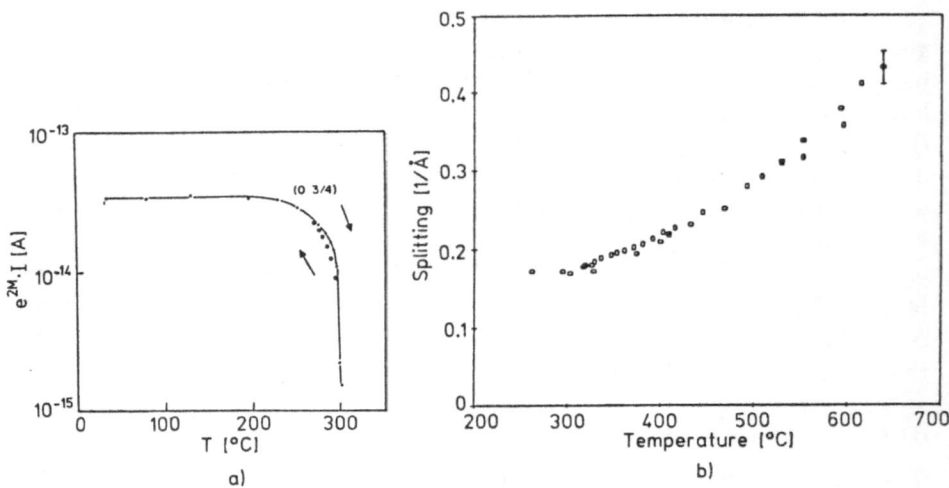

**Fig. 4.19a,b.** Ge(111)-c$(2 \times 8) \leftrightarrow$ I$(2 \times 2)$ transition. (a) Debye-Waller factor corrected intensity of fractional order beam with slight hysteresis as a function of T. (b) Peak splitting due to "wall lattice" ($\sim$ inverse wall distance) in I phase as a function of T [4.224]

[4.224]. This difference in the disordering ("melting") process between the two cases is probably due to the different number of states ($4 = 2^2$ instead of 3, see [4.88] and Sect. 4.2.6) and the existence of one strongly preferred type of wall. No theoretical work has been done on this transition.

For kinked vicinal surfaces theory predicts phase transitions in the Ising, the XY model with cubic anisotropy and – in special cases – the Potts universality class. The diffraction from the ordered phases and the pair correlation function have been calculated [4.225] but no systematic experimental studies have been made up to now.

### 4.4.2 Chemisorbed Layers with Repulsive Nearest-Neighbour Interactions

#### a) Two-Dimensionally Densely Packed Surfaces

There are three kinds of these surfaces on crystals with simple structures: (i) The (111) face of fcc metals and the (0001) face of hcp metals. They contain three densely packed directions in the hexagonal array of atoms and are the smoothest surfaces on which surface diffusion can occur most easily. (ii) The (110) face of bcc metals which contains two densely packed directions and may be considered as a distorted hexagonal structure. (iii) The (100) face of fcc metals which also contains two densely packed rows. Due to the differences in symmetry between the three types of faces transitions in different universality classes should occur on them.

### Hexagonal Surfaces (p6mm, p3m1)

There have been a number of specific calculations for adsorbates on this surface which will be illustrated by two examples. One is a Monte-Carlo calculation for repulsive $J_1(\varphi_1 > 0)$ and attractive $J_2(\varphi_2 < 0)$ interactions in which phase diagram and critical exponents were obtained by finite size scaling of $L \times L$ lattices with $12 \leq L \leq 90$ [4.226]. The ratio $J_2/J_1$ was varied from $-1/16$ to $-\infty$. The two $(\sqrt{3} \times \sqrt{3})$ ordered phases shown in Fig. 4.20b,c were found. They are separated by a coexistence region and have – as expected from Table 4.2 – 3-state Potts critical exponents. No "chiral" behaviour (Sects. 4.2.4c,7) is seen. Figure 4.20f shows the phase diagram for $J_2/J_1 = -1$ which is symmetric about $\theta = 1/2$ (H = 0). Dashed and solid phase boundaries indicate second- and first-order transitions, respectively, G and L gas and liquid and $T_1$, $T_2$ are the limits of an XY-like line of critical points (Sect. 4.2.3). The other example is a renormalization group theory study [4.227] specifically aimed at understanding the asymmetric and broad $(2 \times 2)$ phase region seen in experiment for hydrogen on Ni(111) [4.228]. Assuming all threefold sites on the surface to be equivalent (Fig. 4.20a) makes neighbouring atoms in the p$(2 \times 2)$ structure (Fig. 4.20d) 6$^{th}$ nearest neighbours which have to be taken into account in the description of the interactions $J_i$ between the adsorbed atoms. This leads to a number of high-symmetry ordered structures whose relative stability depends upon relative magnitude and sign of the interactions. The observed

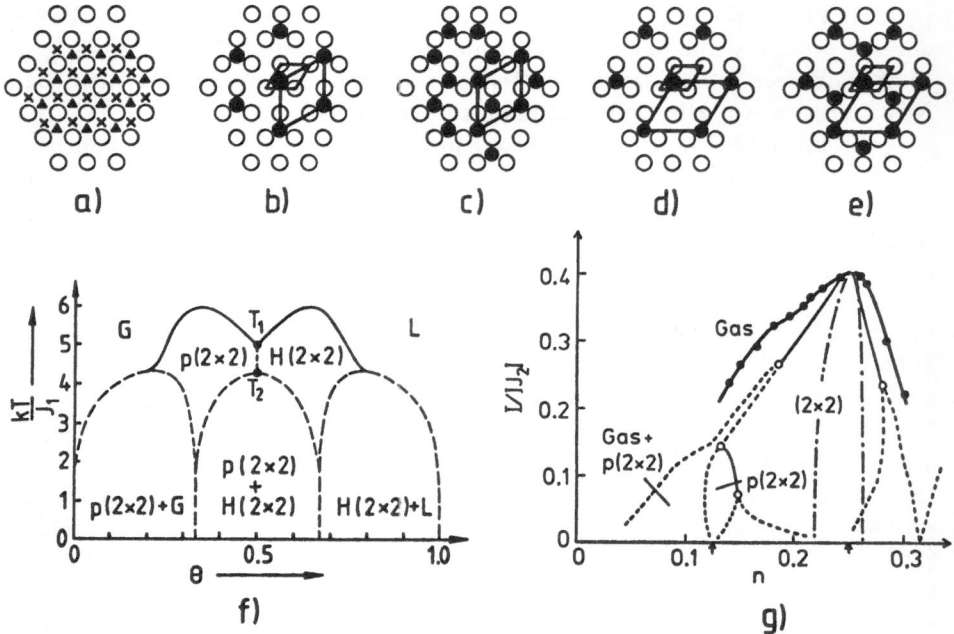

**Fig. 4.20a–g.** Submonolayer phases and phase transitions on hexagonal surfaces. **(a)** Equivalent adsorption sites without (triangles and crosses, p6mm) and with (triangles only, p3m1) influences of layers below the topmost layers ("crystal field"). **(b–e)** $p(\sqrt{3} \times \sqrt{3})$ $(\theta = 1/3)$, honeycomb $(\sqrt{3} \times \sqrt{3})$ $(\theta = 2/3)$, $p(2 \times 2)$ $(\theta = 1/4)$ and honeycomb $(2 \times 2)$ $(\theta = 1/2)$ structure, respectively. Referred to a substrate without crystal field the coverages are only half of those quoted. **(f)** Phase diagram for $J_2/J_1 = -1$, $J_n = 0$ $(n>2)$, $J_1>0$ (repulsive) [4.226]. **(g)** Phase diagram for $J_3/J_2 = 0.4$, $J_4/J_2 = 0.8$, $J_5/J_2 = 0.1$, $J_6/J_2 = -0.02$, $J_n = 0$ $(n>6)$, $J_1 = +\infty$, $J_2>0$ [4.227]

$(2 \times 2)$ phase boundary (heavy line in Fig. 4.20g) can be approximated best with the $J_i$ values given in the caption of Fig. 4.20g. Note that the interactions are nonmonotonic ($|J_4|>|J_3|$) as a consequence of indirect electronic interactions [4.229]. This example illustrates how involved calculations have to be in order to give a realistic description of nature. More recent work on this system [4.229a] points out further problems in previous calculations [4.227,229].

The only adsorbate system on a hexagonal surface for which not only the phase diagram has been studied but also the critical exponents have been determined is oxygen on Ni(111). The experimental phase diagram which has two ordered phases, $p(2 \times 2)$ $(\theta = 1/4)$ and $(\sqrt{3} \times \sqrt{3})$ $(\theta = 1/3)$ could be simulated with $\varphi_3/\varphi_2 = 2/3$, $\varphi_4/\varphi_2 = 1/6$, $\varphi_5/\varphi_2 = 1/20$, $\varphi_6/\varphi_2 = -0.06$, $\varphi_1 = +\infty$ and $\varphi_2 = 1/20$, referred to the sites shown in Fig. 4.20a [4.230]. The continuous transition near $\theta = 1/4$ could, however, be only reproduced by $\varphi_2 \approx \varphi_1$. A detailed analysis of this transition as outlined in Sect. 4.3.2 gave critical exponents which agreed within the limits of error with the 2d Ising values while Table 4.2 predicts 4-state Potts exponents. This discrepancy could be explained by assuming that the influence of the layers below the topmost layer on the transition was negligible. This puts the resulting $p(2 \times 2)$ structure into the universality class of the Heisenberg model with corner cubic anisotropy (Table 4.2) which has the same critical exponents as the Ising model [4.69,231]. It should be mentioned that this was the first determination of critical exponents in chemisorbed layers and that the experiments were carried out under difficult experimental conditions, both from the point of view of the instrument and of the specimen: a display-type system was used (Sect. 4.3.2) and oxygen goes into solid solution in Ni close to the transition temperature.

The system S/Pt(111) which was also studied in a display-type system has the same ordered phases as O/Ni(111). Their transitions at $T_c = 532$ K ($\theta = 1/4$) and $T_c = 665$ K ($\theta = 1/3$) can be simulated well by $\varphi_1 = +117$ meV (repulsive), $\varphi_2 = -26$ meV and $\varphi_3 = -30$ meV [4.232]. The system CO/Ru(0001) shows a complex phase transition behaviour which has been the subject of several studies [4.233,234]. Finally, the rotational transition of Na on Ru(0001) [4.235] should be mentioned as an example in connection with Sect. 4.2.6.

**Centered Rectangular Surfaces (c2mm)**
Detailed model calculations have been made with the intent of simulating the systems H/Fe(110) [4.51,130,132a,236–239], O/W(110) [4.51,239,240] and Si/W(110) [4.241,242]. In addition, theoretical interaction parameters have become available recently for the system H/Fe(110) and have been used [4.243] to calculate the stability of the ordered structures ($c(2 \times 2)$ at $\theta = 1/2$ and $(3 \times 3)$ at $\theta = 2/3$) proposed on the basis of the original experimental work [4.244]. These calculations show that the $c(2 \times 2)$ structure is compatible with the calculated interaction parameters but that the $(3 \times 3)$ structure requires much larger repulsive interactions than the calculated ones. Furthermore, the trio interactions are calculated to be much weaker than those necessary in the lattice gas computer simulations [4.15,237]. Nevertheless these simulations show very nicely the dependence of the possible ordered structures upon the interaction parameters and the variety of transitions (CI transitions, disordering lines, wetting transitions) which can occur in this system. The phase diagram of [4.244] could be described in a satisfactory way with repulsive pair interactions $\varphi_1 = 13$ meV, $\varphi_2 = 83$ meV and $\varphi_3 = 22$ meV and attractive trio interactions $\varphi_t = -35$ meV. In view of the fact that the binding sites assumed in the original work have been revised on the basis of a LEED structure analysis [4.245] – changing the symmetry of the problem – and that new calculations for the new models are in progress it is premature to discuss this system more thoroughly. It should be mentioned, however, that it promises to be a good model system for the problems discussed in Sects. 4.2.6–8.

In the system H/W(110) the critical exponent $\beta$ of the order parameter has been determined with a display-type system at and close to $\theta = 1/2$ ($p(2 \times 1)$ phase) and $\theta = 3/4$ (($2 \times 2$) phase). The transitions were smeared out which was attributed to a distribution $\Delta T$ of $T_c$ values due to finite-size effects. Taking these into account the analysis leads to $\beta = 0.13\pm0.04$, $T_c = 201$ K, $\Delta T = 8$ K for the $p(2 \times 1)$ phase and to $\beta = 0.25\pm0.07$, $T_c = 247$ K, $\Delta T = 10$ K for the

$(2 \times 2)$ phase [4.246]. Both transitions belong into the universality class of the XY model with cubic anisotropy (Table 4.2) which has nonuniversal critical exponents so that no comparison with theory is possible at present.

The system $O/W(110)$ has been much discussed in the literature although only a rather limited amount of experimental information on its phase transitions is available [4.247,248]. There has, however, been a considerable amount of modelling of this system ([4.240,241] and references therein). The most elaborate calculations take into account $\varphi_1, \varphi_5, \varphi_t < 0$ (attractive), $\varphi_2, \varphi_3 > 0$. With $\varphi_2/\varphi_1 = \varphi_3/\varphi_1 = -1.0$, $\varphi_t/\varphi_1 = 1.25$, $\varphi_5/\varphi_1 = 0.01$ good qualitative agreement with experiment was obtained [4.241]. It should be noted that for $\varphi_5 = 0$ only second-order phase boundaries are obtained but that the weak $\varphi_5$ interaction is already sufficient to produce the experimentally observed coexistence region between $p(2 \times 1)$ structure and empty lattice ($p(2 \times 1)$ islands). For $\varphi_5 = 0$ the (nonuniversal) critical exponents $\nu$ and $\eta$ have been calculated as a function of coverage [4.240] but no experimental data are available at present.

The system $O/Mo(110)$ differs significantly from $O/W(110)$. It does not form a $p(2 \times 1)$ structure at $\theta = 1/2$ but rather a $p(2 \times 2)$ structure at $\theta = 1/4$ and develops above $\theta = 0.3$ complex structures. For the $p(2 \times 2)$ structure $\beta$ has been determined with a high quality diffractometer to be $0.12 \pm 0.05$ with $T_c = 684$ K [4.249]. The system $S/Mo(110)$ has at $\theta = 1/4$ also a $(2 \times 2)$ structure but disorders in a different manner: the $\beta$ value of the $(1/2, 1/2)$ reflection is zero within the limits of error, the $(1/2, 0)$ reflection $\beta = 0.096$, close to the 4-state Potts value (Fig. 4.21) [4.249]. At higher coverages various C and I structures are seen. The comparison of O on $W(110)$ and $Mo(110)$ shows that even minor changes in the electronic structure of the substrate can change phase diagram and phase transitions of adsorbates drastically. No calculations are available for O and S on $Mo(110)$.

The Monte-Carlo and transfer-matrix calculations for the system $Si/W(110)$ using interaction parameters obtained by field ion microscopy [4.250] have produced a rich phase diagram with three commensurate phases ($p(2 \times 1)$ with $\theta = 1/4$, $(5 \times 1)$ with $\theta = 2/5$ and $(6 \times 1)$ with $\theta = 1/2$), CI transitions and I melting transitions for the $(6 \times 1)$ and $(5 \times 1)$ structures and a first-order order-disorder transition for the $p(2 \times 1)$ phase [4.241,242]. The interactions are characterized by a strongly attractive $\varphi_3$ and strongly repulsive $\varphi_2, \varphi_4$ but are apparently incorrect at higher coverages because the experimental phase diagram determined by LEED disagrees with the calculated one, except for the $p(2 \times 1)$ structure [4.251,252].

Other phase transitions studies on W and $Mo(110)$ surfaces are those of the $p(2 \times 1) \leftrightarrow p(2 \times 2)$ antiphase transition in Te on $W(110)$ ($p(2 \times 1)$ is the "$(4 \times 2)$" of [4.253]) and of the CI and distortional transitions in this system. The critical exponent of the order parameter has been determined for the $(3 \times 2)$ phase of $Ba/W(110)$ ($\beta = 0.16$, $T_c = 130$ K), for the $c(1 \times 3)$ phase of $Li/W(110)$ and for the $c(3 \times 3)$ phase of Sr on $Mo(110)$ which both have $\beta = 0.14$ [4.254]. These three phases actually should not have a simple con-

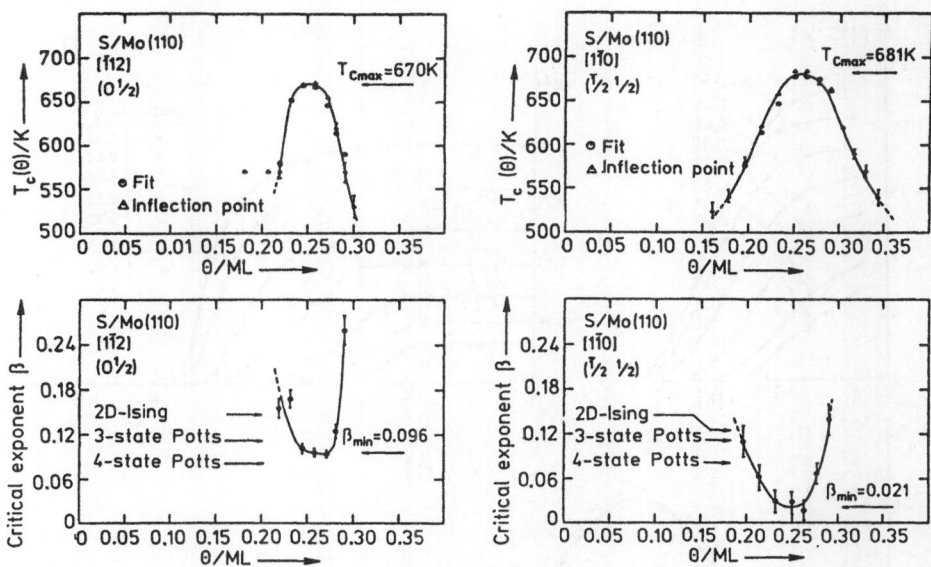

**Fig. 4.21.** S/Mo(110)-(2 × 2) ↔ (1 × 1) transition. Critical parameters T$_c$, $\beta$ as obtained from the peak intensities of the (0,1/2) and (1/2,1/2) reflections. $\beta$ is Fisher-renormalized (constant $\theta$ measurement) [4.195,249]

tinuous order-disorder transition (Table 4.2) but are governed by long-range interactions which have been insufficiently studied theoretically (Sect. 4.2.9).

## Square Surfaces (p4mm)

This is the surface which in the simplest case – when only $\varphi_1 \neq 0$ – is described by the classical Ising model. Because first-nearest-neighbour interactions only are not sufficient to describe real adsorption systems there have been many calculations for more then one $\varphi_i$ ([4.236,255–261] and references therein). Some of them were aimed at understanding specific adsorption systems such as H/Pd(100) [4.236,256] or O/Ni(100) [4.260]; most of them are concerned with the influence of number, sign and magnitude of interaction energies and of the magnetic field h (or $\Delta\mu$) on the phase diagrams, type of transitions and critical exponents which are now awaiting experimental verification. The system O/Ni(100) unfortunately cannot be studied reliably because of the exchange of O between adsorption layer and the bulk in the temperature range of the order-disorder transition [4.206]. The system H/Pd(100), however, could be studied quite successfully experimentally [4.262] and modelled with reasonable agreement by Monte-Carlo calculations [4.236,256]. Figure 4.22 compares experimental and theoretical order parameter curves (a, b) and apparent phase boundaries obtained from (c) which indicate $\varphi_1 > 0$ (repulsive), $-0.8 < \varphi_2/\varphi_1 < -0.6$ and $-0.3 < \varphi_3/\varphi_1 < -0.2$ (attractive) without noticeable trio interactions $\varphi_t$ [4.256]. It should be noted that the c(2 × 2) region determined experimentally from T$_{1/2}(\theta)$ (Fig. 4.22) could be simulated

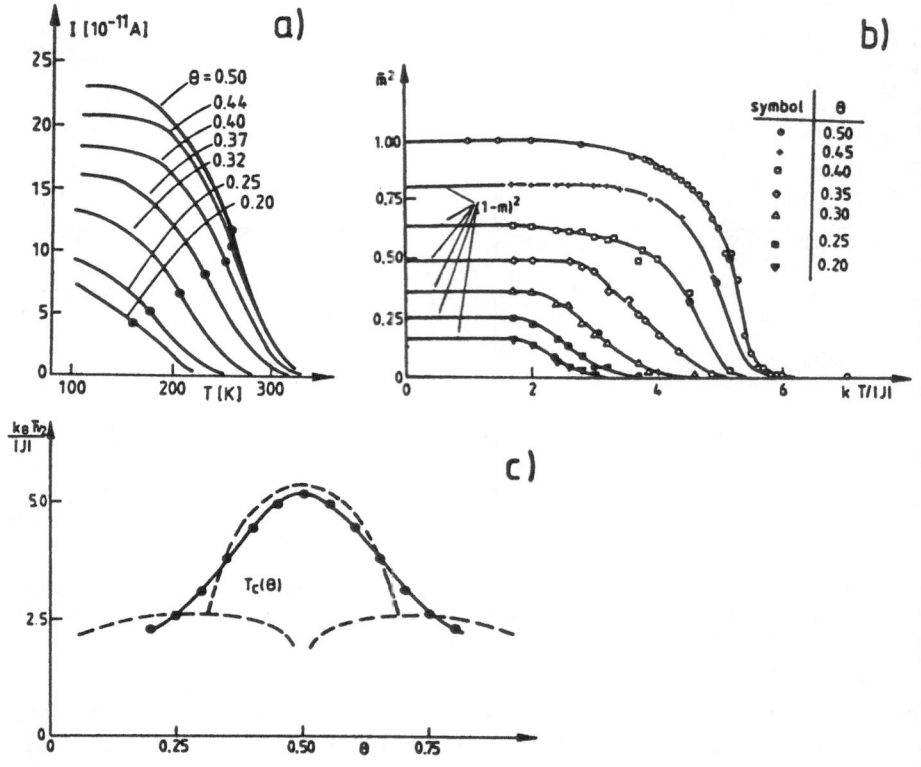

**Fig. 4.22a–c.** H/Pd(100)-c(2 × 2) ↔ (1 × 1) transition. Temperature dependence of (a) the intensity of superstructure spot ( ∼ (order parameter)$^2$) [4.262], (b) the square of the magnetization ((order parameter)$^2$) [4.256], for various coverages. *Solid circles:* $T_{1/2}$ points, defined by $I(T) = I(0)/2$. (c) Comparison of theoretical phase diagram (– – –) with experimental phase boundary as determined by the $T_{1/2}(\theta)$ points in (a) and (b) [4.256]

only by assuming rather small lattice dimensions which is attributed to strong specimen- or instrument-determined finite-size effects (Sect. 4.3.2).

The system Cl/Ag(100) has been studied in considerably more detail [4.263] with the following results. Adsorption saturates at $\theta = 1/2$ with a c(2 × 2) structure which forms at a critical coverage of 0.394±0.007 at 300 K. Disordering is pre-empted by desorption. The dipole moment is 0.74 D which implies $\varphi_1 = 28$ meV and $\varphi_2 = 10$ meV. The data could be Monte-Carlo simulated very well with a hard-square model which has a critical coverage $\theta_c = 0.7355/2 \approx 0.368$ (Sects. 4.2.3,4a). The critical behaviour of the square of the order parameter, the superstructure reflection intensity $I$ close to $\theta_c$ could be fitted well by $I/\theta^2 \sim (\theta - \theta_c)^{2\beta'}$ with $\beta' = \beta/(1-\alpha) = 0.12\pm0.03$ being the Fisher-renormalized $\beta$ value (because of $\theta = $ const instead of $\mu = $ const). The $\beta/(1-\alpha)$ value is in agreement with the 2d Ising exponents $\beta = 1/8$ and $\alpha = 0$ (Table 4.1), the disagreement of the $\theta_c$ values is attributed to the nonzero $\varphi_i$ values for i>1 which are assumed to be zero in the hard-square model.

The system Se/Ni(100) has also been the subject of a joint experimental-theoretical study [4.264]. It has two ordered phases, a $p(2 \times 2)$ phase at $\theta = 1/4$ with $T_c = 500\,\text{K}$ and a $c(2 \times 2)$ phase at $\theta = 1/2$ with $T_c > 600\,\text{K}$. The system can be described by the Ashkin-Teller model (Sects. 4.2.3,4). There are many other adsorption systems whose phases have been studied qualitatively but cannot be discussed here. Only the rotational transition of K on Cu(100) [4.265] should be mentioned in connection with Sect. 4.2.6.

### b) One-Dimensionally Densely Packed Surfaces (p2mm)

The fcc metal (110), bcc metal (211) and hcp metal (10$\bar{1}$0) surfaces belong to this category. They consist of densely packed atomic rows parallel to the [1$\bar{1}$0], [$\bar{1}$11] and [0100] directions, respectively. These are separated by troughs whose bottom is also densely packed so that diffusion and ordering in this direction is easy.

The first phase-transition study [4.266] and Monte-Carlo modelling [4.267] were actually done on one of these surfaces, Pd(110) with an oxygen adsorption layer. Since then, this system has been subject of additional theoretical studies [4.52,132a,268], particularly in connection with asymmetric models and the interesting phase transitions of the $(1 \times 2)$ phase and between $(1 \times 2)$ and $(1 \times 3)$ phases predicted for them (Sects. 4.2.6,8, Fig. 4.9) [4.52]. Experimental data checking these predictions are lacking, however. The system $N_2$/Ni(110) in which $N_2$ is weakly chemisorbed in a "standing up" position has been reported to have a fluid region between the C and I phase as predicted for some statistical models [4.269]. A detailed study of this interesting system is still missing. Another interesting system from the point of view of CI transitions (Sect. 4.2.6) is H/Ni(110) which has a continuous sequence of ordered structures for coverages $1/3 < \theta < 1$ [4.270].

O/W(211) is the system in this category which has been studied in most detail [4.271,272]. It is a classical case in the sense that it forms a $p(2 \times 1)$ phase at $\theta = 1/2$ and a $(1 \times 1)$ phase at $\theta = 1$ without any intermediate phases. The boundary of the $p(2 \times 1)$ phase was determined from the inflection point of the DW factor-corrected I(T) curves of one superstructure reflection for various coverages $\theta$ and is shown in Fig. 4.23a. The asymmetry shows that the system is not a pure classical Ising system. For $\theta = 1/2$ a detailed peak shape analysis was made as a function of temperature in order to determine the critical exponents (Sect. 4.3.2). The I(T) data were fitted for $0.016 \leq |t| \leq 0.083$. The region $|t| \leq 0.016$ had to be excluded because of finite-size effects of the order $L \approx 20\,\text{nm}$ which also made it necessary to assume a distribution of critical temperatures with Gaussian half width $\Delta T = 4\,\text{K}$ in order to fit the I(T) curves over this t range. This resulted in $\beta = 0.13 \pm 0.01$ and $T_c = 899\,\text{K}$. The data above $T_c$ were fitted with a Lorentzian for $0 < t < 0.11$ (Fig. 4.23b) in order to determine $\gamma$ and $\nu$ from peak height and half width which gave $\gamma = 1.79 \pm 0.14$ and $\nu = 1.09 \pm 0.11$ in agreement with the 2d Ising values.

Concluding this section it should be mentioned that alkali, alkaline-earth and rare-earth adsorbates on the (112) surface of W and Mo and the (1010)

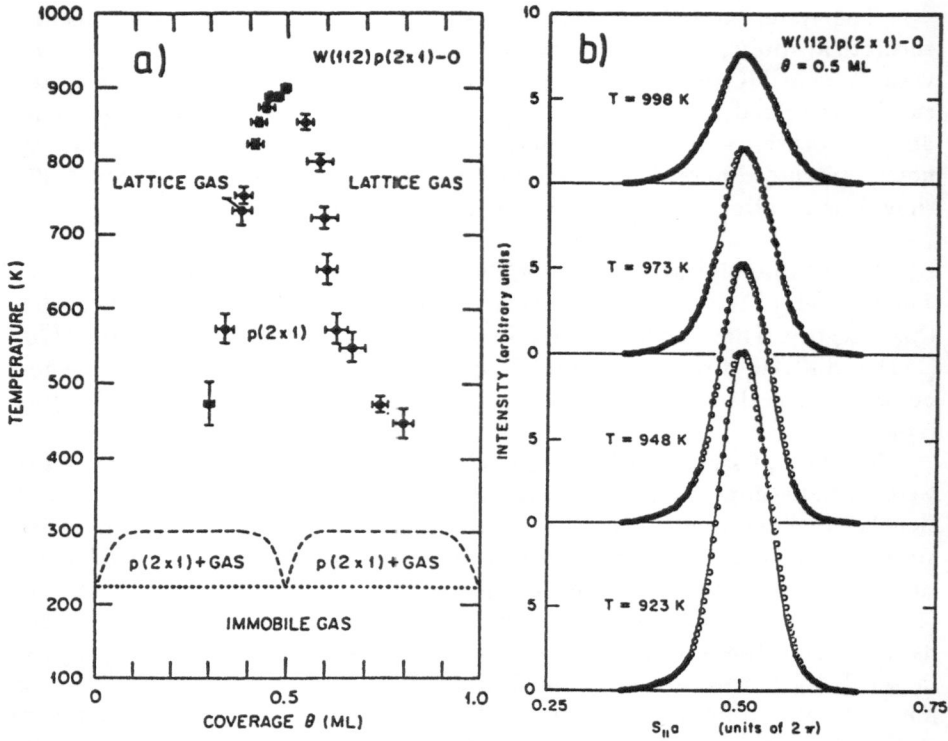

**Fig. 4.23a,b.** O/W(211)-p(2 × 1) ↔ (1 × 1) transition [4.271]. (a) Phase diagram; (– – –): possible first-order boundaries. (b) Lorentzian shape of peak profiles above $T_c$

surface of Re are the systems in which the nonmonotonic disordering over a wide temperature range is seen best [4.133,139,273,274].

### c) Open Surfaces

They contain no densely packed rows of atoms. Typical representatives are the (100) and (111) surfaces of bcc metals. Due to their openness they lack the stability of the densely packed planes and are easily reconstructed by adsorption. Thus oxygen adsorption reconstructs the W(100) surface irreversibly at 600 K [4.275], the Mo(100) surface already at 300 K [4.276]. Once irreversibly reconstructed into a p(2 × 1) structure the W(100) surface has a reversible p(2 × 1) ↔ (1 × 1) phase transition which, however, has not been studied in detail [4.277].

H has a less serious but much discussed influence on the two surfaces: by bonding to two substrate atoms it displaces them to form a pair. The pairs form a c(2 × 2) structure with c2mm symmetry which is different from the low-temperature structure of the clean surface. It transforms into an incommensurate phase with increasing coverage and finally into a (1 × 1) phase with 2 H atoms per W atom occupying all bridge sites. The phase diagram and transitions have been reviewed [4.33,73] and discussed in many theoretical papers

[4.28,30a,218,220,221,278,279]. In particular, the mechanism of the H-induced displacive transition and the CI transition via a striped phase of domain walls (Sect. 4.2.6) has received much attention. Except for the determination of the phase boundaries the main experimental effort has been directed at an understanding of finite-size effects on the formation and stability of the c(2 × 2) and I phases [4.280].

In addition to H, there are many other adsorbates, including metal atoms, which do not reconstruct W and Mo(100) surfaces but only cause displacements of the topmost atoms. This is the case for N [4.281] and probably also for Cl [4.282], S, Se and Te [4.283] on W(100). The phase diagrams of the chalkogenes up to $\theta = 1/2$ are very similar to that of Se on Ni(100) but above $\theta = 1/2$ complex phases are formed in which the atoms cannot be accommodated any longer in equivalent sites unless a high density of superheavy domain walls is accepted. Metal atom adsorption on bcc(100) surfaces which occurs at least at low coverages with repulsive interactions has been reviewed recently [4.284].

### 4.4.3 Chemisorbed Layers
### with Attractive Nearest-Neighbour Interactions

This class of systems has attracted little attention in the past because it is rather uninteresting from the point of view of critical phenomena: the attractive interactions ensure the formation of a 2d condensed phase which coexists with a 2d vapour phase at low coverages and temperatures and which is separated from the high temperature single phase vapour and the high coverage single phase condensate by a first-order transition line. Only the critical point on this line and its environment appear worthy of attention. It is just this region which is difficult experimentally so that only a few studies with new techniques (Sects. 4.3.3,4) were made. From the point of view of surface physics and chemistry these systems are, however, very interesting: 2d metals have physical properties quite different from 3d metals and − to give an example from chemistry − the state of dispersion of bimetallic catalysts (2d vapour or condensate) is expected to have a strong influence on its catalytic properties.

Adsorbates with attractive interactions have another property which is not found in the case of repulsive interactions: they can form adsorbed multilayers. This phenomenon may be called partial or incomplete wetting. Complete wetting, which has been widely discussed for physisorbed layers in recent years (Sect. 4.2.8) cannot occur in chemisorption for two reasons: (i) With increasing thickness the strain energy in the layer becomes so large that it is energetically more favourable to form a 3d phase, in addition to a double or triple layer. (ii) Due to the strong interaction with the substrate the electronic structure of the first layer is sometimes so strongly modified that it cannot bond strongly enough to a second layer to stabilize it against formation of a 3d phase (for references, see [4.285]). In these double or triple layers quasi-2d phase transitions can take place such as the one illustrated in Fig. 4.24. Here the misfit wall orientation in a Cu double layer on a Mo(110) surface changes from walls parallel to the [1$\bar{1}$0] direction at low temperature in a first-order transition to

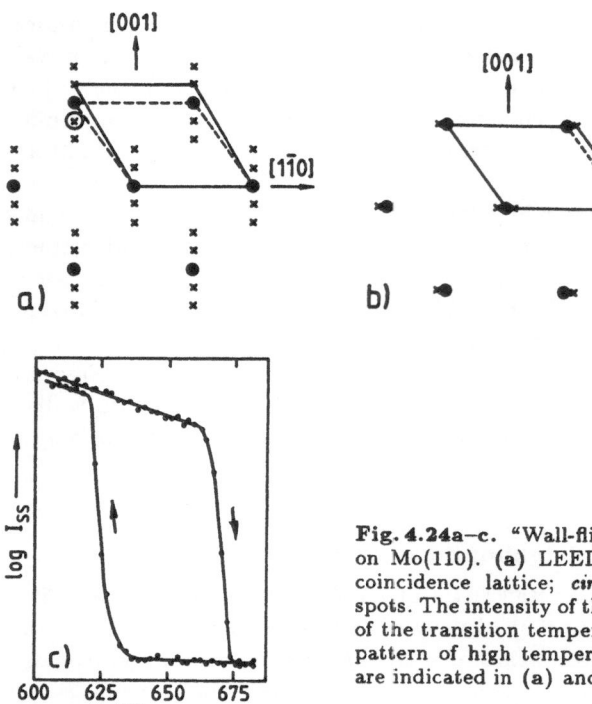

**Fig. 4.24a–c.** "Wall-flip" transition of Cu double layer on Mo(110). (a) LEED pattern of low temperature Cu coincidence lattice; *circles:* substrate, *crosses:* overlayer spots. The intensity of the marked spot in the environment of the transition temperature is shown in (c). (b) LEED pattern of high temperature structure. The unit meshes are indicated in (a) and (b) [4.201]

walls parallel to the [001] direction at high temperatures ("wall-flip") [4.201]. Practically no work has been done up to now in this area. For reviews on metal adsorption see [4.202,284,286,287].

## 4.5 Concluding Remarks and Outlook

There are several aspects of phase transitions on surfaces which had to be excluded from this review for lack of space and time: (i) Dynamical critical phenomena and the kinetics of phase transitions in general such as domain growth or 2d vaporization, although these phenomena are now accessible to experiments as discussed in Sects. 4.3.2,3 and 5. (ii) Interfacial delocalization transitions in three dimensions such as wetting, surface roughening and surface-induced disorder or order transitions. (iii) Phase transitions in physisorption systems which were mentioned only occasionally as motivation for the theory and as a link between theory and chemisorption.

In spite of this limitation the review could not treat any of the subjects in depth so that its aims had to be modest: (i) to give the experimentalist a shorthand introduction into the theory which could not be more than an annotated incomplete dictionary, (ii) to confront the theorist with the possibilities and difficulties of the experimentalist, and (iii) to summarize what has been done already in order to indicate what still needs to be done. The literature references are only the tip of an iceberg and limited in general to the most recent papers whose selection was dictated by availability at the time of writing, the usefulness for the aim of the review and last but not least by the understandability to the author. Hardly any of the pioneering work has been referenced, but this can be found in the references quoted.

The result of the last aim is rather evident: the field is still in its infancy on the experimental side. The chances that it may grow rapidly are good now due to the development of powerful experimental techniques such as microscopy and diffraction techniques with transfer widths in the 100 nm range (Sects. 4.3.2 and 5). If the problems of specimen preparation can be solved and the coordination between theory and experiment can be improved then considerable progress should be made in the next few years. Chemisorption systems may turn out to be models for critical phenomena superior to physisorption systems, in particular as it is becoming increasingly clear that physisorption systems do not mimic a structureless substrate either.

*Acknowledgements and Apologies.* The author wishes to thank L.D. Roelofs for valuable suggestions and for polishing off the worst theoretical corners; K. Binder, D.P. Landau, W. Selke and J. Villain for eliminating some of the remaining flaws and Mrs. B. Schmidt for her dedicated and excellent typing and error search. Particular thanks are due to D.S.Y. Tong and R.W. Vanselow for their hospitality at the Laboratory for Surface Studies, University of Wisconsin at Milwaukee, during the 1985 E.W. Müller lectureship in which the review was started. The author wishes to apologize to his "students" in these lectures who shared his suffering during his struggle through the jungle of theoretical literature. Apologies are also due to the many authors whose work was not mentioned or not presented in the right perspective.

# References

4.1   H.E. Stanley: *Introduction to Phase Transitions and Critical Phenomena* (Clarendon, Oxford 1971)

4.2   S.K. Ma: *Modern Theory of Critical Phenomena* (Benjamin, Reading, Mass. 1976)

4.3   P. Pfeuty, G. Toulouse: *Introduction to the Renormalization Group and Critical Phenomena* (Wiley, New York 1977)

4.4   A.Z. Patashinskii, V.L. Pokrovskii: *Fluctuation Theory of Phase Transitions* (Pergamon, Oxford 1979)

4.5   C. Domb, M.S. Green (eds.): *Phase Transitions and Critical Phenomena*, Vols. 1–9 (Academic, New York 1972–1984)

4.6   J.G. Dash, J. Ruvalds (eds.): *Phase Transitions in Surface Films* (Plenum, New York 1980)

4.7   T. Riste (ed.): *Ordering in Strongly Fluctuating Condensed Matter Systems* (Plenum, New York 1980)

4.8   K. Sinha (ed.): *Ordering in Two Dimensions* (North Holland, New York 1980)

4.9   W. Gebhardt, U. Krey: *Phasenübergänge und kritische Phänomene* (Vieweg, Braunschweig 1980)

4.10  A.D. Bruce, R.A. Cowley: *Structural Phase Transitions* (Taylor and Francis, London 1981)

4.11  M. Lévy, J.-C. LeGuillon, J. Zinn-Justin: *Phase Transitions Cargèse 1980* (Plenum, New York 1982)

4.12  R.J. Baxter: *Exactly Solved Models in Statistical Mechanics* (Academic, London 1982)

4.13    F.W. Hahne (ed.): *Critical Phenomena*, Lect. Notes Phys., Vol. 186 (Springer, Berlin, Heidelberg 1983)
4.14    R. Pynn, A. Skjeltorp (eds.): *Multicritical Phenomena*, NATO ASI Series B Physics, Vol. 106 (Plenum, New York 1984)
4.15    A. Pekalski, J. Sznajd (eds.): *Static Critical Phenomena in Inhomogeneous Systems*, Lect. Notes Phys., Vol. 206 (Springer, Berlin, Heidelberg 1984)
4.16    E.G.D. Cohen (ed.): *Fundamental Problems in Statistical Mechanics, Vol. 6* (Elsevier, New York 1985)
4.17    J.G. Dash: *Films on Solid Surfaces* (Academic, New York 1975)
4.18    M. Bienfait: "Two-dimensional phase transitions of simple molecules adsorbed on graphite", in Current Topics in Materials Science **4**, 361 (North Holland, Amsterdam 1980)
4.19    A.Thomy, X. Duval, J. Regnier: Surf. Sci. Rpt. **1**, 1 (1981)
4.20    S.C. Fain, jr: "Low Energy Electron Diffraction Studies on Physically Adsorbed Films", in *Chemistry and Physics of Solid Surfaces IV*, ed. by R. Vanselow, R. Howe, Springer Ser. Chem. Phys., Vol. 20 (Springer, Berlin, Heidelberg 1982) p.203
4.21    R. Marx: Phys. Rpts. **125**, 1 (1985); Phys. Bl. **42**, 97 (1986)
4.22    E. Bauer: "Chemisorbed Phases" in [4.6]
4.23    L.D. Roelofs: Appl. Surf. Sci. **11/12**, 425 (1982)
4.24    T.L. Einstein: "Critical Phenomena of Chemisorbed Overlayers", in *Chemistry and Physics of Solid Surfaces IV*, ed. by R. Vanselow, R. Howe, Springer Ser. Chem. Phys., Vol. 20 (Springer, Berlin, Heidelberg 1982) p. 251
4.25    L.D. Roelofs, P.J. Estrup: Surf. Sci. **125**, 51 (1983)
4.26    W.H. Weinberg: Ann. Rev. Phys. Chem. **34**, 217 (1983)
4.27    S. Ostlund: Phys. Rev. B**23**, 2235 (1981)
4.28    E. Tosatti: "Displacive Reconstruction Phase Transitions of Clean Transition Metal and Semiconductor Surfaces", in *Modern Trends in the Theory of Condensed Matter*, ed. by A. Pekalski, J. Przystawa, Lect. Notes Phys., Vol. 115 (Springer, Berlin, Heidelberg 1980) p. 501
4.29    A.Fasolino, G. Santoro, E. Tosatti: Surf. Sci. **125**, 317 (1983)
4.30    W. Goldammer, W. Ludwig, W. Zierau, C. Falter: Surf. Sci. **141**, 139 (1984)
4.30a   J.E. Inglesfield: Progr. Surf. Sci. **20**, 105 (1985)
4.31    J. Ihm, D.H. Lee, J.D. Joannopoulos, J.J. Xiong: Phys. Rev. Lett. **51**, 1872 (1983)
4.32    A. Saxena, E.T. Gawlinski, J.D. Gunton: Surf. Sci. **160**, 618 (1985)
4.33    R.F. Willis: "Surface Reconstruction Phase Transformations", in *Dynamical Phenomena in Surfaces, Interfaces and Superlattices*, ed. by T. Nizzoli, K.H. Rieder, R.F. Willis, Springer Ser. Surf. Sci., Vol. 3 (Springer, Berlin, Heidelberg 1985) p. 126
4.34    T.L. Einstein: "Theory of Indirect Interaction between Chemisorbed Atoms", in *Chemistry and Physics of Solid Surfaces*, Vol. 2, ed. by R. Vanselow (CRC, Boca Raton, Fl. 1979) p. 181
4.35    F.Y. Wu: Rev. Mod. Phys. **54**, 235 (1982); J. Appl. Phys. **55**, 2421 (1984)
4.36    E. Domany: "Critical Phenomena in Two Dimensions: Theoretical Models and Physical Realizations", in *Interfacial Aspects of Phase Transformations*, ed. by B. Mutaftschiev (Reidel, Dordrecht 1982) p. 119
4.37    M. Blume, V.J. Emery, R.B. Griffiths: Phys. Rev. A**4**, 1071 (1971)
4.38    A.N. Berker, M. Wortis: Phys. Rev. B**14**, 4946 (1976)
4.39    S. Ostlund: Phys. Rev. B**24**, 398 (1981)
4.40    D.A. Huse: Phys. Rev. B**24**, 5180 (1981)
4.41    W. Selke, J.M. Yeomans: Z. Physik B**46**, 311 (1982)
4.42    M. Kardar, A.N. Berker: Phys. Rev. Lett. **48**, 1552 (1982)
4.43    R.G. Caflisch, A.N. Berker, M. Kardar: Phys. Rev. B**31**, 4527 (1985)
4.44    D.A. Huse, M.E. Fisher: Phys. Rev. B**29**, 239 (1984)
4.45    D.A. Huse: Phys. Rev. B**29**, 5031 (1984)
4.46    P. Bak: "Phase Transitions on Surfaces", in *Chemistry and Physics of Solid Surfaces V*, ed. by R. Vanselow, R. Howe, Springer Ser. Chem. Phys., Vol. 35 (Springer, Berlin, Heidelberg 1984) p. 317
4.47    R.M. Hornreich, R. Liebmann, H.G. Schuster, W. Selke: Z. Physik B**35**, 91 (1979)
4.48    M.E. Fisher, W. Selke: Phys. Rev. Lett. **44**, 1502 (1980)
4.49    W. Selke, M.E. Fisher: Z. Physik B**40**, 71 (1980)

4.50  J. Villain, P. Bak: J. Physique **42**, 657 (1981)
4.51  W. Selke, K. Binder, W. Kinzel: Surf. Sci. **125**, 74 (1983)
4.52  R. Rujan, W.Selke, G.V. Uimin: Z. Physik B**53**, 221 (1983)
4.53  E.Domany, B. Schaub: Phys. Rev. B**29**, 4095 (1984)
4.54  A. Aharony, M.E. Fisher: Phys. Rev. B**27**, 4394 (1983)
4.55  M. Suzuki: Progr. Theor. Phys. **51**, 1992 (1974)
4.56  K. Binder (ed.): *Monte Carlo Methods in Statistical Physics*, 2nd ed., Topics Curr. Phys., Vol. 7 (Springer, Berlin, Heidelberg 1986)
4.57  L.D. Roelofs: "Monte Carlo Simulations of Chemisorbed Overlayers", in *Chemistry and Physics of Solid Surfaces IV*, ed. by R. Vanselow, R. Howe, Springer Ser. Chem. Phys., Vol. 20 (Springer, Berlin, Heidelberg 1982) p. 219
4.58  K. Binder (ed.): *Applications of the·Monte Carlo Method in Statistical Physics*, 2nd ed., Topics Curr. Phys., Vol. 36 (Springer, Berlin, Heidelberg 1984)
4.59  D.J. Amit: *Field Theory, the Renormalization Group and Critical Phenomena* (McGraw Hill, New York 1978)
4.60  T.W. Burkhardt, J.M.J. Van Leeuwen (eds.): *Real Space Renormalization*, Topics Curr. Phys., Vol. 30 (Springer, Berlin, Heidelberg 1982)
4.61  P. Nightingale: J. Appl. Phys. **53**, 7927 (1982)
4.62  M. Barber: "Finite Size Scaling" in [Ref. 4.5, Vol. 8 (1983) p. 146]
4.63  H.W.J. Blöte, M.P. Nightingale: Physica **112A**, 405 (1982)
4.64  M. den Nijs: Phys. Rev. B**27**, 1674 (1983)
4.65  B. Nienhuis: J. Phys. A**15**, 199 (1982)
4.66  F.H. Ree, D.A. Chestnut: J. Chem. Phys. **45**, 3983 (1966)
4.67  R.J. Baxter: J. Phys. A**13**, L 61 (1980)
4.68  N.C. Bartelt, T.L. Einstein: Phys. Rev. B**30**, 5339 (1984)
4.69  M. Schick: Progr. Surf. Sci. **11**, 245 (1981)
4.70  C. Rottmann: Phys. Rev. B**24**, 1482 (1981)
4.71  A.D. Bruce, R.A. Cowley: *Structural Phase Transitions* (Taylor and Francis, London 1981)
4.72  S. Deonarine, J.L. Birmann: Phys. Rev. B**27**, 2855 (1983)
4.73  D.A. King: Physica Scripta T**4**, 34 (1983)
4.74  D.J. Bergman, B.I. Halperin: Phys. Rev. B**13**, 2145 (1976)
4.75  P. Bak: Rpts. Progr. Phys. **45**, 587 (1982)
4.76  S. Howes, L.P. Kadanoff, M. den Nijs: Nuclear Physics B**215**, FS7, 169 (1983)
4.77  S.F. Howes: Phys. Rev. B**27**, 1762 (1983)
4.78  J. Villain, M.B. Gordon: Surf. Sci. **125**, 1 (1983)
4.79  M.E. Fisher: "Walks, Walls and Ordering in Low Dimensions", in [Ref. 4.16, p. 1]; J. Statist. Phys. **34**, 667 (1984)
4.80  H.J. Schulz, B.I. Halperin, C.L. Henley: Phys. Rev. B**26**, 3797 (1982) and references therein
4.81  M.H. Jensen, P. Bak: Phys. Rev. B**29**, 6280 (1984)
4.82  F.D.M. Haldane, J. Villain: J. Physique **42**, 1673 (1981)
4.83  A.L. Talapov: Sov. Phys. JETP **56**, 241 (1982)
4.84  F.F. Abraham, W.R. Rudge, D.J. Auerbach, S.W. Koch: Phys. Rev. Lett. **52**, 445 (1984); for more details see [4.82,83]
4.85  S.W. Koch, W.E. Rudge, F.F. Abraham: Surf. Sci. **145**, 329 (1984)
4.86  M. Schöbinger, F.F. Abraham: Phys. Rev. B**31**, 4590 (1985)
4.87  E.D. Specht, M. Sutton, R.J. Birgeneau, D.E. Moncton, P.M. Horn: Phys. Rev. B**30**, 1589 (1984)
4.88  M. Kardar, R. Shankar: Phys. Rev. B**31**, 1525 (1985)
4.89  A.D. Novaco, J.P. McTague: Phys. Rev. Lett. **38**, 1286 (1977): J. Physique **38**, C4 − 116 (1977)
4.90  J.P. McTague, A.D. Novaco: Phys. Rev. B**19**, 5299 (1979)
4.91  H. Shiba: J. Phys. Soc. Japan **46**, 1852 (1979); **48**, 211 (1980)
4.92  C.R. Fuselier, J.R. Raich, N.S. Gillis: Surf. Sci. **92**, 667 (1980)
4.93  M.B. Gordon, J. Villain: J. Phys. C**15**, 1817 (1982)
4.94  K.L. D'Amico, D.E. Moncton, E.D. Specht, R.J. Birgeneau, S.E. Nagler, P.M. Horn: Phys. Rev. Lett. **53**, 2250 (1984)
4.95  A.J. Dahm: Phys. Rev. B**29**, 484 (1984)

4.96   Y. Saito: Phys. Rev. Lett. **48**, 1114 (1982): Phys. Rev. B**26**, 6239 (1982); Surf. Sci. **125**, 285 (1983)
4.97   S.T. Chui: Phys. Rev. B**28**, 178 (1983)
4.98   H. Müller-Krumbhaar, Y. Saito: Surf. Sci. **144**, 84 (1984)
4.99   S. Ostlund, B.I. Halperin: Phys. Rev. B**23**, 335 (1981)
4.100  E. Domany, M. Schick, R.H. Swendsen: Phys. Rev. Lett. **52**, 1535 (1984)
4.101  J.E. van Himbergen: Phys. Rev. Lett. **53**, 5 (1984)
4.102  H.J.F. Knops: Phys. Rev. B**30**, 470 (1984)
4.103  J.E. van Himbergen: Solid State Commun. **55**, 289 (1985)
4.104  D.A. Huse: Phys. Rev. B**30**, 3908 (1984)
4.105  M. den Nijs: Phys. Rev. B**31**, 266 (1985)
4.106  M.S.S. Challa, D.P. Landau: Phys. Rev. B**33**, 437 (1986)
4.107  M.Y. Choi, S. Doniach: Phys. Rev. B**31**, 4516 (1985)
4.108  M.Y. Choi, D. Stroud: Phys. Rev. B**32**, 5773 (1985)
4.109  D.H. Lee, J.D. Joannopoulos, J.W. Negele, D.P. Landau: Phys. Rev. B**33**, 350 (1986)
4.110  M.L. Cardy, S. Ostlund: Phys. Rev. B**25**, 6899 (1982)
4.111  B.J. Minschau, R.A. Pelcovits: Phys. Rev. B**32**, 3081 (1985)
4.112  L.M. Sander, J. Hautman: Phys. Rev. B**29**, 2171 (1984)
4.113  K.J. Niskanen, R.B. Griffiths: Phys. Rev. B**32**, 5858 (1985)
4.114  P. Dimon, P.M. Horn, M. Sutton, R.J. Birgeneau, D.E. Moncton: Phys. Rev. B**31**, 437 (1985)
4.115  W. Schommers, P. von Blanckenhagen (eds.): *Structure and Dynamics of Surfaces I*, Topics Curr. Phys., Vol. 41 (Springer, Berlin, Heidelberg 1986)
4.116a F.F. Abraham: Phys. Rev. B**29**, 2606 (1984)
4.116b F.F. Abraham: Phys. Rev. B**28**, 7338 (1983)
4.117  N.J. Colella, R.M. Suter: Phys. Rev. B**34**, 2052 (1986)
4.118  A.D. Migone, Z.R. Li, M.H.W. Chan: Phys. Rev. Lett. **53**, 810 (1984)
4.119  S.B. Hurlbut, J.G. Dash: Phys. Rev. Lett. **53**, 1931 (1984); **55**, 2227 (1985)
4.120  B. Joos, M.S. Duesbery: Phys. Rev. Lett. **55**, 1997 (1985); Phys. Rev. B**33**, 8632 (1986)
4.121  D.B. Abraham: Phys. Rev. Lett. **44**, 1165 (1980)
4.122  J. Dudowicz, J. Stecki: Phys. Rev. B**26**, 383 (1982)
4.123  I. Schmidt, W. Pesch: Z. Physik B**58**, 63 (1984)
4.124  D.A. Huse, M.E. Fisher: Phys. Rev. Lett. **49**, 793 (1982)
4.125  W. Selke, W. Pesch: Z. Physik B**47**, 335 (1982)
4.126  W. Selke, J. Yeomans: J. Phys. A**16**, 2789 (1983)
4.127  W. Selke: Surf. Sci. **144**, 176 (1984)
4.128  W. Selke: "Interfacial adsorption in multi-state models" in [Ref. 4.15, p. 191]
4.129  W. Selke, D.A. Huse, D.M. Kroll: J. Phys. A**17**, 3019 (1984)
4.130  I. Sega, W. Selke, K. Binder: Surf. Sci. **154**, 331 (1985)
4.131  D.M. Kroll, R. Lipowsky, R.K.P. Zia: Phys. Rev. B**32**, 1862 (1985)
4.132  D.B. Abraham, P.M. Duxbury: J. Phys. A**19**, 385 (1986)
4.132a W. Selke: Ber. Bunsenges. Phys. Chem., **90**, 232 (1986)
4.133  A.G. Naumovets: Sov. Sci. Rev. A Phys. **5**, 443 (1984)
4.134  V.K. Medvedev, I.N. Yakovkin: Sov. Phys. Solid State **19**, 1515 (1977)
4.135  M. Kaburagi, J. Kanamori: Jap. J. Appl. Phys. Suppl. 2, Pt. 2, 145 (1974); J. Phys. Soc. Japan **43**, 1686 (1977)
4.136  K. Shinjo, T. Sasada: In *Dynamical Processes and Ordering on Solid Surfaces*, ed. by A. Yoshimori, M. Tsukada, Springer Ser. Solid-State Sci., Vol. 59 (Springer, Berlin, Heidelberg 1985) p. 174; J. Phys. Soc. Japan **54**, 1469 (1985)
4.137  V.L. Pokrovsky, G.V. Uimin: J. Phys. C**11**, 3535 (1978)
4.138  O.M. Braun: Sov. Phys. Solid State **23**, 1626 (1982)
4.139  I.F. Lyuksyutov, V.K. Medvedev, I.N. Yakovkin: Sov. Phys. JETP **53**, 1284 (1981)
4.140  L.A. Bol'shov: Sov. Phys. Solid State **13**, 1404 (1971)
4.141  L.A. Bol'shov, A.P. Napartovich: Sov. Phys. JETP **37**, 713 (1973)
4.142  H.K. Kim, M.H.W. Chan: Phys. Rev. Lett. **53**, 170 (1984)
4.143  J.R. Klein, M.H.W. Chan, M.W. Cole: Surf. Sci. **148**, 200 (1984)
4.144  J. Kolaczkiewicz, E. Bauer: Phys. Rev. Lett. **53**, 458 (1984); Surf. Sci. **151**, 333 (1985)
4.145  S. Jüngst, B. Knuth, F. Hensel: Phys. Rev. Lett. **55**, 2160 (1985)
4.146  E.A. Guggenheim: *Thermodynamics* (North Holland, Amsterdam 1977) p. 138

4.147   H. Gollisch: Surf. Sci. **166**, 87 (1986)
4.148   A. Milchev: J. Chem. Phys. **78**, 1994 (1983)
4.149   A. Milchev, K. Binder: Surf. Sci. **164**, 1 (1985)
4.150   L.C.A. Stoop: Thin Solid Films **103**, 375 (1983)
4.151   A. Milchev, M. Paunov: Surf. Sci. **108**, 25, 38 (1981)
4.152   M. Paunov, A. Milchev: phys. stat. sol. (a) **73**, 339 (1982)
4.153   H. Asada: Surf. Sci. **133**, 279 (1983); **137**, 412 (1984)
4.154   P. Cavalotti: Surf. Sci. **83**, 325 (1979)
4.155   J. Kolaczkiewicz, E. Bauer: Surf. Sci. **144**, 495 (1984)
4.156   J. Lajzerowicz, J. Sivardière: Phys. Rev. A**11**, 2079 (1975)
4.157   E. Helfand, H.L. Frisch, J.L. Lebowitz: J. Chem. Phys. **34**, 1037 (1961)
4.158   D. Henderson: Molec. Phys. **30**, 971 (1975)
4.159   F.H. Ree, W.G. Hoover: J. Chem. Phys. **40**, 939 (1964); **46**, 4181 (1967)
4.160   E. Bauer: unpublished
4.161   D. Henderson: Molec. Phys. **34**, 301 (1977)
4.162   J.G. Briano, E.D. Glandt: J. Chem. Soc. Faraday Trans. II **76**, 812 (1980)
4.163   J.A. Barker: Proc. Roy. Soc. Lond. A**377**, 425 (1981)
4.164   J.A. Barker, D. Henderson, F.F. Abraham: Physica **106A**, 226 (1981)
4.165   R. Tenne, E. Bergmann: Phys. Rev. B**22**, 702 (1980)
4.166   W.C. Marra, P.H. Fuoss, P.E. Eisenberger: Phys. Rev. Lett. **49**, 1169 (1982)
4.167   B. Poelsema, L.K. Verheij, G. Comsa: Phys. Rev. Lett. **51**, 2410 (1983)
4.168   R. Miranda, E.V. Albano, S. Daiser, G. Ertl, K. Wandelt: Phys. Rev. Lett. **51**, 782 (1983); J. Chem. Phys. **80**, 2931 (1984)
4.168a  G. Guillot, M.C. Desjonquères, D. Chauveau, G. Treglia, J. Lecante, D. Spanjaard, Tran Minh Duc: Solid State Commun. **50**, 393 (1984)
4.169   S.F. Alvarado, M. Campagna, F. Ciccacci, H. Hopster: J. Appl. Phys. **53**, 7920 (1982)
4.170   J. Lapujoulade, J. Perreau, A. Kara: Surf. Sci. **129**, 59 (1983); Surf. Sci. **178**, 406 (1986)
4.171   M. den Nijs, E.K. Riedel, E.H. Conrad, T. Engel: Phys. Rev. Lett. **55**, 1689 (1985); **57**, 1279 (1986)
4.172   E.H. Conrad, R.M. Aten, D.S. Kaufman, L.R. Allen, T. Engel, M. den Nijs, E.K. Riedel: J. Chem. Phys. **84**, 1015 (1986)
4.173   M.G. Lagally: Appl. Surf. Sci. **13**, 260 (1982); "Diffraction Techniques", in *Methods of Experimental Physics. Surfaces*, Vol. 22, ed. by R.L. Park, M.G. Lagally (Academic, Orlando 1985) p. 237
4.174   D.P. Woodruff, G.-C. Wang, T.-M. Lu: "Surface Structure and Order-Disorder Phenomena", in *The Chemical Physics of Solid Surfaces and Heterogeneous Catalysis*, Vol. 2, ed. by D.A. King, D.P. Woodruff (Elsevier, Amsterdam 1983) p. 259
4.175   M. Henzler: Appl. Surf. Sci. **11/12**, 450 (1982); Appl. Phys. A**34**, 205 (1984); "Defects at Surfaces", in *Dynamical Phenomena at Surfaces, Interfaces and Superlattices*, ed. by F. Nizzoli, K.H. Rieder, R.F. Willis, Springer Ser. Surf. Sci., Vol. 3 (Springer, Berlin, Heidelberg 1985) p. 14
4.176   W. Moritz, M.G. Lagally: Phys. Rev. Lett. **56**, 865, 2882 (1986)
4.177   N.C. Bartelt, T.L. Einstein, L.D. Roelofs: Phys. Rev. Lett. **56**, 2881 (1986)
4.178   G. Meissner, K. Binder: Phys. Rev. B**12**, 3948 (1975)
4.179   T. Schneider, E. Stoll: Phys. Rev. B**13**, 1216 (1976)
4.180   K. Binder, G. Meissner, H. Mais: Phys. Rev. B**13**, 4890 (1976)
4.181   M.B. Webb, M.G. Lagally: Solid State Phys. **28**, 305 (1973)
4.182   H. Froitzheim, U. Köhler, H. Lammering: Phys. Rev. B**30**, 5771 (1984)
4.183   D.K. Saldin, J.B. Pendry, M.A. van Hove, G.A. Somorjai: Phys. Rev. B**31**, 1216 (1985)
4.184   W. Telieps, E. Bauer: Ultramicroscopy **17**, 51 (1985)
4.185   P.M. Horn, R.J. Birgeneau, P. Heiney, E.M. Hammonds: Phys. Rev. Lett. **41**, 961 (1978)
4.186   P. Kleban: "Finite Size Effects, Surface Steps and Phase Transitions", in *Chemistry and Physics of Solid Surfaces* V, ed. by R. Vanselow, R. Howe, Springer Ser. Chem. Phys., Vol. 35 (Springer, Berlin, Heidelberg 1984) p. 339
4.187   K. Binder, D.P. Landau: Phys. Rev. B**30**, 1477 (1984)
4.188   P. Kleban, G. Akinci, R. Hentschke, K.R. Brownstein: J. Phys. A**19**, 437 (1986); Surf. Sci. **166**, 159 (1986)

4.188a G. Grinstein: J. Appl. Phys. **55**, 2371 (1984)

4.189  M.G. Lagally, J.A. Martin: Rev. Sci. Instrum. **54**, 1273 (1984)

4.190  J.A. Martin, M.G. Lagally: Scanning Electron Microscopy 1985 Part III, p. 1357 (SEM Inc., AMF O'Hare, Chicago)
L. Reimer: *Scanning Electron Microscopy*, Springer Ser. Opt. Sci., Vol. 45 (Springer, Berlin, Heidelberg 1985)

4.191  K.D. Gronwald, M. Henzler: Surf. Sci. **117**, 180 (1982)

4.192  U. Scheithauer, G. Meyer, M. Henzler: Surf. Sci. **178**, 441 (1986)

4.193  A.R. Kortan: Ph. D. thesis, Univ. of Maryland (1980)

4.194  P.A. Bennett: Ph. D. thesis, Univ. of Wisconsin, Madison (1980)

4.195  W. Witt: Dr. dissertation, TU Clausthal (1984)

4.196  M.E. Fisher: Phys. Rev. **176**, 257 (1968)

4.197  N.C. Bartelt, T.L. Einstein, L.D. Roelofs: Surf. Sci. **149**, L 47 (1984); Phys. Rev. B**32**, 2993 (1985)

4.198  H. Jónsson, J.H. Weare, A.C. Levi: Phys. Rev. B**30**, 2241 (1984); Surf. Sci. **148**, 126 (1984)

4.199  K. Kern, R. David, R.L. Palmer, G. Comsa: Phys. Rev. Lett. **56**, 620 (1986)

4.200  J.W.M. Frenken, J.K. van der Veen: Phys. Rev. Lett. **54**, 134 (1985); Surf. Sci. **178**, 382 (1986)

4.201  M. Tikhov, M. Stolzenberg, E. Bauer: Surf. Sci. **175**, 508 (1986)

4.202  E. Bauer, J. Kolaczkiewicz: Proc. IXth Intern. Vacuum Congr. and Vth Intern. Conf. Solid Surfaces, Madrid (1983) p. 363

4.203  J. Kolaczkiewicz, E. Bauer: Surf. Sci. **175**, 508 (1986)

4.204  N. Osakabe, Y. Tanishiro, K. Yagi, G. Honjo: Jap. J. Appl. Phys. **19**, L 309 (1980); Surf. Sci. **109**, 353 (1981)

4.205  W. Telieps, E. Bauer: Surf. Sci. **162**, 163 (1985); Ber. Bunsenges. Phys. Chem., **90**, 197 (1986)

4.206  D.E. Taylor, R.L. Park: Surf. Sci. **125**, L 73 (1983)

4.207  P.A. Bennett, M.W. Webb: Surf. Sci. **104**, 74 (1981)

4.208  Y. Tanishiro, K. Takayanagi, K. Yagi: Ultramicroscopy **11**, 95 (1983)

4.209  W. Witt, E. Bauer: Surf. Sci., to be published

4.210  W. Telieps: personal communication

4.211  J. Möller, H. Niehus, W. Heiland: Surf. Sci. **166**, L 111 (1986) and references therein

4.212  J.C. Campuzano, M.S. Foster, G. Jennings, R.F. Willis, W. Unertl: Phys. Rev. Lett. **54**, 2684 (1985)

4.213  S. Pick, M. Tomásek: Czech. J. Phys. B**35**, 183 (1985) and references therein

4.214  K. Masuda-Jindo, N. Hamada, K. Terakura: J. Phys. C**17**, L 271 (1984)

4.215  J.E. Inglesfield, A. Tagliacozzo: J. Phys. C**17**, 5227 (1984)

4.216  H. Krakauer: Phys. Rev. B**30**, 6834 (1984)

4.217  C.L. Fu, A.J. Freeman, E. Wimmer, M. Weinert: Phys. Rev. Lett. **54**, 2261 (1985)

4.218  S.C. Ying, L.D. Roelofs: Surf. Sci. **125**, 128 (1983)

4.219  L.D. Roelofs, G.Y. Hu, S.C. Ying: Phys. Rev. B**28**, 6369 (1983)

4.220  L.D. Roelofs, S.C. Ying: Surf. Sci. **147**, 203 (1984)

4.221  G.Y. Hu, S.C. Ying: Surf. Sci. **150**, 47 (1985)

4.221a L.D. Roelofs, J.F. Wendelken: Phys. Rev. B**34**, 3319 (1986)

4.221b L.D. Roelofs: Phys. Rev. B**34**, 3337 (1986); Surf. Sci. **178**, 396 (1986)

4.222  J.F. Wendelken, G.-C. Wang: Phys. Rev. B**32**, 7542 (1985)

4.223  K. Müller: Ber. Bunsenges. Phys. Chem., **90**, 184 (1986)

4.224  R.J. Phaneuf, M.B. Webb: Surf. Sci. **164**, 167 (1985)

4.225  A. Saxena, T. Ala-Nissilä, J.D. Gunton: Surf. Sci. **169**, L231 (1986)

4.226  D.P. Landau: Phys. Rev. B**27**, 5604 (1983)

4.227  N. Nagai: Surface Sci. **136**, L 14 (1984); Phys. Rev. B**30**, 1461 (1984)

4.228  K. Christmann, R.J. Behm, G. Ertl, M.A. van Hove, W.H. Weinberg: J. Chem. Phys. **70**, 4168 (1979)

4.229  J.-P. Muscat: Progr. Surf. Sci. **18**, 59 (1985); Surf. Sci. **110**, 85 (1981); **105**, 503 (1981); Phys. Rev. B**33** (1986)

4.229a L.D. Roelofs, T.L. Einstein, N.C. Bartelt, J.D. Shore: Surf. Sci. **176**, 295 (1986)

4.230  L.D. Roelofs, A.R. Kortan, T.L. Einstein, R.L. Park: J. Vac. Sci. Technol. **18**, 492 (1981); Phys. Rev. Lett. **46**, 1465 (1981)

4.231 M. Schick: Phys. Rev. Lett. **47**, 1347 (1981)
4.232 M. Auer, H. Leonhard, K. Hayek: Appl. Surf. Sci. **17**, 70 (1983)
4.233 E.D. Williams, W.H. Weinberg, A.C. Sobrero: J. Chem. Phys. **76**, 1150 (1982)
4.234 H. Pfnür, D. Menzel: Surf. Sci. **148**, 141 (1984); Ber. Bunsenges. Phys. Chem. **90**, 272 (1986)
4.235 D.L. Doering, S. Semancik: Phys. Rev. Lett. **53**, 66 (1984)
4.236 K. Binder, W. Kinzel, D.P. Landau: Surf. Sci. **117**, 232 (1982)
4.237 K. Kinzel, W. Selke, K. Binder: Surf. Sci. **121**, 13 (1982)
4.238 W. Kinzel: Phys. Rev. Lett. **51**, 996 (1983)
4.239 K. Kaski, W. Kinzel, J.D. Gunton: Phys. Rev. B**27**, 6777 (1983)
4.240 P.A. Rikvold, K. Kaski, J.D. Gunton, M.C. Yalabik: Phys. Rev. B**29**, 6285 (1984)
4.241 J. Amar, S. Katz, J.D. Gunton: Surf. Sci. **155**, 667 (1985)
4.242 J. Amar, J.D. Gunton: Phys. Rev. B**32**, 7250 (1985)
4.243 J.-P. Muscat: Surf. Sci. **139**, 491 (1984)
4.244 R. Imbihl, R.J. Behm, K. Christmann, G.Ertl, T. Matsushima: Surf. Sci. **117**, 257 (1982)
4.245 W. Moritz, R. Imbihl, R.J.Behm, G.Ertl, T. Matsushima: J. Chem. Phys. **83**, 1959 (1985)
4.246 I.F. Lyuksyutov, A.G. Fedorus: Sov. Phys. JETP **53**, 1317 (1981)
4.247 G.-C. Wang, T.-M. Lu, M.G. Lagally: J. Chem. Phys. **69**, 479 (1978)
4.248 M.G. Lagally, G.-C. Wang, T.-M. Lu: "Chemisorption: Island Formation and Adatom Interactions", in *The Chemistry and Physics of Solid Surfaces*, Vol. 2, ed. by R. Vanselow (CRC, Boca Raton, Fl. 1979) p. 153
4.249 W. Witt, E. Bauer: Ber. Bunsenges. Phys. Chem. **90**, 248 (1986)
4.250 T.T. Tsong: Physica Scripta T**4**, 17 (1983)
4.251 B.A. Boiko, D.A. Gorodetskii, A.A. Yas'ko: Sov. Phys. Solid State **15**, 2101 (1974)
4.252 A. Pavlovska, E. Bauer, H. Poppa: to be published
4.253 Ch. Park, E. Bauer, H.M. Kramer: Surf. Sci. **119**, 251 (1982)
4.254 A.G. Fedorus, V.V. Gonchar: Surf. Sci. **140**, 499 (1984)
4.255 W. Kinzel, M. Schick: Phys. Rev. B**24**, 324 (1981)
4.256 K. Binder, D.P. Landau: Surf. Sci. **108**, 503 (1981)
4.257 P.A. Rikvold, W. Kinzel, J.D. Gunton, K. Kaski: Phys. Rev. B**28**, 2686 (1983)
4.258 R.G. Caflisch, A.N. Berker: Phys. Rev. B**29**, 1279 (1984)
4.259 J. Amar, K. Kaski; J.D. Gunton: Phys. Rev. B**29**, 1462 (1984)
4.260 D.P. Landau, K. Binder: Phys. Rev. B**31**, 5946 (1985)
4.261 I. Harada, K. Takasaki: Jap. J. Appl. Phys. **54**, 2210 (1985)
4.262 R.J. Behm, K. Christmann, G. Ertl: Surf. Sci. **99**, 320 (1980)
4.263 D.E. Taylor, E.D. Williams, R.L.Park, N.C. Bartelt, T.L. Einstein: Phys. Rev. B**32**, 4653 (1985)
4.264 P. Bak, P. Kleban, W.N. Unertl, J. Ochab, G. Akinci, N.C. Bartelt, T.L. Einstein: Phys. Rev. Lett. **54**, 1539 (1985)
4.265 T. Aruga, H. Tochihara, Y. Murata: Phys. Rev. Lett. **52**, 1794 (1984)
4.266 G. Ertl, P. Rau: Surf. Sci. **15**, 443 (1969)
4.267 G. Ertl, J. Küppers: Surf. Sci. **21**, 61 (1970)
4.268 P.H. Kleban: Surf. Sci. **83**, L 335 (1979)
4.269 M. Grunze, P.H. Kleban, W.N. Unertl, F.S. Rys: Phys. Rev. Lett. **51**, 582 (1983)
4.270 V. Penka, K. Christmann, G. Ertl: Surf. Sci. **136**, 307 (1984)
4.271 G.-C. Wang, T.-M. Lu: Phys. Rev. Lett. **50**, 2014 (1983); Phys. Rev. B**28**, 6795 (1983); B**31**, 5918 (1985)
4.272 G.-C. Wang, J.M. Pimbley, T.-M. Lu: Phys. Rev. B**31**, 1950 (1985)
4.273 S. Pick: Czech. J. Phys. B**31**, 1401 (1981)
4.274 I.F. Lyuksyutov: Sov. Phys. JETP **55**, 737 (1982)
4.275 E. Bauer, H. Poppa, Y. Viswanath: Surf. Sci. **58**, 517 (1976)
4.276 E. Bauer, H. Poppa: Surf. Sci. **88**, 31 (1979)
4.277 H.M. Kramer, E. Bauer: Surf. Sci. **92**, 53 (1980); **93**, 704 (1980)
4.278 R.F. Willis: In *Many-Body Phenomena at Surfaces*, ed. by D. Langreth, H. Suhl (Academic, New York 1984) p. 297; Ber. Bunsenges. Phys. Chem. **90**, 190 (1986)
4.279 J.P. Woods, J.L. Erskine: J. Vac. Sci. Technol. A**4**, 1414 (1986)
4.280 J.F. Wendelken, G.-C. Wang: Surf. Sci. **140**, 425 (1984); J. Vac. Sci. Technol. A**2**, 888 (1984)

4.281 K. Griffiths, C. Kendon, D.A. King, J.B. Pendry: Phys. Rev. Lett. **46**, 1548 (1981)
4.282 H.M. Kramer, E. Bauer: Surf. Sci. **107**, 1 (1981)
4.283 C. Park, H.M. Kramer, E. Bauer: Surf. Sci. **115**, 1 (1982); **116**, 456, 467 (1982)
4.284 E. Bauer: "Metals on Metals", in *The Physics and Chemistry of Heterogeneous Catalysis*, Vol. 3a, ed. by D.A. King, D.P. Woodruff (Elsevier, Amsterdam 1984) p. 1
4.285 E. Bauer. J.H. van der Merwe: Phys. Rev. B**33**, 3657 (1986)
4.286 L.A. Bol'shov, A.P. Napartovich, A.G.N. Naumovets, A.G. Fedorus: Sov. Phys. Usp. **20**, 432 (1977)
4.287 E. Bauer: "Chemisorption of Metals on Metals and Semiconductors", in *Interfacial Aspects of Phase Transitions*, ed. by B. Mutaftschiev (Reidel, Dordrecht 1982) p. 411

# Additional References with Titles

Ala-Nissila, T., Amar, J., Gunton, J.D.:Wetting in the two-dimensional ANNNI model. J. Phys. A**19**, L41 (1986). [4.2.8]

Binder, K.: Theory of first-order phase transitions. Repts. Progr. Phys. (in print); Recent progress in the theory of first-order phase transitions: statics and dynamics, in: Recent Advances in Phase Transitions and Disorder Phenomena (World Scientific Publ., Singapore, in print). [4.2]

Bartelt, N.C., Einstein, T.L., Roelofs, L.D.: Structure factors associated with the melting of a $(3 \times 1)$ ordered phase on a centered-rectangular lattice-gas: effective scaling in a three-state chiral clock-like model. Phys. Rev. B**35**, 4812 (1987) [4.2.4, 4.4.2]

Bartelt, N.C., Einstein, T.L., Roelofs, L.D.: Structure factors associated with the continuous melting of two-dimensional lattice gases: models with $(\sqrt{3} \times \sqrt{3})$ R30° and p$(2 \times 2)$ ordered states on triangular nets. Phys. Rev. B**35**, 1776 (1987) [4.2.4, 4.4.2]

Bartelt, N.C., Einstein, T.L., Roelofs, L.D.: Transfer-matrix approach to estimating coverage discontinuities and multi-critical-point positions in two-dimensional lattice-gas phase diagrams. Phys. Rev. B**34**, 1616 (1986). [4.2.4]

Berge, B., Diep, H.T., Ghazali, A., Lallemand, P.: Phase transitions in two-dimensional uniformly frustrated XY spin systems. Phys. Rev. B**34**, 3177 (1986). [4.2.7]

Brandt, U., Stolze, J.: Ground states of the triangular Ising model with two- and three-spin interactions. Z. Phys. B**64**, 481 (1986). [4.2.4]

Challa, M.S.S., Landau, D.P., Binder, K.: Finite-size effects at temperature-driven first-order transitions. Phys. Rev. B**34**, 1841 (1986). [4.2]

Christmann, K.: Phase transitions in chemisorbed hydrogen layers. Ber. Bunsenges. Phys. Chem. **90**, 307 (1986). [4.4.2]

Chudnovsky, E.M.: Structure of a solid film on an imperfect surface. Phys. Rev. B**33**, 245 (1986). [4.2.7]

Fisher, M.E., Milton, G.W.: Classifying first-order phase transitions. Physica **138A**, 22 (1986). [4.2]

Francis, S.M., Richardson, N.V.: Observation of an order-disorder phase transition on the Pd(110) surface. Phys. Rev. B**33**, 662 (1986). [4.4.1]

Glaus, U.: Correlations in the two-dimensional random-field Ising model. Phys. Rev. B**34**, 3203 (1986). [4.2.4]

Grest, G.S., Soukoulis, C.M., Levin, K.: Comparative Monte Carlo and mean-field studies of random-field Ising systems. Phys. Rev. B**33**, 7659 (1986). [4.2.4]

van Himbergen, J.E.: A new phase transition scenario for chiral and algebraic order in a generalized planar model. Phys. Rev. B**34**, 6567 (1986). [4.2.4]

Houlrik, J.M., Knak Jensen, S.J.: Phase diagram of the three-state chiral clock model studied by Monte Carlo renormalization-group calculations. Phys. Rev. B**34**, 325 (1986). [4.2.4]

Indekeu, J.O., Nightingale, M.P., Wang, W.V.: Finite-size interaction amplitudes and their universality: Exact, mean-field, and renormalization-group results. Phys. Rev. B**34**, 330 (1986). [4.2.4]

Jacobs, A.E.: Intrinsic domain-wall pinning and spatial chaos in continuum models of one-dimensionally incommensurate systems. Phys. Rev. B**33**, 6340 (1986). [4.2.6]

Janke, W., Kleinert, H.: First-order transition in a two-dimensional Laplacian roughening model on a square lattice. Phys. Letters **114A**, 255 (1986). [4.2.7]

Kleban, P., Hentschke, R.: Effects of finite geometry and boundary conditions on scattering functions at two-dimensional critical points. Phys. Rev. B34, 1980 (1986). [4.3.1]

Landau, D.P., Swendsen, R.H.: Monte Carlo renormalization-group study of tricritical behavior in two dimensions. Phys. Rev. B33, 7700 (1986). [4.2.4]

Lindgård, P.-A., Mouritsen, O.G.: Theory and model for martensitic transformations. Phys. Rev. Lett. 57, 2458 (1986). [4.2.5]

Lipowsky, R.: Surface critical phenomena at first-order phase transitions. Ferroelectrics/Bulletin 1, 13 (1986). [4.2]

Lyuksyutov, I.F.: Tetragonal-monoclinic phase transition in two-dimensional crystals. Sov. Phys. JETP 62, 615 (1985). [4.2.6]

Mahanti, S.D., Tang, S.: Phase transitions in diatomic molecular monolayers. Superlattices and Microstructures 1, 517 (1985). [4.2]

Mazzucchelli, G.M, Zeyher, R.: Mean field phase diagram of the discrete Frenkel-Kontorova model. Z. Phys. B62, 367 (1986). [4.2.6]

Milchev, A.: Frenkel-Kontorova model with anharmonic interactions. Phys. Rev. B33, 2062 (1986). [4.2.6]

Milchev, A., Heermann, D.W., Binder, K.: Finite-size scaling analysis of the $\Phi^4$ field theory on the square lattice. J. Stat. Phys. 44, 749 (1986). [4.2]

Niskanen, K.J.: Phase diagrams for submonolayer films on weakly corrugated substrates. Phys. Rev. B33, 1830 (1986). [4.2]

Pearce, P.A.: Multicritical scaling in the magnetic hard-square lattice gas. J. Phys. A20, 447 (1987). [4.2.4]

Pokrovsky, V.L., Talapov, A.L.: Theory of Incommensurate Crystals (Harwood Academic Publ., London, 1984). [4.2.6]

Reiter, G., Moss, S.C.: X-ray scattering from a two-dimensional liquid modulated by its periodic host. Phys. Rev. B33, 7209 (1986) [4.3.1]

Roelofs, L.D., Kriebel, D.. The phase diagram of repulsive dipoles in 2-d with application to Na/W(110). J. Phys. C, in print. [4.2.9]

Rujan, P., Selke, W., Uimin, G.: Wetting phenomena in the two-dimensional ANNNI model in a field. Z. Phys. B65, 235 (1986). [4.2.8]

Saito, Y., Tabe, G.: $\sqrt{3} \times \sqrt{3}$ structure on the triangular lattice. J. Phys. Soc. Jpn. 54, 2955 (1985). [4.2.4]

Selke, W., Szpilka, A.M.: Monte Carlo study of a model for the roughening transition of high-index crystal faces. Z. Phys. B62, 381 (1986). [4.3.1]

Selke, W., Wu, F.Y.: Potts models with competing interactions. J. Phys. A 20, 703 (1987). [4.2.4]

Serota, R.A.: Reentrant melting on an imperfect surface. Phys. Rev. B33, 3403 (1986). [4.2.7]

Speth, W.: Domains induced by elasticity. Z. Phys. B63, 389 (1986). [4.2]

Strandburg, K.J.: Crossover from a hexatic phase to a single first-order transition in a Laplacian-roughening model for two-dimensional melting. Phys. Rev. B34, 3536 (1986). [4.2.7]

Tang, S., Mahanti, S.D.: Vortices and strings: phase transition in anisotropic planar-rotor systems. Phys. Rev. B33, 3419 (1986). [4.2.7]

Uimin, G., Rujan, P.: Wetting in two-dimensional systems with axially competing interactions. Phys. Rev. B34, 3551 (1986). [4.2.8]

Vilfan, I., Galam, S.: Multicritical properties of uniaxial Heisenberg antiferromagnets. Phys. Rev. B34, 6428 (1986). [4.2.4]

Villain, J.: On the ground state of Ising models. Ferroelectrics 66, 143 (1986). [4.2.4]

Villain. J., Grempel, D.R., Lapujoulade, J.: Roughening transition of high-index crystal faces: the case of copper. J. Phys. F15, 809 (1985). [4.3.1]

Wegner, F.: Phasenübergänge und Renormierung. Phys. Bl. 42, 185 (1986). [4.2]

# 5. Solid and Liquid Surfaces Studied by Synchrotron X-Ray Diffraction

J. Als-Nielsen

With 32 Figures

The structure of matter on an atomic length scale is often derived from the diffraction pattern of an incident radiation beam. The beam can be of thermal neutrons, x-ray photons, low energy electrons, neutral He atoms etc., all with wavelengths in the Angstrom range. The former two diffraction probes have the fundamental advantage of having only a weak scattering interaction, so the diffraction pattern can be modelled by simple superposition of spherical waves scattered from each atom, whereas the latter two diffraction probes have stronger and more complicated scattering interactions with a single atom making interpretation of the diffraction pattern in terms of a structure model much more complicated and ambiguous.

The subject of this chapter is *surface* structures implying that the number of diffracting atoms is relatively small and therefore requiring a correspondingly stronger intensity of the incident beam. This has been a serious drawback for the use of neutrons in surface structure work and until recently also for the use of x-ray photons as generated in conventional x-ray tubes. However, with the development of producing very intense and highly collimated x-ray beams, so-called synchrotron radiation, from electron storage rings operating at electron energies in the GeV region new possibilities, including x-ray surface diffraction, have appeared in x-ray physics.

Synchrotron radiation (SR) is generated by the centripetal acceleration of the electron when it passes a bending magnet in the storage ring. As the electron energy $\gamma$ in units of the rest mass energy is typically of the order $10^4$, the emitted radiation is confined to a narrow cone around the instantaneous electron velocity with an opening angle of approximately $\gamma^{-1}$. The wavelength spectrum of SR is continuous with a characteristic wavelength $\lambda_c$ in the Angstrom range. For further details on the properties of SR, see [5.1].

A monochromatic beam may be extracted from the continuous spectrum by Bragg reflection from a single perfect crystal of Si or Ge. The wavelength band is determined by the narrow angular divergence $\gamma^{-1}$ and the small Darwin width of the perfect crystal reflection. The resulting x-ray beam to be diffracted by the surface structure is thus very intense, highly collimated and highly monochromatic allowing very sharp diffraction patterns and concomitantly an effective discrimination between sharp but rather weak Bragg scattering from the surface layer and the inevitable diffuse scattering from the irradiated volume of the bulk.

It is convenient to distinguish between structure, or periodicity, normal to the surface and within the surface plane as indicated in Fig. 5.1. The corresponding diffraction has then a wave-vector transfer between the wave vectors of the incident and scattered beams, which is mainly along the surface normal or within the surface, respectively.

In the first case, depicted in the top part of Fig. 5.1, it is then a semantic question whether one talks about scattering or reflection of the incident beam: The surface normal, the incident and the scattered wave vectors are all in the same plane, with the incident glancing angle $\theta$ being equal to the exit glancing angle. This geometry of rays is denoted specular reflection. In Sects. 5.1–3 we discuss how the reflectivity is related to the variation of density across the surface and review some recent experiments of this kind on liquid surfaces. One deals with the determination of thermal roughness of simple liquids, the others are about the layered structures formed at the surface in liquid crystal materials.

In-plane structures, depicted in the lower part of Fig. 5.1, are discussed in Sect. 5.4. Since the incident and scattered beam must be on the same side of the surface of the sample, assumed to have an infinite extent for $z < 0$, there is a small out-of plane component of the wave vector transfer $Q$, but the main part of $Q$ is within the surface plane. The two-dimensional structure, indicated as a checker-board in Fig. 5.1, is obtained by varying $\omega$ and $2\theta$. This particular geometry is called grazing incidence diffraction. It has interesting possibilities in obtaining particular surface sensitivity of the rather penetrating x-ray beam by letting $\alpha_i$ and $\alpha_f$ be close to the critical angle for total reflection as will be discussed. In Sect. 5.4.3 we shall be concerned with the reconstructed surface

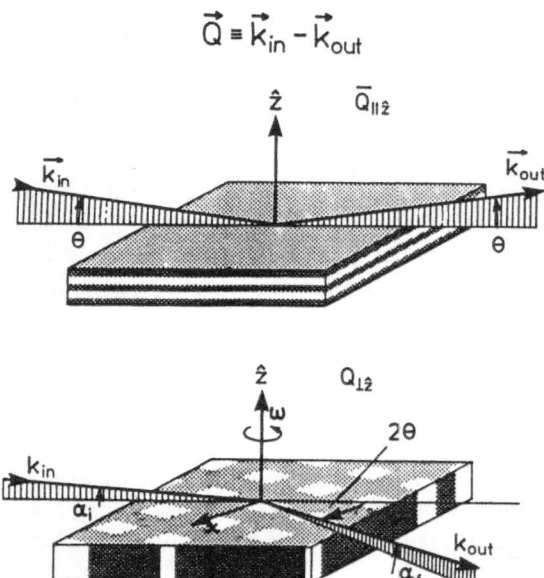

$$\vec{Q} \equiv \vec{k}_{in} - \vec{k}_{out}$$

**Fig. 5.1.** *Top:* Specular reflection measures the density variation across the surface. *Bottom:* Grazing incidence diffraction measures the structure within the surface plane

structure of a bulk, single crystal. The diffraction principles for overlayers on a bulk crystal are similar, and we have chosen to give a tutorial presentation of one single example of a reconstructed surface determination rather than to review the numerous studies of this kind which have been carried out recently. Reference to these studies is given in tabular form in Sect. 5.4.4.

# 5.1 Reflectivity and Density Profile

## 5.1.1 Master Formula

In Fig. 5.2 we consider the specular reflection of monochromatic x-rays incident on the interface at glancing angle $\theta$. Specular reflection means that the reflected ray is in the plane spanned by the incident wave-vector $k_{in}$ and the surface unit normal, $\hat{n}$, and that the reflected ray wave-vector $k_{out}$ is also at the angle $\theta$ with respect to the surface. It turns out to be convenient to use the wave-vector transfer $Q \equiv k_{out} - k_{in}$ as the independent variable rather than the glancing angle $\theta$. The condition of specular reflection means that $Q = 2k \sin \theta \, \hat{n}$.

In the right part of Fig. 5.2 is sketched the density profile across the interface (Chap. 6). The density $\varrho(z)$ has been normalized to unity deep in the liquid (or substrate) and thus is dimensionless. We seek the relation between the reflected intensity I(Q) and $\varrho(z)$.

Had $\varrho(z)$ been a step function, that is zero for z<0 and unity for z>1, the reflected intensity would follow the Fresnel law $I_F(Q)$ which we discuss in the following section. Here we state the master formula relating I(Q) and $\varrho(z)$ :

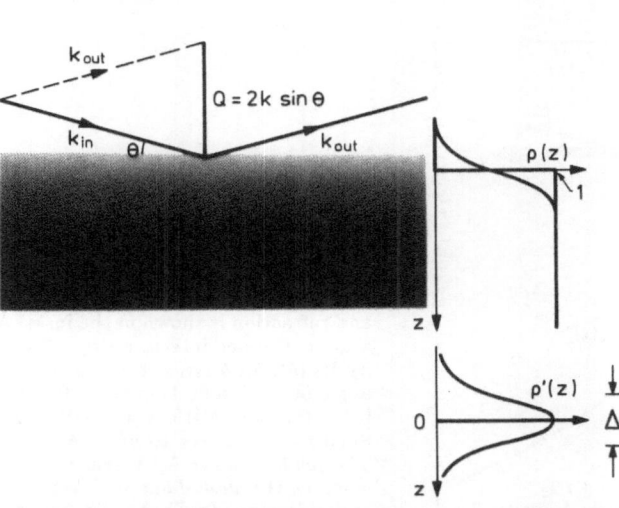

**Fig. 5.2.** The density variation across a surface is indicated by shading in the *left part* and more quantitatively by the function $\varrho(z)$ in the *right part*. The reflectivity versus wave-vector transfer Q is related to the Fourier transform of the gradient of the density, $\varrho'(z)$

$$\frac{I(Q)}{I_F(Q)} = \left| \int \varrho'(z)\exp[iQz]dz \right|^2 \tag{5.1}$$

where $\varrho'(z)$ denotes the gradient of $\varrho(z)$. The gradient $\varrho'(z)$ is also sketched in Fig. 5.2; it has a typical width which we denote $\Delta$. Equation (5.1) is certainly a plausible result. If $\varrho(z)$ is a step function, the gradient $\varrho'(z)$ is a delta function and the Fourier transform is unity, i.e. $I(Q) = I_F(Q)$ as required. After discussing $I_F(Q)$ below we shall give a heuristic derivation of (5.1).

### 5.1.2 Fresnel Reflectivity

In the inset of Fig. 5.3 is shown the usual optical refraction where an electromagnetic wave is incident at glancing angle $\theta$ on a sharp interface where the index of refraction changes from unity to n. For x-rays, n is less than unity,

$$n = 1 - \lambda^2 \varrho_{el} r_0/(2\pi) \ . \tag{5.2}$$

Here $\lambda$ is the x-ray wavelength, $\varrho_{el}$ is the electron density and $r_0$ the classical radius of the electron $e^2/mc^2 = 2.82 \times 10^{-13}$ cm. The deviation of n from unity is of the order $(10^{-8})^2 10^{24} \times 10^{-13} \sim 10^{-5} \ll 1$ and Snell's law relating $\theta$ and $\theta'$ is accurately given by expanding $\cos \theta$ and $\cos \theta'$ to second order:

$$\theta^2 = \theta'^2 + \theta_c^2 \ , \tag{5.3}$$

$$\theta_c^2 = \lambda^2 \varrho_{el} r_0/\pi \ . \tag{5.4}$$

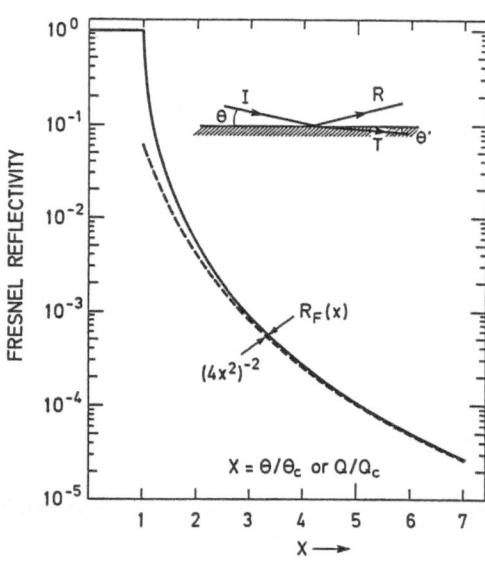

Fig. 5.3. The geometry of reflection and refraction is shown in the inset. At a discontinuous interface the reflectivity $R_F(\theta)$, for $\theta$ larger than the critical angle of total reflection $\theta_c$, is $R_F(\theta) = \left| (\theta - \theta')/(\theta + \theta') \right|^2$. For small angles Snell's law reduces to $\theta^2 = \theta'^2 + \theta_c^2$. The *full line* shows $R_F$ versus $x = \theta/\theta_c$ whereas the *dashed line* represents the approximation $(4x^2)^{-2}$ valid for $x \gg 1$

Here $\theta_c$ denotes the critical angle for total reflection: when $\theta < \theta_c$ the incident beam is totally reflected from the interface. For $\theta > \theta_c$ the standard derivation from optics [5.2] leads to the Fresnel law

$$R_F(\theta) = [(\theta - \theta')/(\theta + \theta')]^2 \ , \quad \theta > \theta_c \ . \tag{5.5}$$

The Fresnel reflectivity is shown in a semi-log plot in Fig. 5.3 versus $\theta/\theta_c$. When $\theta \gg \theta_c$, (5.5) is readily expanded to give the even simpler result

$$R_F(\theta) \simeq (2\theta/\theta_c)^{-4} = (2Q/Q_c)^{-4} \ , \quad \theta \gg \theta_c \tag{5.5a}$$

which is also shown in Fig. 5.3 as the dashed curve.

As mentioned above the wave-vector transfer $Q = 2k \sin \theta = (4\pi/\lambda) \sin \theta$ is a more convenient independent variable than the incident angle $\theta$. The wave-vector transfer corresponding to total reflection is

$$Q_c \equiv 2k \sin \theta_c \simeq 4(\varrho_{el} r_0 \pi)^{1/2} \ . \tag{5.6}$$

In the small angle approximation the scales $\theta/\theta_c$ and $Q/Q_c$ are identical.

In the formula for $R_F(\theta)$, (5.5), we have neglected absorption of the refracted ray. The effect of absorption, which can easily be incorporated, is a rounding of the kink at $\theta = \theta_c$. In practice, the rounding effect is negligible for $\theta \geq 1.2\theta_c$ and, as we shall mainly be concerned with the region $\theta \gg \theta_c$, absorption is irrelevant for our discussion.

Anticipating the master formula, (5.1), we do not find any significant deviation from the Fresnel law until Q is of the order of the inverse of the width $\Delta$ of the smearing of the interface. Let us estimate the magnitude of $\Delta \cdot Q_c$. The smearing region will in general be of the order of the molecular radius R. The electron density in (5.6) for $Q_c$ is of the order $Z(4/3\pi R^3)^{-1}$, Z being the number of electrons per atom or molecule. We then find

$$\Delta \cdot Q_c \simeq R \cdot 4(\varrho_{el} r_0 \pi)^{1/2} \simeq 4(r_0 Z/R)^{1/2} \simeq 10^{-1} \ . \tag{5.7}$$

This estimate shows that it is not until $Q/Q_c$ is of the order of 10 that significant deviation from the Fresnel reflectivity can be expected. The neglect of absorption effects for $Q \simeq Q_c$ is therefore a reasonable simplification. The interesting region of reflectivities is quite small, of the order $(4 \times 10^2)^{-2} \sim 10^{-5}$ to $10^{-6}$, as seen from (5.5a), and requires a strong source as well as effective background discrimination to obtain $\varrho(z)$ with sufficient accuracy.

### 5.1.3 Derivation of Master Formula

The basic idea in deriving (5.1) is to think of the reflected ray as a superposition of reflections from infinitesimal layers at various depths z.

Consider therefore first in Fig. 5.4 the x-ray reflectivity of a thin plate of thickness $\Delta z$ much smaller than the wavelength $\lambda$. A ray from the source point S is Thomson scattered from the electrons in the plate and the scattered, or

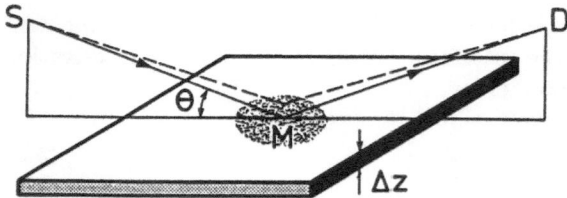

**Fig. 5.4.** Reflectivity of a thin slab. X-ray photons from the source point S are scattered coherently to the detection point D by the electrons in the shaded region around the midpoint M

reflected, wave at the detection point D will be proportional to the scattering length $r_0$ of one electron and to the density of electrons perpendicular to the beam $(\varrho_{el}/\sin\theta)\Delta z$. The ratio of amplitudes of the scattered and incident ray, $\mathcal{E}_D/\mathcal{E}_S$, is of course dimensionless and, as argued above, it must be proportional to $r_0(\varrho_{el}/\sin\theta)\Delta z$ which has the dimension of inverse length. The "missing length" can then only be the x-ray wavelength and this argument leads to

$$\frac{\mathcal{E}_D}{\mathcal{E}_S} = Cr_0\varrho_{el}(z)\Delta z\lambda/\sin\theta \ . \tag{5.8}$$

To derive the complex constant C one must integrate over rays reflected from each area element of the plate with the appropriate phase factors. This is for instance given in [5.3] Appendix 5.A. The result is that $C = \exp(i\pi/2) = i$.

Next we superimpose rays reflected at different depths z. This will involve a phase factor $\exp[iQz]$. In addition, the electron density varies with z. This variation is given by the product of the bulk density $\varrho_{el}$ used in (5.1,4 and 6) and the normalized density profile $\varrho(z)$ in (5.1). Finally, we note that the factor $\lambda/\sin\theta$ in (5.8) is simply $4\pi/Q$ so that the reflectivity involving the *squared* amplitude ratio becomes

$$R(Q) = |i4\pi Q^{-1}\varrho_{el}r_0 \int \varrho(z)\exp(iQz)dz|^2 \tag{5.9}$$

$$= (2Q/Q_c)^{-4}| \int \varrho'(z)\exp(iQz)dz|^2 \ . \tag{5.9a}$$

Equation (5.9a) follows from (5.9) by integration by parts and utilizing $Q_c = 4(\varrho r_0\pi)^{1/2}$.

We note that the prefactor $(2Q/Q_c)^{-4}$ is indeed the Fresnel law of (5.5a) consistent with the neglect in this derivation of refraction phenomena, through the facit assumption that $\theta\gg\theta_c$. The master formula is an improvement of (5.9a) in this respect because $(2Q/Q_c)^{-4}$ has been changed to the more general experession $R_F(Q)$.

### 5.1.4 Traditional Calculation Scheme for R(Q)

The reflection of an electromagnetic wave from an interface is, of course, a classical problem in optics. The graded density profile, or index of refraction, can be approximated by a series of small discontinuities, and the optics problem is then to calculate the reflectivity of a (large) number N of stratified layers.

The basic requirement is continuity of the lateral component of the electric field vector at each interface. The explicit formulae have been worked out in a classical paper [5.4] in 1954. We shall here for the sake of completeness outline this method. For each layer n one assumes a certain thickness $d_n$ and complex refractive index $(1 - \delta_n - i\beta_n)$. The layer index n = 1 corresponds to vacuum and n = N to the substrate, both of semi-infinite extent. For a given incident angle $\theta$ one calculates for each layer the complex numbers $f_n(\theta) = (\theta^2 - 2\delta_n - 2i\beta_n)^{1/2}$ and $a_n = \exp(-ikf_nd_n/2)$, where k is the wave vector in vacuum. The absorption is given by the parameter $\beta_n \equiv (4\pi)^{-1}\lambda\mu_n$ where $\mu_n$ is the linear absorption coefficient of the n'th layer. Note that for $\beta_n = 0$, $f_n(\theta)$ is nothing but the grazing angle inside medium n, $\theta_n'$ (Fig. 5.3). The reflectivity between layer (n−1) and n is expressed as $|F_{n,n-1}|^2$ where $F_{n,n-1}$ is the natural generalization of (5.5):

$$F_{n,n-1} = (f_{n-1} - f_n)/(f_{n-1} + f_n) \ . \tag{5.10}$$

The required reflectivity, $R(\theta)$, follows from N recursive calculations:

$$R(\theta) = |A_{1,2}|^2 \tag{5.11}$$

$$A_{1,2} = (A_{2,3} + F_{1,2})/(A_{2,3}F_{1,2} + 1)$$

$$\ldots$$

$$A_{n-1,n} = a_{n-1}^4(A_{n,n+1} + F_{n-1})/(A_{n,n+1}F_{n-1,n} + 1)$$

$$\ldots$$

$$A_{N,N+1} = 0 \ .$$

At the time *Parratt* derived these equations digital computing with complex numbers was in its infancy and noted: "The algebra is straightforward but very tedious for N>3. Evaluations up to N = 6 have been carried out."

This method has, of course, the advantage of being exact, taking both refraction and absorption properly into account, which might be important in studying the reflectivity of an interface for x-ray wavelengths on both sides of an absorption edge. As was already mentioned in Sect. 5.1.2, absorption is irrelevant for our discussion. In this case one can use the method oulined in Sects. 5.1.1–4; it is more transparent in relating $\varrho(z)$ to $R(\theta)$ than the *Parratt* method.

## 5.2 Experimental Reflectivity Set Up

The experimental set-up to be described in this section assumes that the surface to be studied is the free surface of a liquid which must be kept horizontal in scanning the wave-vector transfer $Q$. For solid surfaces, as for example Langmuir-Blodgett films on a solid substrate, this condition is not necessary and the spectrometer can be somewhat simpler.

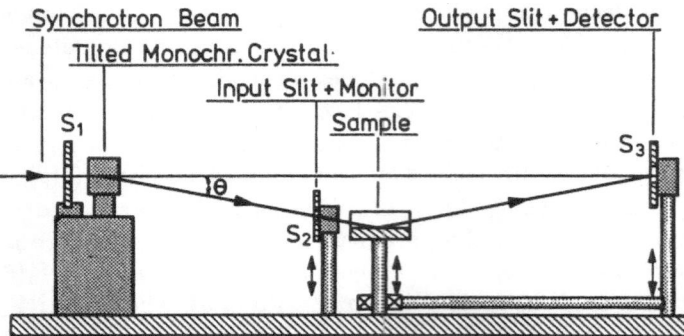

**Fig. 5.5.** Side view of the vertical, scattering plane. Beam directions are defined by slits. The monochromatic beam is bent down towards the sample by tilting the monochromator crystal. The incident beam intensity is monitored after slit $S_2$

Figure 5.5 is a side view of the vertical plane through the monochromatic beam. The horizontal, white synchrotron beam incident from the left in the figure through a slit $S_1$ is monochromatized by Bragg reflection from a pefect Si or Ge crystal, and bent downwards by tilting the reflecting planes out from the vertical plane (Fig. 5.10). The monochromatic beam, at an angle $\theta$ with the horizontal, passes through a narrow slit $S_2$ in front of the sample to provide on the surface a footprint which is shorter along the beam direction than the surface dimension. The slit $S_2$ is mounted on an elevator so that the incident angle $\theta$ can be scanned by a coupled (computer-controlled) motion of the monochromator tilt and the slit height. Also the sample is mounted on an elevator so that the central ray hits the center of the sample at all incident angles. The specularly reflected beam passes through a slit $S_3$ in front of the detector. When $S_3$ is situated at the same distance from the sample axis as the monochromator axis, the vertical position of $S_3$ is stationary during a $\theta$-scan for specular reflection. Nevertheless, it is convenient to mount $S_3$ with the detector on an elevator for a number of reasons given below.

### 5.2.1 General Wave-Vector Transfer

Specular reflection means that the wave-vector transfer, $Q \equiv k_{in} - k_{out}$, is along the surface normal $\hat{n}$. Let this vertical direction be the z-direction, with the transverse, horizontal directions being denoted y in the plane spanned by $k_{in}$ and $\hat{n}$, and x along the normal to this plane.

Figure 5.6 illustrates that a transverse component $Q_y$ is obtained by moving $S_3$ vertically with respect to the height of $S_1$. The slit $S_3$ and detector are mounted on an arm which can be turned around the sample axis. In this way it is also possible to move $S_3$ horizontally with respect to $S_1$, thereby obtaining a transverse component $Q_x$. It is straightforward to calculate the wave-vector transfer $Q = (Q_x, Q_y, Q_z)$ corresponding to any settings of the slits. Conversely, by a suitable algorithm, it is also possible to calculate the slit settings corresponding to any input wave-vector transfer.

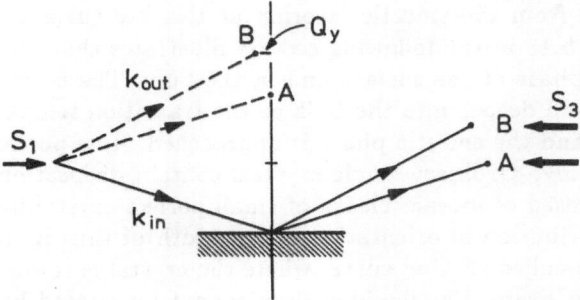

**Fig. 5.6.** A transverse wave-vector component may be obtained without rotating the sample by having the incident glancing angle different from the reflected angle. With slit $S_3$ in position A the longitudinal wave-vector component is $Q_z(A)$ and the transverse wave-vector component is zero. With slit $S_3$ in position B, $Q_z(B) > Q_z(A)$ and $Q_y(B) \neq 0$

The usefulness of being able to scan the detector height, or the general wave-vector transfer $Q$, is illustrated by a few examples. In the next section we shall discuss these and other examples with the main emphasis on the physics these experiments have revealed, but here we shall only describe the techniques of measurements.

The first example is from the study of the free surface of water [5.7]. A good water surface was obtained by applying a thin (0.3 mm) water layer of approximately 60 mm diameter onto a somewhat larger glass flat thoroughly cleaned in a mixture of hot chromic and sulphuric acids. The macroscopic flatness of the water surface is apparent from Fig. 5.7 where the direct beam profile in a detector height scan is compared to the reflected beam profile at an incident angle of $\theta = 0.00244$ radian $= 0.924\,\theta_c$ for water. The reflected beam profile is almost identical to the incident beam profile. An upper limit of the intrinsic meniscus curvature of the irradiated $0.6 \times 16\,\text{mm}^2$ "footprint" strip on the surface of no more than $6 \times 10^{-6}$ radians could be deduced from the data of Fig. 5.7.

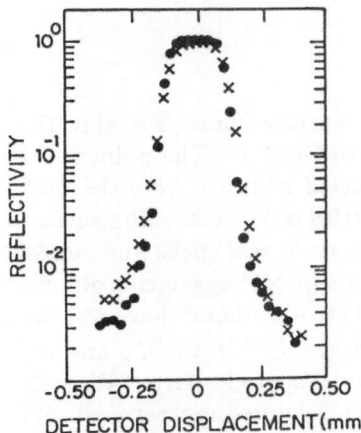

**Fig. 5.7.** Detector height scans of the direct beam *(solid circles)* and the beam reflected from a water surface *(crosses)*

The second example is from the smectic layering at the free surface of liquid crystals [5.6]. Figure 5.18 in the following section illustrates the smectic layering in the nematic phase at the surface and in the bulk. The surface layering penetrates deeper and deeper into the bulk as the transition temperature between the nematic and the smectic phase is approached. How perfect is this surface-induced layering? Ordinary single crystals contain dislocations and are thought of as composed of mosaic blocks of small perfect crystallites with a so-called mosaic distribution of orientations. The width of this distribution is determined by a so-called rocking curve, where the crystal is rotated through a stationary incident beam. The liquid surface cannot be rotated but the equivalent scan can be carried out as a $Q_y$-scan (Fig. 5.6). Figure 5.8 shows the result when the liquid crystal is held only a few millidegrees from the transition temperature. The half width at half maximum is only $2.4 \times 10^{-4}\,nm^{-1}$, and a substantial fraction of this is the instrumental broadening. One can conclude that the smectic layering is essentially perfect, with no dislocations over correlated distances of at least 100 000 Angstrom or 0.1 mm, an amazing result when it is borne in mind that it is obtained by diffraction of radiation in the 0.1 nm range.

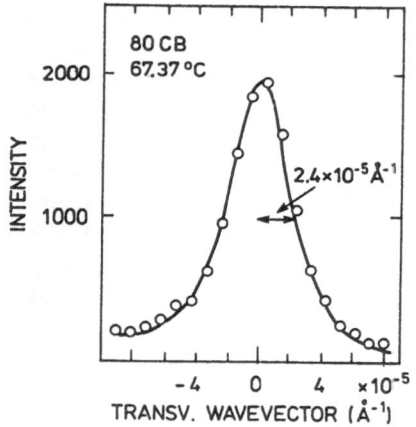

Fig. 5.8. Transverse scan equivalent to a rocking curve for smectic A layering a few millidegrees from the nematic to smectic A transition temperature ($1\,Å = 10\,nm^{-1}$)

The third example derives also from the free surface liquid crystal in the nematic phase [5.8] and reference is again made to Fig. 5.18. The point to be made is how one can distinguish the surface *reflected* intensity from the bulk *scattered* intensity. The answer lies in the symmetries of Fig. 5.18. The surface layers have an infinite extent in the lateral directions in real space and consequently the scattering is a delta function in the reciprocal space coordinates $Q_x$ and $Q_y$. The smectic regions in the bulk, on the other hand, have a finite extent $\xi_\perp$ in the lateral directions and thus a width of $\xi_\perp^{-1}$ in the $Q_x$ and $Q_y$ directions. This is indeed borne out by the $Q_x$ scan in the right part of Fig. 5.9. Notice the logarithmic intensity scale; the resolution-limited central peak de-

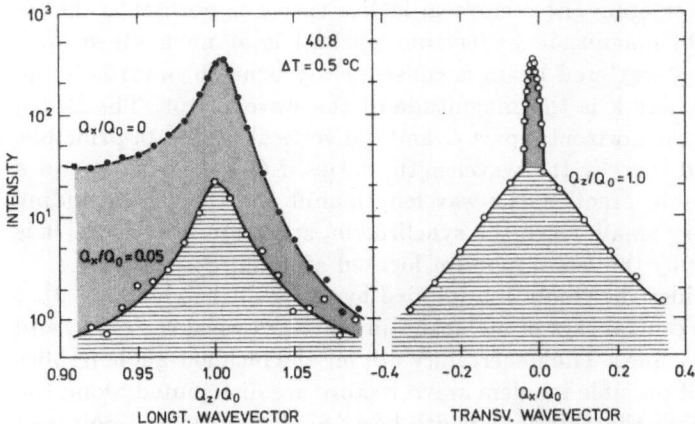

**Fig. 5.9.** Scans illustrating scattering from the bulk *(cross-hatched)* and the surface *(shaded)* smectic layering of a liquid crystal in the nematic phase

rives from the surface layers and is ten times as intense as the bulk signal. The left part of Fig. 5.9 illustrates that the longitudinal bulk correlation range $\xi_\parallel$ (Fig. 5.18) is very nearly the same as the surface layering penetration depth, since the two peaks obtained respectively at $Q_x = 0.05\,Q_0$ and $Q_x = 0$ have identical widths. In the same study it was also shown that the ratio of surface to bulk scattering depends on the x-ray wavelength; the longer the wavelength the less the x-ray beam penetrates into the bulk.

## 5.2.2 Resolution

We shall now consider the finer details of the principles outlined in Fig. 5.5.

Let us first return to the way the monochromatic beam is bent downwards towards the sample surface at glancing angle $\theta$ (Fig. 5.10). The horizontal synchrotron beam is incident from the left on the slit $S_1$. It then hits the net-planes

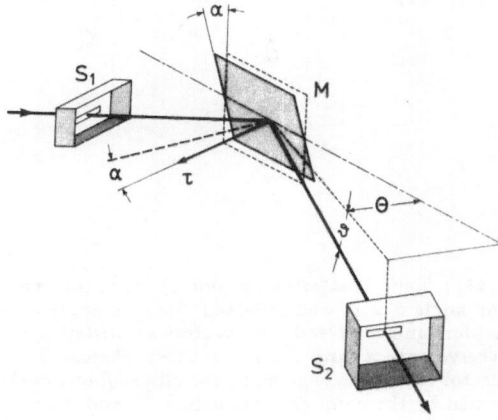

**Fig. 5.10.** The white synchrotron radiation beam passes through a slit $S_1$ and an ionization monitor behind it. A monochromatic beam is extracted by Bragg reflection from a single crystal M of Si or Ge. The monochromatic beam is bent downwards at angle $\theta$ by tilting the reflecting planes an angle $\alpha$ from vertical. This beam passes through a slit $S_2$ with a beam monitor before hitting the sample surface (not shown)

of a perfect Si or Ge crystal. The reciprocal lattice vector $\tau$, normal to the reflecting planes and of magnitude $2\pi/$(plane spacing) is at an angle $\alpha$ with horizontal. The Bragg scattered beam is consequently bent downwards by an amount $\theta \simeq (k/\tau)\alpha$, where k is the magnitude of the wave vector. The Bragg angle is composed of the horizontal part $\Theta$ and the vertical part $\theta$. In principle, this Bragg angle, and thereby the wavelength, varies if $\Theta$ is kept fixed and $\theta$ is varied, but as $\Theta \gg \theta$ in practice, the wavelength shift and the corresponding intensity shift are very small, since the synchrotron spectrum is smooth. It is anyhow corrected for by the beam monitor located after slit $S_2$.

Next let us consider the resolution implied by finite slit heights [5.8–9].

In Fig. 5.11b the central rays of incident and scattered wave vectors $k_i$ and $k_s$ are shown by heavy lines. The uncertainty $\delta\phi_i$ for the incident angle implies that the end points of possible incident wave vectors are distributed along the vector $X_i$ with a certain characteristic width $k\delta\phi_i$. Similarly the end points of possible scattered wave vectors $k_s$ are located along $X_s$ and crudely speaking the actual wave-vector transfers are distributed within the shaded parallelogram spanned by $X_i$ and $X_s$. To be more precise, if the $k_i$ and $k_s$ distributions are box-like, that is constant along $X_i$ and $X_s$, respectively, and zero outside, then any wave-vector transfer to a point inside the parallelogram is equally likely, but the probability outside the parallelogram is zero. Alternatively, one might assume Gaussian distributions along $X_i$ and $X_s$ for $k_i$ and $k_s$. In that case the wave-vector transfer distribution will be a 2-dimensional Gaussian characterized by an equi-probability ellipse where $X_i$ and $X_s$ are conjugate diameters [5.10–11].

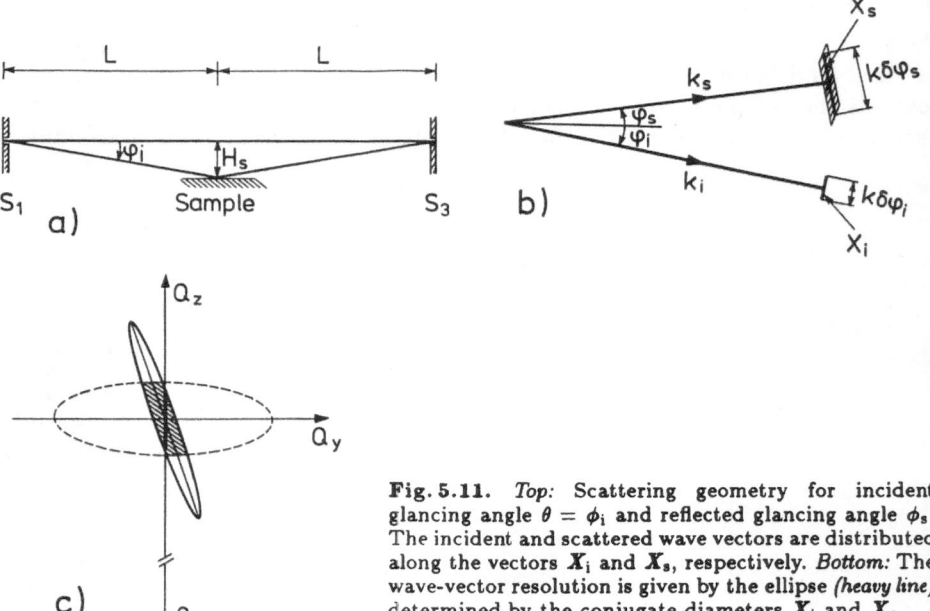

**Fig. 5.11.** *Top:* Scattering geometry for incident glancing angle $\theta = \phi_i$ and reflected glancing angle $\phi_s$. The incident and scattered wave vectors are distributed along the vectors $X_i$ and $X_s$, respectively. *Bottom:* The wave-vector resolution is given by the ellipse *(heavy line)* determined by the conjugate diameters $X_i$ and $X_s$

This resolution ellipse is shown in the bottom part of Fig. 5.11. The overlap between the resolution ellipse and the cross section determines the registered intensity. For specular reflection the cross section is confined to the $Q_z$-axis and the effective width is therefore given by the $Q_z$ diameter of the ellipse rather than the $Q_z$ projection. The cross-hatched region indicates a possible overlap between the bulk cross section, as given by the half-value contour (dashed line) ellipse, and the resolution ellipse.

In considering Fig. 5.10 again one may wonder why the slit $S_2$ is necessary at all – the synchrotron beam is well collimated and will continue to be so after Bragg reflection from a perfect crystal, and the glancing angle $\theta$ is completely determined by the monochromator tilt $\alpha$. In practice, the slit $S_2$ is useful for two reasons. First, since it is located only a few centimeters from the sample surface it provides a "foot-print" length along the beam which is only determined by the slit height of $S_2$ (and $\theta$), whereas with slit $S_1$ only there will be a sizeable contribution from angular divergence, because $S_1$ is located almost 1 meter from the sample axis. Second, the height of $S_2$ can be set and reproduced to define the glancing angle $\theta$ more accurately than can be provided by the monochromator $\alpha$-arc. If $\alpha$ does not track exactly with the height of $S_2$ it will only prolong the counting time which is determined by a preset count in the beam monitor located *after* $S_2$.

A convenient monitor-detector arrangement is shown in Fig. 5.12. The NaI scintillation detector can be placed in either the monitor or detector position. In the monitor position a small fraction of the photons in the beam is scattered by a plastic foil inserted in the slot S into the scintillation detector. Pulse-height discrimination allows separation of the fundamental wavelength in the monochromatic beam from higher-order contamination – a feature which cannot be obtained by an ionization-chamber monitor. In the detector mode the slot S is empty and all photons in the beam hit the detector. A dynamic range of $1:10^8$ is typically required for the type of reflectivity measurements we are dealing with. The high intensity part of this range is conveniently measured by the detector in monitor mode where the efficiency may be of the order of $10^{-4}$. Another practical feature of the system in Fig. 5.12 is to scan the wavelength with a suitable absorption foil inserted in the slot S, and to notice the jump in fluorescent yield in the monitor mode corresponding to the K absorption edge of the foil for wavelength calibration.

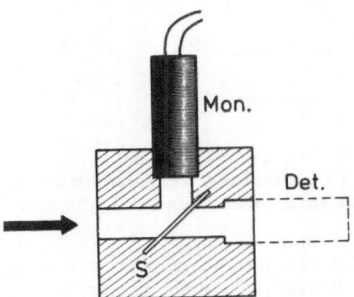

**Fig. 5.12.** A NaI scintillation detector may be used as a wavelength discriminating beam monitor by inserting a plastic foil in the slot S to scatter a small fraction of the beam into the detector

The higher-order contamination of the monochromatic beam may actually be utilized for determining the thickness of the liquid layer on the substrate. The latter is usually opaque for all wavelengths in the x-ray beam whereas the liquid may be at least partly transparent for the higher-order wavelengths but opaque for the fundamental wavelength. A sample height scan through the direct beam, registering photon intensities of both the fundamental wavelength and higher order wavelengths, will exhibit jumps corresponding to the liquid surface as well as the substrate surface passing through the direct beam, and provides therefore the liquid thickness as well as a line up of the sample elevator.

## 5.3 Examples on Reflectivity Measurements

### 5.3.1 Thermal Roughness

In the previous section we mentioned briefly reflectivity measurements on a rather thin water layer [5.7] on a glass flat and found (Fig. 5.7) that the water surface was very flat with a reflectivity of more than 95 % for $\theta < \theta_c$. Data extending to $\theta \gg \theta_c$ are given in Fig. 5.13 in a semi-log plot of reflectivity relative to Fresnel reflectivity versus the *square* of the wave-vector transfer. This plot shows that the data empirically obeys Gaussian smearing, which is analogous to the Debye-Waller factor in crystallography, where $\sigma^2$ is interpreted as the thermal mean-squared displacement $\langle u^2 \rangle$. It is therefore tempting to estimate $\langle u^2 \rangle$ for a liquid surface.

In a classical continuum model we consider an area of $L \times L$ units with a sinusoidal perturbation of lateral wave vector $q$ and amplitude $u_q$ (Fig. 5.14).

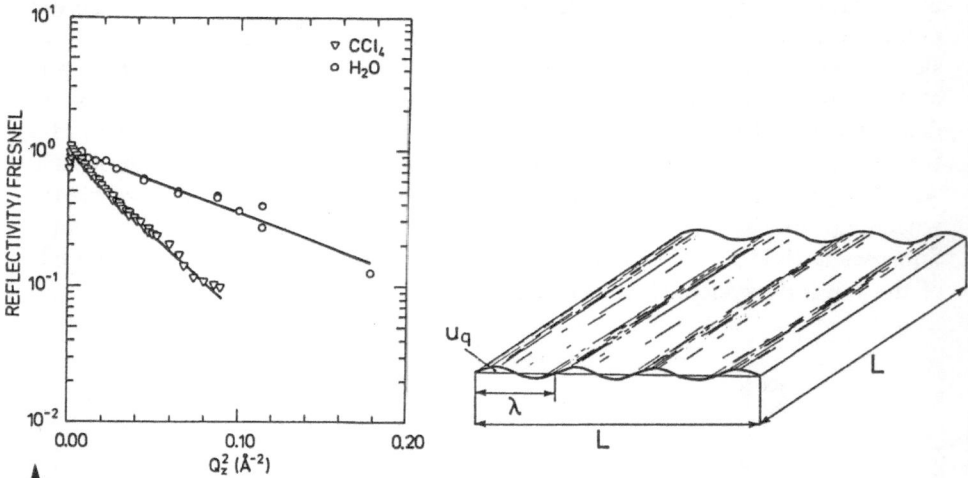

**Fig. 5.13.** Reflectivity data versus the squared wave-vector transfer for water *(open circles)* and carbon tetrachloride *(triangles)*. The slope expresses the thermal mean squared displacement of the atoms from the ideal flat surface

**Fig. 5.14.** A sinusoidal perturbation of a flat surface of $L \times L$ units has a potential energy from surface tension and gravity. The latter contribution is negligible in the present context

The energy, $\varepsilon_{\boldsymbol{q}}$, of this $\boldsymbol{q}$-mode is proportional to $u_{\boldsymbol{q}}^2$, as we shall see, and the thermal average value, $\langle u_{\boldsymbol{q}}^2 \rangle$, follows from the equi-partition principle $\langle \varepsilon_{\boldsymbol{q}} \rangle = k_B T/2$ where $k_B$ is Boltzmann's constant and T the absolute temperature.

The perturbed, $\boldsymbol{q}$-mode surface has a larger surface than the unperturbed flat surface. It is an elementary exercise to see that the excess surface area is $(1/4)u_{\boldsymbol{q}}^2 q^2 L^2$ implying a contribution to $\varepsilon_{\boldsymbol{q}}$ from surface tension of

$$\varepsilon_{\boldsymbol{q}}^s = \gamma u_{\boldsymbol{q}}^2 q^2 L^2/4 , \tag{5.12}$$

where $\gamma$ is the surface tension constant. There is also a contribution to $\varepsilon_{\boldsymbol{q}}$ from gravity effects, $\varepsilon_{\boldsymbol{q}}^g$. This contribution turns out to be negligible compared to $\varepsilon_{\boldsymbol{q}}^s$ when the water film thickness is only 0.3 mm. The thermal average of $u_{\boldsymbol{q}}^2$ is therefore determined by $\langle \varepsilon_{\boldsymbol{q}}^s \rangle = k_B T/2$ or

$$\langle u_{\boldsymbol{q}}^2 \rangle = 2k_B T/(\gamma q^2 L^2) \tag{5.13}$$

and the general mean-squared displacement $\langle u^2 \rangle$ is found by summing over all wave vectors. Turning the sum into a two-dimensional integral utilizing the phase space density $(2\pi/L)^{-2}$ yields

$$\langle u^2 \rangle = (1/2) \sum_{\boldsymbol{q}} \langle u_{\boldsymbol{q}}^2 \rangle = (2\pi)^{-2} k_B T/\gamma \int_{q_{min}}^{q_{max}} q^{-2} d^2 q . \tag{5.14}$$

The upper integration limit $q_{max}$ must be of the order of $2\pi/$(molecular diameter). The lower integration limit is determined by the instrumental resolution in the two lateral wave-vector components. Had the instrumental resolution been isotropic in the lateral components, the integration in (5.14) could be carried out analytically to yield $\langle u^2 \rangle = (2\pi)^{-1} k_B T/\gamma \log(q_{max}/q_{min})$. In the experimental case the integration must be carried out numerically. The result for water is $\langle u^2 \rangle_T^{H_2O} \simeq (0.28\,\text{nm})^2$. Including the estimate that the electron cloud of a water molecule has a mean-square radius of the order of $(0.19\,\text{nm})^2$, we calculate the resulting mean-square displacement of the water surface as determined with the applied resolution to be $\langle u^2 \rangle_{res}^{H_2O} \simeq (0.34\,\text{nm})^2$. The slope of the straight line through the water data in Fig. 5.13 yields $\langle u^2 \rangle_{exp}^{H_2O} \simeq (0.32\,\text{nm})^2$ which is in remarkable agreement with the theoretical estimate. Carbon tetrachloride is another simple liquid, but with a smaller surface tension than water by a factor of 3. Indeed the data for $CCl_4$ in Fig. 5.13 exhibit a slope about three times larger than that for water. These data (unpublished) strengthen significantly the interpretation that $\langle u^2 \rangle$ is due to thermal roughness. A theoretical discussion of $\langle u^2 \rangle$ from a liquid surface including quantum effects and gravity was given by *Cole* in 1980 [5.12].

## 5.3.2 Smectic Layering on Liquid Crystal Surfaces

Recent x-ray reflectivity measurements on liquid crystal surfaces have revealed a rich variety of phenomena induced by the boundary condition of the free surface. While the remaining examples of synchrotron x-ray reflectivity studies are all chosen from liquid crystals, the intention is more to illustrate the variety of surface spectra and their interpretation than to review the physics of the liquid-crystal surfaces. It is convenient though that all three examples require the same background knowledge as a prerequisite for their interpretation and we shall therefore first briefly recall the basic features of the structure of liquid crystal materials. For a general reference, see [5.13].

Liquid crystals consist of long molecules with a typical length-to-diameter ratio of 5 to 1. In describing the structures, the molecules are considered as rigid rods. The variety of structures or phases is due to the combination of order/disorder between the *position* of molecules and their *orientation*. For the present purpose it suffices to recall three liquid phases: the *isotropic* phase where both position and orientation are disordered as in an ordinary simple liquid, the *nematic* phase where the position is disordered but all molecules have the same spontaneous average direction, and the *smectic A* phase where the common orientation is maintained but, in addition, the molecules are positioned in layers perpendicular to their long axis with a well-defined repetition distance between layers, but with positional disorder of molecules within the same layer. Different sequences of transitions between these phases may occur. By decreasing temperatures the high temperature isotropic phase may be followed directly by the smectic A phase, or a nematic phase at intermediate temperatures may intervene between the isotropic and smectic A phases. The transition from the isotropic phase is always discontinuous or first order, whereas the nematic to smectic A phase may be first order or continuous (second order). In the latter case critical fluctuations of short-range order smectic A regions in the nematic matrix become more and more pronounced as the nematic to smectic A phase transition temperature is approached. An example of the molecular structure is the so-called nCB molecules

$$C_nH_{2n+1}\text{---}\langle\rangle\langle\rangle\text{---}CN$$
$$NC\text{---}\langle\rangle\langle\rangle\text{---}C_nH_{2n+1} \tag{5.15}$$

The upper molecule in the dashed frame is shown with its aliphatic tail $C_nH_{2n+1}$ to the left and the polar cyano head CN to the right. The intramolecular interactions are so strong that this molecule pairs with one of opposite orientation as shown, and the unit rod should be considered as this pair of molecules within the dashed frame.

Another example of a liquid crystal molecule is

$$C_nH_{2n+1}\text{—}\bigcirc\text{—O–CO—}\bigcirc\text{—OCH}_2\text{—}\bigcirc\text{—CN} \tag{5.16}$$

Here the molecular interaction is weaker, so that a neighboring molecule may or may not have the opposite head-tail orientation.

### a) Smectic Layering in the Isotropic Phase

The data in this example [5.14] derives from the material 12CB with the molecular structure given in (5.15) with n = 12. This material exhibits a strong first-order transition from the isotropic phase directly to the smectic A phase at a transition temperature $T_{IA} = 57.7°C$. The lattice spacing in the smectic A phase gives a Bragg peak at wave-vector transfer $Q_0 = 1.62\,\text{nm}^{-1}$. The reflected intensity at a slightly smaller wave-vector transfer of $Q = 1.5\,\text{nm}^{-1}$ exhibits a remarkable temperature dependence as shown in Fig. 5.15. Each time the temperature difference from $T_{IA}$ is diminished by an almost constant factor of four the intensity changes discontinuously. Each time this is a result of creation of an additional smectic A layer at the surface.

The reflectivity as function of the wave vector at different temperatures is shown in Fig. 5.16. The full lines are calculated from the following model. The density is composed of two contributions. The first is a step function smeared with a Gaussian, giving the contribution

$$\phi_1(Q) = \exp(-Q^2\sigma_s^2/2) \ . \tag{5.17}$$

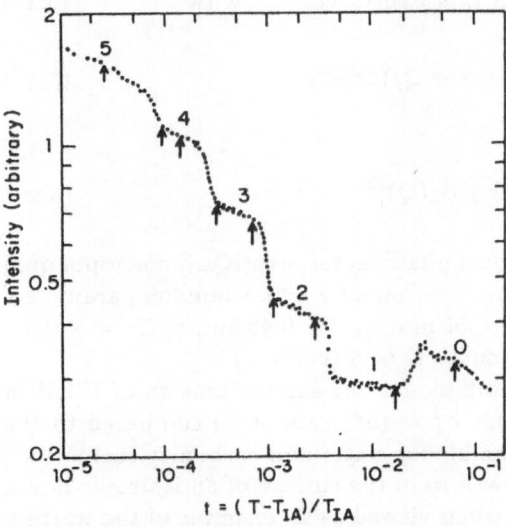

**Fig. 5.15.** The reflected intensity from 12CB in the isotropic phase changes in discrete steps as the temperature is lowered towards the transition temperature to the smectic A phase. The numbers indicate the number of smectic A layers at the surface

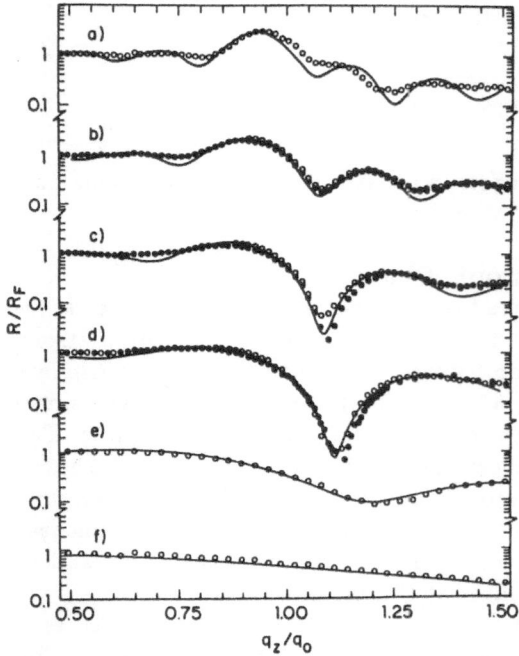

**Fig. 5.16.** Reflectivity relative to Fresnel reflectivity versus wave vector at different temperature intervals $\Delta T$ above the transition to the smectic A phase. The reflectivity scale is logarithmic and each spectrum is displaced two decades for clarity. The temperature intervals relative to the transition temperature $T_{IA}$ are *(a)* t = $3 \times 10^{-5}$, *(b)* t = $8 \times 10^{-5}$ *(open circles)*, t = $1.4 \times 10^{-4}$ *(closed circles)*, *(c)* t = $3 \times 10^{-4}$ *(open circles)*, t = $8.3 \times 10^{-4}$ *(closed circles)*, *(d)* t = $1.1 \times 10^{-3}$ *(open circles)*, t = $3 \times 10^{-3}$ *(closed circles)*, *(e)* t = $1.9 \times 10^{-2}$, and *(f)* t = $6.1 \times 10^{-2}$. The *solid line* is for a density model with a sinusoidal modulation terminated after an integral number *(a)* 5, *(b)* 4, *(c)* 3, *(d)* 2, *(e)* 1 and *(f)* 0 of periods

The second is a sinusoidal density wave of amplitude $A_0$ between $z = 0$ and $z = NL$, where L is the layer spacing and N an integer number of layers. Also this part is smeared by a Gaussian and the corresponding contribution is

$$\phi_2(Q) = A(x)\exp(-Q^2\sigma_m^2/2)\sin(\pi N)\exp(i\pi N) \qquad \text{with} \qquad (5.18)$$

$$A(x) = 2A_0 x/[(x+1)(x-1)] \ , \quad x \equiv Q/(2\pi/L) \ . \qquad (5.19)$$

The reflectivity is then

$$R(Q)/R_F(Q) = |\phi_1(Q)\exp(iQz_0) + \phi_2(Q)|^2 \ . \qquad (5.20)$$

We have introduced a phenomenological phase factor, $\exp(iQz_0)$ corresponding to the step function being at $z = z_0$ rather than at $z = 0$. Common parameters for all curves in Fig. 5.16 are $\sigma_s = 0.55\,\text{nm}$, $\sigma_m = 0.45\,\text{nm}$, $z_0/L = -0.35$, $A_0 = 0.12$ whereas N varies from 0 (curve f) to 5 (curve a).

At present there are no data available for the surface tension of 12CB in the isotropic phase, so the number for $\sigma_s^2 = \langle u^2 \rangle$ cannot be compared to the thermal roughness model, as was possible for water and carbon tetrachloride.

These data of layer by layer growth from the surface of an isotropic liquid crystal are interesting, in particular when viewed as an example of the wetting problem [5.14].

## b) Layer Penetration in the Nematic Phase

All it takes in nCB molecules to obtain a nematic phase between the isotropic and smectic A phases is a shortening of the aliphatic tail from n = 12 to say n = 8. The surface layering is now quite different and so is the reflectivity data. In the isotropic phase a few smectic layers are formed on the surface, but now with a decreasing amplitude, as shown in the top of Fig. 5.17, in contrast to 12CB where the model calculation of Fig. 5.16 assumed a sinusoidal density variation of constant amplitude up to N periods. The most remarkable difference is however in the nematic phase. The reflected intensity versus $Q_z/Q_0$ is shown in the bottom left part of Fig. 5.17 and the corresponding model density in the right bottom part of Fig. 5.17.

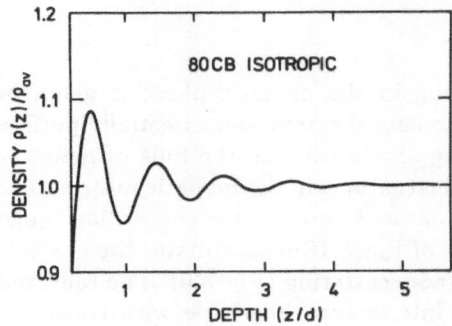

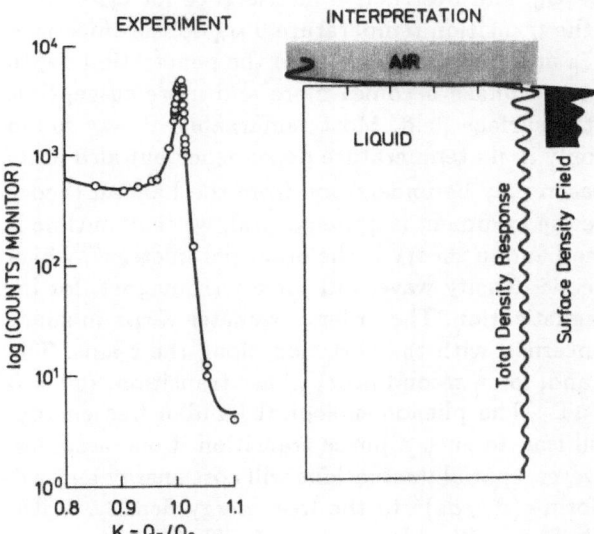

**Fig. 5.17.** *Top:* Density profile in the isotropic phase slightly above the transition temperature to the nematic phase of the liquid crystal 80CB. *Bottom:* Reflected intensity versus $Q_z/Q_0$ in the nematic phase *(left)* and the corresponding density *(right)*. This can be considered as the response in the nematic phase to smectic layering imposed by the first top layers

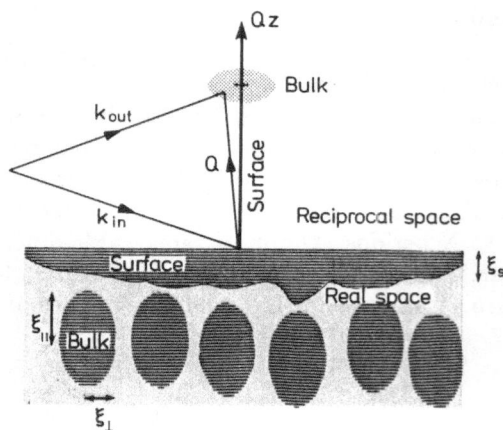

**Fig. 5.18.** Smectic layering in the nematic phase. The boundary conditions at the surface impose smectic layering of infinite lateral extent at the surface, decaying exponentially in going into the bulk. Here spontaneous critical fluctuations have correlation ranges $\xi_\parallel$ and $\xi_\perp$ along and perpendicular to the molecular axis, respectively. The top part shows scattering geometries to separate bulk and surface signals (Fig. 5.6)

The total picture of smectic layering in the nematic phase is given in Fig. 5.18. The surface layers are of infinite lateral extent and essentially perfect (Fig. 5.8), with an exponentially decaying amplitude into the bulk of penetration depth $\xi_s$. Deep in the bulk smectic fluctuations in the nematic matrix have one correlation range $\xi_\parallel$ along the molecular axis and another correlation range $\xi_\perp$ in the lateral direction. The top part of Fig. 5.18 recapitulates the scattering diagram for separating bulk and surface scattering (Fig. 5.9). The reflected intensity in Fig. 5.17 displays a typical interference lineshape with constructive interference between the ordinary Fresnel wave and the wave scattered from the surface layers for $Q<Q_0$, and destructive interference for $Q>Q_0$. As the temperature approaches the transition temperature $T_{NA}$ to the smectic A phase the correlation ranges $\xi_\parallel$ and $\xi_\perp$ diverge and also the penetration depth $\xi_s$ increases, because the nematic phase becomes more and more susceptible to the layering imposed by the surface field. Most remarkably, it was found that $\xi_s$ is identical to $\xi_\parallel$ not only in its temperature dependence but also in its numerical value [5.5]. The reason may be understood from the Landau theory of phase transitions and, since the argument is quite general, we shall outline it here. The basic quantity in the Landau theory is the order parameter $\psi$, which in the present case is the smectic density wave, but for a ferromagnet, for instance, would be the local magnetization. The order parameter varies in space and we shall here only be concerned with the variation along the z-axis. The average value is denoted $\langle\psi\rangle$ and, for a second-order phase transition, $\langle\psi\rangle = 0$ for $T \geq T_c$ but is finite for $T < T_c$. The phenomenological Landau free energy density $f = a\langle\psi\rangle^2 + b\langle\psi\rangle^4$ will lead to such a phase transition if one assumes $a = a_0(T - T_c)$ and $b>0$. However, spatial fluctuations will cost energy and we therefore add a term of the form $c(\partial\psi/\partial z)^2$ to the free energy density. With this expression for f one readily finds critical fluctuations for $T>T_c$ with a correlation range $\xi_\parallel = (c/a)^{1/2}$, assuming an infinite system. Now we consider the penetration from the surface, requiring a finite value $\psi_0$ of the order parameter at $z = 0$. The order parameter must decay as z increases to reach its bulk average value of zero. On the other hand, any spatial change costs energy, cf.

the term $c(\partial\psi/\partial z)^2$, so there must be an optimum way for $\psi$ to decay in the sense that the total energy obtained by integrating the energy density becomes minimal, i.e. a typical problem of variational calculus. Explicitly one finds that the optimal decay is exponential with a decay length $\xi_s = (c/a)^{1/2}$, i.e. $\xi_s = \xi_{\parallel}$.

## c) Antiferroelectric Surface Layers

The final example of surface structure of liquid crystals is also the most complex and it illustrates the degree of detail that x-ray reflectivity measurements can provide. The liquid crystal molecule in this example is given in (5.16). Although the molecular interactions favor that the neighboring molecules along the molecular axis are oriented head to head rather than head to tail, this interaction is not nearly as strong as in the nCB molecules of (5.15), where the head to head interaction implies bi-layering and a symmetrical unit cell as shown in (5.15). Since the molecule, or unit cell, of (5.16) is polar, it is necessary to consider both layering irrespective of molecular orientation, and polarized layering where the molecules in a layer have predominantly heads up or heads down.

Reflectivity data at various temperatures in the nematic phase are shown in Fig. 5.19. The peak around $Q = Q_0 = 2\pi/(\text{molecular length})$ becomes gradually more and more intense and narrow as the transition temperature $T_{NA}$ is approached, corresponding to deeper and deeper response to the surface layering field, just as in the previous example. In addition, the reflectivity spectra

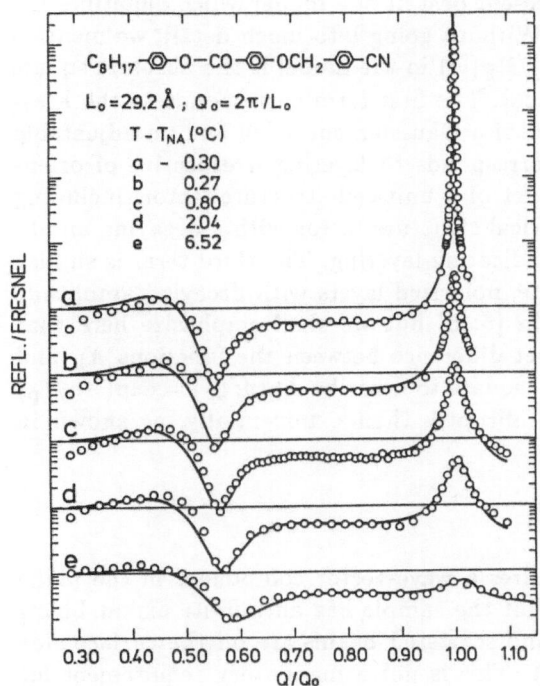

Fig. 5.19. Reflectivity data showing structure both around $Q = Q_0$ and $Q = Q_0/2$, corresponding to smectic layering irrespective of molecular orientation and polarized smectic layering, respectively

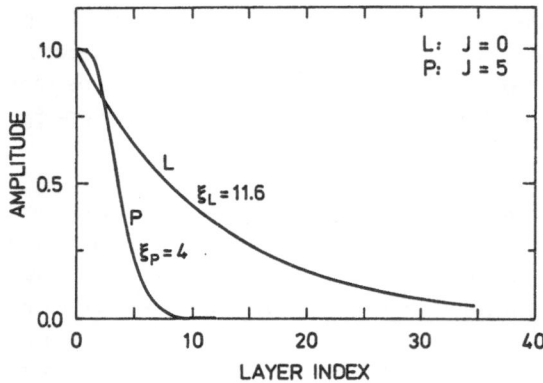

**Fig. 5.20.** The decay of polarized layers (P) and of smectic layers (L) corresponding to the curve labelled d in Fig. 5.19

show structure around $Q = Q_0/2$ in reciprocal space, corresponding to a doubling of the unit cell in real space. Qualitatively this may easily be understood. In the top layer it is favorable to have the aliphatic tails sticking out from the surface, and this first layer is therefore polarized with heads down. Since head-head orientation is favoured over head-tail orientation, the next layer will predominantly have heads up and so on. However, in contrast to the response to layering *per se* which diverges as T approaches $T_{NA}$, there is no particularly large or temperature-dependent response to a polarization field in the nematic phase. The structure around $Q = Q_0/2$ is therefore rather broad and not very temperature dependent.

The full lines in Fig. 5.19 represent best fit of a model which quantifies the qualitative picture just outlined. Without going into much detail, we mention that the general structure of $R(Q)/R_F(Q)$ in the model is the absolute square of the sum of three terms, see (5.20). The first term corresponds to the Fresnel wave and includes the product of a Gaussian smearing and an adjustable phase factor. The second term corresponds to layering irrespective of orientation and is written as the product of a unit-cell structure factor, including Gaussian smearing, and a geometrical structure factor with a decaying amplitude $A_L(n; \xi_L)$ with subscript L indicating layering. The third term is similar but now corresponding to alternate, polarized layers with decaying amplitude $A_P(n; \xi_P)$. Details can be found in [5.15] but we shall emphasize here that the data analysis showed a distinct difference between the functions $A_L$ and $A_P$. Simple exponential decay is adequate to describe $A_L(n; \xi_P) = \exp(-n/\xi_P)$ whereas $A_P$ vs. n falls off more abruptly than exponentially, as shown in Fig. 5.20.

## 5.4 In-Plane Structures

In-plane structure diffraction requires a wave-vector component in the plane of the surface. We shall assume that the sample has an infinite extent below the surface so that the incident and scattered beams are on the surface side of the crystal as shown in Fig. 5.1. This is not a mandatory requirement for surface studies. Overlayers on a graphite substrate in transmission geometry [5.16] and freely suspended thin films of liquid crystals [5.17] have been studied very successfully but will not be discussed here.

For the reflection geometry grazing incidence may be used to enhance surface sensitivity. This is discussed in Sect. 5.4.1. We shall restrict the discussion only to deal with single crystalline in-plane structure. The surface geometry implies a particular choice of unit cell and corresponding reciprocal lattice. This, together with interpretation of data using the Patterson function method, is dealt with in Sect. 5.4.2. An example illustrating these principles is described in Sect. 5.4.3. References to x-ray diffraction studies of this kind are given in Sect. 5.4.4.

### 5.4.1 Grazing Incidence Geometry

The grazing incidence geometry is depicted in the bottom part of Fig. 5.1. When the grazing angles $\alpha_i$ and/or $\alpha_f$ is close to the critical angle of total reflection, $\alpha_c$, the penetration depth of the radiation as well as the field strength vary in a non-trivial way. These effects were first utilized in x-ray surface crystallography by *Eisenberger* and *Marra* [5.18].

#### a) Penetration Depth

In this subsection we shall discuss how the effective penetration depth $\Lambda_{\mathrm{eff}}$ depends on the incident and exit angles $\alpha_i$ and $\alpha_f$, on the wave number k and on the linear absorption coefficient $\mu(k)$. The scale of angles is the critical angle for total reflection, $\alpha_c$, see (5.6):

$$k\alpha_c = (4\pi\varrho_{\mathrm{el}}r_0)^{1/2} \; , \tag{5.21}$$

where $\varrho_{\mathrm{el}}$ is the number density of electrons in the bulk and $r_0$ the electron radius $e^2/mc^2$.

The effective penetration depth $\Lambda_{\mathrm{eff}}$ is defined in Fig. 5.21. The total path length of the incident and reflected ray depends on the depth of reflection. The depth where the *intensity* (not wave amplitude) attenuation of the incident and reflected ray together is $1/e$ defines $\Lambda_{\mathrm{eff}}$.

Let us first assume that $\alpha_i$ and $\alpha_f$ are both much larger than $\alpha_c$. In that case refraction can be neglected, and the absorption is solely determined by the

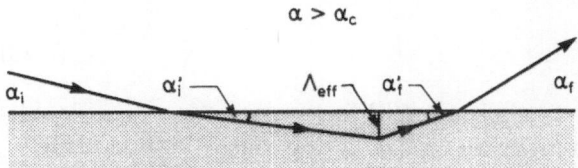

**Fig. 5.21.** The penetration depth $\Lambda_{\mathrm{eff}}$ is determined by the sum of the path lengths of the incident and exit beams inside the crystal being equal to $\mu^{-1}$, the linear absorption coefficient

linear absorption coefficient $\mu$. The definition of $\Lambda_{\text{eff}}$ gives $(\Lambda_{\text{eff}}/\alpha_i + \Lambda_{\text{eff}}/\alpha_f)\mu = 1$ or

$$\alpha_{i,f} \gg \alpha_c \quad : \quad \Lambda_{\text{eff}}^{-1} = \mu/\alpha_i + \mu/\alpha_f \ . \tag{5.22}$$

This relation is of the general form

$$\Lambda_{\text{eff}}^{-1} = \Lambda^{-1}(\alpha_i) + \Lambda^{-1}(\alpha_f) \tag{5.23}$$

which explicitly shows the symmetric contribution of the incident and exit rays caused by the fundamental theorem of reciprocity [5.19]. We shall therefore only need to consider the function $\Lambda(\alpha)$ independent of whether the argument $\alpha$ is $\alpha_i$ or $\alpha_f$. The limit $\alpha \gg \alpha_c$ is denoted $\Lambda_\infty(\alpha)$, and as (5.22) shows

$$\Lambda_\infty(\alpha) = \alpha\mu^{-1} \ . \tag{5.24}$$

The direction of rays *inside* the sample is denoted $\alpha'$. Snell's law in the small angle approximation gives the relation

$$\alpha^2 = \alpha'^2 + \alpha_c^2 \ . \tag{5.25}$$

The part of the wave describing the z-dependence inside the crystal, z being the inward normal to the surface, is $\psi_z = \exp(ik\alpha'z)$. We neglect then the change in the modulus of the wave vector by entering the crystal since the index of refraction for x-rays is so close to unity. When $\alpha$ is less than $\alpha_c$, (5.25) shows that $\alpha'$ becomes imaginary. For $\alpha \ll \alpha_c$ one can neglect the normal absorption contribution to attenuation and one finds $\psi_z = \exp(-k\alpha_c z)$. The intensity is $\psi_z^2$, so in this limit we find $\Lambda = \Lambda_0$

$$\alpha \ll \alpha_c \quad : \quad \Lambda_0 = (2k\alpha_c)^{-1} \ . \tag{5.26}$$

For the general case we state the result without proof

$$\Lambda^{-1}(\alpha) = \sqrt{2}k \left[ \sqrt{\alpha'^4 + (\mu/k)^2} - \alpha'^2 \right]^{1/2} \ . \tag{5.27}$$

The reader may check that (5.27) has the correct limiting values for $\alpha \ll \alpha_c$ and $\alpha \gg \alpha_c$, and furthermore we note that $\Lambda(\alpha_c)$ is the geometrical mean of $\Lambda_0$ and $\Lambda_\infty(\alpha_c)$

$$\Lambda(\alpha_c) = [\Lambda_\infty(\alpha_c)\Lambda_0]^{1/2} \ . \tag{5.28}$$

Figure 5.22 illustrates (5.27) in showing $\Lambda(\alpha)/\Lambda_0$ versus $\alpha/\alpha_c$ in a double-logarithmic plot for the material InSb at an x-ray wavelength of $0.1119\,\text{nm}$.

In summarizing, for a given material and x-ray wavelength the effective penetration depth may be quickly estimated for any combination of $(\alpha_i, \alpha_f)$ by first calculating $\alpha_c$ and $\Lambda_0$ from (5.21 and 26), respectively, then drawing

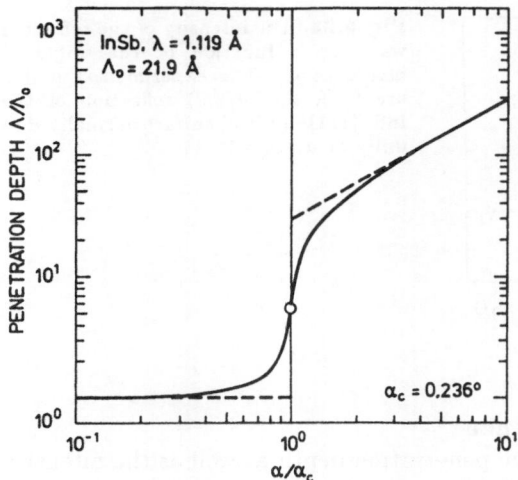

**Fig. 5.22.** Penetration depth $\Lambda(\alpha)$ relative to its minimal value $\Lambda_0$ versus glancing angle relative to critical angle for total reflection. Note that the value at $\alpha = \alpha_c$ is the geometrical mean of the asymptotes, shown as *dashed lines*, extrapolated to $\alpha = \alpha_c$ (1 Å = 0.1 nm)

the asymptotes $\Lambda_0$ and $\Lambda_\infty(\alpha)$ versus $\alpha$ on a double-log plot together with the geometrical mean point $\Lambda(\alpha_c)$, then sketching the full curve $\Lambda(\alpha)$ and finally using the "parallel resistor law", (5.23), to find $\Lambda_{\text{eff}}$.

### b) Field Strength
Let the electrical field strengths of the incident, reflected and refracted wave be $E_i$, $E_r$ and $E_i'$, respectively. Continuity across the interface implies

$$E_i' = E_i + E_r \ , \tag{5.29}$$

and for $\mu = 0$ one finds the simple relation, see (5.5),

$$E_r = (\alpha_i - \alpha_i')/(\alpha_i + \alpha_i')E_i \ . \tag{5.30}$$

An interesting consequence occurs for $\alpha \simeq \alpha_c$ because then $E_r = E_i$ and by (5.29) $E_i' = 2E_i$, i.e. the evanescent wave intensity is 4 times the incident wave intensity! For $\alpha_i \ll \alpha_c$, the incident and reflected wave have opposite signs, $E_r = -E_i$, i.e. they interfere destructively and the evanescent wave intensity tends to zero. Striking experimental confirmation of these effects has been reported by *Feidenhans'l* et al. [5.20]. In a study of the $(3 \times 3)$ InSb($\bar{1}\bar{1}\bar{1}$) surface the intensity, solely due to the reconstructed surface and therefore independent of the penetration depth $\Lambda(\alpha_i, \alpha_f)$, was measured versus the incident angle $\alpha_i$. The result is shown in Fig. 5.23. The full line represents $(E_i'/E_i)^2$ as discussed above. The value is not 4 at $\alpha_i/\alpha_c = 1.0$ because the effect of a finite absorption coefficient $\mu$ is included. One finds perfect agreement with the measured intensity variation with no adjustable parameters except a scale factor fixing the intensity to unity for $\alpha_i/\alpha_c \gg 1$.

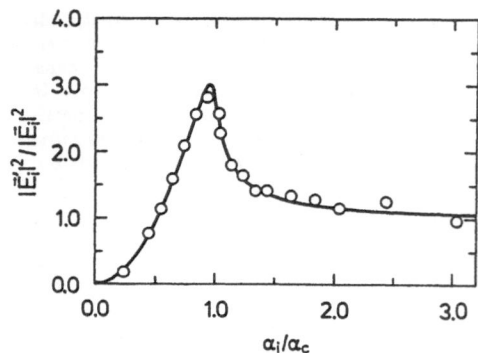

**Fig. 5.23.** The intensity of the refracted wave as a function of the angle of incidence $\alpha_i$. The experimental points are from the (4/3,0) reflection of the InSb($\bar{1}\bar{1}\bar{1}$)- (3 × 3) surface normalized to unity at $\alpha_i/\alpha_c \gg 1$

### c) Summary of Grazing Incidence

At grazing incidence and/or exit the penetration depth as well as the intensity of the wave inside the sample (which can be considered as the wave incident on the diffracting surface atoms) depend drastically on the grazing angle $\alpha$. Incident and exit angles enter in a symmetric way due to the reciprocity principle. In particular,

i)   the penetration depth is only a few atomic spacings when $\alpha_i$ and $\alpha_f$ is somewhat less than the critical angle for total reflection (Fig. 5.22), and

ii)  the evanescent wave for $\alpha_i \simeq \alpha_f \simeq \alpha_c$ is considerably enhanced relative to the incident wave, but angular divergence and normal absorption diminishes the effect.

Although these effects may be advantageous for a given experiment they also imply that the line-up must be accurate to ensure constant $\alpha_i$ and $\alpha_f$ for the different Bragg orientations of the crystal, so that reliable structure factors can be obtained from integrated intensities.

### 5.4.2 Principles of Surface Crystallography

### a) Bragg Scattering

The basic advantage of diffracting x-rays rather than low-energy electrons, the conventional diffraction probe for surface structures due to the penetration depth of only a few atomic layers, is the relatively weak interaction between the x-ray photon and an atom. This fact allows us to calculate the wave scattered from a small crystal by superposition of the scattering of

i)    the electrons of each atom,
ii)   the atoms of each unit cell, and
iii)  the unit cells of the crystal lattice.

We consider a scattering process with wave-vector transfer $Q \equiv k_{in} - k_{out}$, where $k_{in}$ and $k_{out}$ are the wave vectors of the incident plane wave with amplitude $A_0$ and the scattered wave with amplitude $A(Q)$ at a large distance R from the diffracting object.

The scattering amplitude of a single electron is $r_0 = e^2/mc^2$, so the scattering amplitude of an atoms of type $\eta$ with electron density $\varrho_\eta(r)$ becomes

$$f_\eta(Q) = r_0 \int \varrho_\eta(r) \exp(iQ\cdot r)dr \ . \tag{5.31}$$

The scattering amplitude from a unit cell with atoms of type $\eta$ located at position $\eta$ is then

$$F(Q) = \sum_\eta \exp[-B_\eta(Q/4\pi)^2]f_\eta(Q)\exp(iQ\cdot\eta) \ . \tag{5.32}$$

Here $\exp[-B_\eta(Q/4\pi)^2]$ is due to the atomic vibrations of the atom labelled $\eta$. The $4\pi$ is only crystallographic convention.

Finally, the scattering amplitude of a lattice composed of unit cells located at lattice vectors $n$ involves the summation of phase factors

$$G(Q) = \sum_n \exp(iQ\cdot n) \ . \tag{5.33}$$

We then find

$$A(Q) = A_0 R^{-1}F(Q)\cdot G(Q) \ . \tag{5.34}$$

The measured intensity is $|A(Q)|^2$ properly integrated over the experimental resolution volume.

The aim of the diffraction study is then to determine the atomic positions $\eta$ in the unit cell from the measured intensities.

Below we shall evaluate $G(Q)$ for the special geometry used in surface investigations, that is a semi-infinite crystal with the surface parallel to a certain set of atomic planes in the bulk, possibly with an overlayer of adsorbed or chemisorbed atoms or with a reconstructed surface with a different unit cell than the bulk. In conventional crystallography evaluation of $G(Q)$ for a small crystal leads to Bragg's selection rule that $Q$ must coincide with a lattice vector of the reciprocal lattice forming an array of Bragg *points*. For the surface geometry we shall see that selection rules also involve Bragg *rods*, partly from the termination of the bulk crystal and partly from the overlayer or the reconstructed surface layer.

The choice of unit cell is not unique. For the surface study geometry it is particularly convenient to chose a unit cell with two basis vectors in the plane of the surface and the third basis vector pointing into the bulk. This is probably best illustrated by way of an example and we have chosen the lattice of InSb terminating in the (111) plane, that is with a surface normal along the cube diagonal of the conventional cubic unit cell.

## b) Unit Cell in Surface Geometry, InSb(111)

In Fig. 5.24 is shown the 2-dimensional honeycomb lattice. The smallest repeat unit or *unit cell* is shown as the shaded parallelogram. Denoting every other of the six corners in the honeycomb as being A sites and the remaining three corners as being B sites we see that there is one B site inside the unit cell and an A site on each of the four corners so that each A site is shared by 4 unit cells, i.e. one A site and one B site per unit cell.

The A sites of the honeycomb form an A lattice, as shown in the top part of Fig. 5.25. The triangular A lattice has the same unit cell as the honeycomb lattice but of course with only one site per unit cell. The B atoms occupy the center of every other triangle in the A lattice and a stack of alternate A and B layers on top of each other forms therefore a possible closest packing of spheres. Such a 3-dimensional lattice is called the hcp lattice (*h*exagonal *c*losest *p*acked). Another possible closest packing stack of triangular lattice layers is the sequence ABC ABC ..., where C denotes the center points of the honeycombs in Fig. 5.25. This stacking sequence occurs in the fcc lattice (*f*ace *c*entered *c*ubic) viewed along the cube diagonal as the stacking direction.

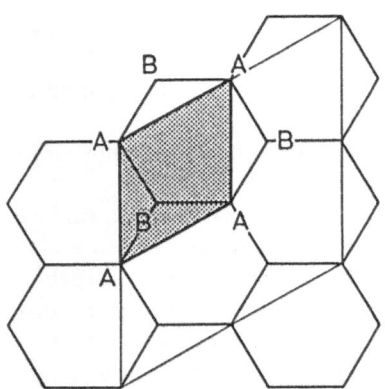

**Fig. 5.24.** The unit cell of a honeycomb lattice is the shaded parallelogram with two types of sites (A and B) per cell

**Fig. 5.25.** *Upper part:* Top view along the cube diagonal of an fcc lattice. The unit cell *(shaded)* contains three atoms, A, B and C sites. *Lower part:* Side view of the fcc lattice with stacking sequence A, B, C, A ... with another interpenetrating fcc lattice a, b, c, a ... displaced 1/4 of a cube diagonal (the diamond lattice)

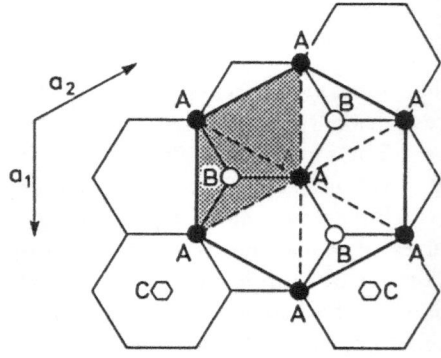

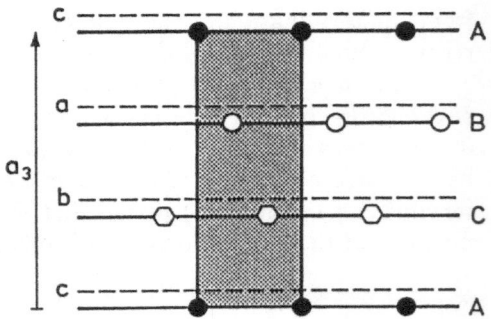

Figure 5.25 can therefore represent a top view of an fcc lattice cut normal to the cube diagonal so the A atoms are the top layer, the B atoms the second layer, the C atoms the third layer, and so on.

This stacking is shown in a side view in the lower part of Fig. 5.25, the vertical direction being the cube diagonal in the description of the lattice with a fcc unit cell. This side view shows in addition to A, B, C planes of atoms also a, b, c planes. The a atoms are supposed to be exactly below the A atoms, the b atoms below the B atoms, etc. The distance between the A plane and the a plane is 1/4 of the repeat distance A to A and the total assembly of A, B, C and a, b, c atoms therefore represents *two* interpenetrating fcc lattices, one with atoms denoted by capital letters and the other with atoms denoted by small letters. These two lattices are displaced 1/4 of a cube diagonal with respect to each other. This is the so-called diamond lattice and it is typical for semiconductor crystals such as Si, Ge and GaAs, InSb etc. In the latter case In atoms occupy one fcc lattice, Sb atoms the other.

The a-B, b-C and c-A layers are particularly close together (1/12 of the cube diagonal) and such bi-layers may be considered as one composite layer. Thus the surface layer of an InSb(111) crystal would be a honeycomb lattice, as shown in Fig. 5.24 with In alternating with Sb around the corners of the honeycomb if the bulk structure terminated undistorted at the surface. The resulting unit cell is spanned by the three basis vectors $a_1$, $a_2$ and $a_3$ also shown in Fig. 5.25.

### c) Reciprocal Lattice. Bragg Points

Having discussed in detail the particular choice of unit cell for surface geometry, we can evaluate the lattice summation of $G(Q)$ from (5.33). The lattice vector $n$ is given by the integer coordinates $n_1, n_2, n_3$:

$$n = n_1 a_1 + n_2 a_2 + n_3 a_3 \ . \tag{5.35}$$

The wave vector $Q$ is conveniently expressed by its coordinates $(h, k, \ell)$ in the reciprocal lattice spanned by the vectors $g_1$, $g_2$ and $g_3$:

$$g_1 = 2\pi \frac{a_2 \times a_3}{a_1 \cdot (a_2 \times a_3)} \quad \text{and cyclic permutations} \ . \tag{5.36}$$

The reciprocal lattice points corresponding to the real space 2-dimensional unit cell of Fig. 5.24 are shown in the lower part of Fig. 5.26 as filled circles.

In evaluating $G(Q)$ we get

$$\begin{aligned} Q \cdot n &= (h g_1 + k g_2 + \ell g_3) \cdot (n_1 a_1 + n_2 a_2 + n_3 a_3) \\ &= 2\pi (h n_1 + k n_2 + \ell n_3) \ . \end{aligned} \tag{5.37}$$

Neglecting absorption and extinction and assuming the number of unit cells along $a_1$, $a_2$ and $a_3$ to be large, $G(Q)$ will be non-vanishing only for integer values of h, k and $\ell$.

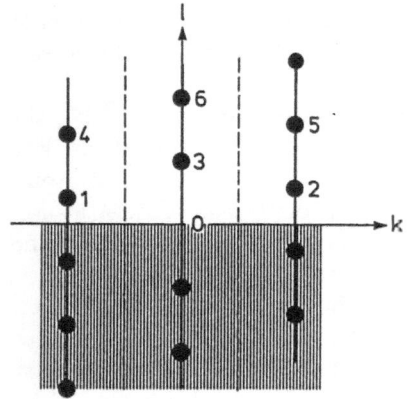

**Fig. 5.26.** Reciprocal lattice *(filled circles)* corresponding to the diamond lattice with surface geometry unit cell. Only lattice points with nonvanishing structure factor are shown. The half-integer lattice points corresponding to a (2 × 2) unit cell are shown as *open circles*. The light-shaded area in the *top part* is inaccessible in glancing incidence geometry

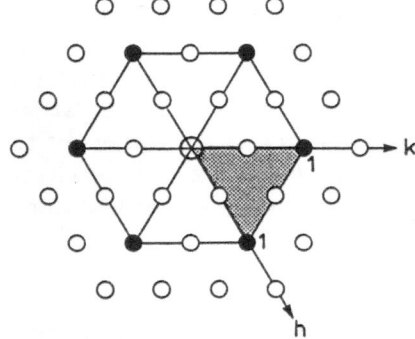

The structure factor (5.32) implies a further restriction on $(h, k, \ell)$. A,B,C stacking of identical layers with the same thermal vibration and atomic form factors (Fig. 5.25) will in the evaluation of $F(Q)$ involve the sum $\exp(iQ \cdot \eta_A) + \exp(iQ \cdot \eta_B) + \exp(iQ \cdot \eta_c)$ where $\eta_A = 0$, $\eta_B = 2/3a_1 + 1/3a_2 + 1/3a_3$, $\eta_c = 1/3a_1 + 2/3a_2 + 2/3a_3$. As $\eta_c = 2\eta_B + a_1$ we find $F(Q) \propto 1 + e^{i\psi} + e^{i2\psi}$ with $\psi = 2\pi(2/3h + 1/3k + 1/3\ell)$ so that $F(Q)$ is non-vanishing only if

$$(2h + k + \ell) = 3j , \quad j \text{ integer} . \tag{5.38}$$

Only such values of $(h, k, \ell)$ are shown as filled circles in the side view (top) and top view (bottom) of the reciprocal lattice in Fig. 5.26.

The surface geometry implies that only positive values of $\ell$ are experimentally accessible since both the incident and scattered wave are on the surface side of the crystal. However, as is apparent from the top part of Fig. 5.26, the inaccessible region for say $k = +1$, $\ell < 0$ is by symmetry equivalent with regard to intensity to the accessible region $k = -1$, $\ell > 0$. More precisely, $|F_{hk\ell}| = |F_{-h-k-\ell}|$ which is known as Friedel's inversion law.

## d) Reconstructed (n × m) Surface Structure. Bragg Rods

Although the reconstructed surface structure or an overlayer structure is often different from the truncated bulk crystal structure they may be related in a commensurate way, that is the surface unit cell is an integer multiple of the truncated bulk unit cell. When the basis vectors of the surface unit cell $a_1^s$ and $a_2^s$, are co-linear with the basis vectors of the truncated bulk unit cell, i.e. $a_1^s = na_1$ and $a_2^s = ma_2$, the convention of the "(n × m)" surface structure is used. This convention can be extended to the case where n and m are not integers. For example, it is apparent from Fig. 5.25 that 3 honeycombs have the same area as the full-line hexagon, which however is turned 30° with respect to the honeycomb. In that case the hexagon is denoted as "$(\sqrt{3} \times \sqrt{3})$ R30" reconstruction of the honeycomb. In our prototype case of InSb(111) the surface structure is (2 × 2).

In the reciprocal lattice of Fig. 5.26 the (2 × 2) surface structure implies that both integer and half-integer values of h and k will lead to a non-vanishing lattice sum $G(Q)$. Furthermore, as the summation over $n$ along the surface normal only involves the surface layer itself there will be no dependence on $\ell$ in $G(Q)$ from the surface layer. A non-vanishing $G(Q)$ is therefore not a lattice of Bragg *points* but rather a lattice of Bragg *rods* perpendicular to the surface as also shown in Fig. 5.26 with $G(Q) = G_{sr}(Q)$, the index sr indicating "surface rod". With $N_1$ unit cells along $a_1$ and $N_2$ unit cells along $a_2$ we get

$$G_{sr}(Q) = N_1 N_2 . \tag{5.39}$$

## e) Bragg Rods from Termination at Surface

In discussing Bragg scattering from the bulk crystal, see (5.37), it was assumed that absorption and extinction could be neglected. On the other hand, in Sect. 5.4.1 we discussed the anomalous absorption and the associated surface sensitivity which could be obtained at grazing incidence and/or exit beams.

In this subsection we shall look again on the Bragg scattering from a semi-infinite crystal terminating at the surface taking absorption into account. For grazing incidence at angle $\alpha$ only absorption along the inward normal needs to be considered as the in-plane penetration length is $\alpha^{-1}$ times $\Lambda_{\text{eff}}$ when $\alpha_c \ll \alpha \ll 1$ (Fig. 5.21), and even a larger factor bigger than $\Lambda_{\text{eff}}$ when $\alpha < \alpha_c$ (Fig. 5.22).

In this approximation the lattice sum corresponding to (5.33) becomes

$$G(Q) = \sum_{n_1=1}^{N_1} \exp(i2\pi h n_1) \sum_{n_2=1}^{N_2} \exp(i2\pi k n_2)$$
$$\times \sum_{n_3=1}^{N_3} \exp(i2\pi \ell n_3) \exp(-\nu n_3) . \tag{5.40}$$

In the summation over $n_3$ the attenuation of wave *amplitude* is $\exp(-\nu)$ per layer, that is $\exp(-2\nu)$ in *intensity* per layer, or in terms of $\Lambda_{\text{eff}}$, $\exp(-a_3 \Lambda_{\text{eff}})$. The dimensionless absorption parameter $\nu$ in (5.40) is thus

$$\nu = \tfrac{1}{2} a_3 / \Lambda_{\text{eff}} \ . \tag{5.41}$$

Even for $\alpha \ll \alpha_c$, Fig. 5.22 indicates that the value of $\nu$ is much less than unity. For the Bragg points in Fig. 5.26, the lattice sum of (5.40) becomes

$$G(\boldsymbol{Q}) = N_1 N_2 \sum_{n_3=1}^{N_3} \exp(-\nu n_3) \simeq N_1 N_2 \nu^{-1} \tag{5.42}$$

utilizing $N_3 \nu \gg 1$ although $\nu \ll 1$, i.e. the effective number of scattering planes along the inward surface normal is $\nu^{-1}$.

When $\ell$ is non-integer one finds readily, still utilizing $N_3 \nu \gg 1$ and $\nu \ll 1$, that

$$G_{\text{br}}(\boldsymbol{Q}) = N_1 N_2 \frac{1 - \exp(2\pi i \ell - \nu N_3)}{1 - \exp(2\pi i \ell - \nu)} \exp(2\pi i \ell - \nu)$$

$$\simeq N_1 N_2 \frac{-\exp(\pi i \ell)}{\exp(\pi i \ell) - \exp(-\pi i \ell)} \quad \text{or} \tag{5.43}$$

$$|G_{\text{br}}(\boldsymbol{Q})| = N_1 N_2 / [2 \sin (\pi \ell)] \ . \tag{5.44}$$

Equation (5.44) shows that the terminated bulk crystal gives rise to Bragg rods, not independent of $\ell$ as for a single monolayer, but of the same order of magnitude in terms of scattered amplitude as a comparison of (5.44) and (5.39) readily shows. The index br in (5.43 and 44) is short-hand for "bulk rod".

The subject of bulk-rods has been discussed extensively, both theoretically and experimentally, by *Robinson* [5.21].

Regardless of the extreme simplicity of the model for the intensity along the truncation Bragg rods, $I(\ell) \sim \sin^{-2}(\pi \ell)$, it agrees remarkably well with the measured variation. This is shown in Fig. 5.27 for the intensity measured vs. $\ell$ along the $(0, \bar{1})$ rod for $\ell > 0$ and along the $(0, 1)$ rod for $\ell < 0$, c.f. top part of Fig. 5.26. The intensity variation (excluding the region around the bulk Bragg point $(0\bar{1}1)$) spans 4 orders of magnitude and the maximum deviation from the $\sin^{-2}(\pi \ell)$ law shown as the dashed line is not more than about a factor of 2. An improved model [5.21] involving one more fitting parameter $\beta$ describing a certain roughness of the truncated crystal surface gives very good agreement, but a more detailed discussion of this model is beyond the scope of this article.

It should be emphasized that the intensity along a so-called truncation rod involves both the wave scattered from the truncated crystal *and* the wave scattered from the reconstructed surface layer or deposited overlayer. The interference of these waves and the resulting intensity depends on the location of the surface unit cell relative to the truncated bulk unit cell as it will be discussed below.

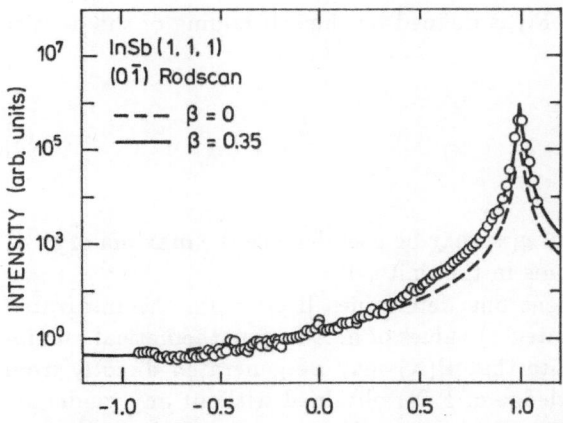

**Fig. 5.27.** Bragg rod due to truncation of an InSb crystal. The surface was contaminated prohibiting the (2×2) reconstructed surface layer. The *dashed line* represent the $\sin^{-2}(\pi\ell)$ law derived in the text including small variations due to atomic form factor and polarization. The *full line* includes a certain roughness model of the surface as described in [5.21]. The data derives from [5.23]

## f) Patterson Function

In our proto-type case of InSb(111) the reconstructed surface structure is (2 × 2). Even for such a relatively small reconstructed unit cell there will a priori be a large number of possible models simply because the (2 × 2) unit cell contains 8 atoms and there is a large number of ways that the position of one or several of 8 atoms may differ from the positions in the (1 × 1) cell including the possibility that one or several atoms may be missing in the (2 × 2) cell. The use of the Patterson function may be very efficient in limiting the number of possible models as pointed out for surface crystallography by *Robinson* [5.22] in his study of the Au(110)–(2 × 1) surface.

In the exposition of electron density in Sect. 5.4.2a a separation into the density of each atom (5.31), and into the thermal vibrations of the atom together with its location in the unit cell (5.32) was made. Although this separation may be useful, it is clear that the structure factor simply expresses the Fourier transform of the electron density within the unit cell irrespective of whether the electrons "belong" to one or another atom in the unit cell.

For simplicity, we consider a one-dimensional unit cell [5.3]. According to the remarks above, the structure factor $F_h$ for a unit cell dimension a is

$$F_h = \int_0^a \varrho(x)\exp[i(2\pi/a)hx]dx \qquad (5.45)$$

irrespective of how many atoms there might be in the unit cell and where they are located.

The Patterson function (P.F.) is defined by the self-folding of the density function:

$$P(X) \equiv \int_0^a \varrho(x)\varrho(x + X)dx .$$  (5.46)

Here the atomic interpretation of $\varrho(x)$ may be useful, since the maxima of $P(X)$ show the distances between atoms in the unit cell.

In the diffraction experiment one determines $|F_h|^2$ from the integrated intensities, in principle for all (integer) values of h. A few mathematical manipulations given below demonstrate that $P(X)$ may be generated directly from the data, see (5.47). This knowledge of $P(X)$ obtained without any model assumptions whatsoever puts very stringent restrictions on possible atomic arrangements in the unit cell and may provide one unique model allowing only refinement of a few parameters in the final interpretation of the data as we shall exemplify in the following section.

A Fourier series expansion of $\varrho(x)$

$$\varrho(x) = \sum_{n=-\infty}^{\infty} C_n exp[-i2\pi(x/a)n]$$

leads to the expansion coefficients $C_n$

$$F_h = \sum_{n=-\infty}^{\infty} C_n \int_0^a exp[i(2\pi/a)(h-n)]dx = aC_h .$$

Introducing the Fourier-transform representation of $\varrho(x)$ with expansion coefficients $C_h = a^{-1}F_h$ into the defining equation for $P(X)$ gives

$$\begin{aligned}
P(X) &= a^{-2} \sum_{h,n} F_n F_h exp(-i2\pi X/a) \int_0^a exp[-i2\pi(x/a)(n+h)]dx \\
&= a^{-1} \sum_h F_{-h} F_h exp(-i2\pi X/a) \\
&= a^{-1} \sum_h |F_h|^2 exp(-i2\pi X/a) .
\end{aligned}$$  (5.47)

Generalization from one to two dimensions is straightforward.

In practise only a finite set of squared structure factors are available for generating P.F. using (5.47) and the data-generated P.F. is therefore not the true P.F. Implications of the necessarily incomplete data-set are discussed in the thesis of *Feidenhans'l* [5.23].

The point is that in x-ray diffraction the squared structure factors $|F_{hk}|^2$ are determined independent of any model from the intensities $I_{hk}$ and one can readily calculate the Patterson function $P(X)$ from the data. We shall now see explicitly how the method was used [5.24] in determining the InSb(111) $2 \times 2$ surface structure.

### 5.4.3 Reconstructed Surface of InSb(111)–(2 × 2)

#### a) Diffractometer Set Up

A typical, modern beam line for surface crystallography is the Risø group instrument at the wiggler beam line W1 in HASYLAB, Hamburg. An overall view is shown in Fig. 5.28.

The x-ray source point is a 32 pole wiggler providing about 32 times the intensity from a bending magnet within an opening angle of 2 mrad in the horizontal plane. The S.R. collimation in the vertical plane is about $mc^2/E \sim 0.1$ mrad. A monochromatic beam is extracted by 2 successive Bragg reflections from two flat Ge crystals in (111) orientation. The first crystal must be water-cooled to remove the heat load of several hundred watts in the S.R. beam.

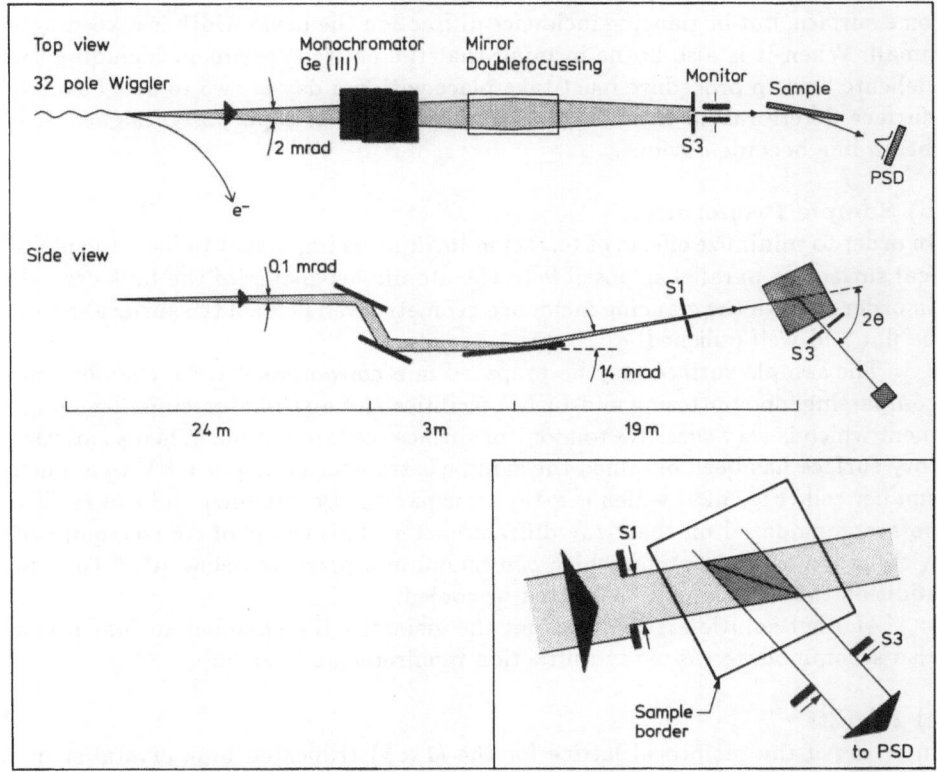

**Fig. 5.28.** Beam line geometry for surface crystallography. PSD stands for position sensitive detector. The inset shows that the diffracting surface area is determined by the overlap of the incident beam defined by slit S1 and the diffracted beam defined by slit S3

The horizontal monochromatic beam is reflected 0.8° or 14 mrad upwards by a gold-coated, toroidal mirror. In addition to providing focussing, the mirror removes higher harmonics from the monochromatic beam.

Since the narrow collimation is in the vertical plane it is advantageous to have this as the scattering plane of the diffractometer. In addition, there are practically no corrections for x-ray polarization in this plane.

The detector is a linear position sensitive proportional counter with the position sensitive direction perpendicular to the scattering plane. The position sensitive detector then measures a typical rod pattern analogous to that given in Fig. 5.23 and provides a convenient on-line monitoring of the quasi-two-dimensional origin of the scattering.

The detailed geometry around the sample is shown in the inset. The diffracting surface area is determined by the overlap of the incident beam and the scattered beam as determined by slits. This area depends (trivially) on $2\theta$ for fixed slit widths and this effect must be corrected for. The sample surface is aligned so the incident glancing angle is independent of the sample rotation angle $\omega$. The glancing angle is typically 5 mrad and with a sample dimension around 10 mm the slit width before the sample is only $10 \times 5 \times 10^{-3}$ mm $= 50\,\mu$m. Not only are there relatively few diffracting atoms on a surface, but in glancing incidence diffraction the beam width is exceedingly small. When it is also borne in mind that the entire experiment including the delicate line-up procedure must take place within a day or two to avoid sample surface deterioration the advantage and necessity of a purpose-designed S.R. beam line become obvious.

### b) Sample Preparation

In order to minimize effects of terracing [5.22], it is important to have the physical surface as parallel as possible to the atomic net-planes of the bulk crystal. In order to have the glancing incidence geometry well defined the surface should be flat and well polished.

The sample surface may be prepared in a conventional UHV chamber encompassing ion sputtering and LEED facilities and also photon emission equipment which is very sensitive to monitor surface contamination. When a satisfactory surface has been obtained the sample is transferred under UHV to a much smaller cell (Fig. 5.29) which is x-ray transparent (Be window) and can readily be accommodated on the x-ray diffractometer. This cell is of course equipped with an ion vacuum pump which can maintain a pressure below $10^{-9}$ Torr. In addition the sample may be heated or cooled.

Another solution is to construct the general UHV chamber so that it can also accommodate the x-ray diffraction requirements [5.25,26].

### c) Results

In terms of the reciprocal lattice for the $(1 \times 1)$ truncated bulk crystal structure shown in Fig. 5.26 the $(2 \times 2)$ reconstruction implies Bragg rods for (h,k) integer or half-integer. The half-integer Bragg rods are solely due to the sur-

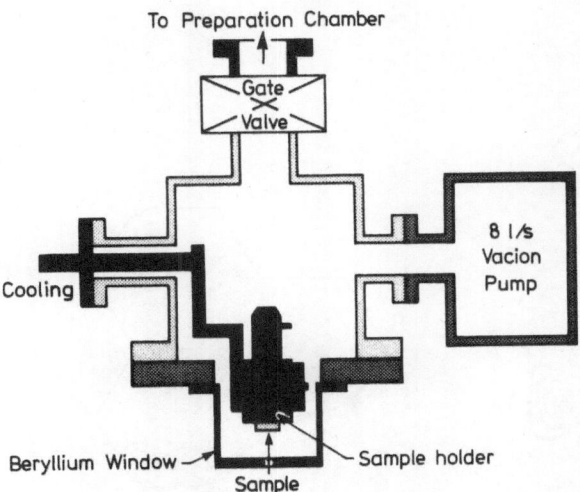

**Fig. 5.29.** X-ray transfer cell. After the sample has been prepared and checked it is transferred into the cell through the gate shown in the top and mounted in the sample holder, the valve is closed, and the cell removed from the preparation equipment. The cell pressure is kept below $10^{-9}$ mbar by constant pumping with a vacuum pump

face layer. When the integrated intensities were corrected for active sample area (sin $2\theta$) and Lorentz factor (sin $2\theta$) 16 non-symmetry related, half-integer structure factors, $|F_{hk}|^2$ were obtained from which the two-dimensional Patterson function (P.F.) was calculated using the two-dimensional analog of (5.47). P.F. in two-dimensions is given as a contour plot in the irreducible part of the real space unit cell, shown in the bottom part of Fig. 5.30. The calculated P.F. is only approximate because only half-integer structure factors were used but it is sufficient to obtain at least a partial model of the correct $(2 \times 2)$ structure. In order to interpret P.F., consider first the undistorted honeycomb in the top part of Fig. 5.30 and imagine one honeycomb lattice sliding on top of an identical lattice. When the sliding direction is along the diameter denoted 4, atomic overlaps between the two lattices occur for the sliding distance being one side (OM) or one diameter (OD = 2 OM) and along this direction P.F. will therefore have peaks at M and D. Another sliding direction is the chord-direction 2. This will lead to overlap and thereby a peak in P.F. at point A. These slidings along the edge, chord and diameter of the hexagon exhaust the possibilities for overlap of the honeycomb lattices and define the irreducible P.F. zone OAD in Fig. 5.30. The observed P.F. maintains the diameter distance OD whereas the chord peak at A apparently is split into a shorter chord (2) and a longer chord (3). The distorted hexagon shown by the full line in the top of Fig. 5.30 is consistent with these 3 peaks and also explains why OM changes direction to O1. The distorted hexagon with its 6 atoms must be an important part of the $(2 \times 2)$ unit cell with maximum 8 atoms, as shown in Fig. 5.31. It does

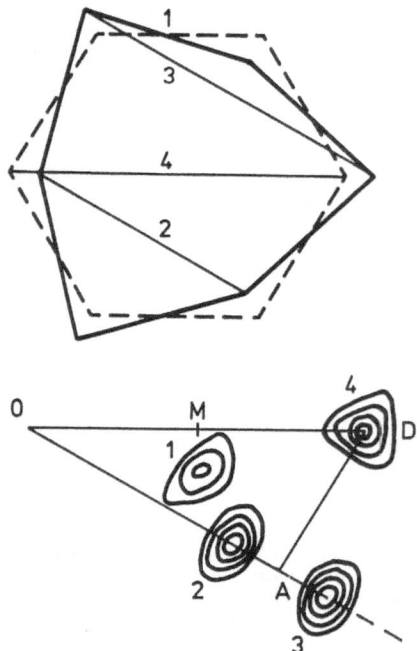

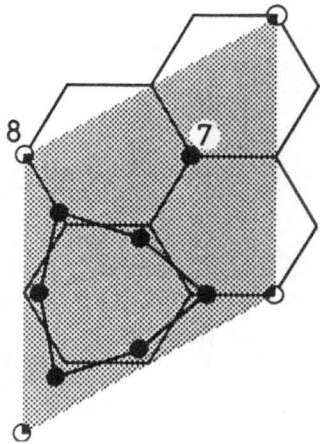

**Fig. 5.30.** The honeycomb gives atomic overlap when sliding along an edge *(1)*, a chord *(2 or 3)* or a diameter *(4)* so the Patterson function has peaks at M, D and A, respectively, in the irreducible cell ODA. The observed Patterson function is indicated by the contours peaking at positions *1, 2, 3* and *4* in the *lower part*. The corresponding distorted honeycomb is shown by *full lines* in the *upper part*

**Fig. 5.31.** The (2 × 2) unit cell of InSb (111) reconstructed surface is the shaded parallelogram. The distorted honeycomb of Fig. 5.30 is entirely embedded by the unit cell. Additional atoms 7 and 8 may or may not belong to the surface unit cell. The data shows that atom 8 is removed and atom 7 remains

not involve the corner atom, marked 8, or the center atom, marked 7. One can readily calculate the discrepancy between observed and calculated $|F_{hk}|^2$, including in the calculation always the distorted hexagon but including or excluding atom Nos. 7 and 8 or both. It turns out that an excellent agreement is obtained only by including atom 7 but excluding the corner atom 8. At this point one can refine the model by parameters desribing the distortion of the hexagon and thermal vibrations. The reconstructed surface structure is thereby solved – almost! Two features have still to be discussed.

One is the registry between the surface unit cell and the truncated bulk unit cell. The easiest way to define the possibilities is to consider the stacking sequence. Let the bulk stacking, read from bulk towards surface, be CBA CBA CB, i.e. terminating with a B layer. Referring to Fig. 5.25 a B layer in the present context is of course a double layer aB, an A layer a double layer cA etc. Will the surface layer then be a distorted A layer, B layer or C layer (3 possibilities) where each possibility can be turned 180°, i.e. 6 possibilities altogether. To settle this question one uses the *integer* reflections, which has contribution from both the truncated bulk crystal, $F_{hk}^{bulk}$, and from the surface layer, $F_{hk}^{surf}$ so that the resulting structure factor becomes

$$F_{hk}^{res} = 4F_{hk}^{bulk} + F_{hk}^{surf}e^{i\phi} \qquad (5.48)$$

to be compared with the measured $|F_{hk}^{res}|^2$. The factor 4 is due to the $(2 \times 2)$ surface unit cell being 4 times larger than the bulk unit cell, and the phase factor $\exp(i\phi)$ is readily calculated for each of the 6 possibilities mentioned. This kind of study was carried out on a GaSb(111)–$(2 \times 2)$ surface [5.23, 28], and led to the conclusion that bulk stacking is maintained also at the surface; in the example above the surface layer is an A layer, not turned 180°.

The other issue is how flat the surface layer is. To this end one can measure the intensity *along* a half-integer Bragg rod. Figure 5.32 taken from [5.23] shows an example along the $(5/2,0)$ rod. The three calculated curves correspond to a layer thickness of 0.0935 nm (the bulk value, full line), zero thickness (dashed curve, $\ell$-variation due to form factor and Debye-Waller factor) and 0.02 nm (dotted curve). Unfortunately, the spectrometer configuration did not allow a wide range of $\ell$ but the data do indicate that the surface layer is very flat, certainly much flatter than the bulk layer value.

In summary, a complete and unambiguous determination of the reconstructed surface of InSb(111) has been possible by means of surface x-ray diffraction using the grazing incidence technique, the Patterson function method to obtain the surface unit cell structure, and the interference between surface and truncated crystal scattering to obtain the registry between the surface unit cell and the bulk crystal unit cell. Finally, the intensity variation along the surface normal measures the thickness of the surface reconstruction.

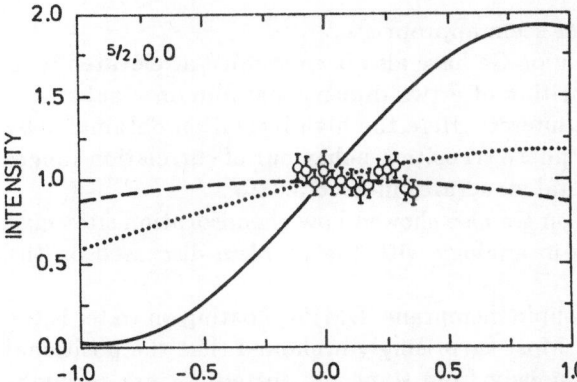

**Fig. 5.32.** Scan along the $(5/2,0)$ rod of the $(2 \times 2)$ InSb(111) reconstructed surface. Models of the surface flatness are indicated by the three curves. The data show that the reconstructed surface is extremely flat

### 5.4.4 Surface Structures Revealed by X-Ray Diffraction

It has been the purpose of Sect. 5.4 up to this point to give a tutorial exposition of x-ray surface crystallography, and to this end one particular study was chosen as an illustrative example.

In this subsection we present a list (Table 5.1) of studies of this kind with references to the original articles when published or to the authors when the work has not yet reached the stage of publication. The list shows that the field is certainly rapidly growing.

**Table 5.1.** X-ray diffraction studies of surface structures

| Class | Material | Year | Ref. |
|---|---|---|---|
| Reconstructed – surfaces | Ge(100)–(2 × 1) | 1981 | 5.18 |
| | Au(110)–(2 × 1) | 1983 | 5.22 |
| | InSb(111)–(3 × 3) | 1985 | 5.27 |
| | InSb(111)–(2 × 2) | 1985 | 5.24 |
| | GaSb(111)–(2 × 2) | 1986 | 5.28 |
| | Si(111)–(7 × 7) | 1986 | 5.29 |
| | W(100) | 1986 | 5.30 |
| | Ge(111)–c(2 × 8) | 1986 | 5.38 |
| Overlayers | Pb/Cu (110) | 1982 | 5.31 |
| | same | 1986 | 5.32 |
| | O/Cu(110) | 1985 | 5.33 |
| | Pb/Ge(111)–($\sqrt{3} \times \sqrt{3}$) | 1986 | 5.34 |
| | Sn/Ge(111)–($\sqrt{3} \times \sqrt{3}$) | 1986 | 5.35 |
| | Sn/Ge(111)–(5 × 5) | 1986 | 5.35 |
| | Sn/Ge(111)–(7 × 7) | 1986 | 5.35 |
| | Phospho Lipid DMPA/$H_2O$ | 1986 | 5.37 |
| Interface | Al/GaAs | 1979 | 5.36 |

A few comments to the list seem appropriate.

The overlayers of Pb on Cu or Ge have also been studied at elevated temperatures to investigate the melting of a two-dimensional film on a substrate, a subject of considerable basic interest. Here the high resolution obtainable by S.R. is important to study the phase transition behaviour of correlation ranges in analogy with the liquid crystal structures in Sect. 5.3.2b.

The study in [5.34] of Pb on Ge also showed how chemisorption sites may be determined unambiguously in analogy with the problem discussed on the previous pages.

The structure of a phospholipid membrane, DMPA, floating on water is the first in-situ study of this kind. Most surprisingly it showed that the positional correlation range grows continuously from some nm to tens of nm throughout the part of the pressure-area phase diagram where fluorescence microscopy had indicated a coexistence of two-dimensional solid and liquid phases with the solid phase patches of order of tens of $\mu$m. The solid phase is thus more subtle than just crystalline. Recently the in-plane structure determined at particular points in surface pressure-area diagram has been supplemented by out-of-plane average structures as determined by reflectivity measurements of the kind described in Sects. 5.1–3.

*Acknowledgements.* It is obvious that the present attempt of a flash-light picture of a new, rapidly growing field has required close collaboration and discussions with many colleagues. A list would have a substantial overlap with the authors in the list of references and is omitted. But, in particular, I would like to thank an older and more experienced scientist than myself, Professor B. Buras, and a younger expert on x-ray surface crystallography, Dr. R. Feidenhans'l, for their scrutinizing reading of the manuscript.

# References

5.1   C. Kunz (ed.): *Synchrotron Radiation*, Topics Curr. Phys., Vol. 10 (Springer, Berlin, Heidelberg 1979)
      E.-E. Koch (ed.): *Handbook on Synchrotron Radiation*, Vol. 1B (North-Holland, Amsterdam 1983)
5.2   M. Born, E. Wolf: *Principles of Optics*, 5. ed. (Pergamon, London 1975)
5.3   B.E. Warren: *X-Ray Diffraction* (Addison-Wesley, Reading, Mass 1969)
5.4   L.G. Paratt: Phys. Rev. **95**, 359 (1954)
5.5   J. Als-Nielsen, F. Christensen, P.S. Pershan: Phys. Rev. Lett. **48**, 1107 (1982)
5.6   J. Als-Nielsen, P.S. Pershan: Nucl. Instr. Methods, **208**, 545 (1983)
      J. Als-Nielsen: Z. Physik B**61**, 411 (1985)
5.7   A. Braslau, M. Deutsch, P.S. Pershan, A.H. Weiss, J. Als-Nielsen, J. Bohr: Phys. Rev. Lett. **54**, 114 (1985)
5.8   P.S. Pershan, A. Braslau, A.H. Weiss, J. Als-Nielsen: (submitted to Phys. Rev.) (1987)
5.9   P.S. Pershan, J. Als-Nielsen: Phys. Rev. Lett. **52**, 759 (1984)
5.10  M. Nielsen, H.B. Möller: Acta Cryst. A**25**, 547 (1969)
5.11  B. Lebech, M. Nielsen: Proc. Neutron Diffraction Conf., Petten, The Netherlands, RCN-234 (1975) p. 466
5.12  M.W. Cole: J. Chem. Phys. **73**, 4012 (1980)
5.13  P.G. de Gennes: *The Physics of Liquid Crystals* (Clarendon, Oxford 1974)
      G. Vertogen, W.H. de Jeu: *Thermotropic Liquid Crystals, Fundamentals*, Springer Ser. Chem. Phys., Vol. 45 (Springer, Berlin, Heidelberg 1987)
5.14  B.M. Ocko, A. Braslau, P.S. Pershan, J. Als-Nielsen, M. Deutsch: Phys. Rev. Lett. **57**, 94 (1986)
5.15  E.F. Gramsbergen, W.H. de Jeu, J. Als-Nielsen: J. Physique, **47**, 711 (1986)
5.16  R.J. Birgeneau, P.M. Horn: Science **232**, 329 (1986)
5.17  D.E. Moncton, R. Pindak: Phys. Rev. Lett. **43**, 701 (1979)
5.18  P. Eisenberger, W.C. Marra: Phys. Rev. Lett. **46**, 1081 (1981)
5.19  L.I. Schiff: *Quantum Mechanics*, 3rd ed. (McGraw Hill, New York) p. 135
      L.D. Landau, E.M. Lifschitz: *Electrodynamics of Continuous Media* (Pergamon, New York 1981) p. 288–289
5.20  R. Feidenhans'l, J. Bohr, M. Nielsen, M. Toney, R.L. Johnson, F. Grey, I.K. Robinson: *Festkörperprobleme* (Adv. Solid State Physics) **25**, 545 (Vieweg, Braunschweig 1985)
5.21  I.K. Robinson: Phys. Rev. B**33**, 3830 (1986)
5.22  I.K. Robinson: Phys. Rev. Lett. **50**, 1145 (1983)
5.23  R. Feidenhans'l: Risø-M-2569 (1986)
5.24  J. Bohr, R. Feidenhans'l, M. Nielsen, M. Toney, R.L. Johnson, I.K. Robinson: Phys. Rev. Lett. **54**, 1275 (1985)
5.25  S. Brennan, P. Eisenberger: Nucl. Instr. Meth. **222A**, 164 (1984)
5.26  P.H. Fuoss, I.K. Robinson: Nucl. Instr. Meth. **222A**, 171 (1984)

5.27  R.L. Johnson, J.H. Fock, I.K. Robinson, J. Bohr, R. Feidenhans'l, J. Als-Nielsen, M. Nielsen, M. Toney: In *The Structure of Surfaces*, ed. by M.A. van Hove, S.Y. Tong, Springer Ser. Surf. Sci., Vol. 2 (Springer, Berlin, Heidelberg 1985) p. 313

5.28  R. Feidenhans'l, M. Nielsen, F. Grey, R.L. Johnson, I.K. Robinson: (submitted to Surf. Sci.)

5.29  I.K. Robinson, W.K. Waskeewicz, P.H. Fuoss, J.B. Stark, P.A. Bennett: Phys. Rev. B33, 7013 (1986)

5.30  M.S. Altman, P.J. Estrup, I.K. Robinson: unpublished

5.31  W.C. Marra, P.H. Fuoss, P. Eisenberger: Phys. Rev. Lett. 49, 1169 (1982)

5.32  S. Brennan, P.H. Fuoss, P. Eisenberger: Phys. Rev. B33, 3678 (1986)

5.33  K.S. Liang, P.H. Fuoss, G.J. Hughes, P. Eisenberger: In *The Structure of Surfaces*, ed. by M.A. van Hove, S.Y. Tong, Springer Ser. Surf. Sci., Vol. 2 (Springer, Berlin, Heidelberg 1985) p. 421

5.34  R. Feidenhans'l, J.S. Pedersen, M. Nielsen, F. Grey, R.L. Johnson: Surf. Sci. 178, 927 (1986)

5.35  J.S. Pedersen, R. Feidenhans'l, M. Nielsen, K. Kjaer, F.Grey, R.L. Johnson (submitted to Surf. Sci.)

5.36  W.C. Marra, P. Eisenberger, A.Y. Cho: J. Appl. Phys. 50, 6927 (1979)

5.37  K. Kjaer, J. Als-Nielsen, C. Helm, L. Laxhuber, H. Möhwald: Phys. Rev. Lett. 58, 2224 (1987)

5.38  R. Feidenhans'l, J.S. Pedersen, M. Nielsen, F. Grey, R.L. Johnson: to be published

5.39  M. Lösche, E. Sackmann, H. Möhwald: Ber. Bunsenges., Phys. Chem. 87, 848–852 (1983)

# 6. Statistical Mechanics of the Liquid Surface and the Effect of Premelting

W. Schommers

With 26 Figures

The *structure* of liquid surfaces is described in terms of *distribution* and *correlation* functions. These functions are introduced on the basis of statistical mechanics. Simple models are discussed for the *single-particle distribution function,* which is important in connection with the determination of the density variation through the liquid-vapour interface. In the description of *thermodynamic functions* (pressure, energy, etc.) of the liquid surface the *two-particle distribution function* $\varrho^{(2)}$ is also of relevance. Approximations are given for $\varrho^{(2)}$, which were often used in the determination of thermodynamics functions.

Since the *harmonic* approximation breaks down in the case of liquids, the phonon density of states $g(\omega)$ is no longer an appropriate quantity for the description of the dynamics, and it is shown that the Fourier transform of the *velocity autocorrelation function* $\psi(t)$, which is more general than $g(\omega)$, becomes important. The behaviour of $\psi(t)$ at the surface is discussed by means of a molecular-dynamics model for liquid argon.

There is strong evidence that some layers near the surface are already in a liquid-like state *below* the melting temperature. This phenomenon is the so-called *effect of premelting.* This effect is discussed at the atomic scale by means of a molecular-dynamics model for a noble-gas (krypton) system, and the few experimental results for different surfaces available till now are briefly described.

## 6.1 The Structure of the Liquid Surface

The structure of *crystalline* solids is given by the geometry of the *unit cell,* and the unit cell is determined by a relatively small number of position vectors. In the case of *disordered* systems (e.g., liquids) a unit cell does not exist and the structure is given by the position vectors of *all* ($\sim 10^{23}$) particles of the system. However, it is not possible to manipulate such a detailed information and, therefore, the structure of disordered systems has to be characterized by certain *averaged* quantities; these quantities are the so-called *distribution functions* and *correlation functions*. In Sects. 6.1.1–5 we will discuss some typical features of these functions for disordered *single-component* systems.

### 6.1.1 General Remarks

In the case of a *classical* system of N atoms in a volume V at temperature T, the probability distribution in the phase space is given by [6.1,2]

$$\exp\left(-\frac{H}{k_B T}\right) , \tag{6.1}$$

where H is the classical Hamiltonian, and $k_B$ is the Boltzmann constant. If we are interested in *distribution functions* in coordinate space we have (6.1) to integrate over all N momenta. The probability

$$P^{(N)}(r_1, r_2, \ldots, r_N) dr_1 dr_2 \ldots dr_N$$

that particle 1 will be found in volume element $dr_1$ around the position $r_1$, particle 2 in $dr_2$ around $r_2$, ..., particle N in $dr_N$ around $r_N$ is given by

$$P^{(N)}(r_1, r_2, \ldots, r_N) dr_1 dr_2 \ldots dr_N$$
$$= \frac{1}{Z_N} \exp\left(-\frac{U(r_1, r_2, \ldots, r_N)}{k_B T}\right) dr_1 dr_2 \ldots dr_N , \tag{6.2}$$

where $U(r_1, r_2, \ldots, r_N)$ is the potential energy and $Z_N$ is the N-body partition function

$$Z_N = \int \ldots \int \exp\left(-\frac{U(r_1, r_2, \ldots, r_N)}{k_B T}\right) dr_1 dr_2 \ldots dr_N . \tag{6.3}$$

The probability that n atoms ($n \leq N$) will be in $dr_1$ around $r_1$, ..., $dr_n$ around $r_n$, regardless of the other N-n particles is obviously expressed by

$$P^{(n)}(r_1, \ldots, r_n) dr_1 \ldots dr_n$$
$$= \left\{ \frac{1}{Z_N} \int \ldots \int \exp\left(-\frac{U(r_1, r_2, \ldots, r_N)}{k_B T}\right) \right.$$
$$\left. dr_{n+1} \ldots dr_N \right\} dr_1 \ldots dr_n . \tag{6.4}$$

In order to avoid the labelling of the N atoms we have to multiply $P^{(n)}$ by the factor $N!/(N-n)!$ and we obtain the function

$$\varrho^{(n)}(r_1, r_2, \ldots, r_n) = \frac{N!}{(N-n)!} P^{(n)}(r_1, r_2, \ldots, r_n) . \tag{6.5}$$

$P^{(n)}(r_1, r_2, \ldots, r_n)$ is normalized to unity and we get

$$\int \ldots \int \varrho^{(n)}(r_1, r_2, \ldots, r_n) dr_1 dr_2 \ldots dr_n = \frac{N!}{(N-n)!} . \tag{6.6}$$

$\varrho^{(n)}(r_1,\ldots,r_n)dr_1\ldots dr_n$ is the probability that one atom of the system will be in $dr_1$ at $r_1$, another in $dr_2$ at $r_2$, .... $\varrho^{(n)}$ is the so-called *n-particle distribution function*.

In the case of *non-interacting* particles we have, see (6.4),

$$P^{(n)}(r_1, r_2,\ldots,r_n) = P^{(1)}(r_1)P^{(1)}(r_2)\ldots P^{(1)}(r_n) \ , \tag{6.7}$$

and with (6.5) $\varrho^{(n)}(r_1, r_2,\ldots,r_n)$ takes for such a *random system* the following form

$$\varrho^{(n)}(r_1, r_2,\ldots,r_n) = \frac{N!}{N^n(N-n)!}\varrho^{(1)}(r_1)\varrho^{(1)}(r_2)\ldots\varrho^{(1)}(r_n) \ . \tag{6.8}$$

Let us now introduce the *n-particle correlation function* $g_n(r_1, r_2,\ldots,r_n)$ by

$$\varrho^{(n)}(r_1, r_2,\ldots,r_n) = [\varrho^{(1)}(r_1)\varrho^{(1)}(r_2)\ldots\varrho^{(1)}(r_n)]g_n(r_1, r_2,\ldots,r_n) \ . \tag{6.9}$$

In other words, there are *correlations* between the N particles of the system (i.e., deviations from a system with randomly distributed N particles) if the function $g_n(r_1, r_2,\ldots,r_n)$ is different from $N!/(N-n)!N^n$ :

$$g_n(r_1, r_2,\ldots,r_n) \neq \frac{N!}{N^n(N-n)!} \ . \tag{6.10}$$

### 6.1.2 Bulk Liquids

The *bulk liquid* forms a *homogeneous* and *isotropic* system. Thus, the single-particle distribution function $\varrho^{(1)}(r_1)$ must be constant, i.e. independent of $r_1$. In particular, we have, using (6.6),

$$\begin{aligned}\varrho^{(1)}(r_1) &= \varrho^{(1)}(r_2) = \ldots = \varrho^{(1)}(r_n) \\ &= \frac{N}{V} = \varrho_L \ .\end{aligned} \tag{6.11}$$

Within the limit of $N\to\infty$, $V\to\infty$, $\varrho_L = N/V$ we obtain for a *random system* the following relations

$$\varrho^{(n)}(r_1, r_2,\ldots,r_n) = \varrho_L^n \ , \tag{6.12}$$

$$g_n(r_1, r_2,\ldots,r_n) = 1 \ . \tag{6.13}$$

In the case of *interacting* particles we have, using (6.11),

$$\varrho^{(n)}(r_1, r_2,\ldots,r_n) = \varrho_L^n g_n(r_1, r_2,\ldots,r_n) \ . \tag{6.14}$$

225

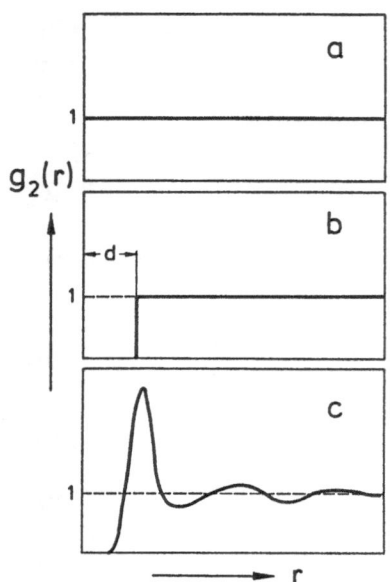

Fig. 6.1a–c. Pair correlation function for several many-particle systems: (a) random system, (b) hard spheres at low density (d is the diameter of the hard spheres), and (c) dense liquid

In the theory of liquids the *two-particle correlation function* $g_2(\mathbf{r}_1, \mathbf{r}_2)$ is of most interest; it depends only on $r_{12} = |\mathbf{r}_1 - \mathbf{r}_2| \equiv r$. $g_2(r)$ is the so-called *pair correlation function* and represents the probability distribution for the distances r of two particles of the system. Examples for $g_2(r)$ are given in Fig. 6.1. Of all the correlation functions only $g_2(r)$ is accessible to experiments [6.3]. Furthermore, within the pair potential approximation

$$U(\mathbf{r}_1, \mathbf{r}_2, \ldots, \mathbf{r}_N) = \frac{1}{2} \sum_{\substack{i,j=1 \\ i \neq j}}^{N} v(r_{ij}), \tag{6.15}$$

where $v(r_{ij})$ is the pair interaction between the particles i and j; the *thermodynamic functions* (energy, pressure, etc.) of the liquid many-particle system are determined by $g_2(r)$ alone, i.e. higher-order correlation functions do not contribute in this case.

**Theoretical Determination of the Pair Correlation Function.** Using (6.4,5,14 and 15) we obtain the following relation between the pair correlation function and the pair potential

$$g_2(\mathbf{r}_1, \mathbf{r}_2) = \frac{V^2}{Z_N} \int \cdots \int \exp\left[-\frac{1}{k_B T} \sum_{i<j} v(r_{ij})\right] d\mathbf{r}_3 \ldots d\mathbf{r}_N . \tag{6.16}$$

Equation (6.16) cannot be used for the determination of $g_2(\mathbf{r}_1, \mathbf{r}_2)$ from a given pair potential $v(r)$, because it is impossible to solve the high-dimensional integrals in (6.16). However, we can eliminate this problem if we consider the spatial variations of $g_2(r_1, r_2)$, rather than the correlation itself.

Let us form $\partial g_2(r_1, r_2)/\partial r_1$. With

$$U(r_1, r_2, \ldots, r_N) = \sum_{j=2}^{N} v(r_{1j})$$

$$+ \text{ terms independent of position vector } r_1 \qquad (6.17)$$

we obtain after some simple manipulations

$$\frac{\partial g_2(r_1, r_2)}{\partial r_1} = -\frac{V^2}{k_B T Z_N} \int \cdots \int \left[ \frac{\partial v(r_{12})}{\partial r_1} + (N-2) \frac{\partial v(r_{13})}{\partial r_1} \right]$$

$$\times \exp\left[ -\frac{1}{k_B T} \sum_{i<j} v(r_{ij}) \right] dr_3 \ldots dr_N \ . \qquad (6.18)$$

With the triplet correlation function

$$g_3(r_1, r_2, r_3) = \frac{V^3}{Z_N} \int \cdots \int \exp\left[ -\frac{1}{k_B T} \sum_{i<j} v(r_{ij}) \right] dr_4 \ldots dr_N \qquad (6.19)$$

we can rewrite (6.18) in the form

$$-k_B T \frac{\partial g_2(r_1, r_2)}{\partial r_1} = \frac{\partial v(r_{12})}{\partial r_1} g_2(r_1, r_2) + \varrho_L \int \frac{\partial v(r_{13})}{\partial r_1} g_3(r_1, r_2, r_3) dr_3 \ . \qquad (6.20)$$

This relation connects $g_2(r_1, r_2)$, $v(r)$ and the triplet correlation function $g_3(r_1, r_2, r_3)$; the high-dimensional integrals do not appear as in the case of (6.16).

Equation (6.20) provides us with a starting point for the theoretical determination of $g_2(r)$ from a given pair potential $v(r)$ provided a model for the triplet correlation function is known. In the literature a number of models are discussed for the triplet correlation function (see, for example, [6.4–14]). Other possibilities for the determination of $g_2(r)$ from $v(r)$ and *vice versa* are indicated, for example, in [6.15]. In this connection it should be emphasized that the determination of $g_2(r)$ from $v(r)$ can be done most reliably by molecular dynamics and Monte-Carlo simulations [6.1,16–19].

### 6.1.3 Liquid Surfaces

#### a) Single-Particle Distribution Function

Let us consider a semi-infinite system with the surface which separates the liquid from the vapour phase defining the x,y-plane. In the surface region the liquid is no longer isotropic and homogeneous as in the bulk liquid and, therefore, the distribution functions and integral relations (for example (6.20)) cannot be described by *scalar* functions of the interatomic separation r, but these

227

quantities and relations depend upon the vector field $r = r_{12} = r_1 - r_2$ and on the distance $z$ as measured along the axis perpendicular to the surface.

In the transition zone between the bulk liquid and the bulk vapour the single-particle distribution function $\varrho^{(1)}(r)$ is no longer a simple constant, see (6.11), but is dependent on the coordinate $z$ (not on x and y):

$$\varrho^{(1)}(r) = \varrho^{(1)}(z) \ . \tag{6.21}$$

Furthermore, the two-particle distribution function must be expressed as

$$\varrho^{(2)}(r_1, r_2) = \varrho^{(2)}(z_1, z_2, |r_1 - r_2|) \tag{6.22}$$

and is not only simply dependent on $r = |r_1 - r_2|$ as in the case of the bulk liquid.

The simplest non-trivial equation relating correlation functions is in the case of *bulk* liquids given by (6.20). On account of the inhomogeneity in the surface zone the single-particle distribution function $\varrho^{(1)}(r)$ is not a constant, see (6.21), as in the case of the bulk liquid and, therefore, the simplest equation of the hierarchy is given by [6.20–24]

$$k_B T \frac{\partial}{\partial r_1} \varrho^{(1)}(z_1) + \int \frac{\partial}{\partial r_1} v(r_{12}) \varrho^{(2)}(z_1, z_2, r_{12}) dr_2 = 0 \ . \tag{6.23}$$

In the derivation of (6.23) it was assumed that the pair potential $v(r)$ remains unchanged as we pass through the surface zone between the liquid and vapour phases. This is certainly a good approximation for noble gases, but probably not an appropriate procedure in a liquid metal where the pair potential is dependent on the density.

Equation (6.23) yields a starting point for the theoretical determination of $\varrho^{(1)}(z)$ from a given pair potential $v(r)$ provided that $\varrho^{(2)}(z_1, z_2, r_{12})$ can be treated reliably. As examples, let us briefly discuss two statistical mechanical models.

*i) Croxton-Ferrier Approach.* With, see (6.9),

$$\varrho^{(2)}(r_1, r_2) = \varrho^{(1)}(r_1) \varrho^{(1)}(r_2) g_2(r_1, r_2) \tag{6.24}$$

and the use of (6.21 and 22), (6.23) takes the form

$$k_B T \frac{1}{\varrho^{(1)}(z_1)} \frac{\partial \varrho^{(1)}(z_1)}{\partial r_1} + \int \frac{\partial}{\partial r_1} v(r_{12}) \varrho^{(1)}(z_2) g_2(z_1, z_2, r_{12}) dr_2 = 0 \ . \tag{6.25}$$

*Croxton* and *Ferrier* replaced the function $\varrho^{(1)}(z_2) g_2(z_1, z_2, r_{12})$ in (6.25) by the bulk properties of the liquid [6.25,26]

$$\varrho^{(1)}(z_2)g_2(z_1, z_2, r_{12}) \longrightarrow \varrho_L g_2(r_{12}) . \tag{6.26}$$

However, instead of the pair interaction $v(r)$, they used an effective one

$$v(r) \longrightarrow \tfrac{1}{2}v(r)\phi(z) . \tag{6.27}$$

Then we have

$$k_B T \frac{1}{\varrho^{(1)}(z_1)} \frac{\partial \varrho^{(1)}(z_1)}{\partial r_1} + \frac{\varrho_L}{2} \int \frac{\partial}{\partial r_1}[v(r_{12})\phi(z_2)]g_2(r_{12})dr_2 = 0 . \tag{6.28}$$

In this way, the explicit appearance of $g_2(r_1, r_2)$ in the presence of the surface is avoided. The solution of (6.28) is given by

$$\varrho^{(1)}(z_1) = \varrho_L \exp\left\{-\frac{\varrho_L}{2k_B T} \int\limits_{-\infty}^{z_1} \int \frac{\partial}{\partial r_1}[v(r_{12})\phi(z_2)]g_2(r_{12})dr_2 dz\right\}, \tag{6.29}$$

where the boundary condition (the z axis is directed from the liquid to the vapour phase)

$$\varrho^{(1)}(-\infty) = \varrho_L \tag{6.30}$$

was used.

The function $\phi(z_2)$ may be readily adjusted to meet any assumed form of "switching off" of the bulk liquid without imposing any unreasonable form on the pair correlation function in the presence of the surface. In [6.25] for $\phi(z_2)$ a Lennard-Jones form was assumed for $z_2 \geq 0$. In particular, $\phi(z_2)$ has been defined in [6.25] as

$$\phi(z_2) = 1.0 , \quad z_2 < 0 \tag{6.31}$$

$$\phi(z_2) = -4\left[\left(\frac{\sigma}{r_0 + z_1 + r_{12}\cos\Theta}\right)^{12} - \left(\frac{\sigma}{r_0 + z_1 + r_{12}\cos\Theta}\right)^6\right] , \quad z_2 \geq 0$$

$$\phi(z_2) = \frac{3(\varrho_L k_B T - p)}{2\pi \varrho_L^2}\left[\int \frac{\partial}{\partial r_1}v(r_{12})g_2(r_{12})r_{12}^3 dr_{12}\right]^{-1} , \quad z_2 \geq z_0$$

where $\sigma$ is the atomic diameter, $r_0$ was chosen to be $1.1224\,\sigma$ and p is the experimental value of the saturated vapour pressure pertaining at the temperature where $g_2(r)$ was measured. The various regions of $\phi(z)$ are shown in Fig. 6.2. The single-particle distribution $\varrho^{(1)}(z)$ obtained from (6.29) for liquid argon at 85 K is shown in Fig. 6.3. The calculation was done with a Lennard-Jones pair potential. It can be seen from Fig. 6.3 that the density profile is *oscillatory* in character.

229

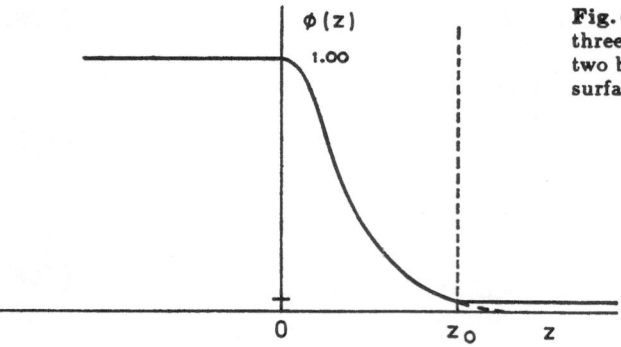

**Fig. 6.2.** The function $\phi(z)$. The three distinct regions refer to the two bulk states $z<0$, $z>z_0$, and the surface zone $z_0 \geq z \geq 0$, see (6.31)

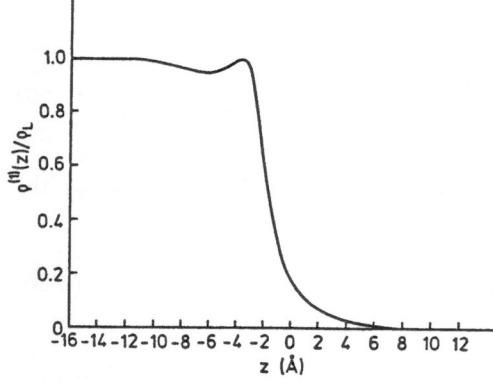

**Fig. 6.3.** The single-particle distribution function $\varrho^{(1)}(z)$ for liquid argon at $T = 85\,K$. $\varrho_L$ is the density in the bulk liquid. Instead of the usual mono-atomic form an oscillation is seen, and the width of the transition zone is approximately two atomic diameters. More details were discussed in [6.25]

*ii) The Model by Nazarian. Nazarian* [6.27] emphasized the importance of statistical mechanical theory of distribution functions in connection with the determination of the density variation through the liquid-vapour interface. He confirmed the *oscillatory* behaviour of $\varrho^{(1)}(z)$ in liquid argon for two specific models

1. $$g_2(r_{12}) = g_2^L(r_{12}) \ , \quad z_1 + z_2 < 0 \tag{6.32}$$

$$g_2(r_{12}) = g_2^V(r_{12}) \ , \quad z_1 + z_2 > 0 \ ,$$

where $g_2^L(r_{12})$ is the pair correlation function in the bulk liquid, and $g_2^V(r_{12})$ in the bulk vapour.

2. The discontinuity is smoothed by taking a linear combination of $g_2^L(r_{12})$ and $g_2^V(r_{12})$ :

$$g_2(z_1, z_2, r_{12}) = g_2^L(r_{12}) + \left[ \frac{z_2}{z_{12}} A(z_2) - \frac{z_1}{z_{12}} A(z_1) \right]$$
$$\times [g_2^V(r_{12}) - g_2^L(r_{12})] \ , \tag{6.33}$$

where $A(z)$ is the unit step function.

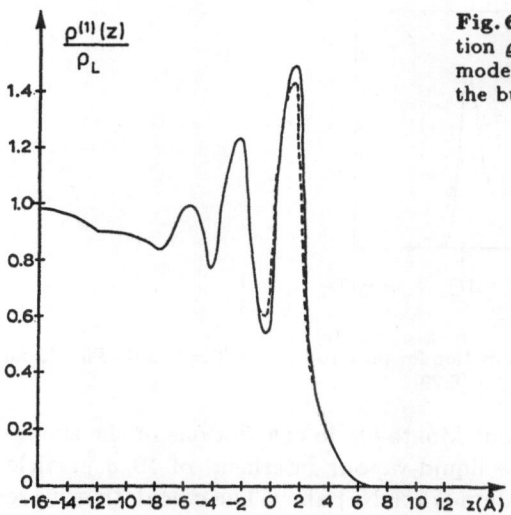

**Fig. 6.4.** The single-particle distribution function $\varrho^{(1)}(z)$ for liquid argon at T = 90 K. (——) model (1); (– – –) model (2). $\varrho_L$ is the density in the bulk liquid

The solution of (6.25) for both cases are shown in Fig. 6.4 for liquid argon. *Nazarian* used a Lennard-Jones potential in the calculations, too. The curves in Fig. 6.4 indicate a crystal-like layer structure in the direction perpendicular to the surface. The oscillations in Fig. 6.4 are much more pronounced than in Fig. 6.3; it is likely that the oscillations in Fig. 6.4 are somewhat overestimated. Nevertheless, the results given in Figs. 6.3 and 4 indicate that the stable density oscillations may develop as a response to the collective constraining field at the liquid surface; note that $\varrho^{(1)}(z)$ is a constant in the bulk liquid.

It should be mentioned that a molecular-dynamics calculation [6.28] for a *two-dimensional* liquid argon system at T = 94.4 K yields a single-particle distribution function with strongly developed density oscillations (Fig. 6.5). For comparison the theoretical density profile obtained from the Croxton-Ferrier approach for three-dimensional liquid argon (Fig. 6.3) is also shown in Fig. 6.5. Although in two-dimensional systems the correlations are expected to be more pronounced than in three-dimensional systems, it is unlikely that the differences between the two curves are entirely due to the different geometries.

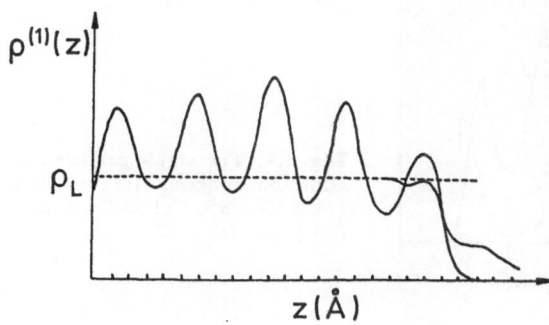

**Fig. 6.5.** A two-dimensional molecular-dynamics study of the single-particle distribution function $\varrho^{(1)}(z)$ for liquid argon at T = 94.4 K. For comparison the result given in Fig. 6.3 (obtained from a three-dimensional analytical calculation) is also shown

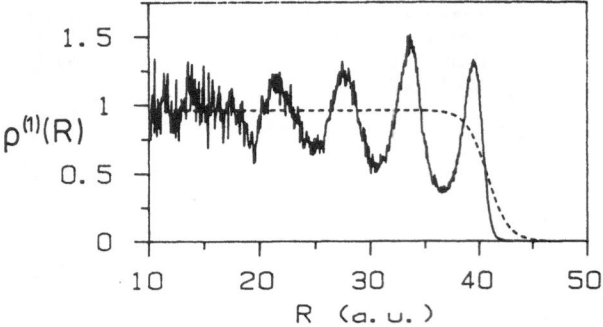

**Fig. 6.6.** The single-particle distribution function for pure sodium at $T = 373$ K. The *dashed curve* denotes the electron reference density n [6.29]

*Gryko* and *Rice* [6.29] carried out Monte-Carlo calculations of the single-particle distribution function at the liquid-vapour interfaces of 1023 particle *clusters* of sodium and a sodium-cesium (75 % Na) alloy. The calculations were performed at a temperature of $T = 373$ K, and the authors used an energy-dependent model pseudopotential introduced by *Woo* et al. [6.30]. The results for $\varrho^{(1)}(R)$, where R is the distance from the surface of the cluster, are shown in Figs. 6.6 and 7. The liquid-vapour transition zone exhibits density oscillations which extend into the bulk of the liquid metal for approximately two to three ion layers. In the case of the sodium-cesium system it was observed that the density of cesium in the surface layer is distinctly higher than in the bulk of the cluster; this effect was explained by the fact that the ionic diameter of cesium is much greater than that of sodium [6.31]. Furthermore, it can be seen from Fig. 6.7 that $\varrho^{(1)}(R)$ for Cs is shifted towards the surface relative to the sodium curve, and *Gryko* and *Rice* [6.29] concluded that the surface of the cluster is composed mainly of cesium ions.

**Experimental Determination of $\varrho^{(1)}(z)$.** It was shown in [6.32–34] that the angular dependence of x-ray reflection at almost grazing angles can be used for the test of density profiles in the liquid-vapour interface. It turned out [6.34]

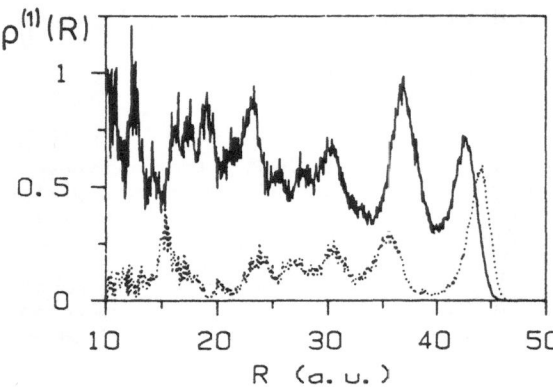

**Fig. 6.7.** The single-particle distribution function for an alloy (75 % Na, 25 % Cs) at $T = 373$ K. (——): Na, (· · ·): Cs

that the density variation of mercury and gallium near the triple point cannot be represented by a *step-like* function. The results suggest *non-monotonic* profiles, and this is consistent with the theoretical results given in Figs. 6.3–7. A detailed study of the structure of the liquid surface by *synchrotron x-rays* is given by Chap. 5 of J. Als-Nielsen.

## b) The Two-Particle Distribution Function

The *thermodynamic functions* of the liquid surface are given in terms of the interatomic potential, the single-particle distribution function $\varrho^{(1)}(z)$, and the two-particle distribution function $\varrho^{(2)}(z_1, z_2, r_{12})$ [6.22]. Thus, not only $\varrho^{(1)}$ is of importance but also $\varrho^{(2)}$. In the following we shall discuss approximations for $\varrho^{(2)}$ which have often been used in the determination of thermodynamic functions.

*Kirkwood* and *Buff* [6.21] assumed in connection with a step-function for $\varrho^{(1)}(z)$

$$\varrho^{(1)}(z) = \varrho_L , \quad z \leq 0 , $$
$$= 0 , \quad z > 0 , \qquad\qquad (6.34)$$

the following ansatz

$$\varrho^{(2)}(z_1, z_2, r_{12}) = \varrho_L^2 g_2^L(r_{12}) , \quad z \leq 0 .$$

In (6.34) the vapour is neglected. A more realistic representation, first proposed by *Green* [6.35] and applied by *Berry* and *Reznek* [6.36] is

$$\varrho^{(2)}(z_1, z_2, r_{12}) = \varrho^{(1)}(z_1)\varrho^{(1)}(z_2)g_2^L(r_{12}) , \qquad\qquad (6.35)$$

where $g_2^L(r_{12})$ is again the pair correlation function in the bulk liquid phase. Clearly, any prescription which retains the bulk liquid distribution throughout cannot accurately describe the gas phase distribution and can lead to incorrect values of the gas density $\varrho^{(1)}(+\infty)$.

*Croxton* and *Ferrier* [6.37,38] expressed the distortion of $\varrho^{(2)}$ in the anisotropic transition zone at the liquid surface in terms of the angular distribution

$$\varrho^{(2)}(z_1, z_2, r_{12}) = \varrho^{(2)}(z_1, r_{12}, \Theta, \phi)$$

$$= \varrho_L^2 g_2(r_{12}) \sum_{\ell \geq}^{\infty} \sum_{m=0}^{L} A_{\ell m}(z_1) P_\ell^m(\cos \Theta)\Phi(m\phi) . \qquad (6.36)$$

The expression (6.36) plays the role of a trial function in the variational determination of the coefficients $A_{\ell m}$. The distribution (6.36) is obviously symmetric with respect to the z-axis and allows one to set $m = 0$ throughout in (6.36). Again, on grounds of symmetry many of the $A_{\ell m}$ values become zero, and *Croxton* and *Ferrier* truncated the series at the $p_z$ harmonic, $\ell = 1$. Then one gets

$$\varrho^{(2)}(z_1, r_{12}, \Theta, \phi) = \varrho_L^2 g_2(r_{12})[A_{00}(z_1)P_0^0(\cos \Theta) + A_{10}(z_1)P_1^0(\cos \Theta)]$$
$$= \varrho_L^2 g_2(r_{12})A_{00}(z_1)[P_0^0(\cos \Theta) + \lambda(z_1)P_1^0(\cos \Theta)] ,$$
$$(6.37)$$

where $P_0^0 = 1$, $P_1^0 = \cos \Theta$ and

$$\lambda(z) = \frac{A_{10}(z)}{A_{00}(z)} ; \qquad (6.38)$$

$\lambda(z)$ is a *hybridization coefficient* between the spherically symmetric bulk modes and the asymmetric modes at the surface. $\lambda(z)$ vanishes for great values of $|z|$.

In order to obtain more flexibility in the trial function (6.37) *Croxton* and *Ferrier* squared the angular term and used the following expression in the analysis

$$\varrho^{(2)}(z_1, r_{12}, \Theta, \phi) = \varrho_L^2 g_2(r_{12})A_{00}(z_1)^2[P_0^0 + \lambda(z_1)P_1^0]^2 . \qquad (6.39)$$

The procedure then consists in minimizing the free energy with respect to this trial function. The *absolute* values of the coefficients $A_{00}$ and $A_{10}$ can be obtained by the identification of the components of (6.39). The result of such an identification procedure is given by [6.37]

$$A_{00}(z_1)^2 = \frac{\varrho^{(1)}(z_1) \int \varrho^{(1)}(z_2)g_2(r)d\mathbf{r}}{2\varrho_L^2[1 + \frac{1}{3}\lambda(z_1)^2] \int\limits_0^\infty g_2(r)r^2dr} . \qquad (6.40)$$

Then, the amplitude of the first harmonic is directly obtained by (6.38).

The hybridization functions $\lambda(z)$ and $A_{00}(z)$ for liquid argon at the triple point are shown in Fig. 6.8. The determination of *thermodynamic* functions on the basis of these results will be discussed in Sect. 6.1.4.

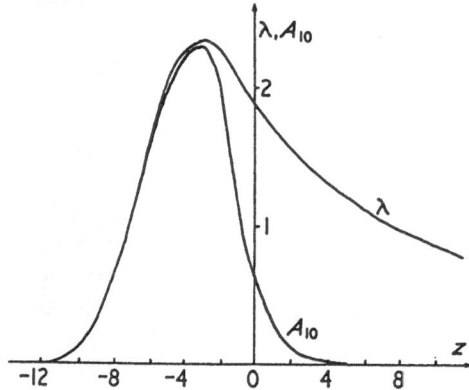

Fig. 6.8. The functions $\lambda = \lambda(z)$ and $A_{10} = A_{10}(z)$ for liquid argon at the triple point [6.38]

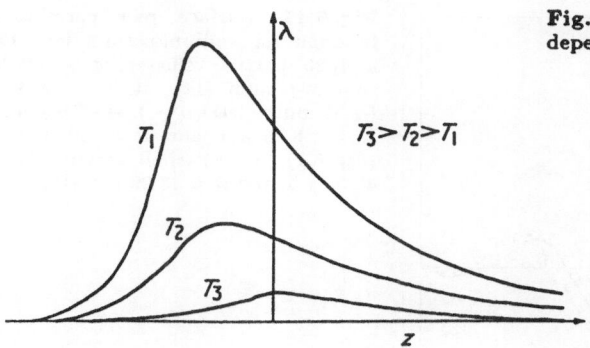

**Fig. 6.9.** The schematic temperature dependence of $\lambda = \lambda(z)$ [6.38]

With increasing *temperature* the liquid surface will *delocalize*, $\varrho^{(1)}(z)$ becoming monotonic, and the local anisotropy will decrease. Consequently, $\lambda(z) \to 0$ as the temperature is increased towards the critical point. In Fig. 6.9 the temperature-dependence of $\lambda(z)$ is shown schematically [6.37,38].

**The Study of Surface Pair Correlations by Monte-Carlo Calculations.** It is instructive to study the pair correlation function $g_2$ for the shells defined by the single-particle distribution function $\varrho^{(1)}$ (see, for example, Fig. 6.6). This was done in [6.29] for the 1023 particle clusters of sodium and a sodium-cesium (75 % Na) alloy which we have already discussed in connection with $\varrho^{(1)}$. In Figs. 6.10 and 11 the surface pair correlation functions at T = 373 K are shown, and the following can be seen:

i)     The first peaks of the surface pair correlation functions have different heights; this effect is obviously due to the fact that the local densities in different shells are not the same. It is well known from bulk studies that the height of the first peak of bulk pair correlation functions depends on the density of the liquid.

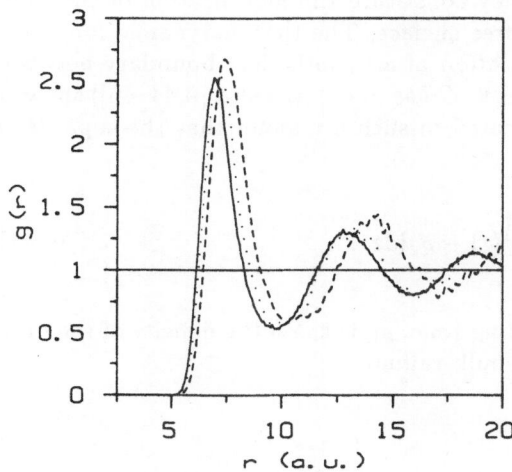

**Fig. 6.10.** Surface pair correlation function for a 1023 particle cluster of sodium at T = 373 K. (——): bulk sodium, ($\cdots$): the shell between R = 33.07 a.u. and R = 35.00 a.u. (Fig. 6.6), (———): the shell between R = 38.86 a.u. and R = 40.79 a.u. (Fig. 6.6)

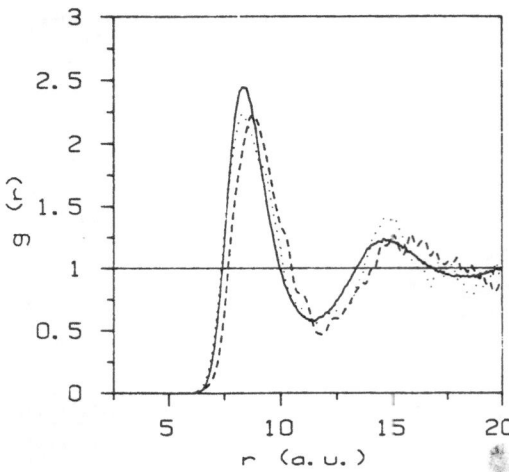

**Fig. 6.11.** Surface pair correlation function for sodium-cesium ions for a 1023 particle cluster of a 75–25 sodium-cesium alloy at $T = 373$ K. (—): bulk metal, ($\cdots$): shell between $R = 38.25$ a.u. and $R = 40.16$ a.u. (Fig. 6.7), (– – –): shell between $R = 42.06$ a.u. and $R = 43.96$ a.u. (Fig. 6.7)

ii)    The position of the first peak of the surface pair correlation function is shifted systematically along the line from the metal interior to the surface. This effect is explained in [6.29] by the finite curvature of the cluster.

### 6.1.4 Surface Thermodynamic Properties

The structure anisotropy in the liquid-vapour transition zone implies anisotropy in the structure-dependent thermodynamic quantities (energy, pressure, etc.). In the statistical mechanical description of the surface thermodynamic properties the distribution functions $\varrho^{(1)}$ and $\varrho^{(2)}$ are of considerable importance. On the other hand, thermodynamic concepts were also used in the calculation of the density variation through the liquid-vapour interface [6.39–42].

**Gibbs Dividing Surface.** The thermodynamic functions at the surface are defined as *excess* quantities, i.e. they constitute the modification of the bulk quantity by the introduction of a free surface. The thermodynamic functions are in general dependent on the *location* of a hypothetical boundary between the bulk liquid and bulk vapour phases. *Gibbs* [6.43] (see also [6.44–46]) showed that this dividing zone must be located in such a manner that the superficial excess density of matter vanishes

$$\int\limits_{-\infty}^{z_0} [\varrho_L - \varrho^{(1)}(z)]\mathrm{d}z = \int\limits_{z_0}^{\infty} [\varrho^{(1)}(z) - \varrho_v]\mathrm{d}z \; , \tag{6.41}$$

where $z_0$ is the location of the dividing zone, $\varrho_L$ is again the density of the bulk liquid, and $\varrho_v$ is the density of the bulk vapour.

**Surface Tension.** The total energy E, the free energy F, and the entropy S are related to each other by the well-known expression

$$E = F + TS .$$ (6.42)

Since the *surface tension* $\gamma$ is identical with the *excess* free energy per unit area (see, for example, [6.46]) we obtain

$$E_s = \gamma + TS_s ,$$ (6.43)

where $E_s$ is the surface excess energy and $S_s$ the surface excess entropy. A similar expression can be obtained if we apply thermodynamic principles to a Carnot cycle:

$$E_s = \gamma - T\frac{d\gamma}{dT}$$ (6.44)

By comparison with (6.43) we get

$$S_s = -\frac{d\gamma}{dT} .$$ (6.45)

Thus, the surface tension and its temperature dependence is of particular interest in the description of thermodynamic properties.

The surface tension $\gamma$ is defined as the work required for the reversible creation of unit area of new surface without change in structure. $\gamma$ can be expressed by [6.21]

$$\begin{aligned} \gamma &= \gamma(T) \\ &= \int_{-\infty}^{\infty} \{p_\perp - p_{||}(z)\}dz , \end{aligned}$$ (6.46)

where $p_\perp$ is the pressure component perpendicular to the surface and $p_{||}$ is the pressure component parallel to the surface. The pressure at a point in an *isotropic* bulk system is given within the pair potential approximation by

$$p = \varrho k_B T - \frac{\varrho^2}{6} \int g_2(r) \frac{dv(r)}{dr} r\, dr ,$$ (6.47)

i.e. the pressure in the bulk is a *constant* and independent of the coordinates. At the liquid surface, however, $g_2$ becomes an *anisotropic* quantity and this results in a *pressure tensor*. Thus, at any point z in the transition zone we have a *normal* pressure component $p_\perp$ and a *tangential* component $p_{||}$. $p_\perp$ and $p_{||}$ are different from each other in the liquid-vapour transition zone, and both quantities are of course identical in the isotropic bulk phases. For a mechanically stable free surface $p_\perp$ must be a constant and identical to the pressure p of the isotropic bulk phase, see (6.47). With $r_{12} = (x_{12}, y_{12}, z_{12})$ the statistical mechanical expressions for $p_\perp$ and $p_\parallel$ take the form [6.46]

$$p_\perp = \varrho^{(1)}(z)k_B T - \frac{1}{2} \int \varrho^{(2)}(z, r)\nabla v(r) \frac{z_{12}^2}{r} d\mathbf{r} , \qquad (6.48)$$

$$p_\parallel = \varrho^{(1)}(z)k_B T - \frac{1}{2} \int \varrho^{(2)}(z, r)\nabla v(r) \frac{x_{12}^2}{r} d\mathbf{r} . \qquad (6.49)$$

Then, the statistical mechanical expression for the surface tension (6.46) becomes

$$\gamma(T) = \frac{1}{2} \iint\limits_{-\infty}^{\infty} \varrho^{(2)}(z, r)\nabla v(r) \frac{x_{12}^2 - z_{12}^2}{r} d\mathbf{r}\, dz . \qquad (6.50)$$

It can be seen from (6.50) that in the case of classical liquids $\gamma(T)$ does not depend explicitly on the single-particle distribution function $\varrho^{(1)}$. Thus, $\gamma(T)$ cannot be used for a sensitive test of model density profiles.

Using (6.34) (KB model) as the model distribution function $\varrho^{(2)}$ we obtain

$$\gamma(T) = \frac{\pi \varrho_L^2}{8} \int\limits_0^{\infty} \nabla v(r) g_2^l(r) r^4 dr . \qquad (6.51)$$

In the case of the model given by (6.37) (CF model) one gets

$$\gamma(T) = -\pi \varrho_L^2 \int\limits_0^{\infty} g_2(r) \frac{\partial v(r)}{\partial r} r^3 dr \int\limits_{-\infty}^{\infty} A_{00}^2(z)\lambda(z) \left[1 + \frac{4}{15}\lambda(z)\right] dz . \qquad (6.52)$$

**Table 6.1.** Surface tension and surface excess energy of liquid argon at the triple point. All calculations have been done with Lennard-Jones pair potentials

|  | KB model (*Shoemaker* et al.) [6.47] | CF model [6.37] | Expt. [6.48] |
|---|---|---|---|
| $\gamma \left[\frac{\text{dyne}}{\text{cm}}\right]$ | 15.6 | 13.48 | 13.45 |
| $E_s \left[\frac{\text{erg}}{\text{cm}^2}\right]$ | 27.08 | 35.35 | 35.01 |

In Table 6.1 numerical results for both models for liquid argon at the triple point are compared with experimental data [6.48]. In Table 6.1 also the numerical results for the surface excess energies are given which were calculated from the following expressions:

*KB model:*

$$E_s = -\frac{\pi \varrho_L^2}{2} \int_0^\infty g_2^L(r) v(r) r^3 dr \tag{6.53}$$

*CF model:*

$$E_s = 2\pi \varrho_L^2 \int_0^\infty v(r) g_2(r) r^2 dr \{ \int_{-\infty}^0 \{A_{00}^2(z)[1 + \tfrac{1}{3}\lambda(z)^2] - 1\} dz$$

$$+ \int_0^\infty A_{00}^2(z)[1 + \tfrac{1}{3}\lambda(z)^2] dz \} \ . \tag{6.54}$$

A more extensive discussion concerning surface thermodynamic properties is given in [6.46].

### 6.1.5 Liquid Metals

In Sects. 6.1.1–4 we have discussed mainly *structural* features of *single-component* systems. Liquid metals consist of *ions* and conduction *electrons* and are therefore at least two-component systems. Let us briefly point out in this subsection how the *statistical mechanics* can be evaluated in the case of such two-component systems.

Since the pioneering work of *Lang* and *Kohn* [6.49] based on the density-functional formalism [6.50,51], surface properties of liquid metals have been studied by several authors (see, for example, [6.52–60]).

In the discussion of the statistical mechanical expressions (Sects. 6.1.1–4) we have assumed that the pair interaction $v(r)$ remains unchanged as we pass through the surface zone between the liquid and the vapour phases. This is certainly a good approximation for noble gases but probably not for liquid metals where $v(r)$ is *density dependent*. Let us briefly discuss some basic characteristics of liquid metal surfaces within the framework of the *pseudo-atom model* for the ions and the electrons. By treating the energy of the electrons as an effective potential for the pseudo-ions, the *statistical mechanics* of the pseudo-ions can be evaluated [6.52,61,62].

The Hamiltonian for the *bulk* liquid of N screened pseudo-ions in a configuration {N} has the form

$$H_{ion} = \sum_{i=1}^N \frac{p_i^2}{2m} + U(\{N\}) \ , \tag{6.55}$$

where the total potential energy is given by

$$U(\{N\}) = Nu(n) + \sum_{i<j} v(r_{ij}, n) \tag{6.56}$$

n is the average density of the conduction electrons ($n = \varrho_L Z$, where Z is the valence of the metal and $\varrho_L$ the density of the ions). The first term u(n) can be interpreted as the self-energy of a pseudo-particle [6.63,64]; u(n) only depends on n and is independent of the arrangement of the ions. u(n) is expressed by the electron-ion pseudopotential and the kinetic, correlation and exchange energies of the electron system. $v(r, n)$ includes the pseudopotential and the dielectric screening function.

From (6.56) follows that the two-component problem is reduced to that of a single-component system, and on the basis of $U(\{N\})$ the system of pseudo-ions is described by classical statistical mechanics (Sect. 6.1.2).

The interaction of the electrons with ions in a configuration with surface will give rise to an averaged electron density profile n(z), where z is again the direction perpendicular to the surface. n(z) will vary from its constant value n in the bulk to zero in the vacuum. By analogy with (6.56) the total potential energy of the pseudo-ions will be of the form described in [6.52–54]:

$$U(\{N\}) = \sum_i u(n(z_i)) + \frac{1}{2} \sum_{i<j} [v(r_{ij}, n(z_i)) + v(r_{ij}, n(z_j))] \ . \tag{6.57}$$

Electrostatic terms have been neglected in (6.57). This can be done if the single-particle distribution function $\varrho_+^{(1)}(z)$ for the ions is identical with n(z)/Z. The second term in (6.57) is constructed so that there is symmetry between the labels i and j [6.52].

Due to the analogy between (6.56 and 57) also in the case of the liquid metal *with* surface the real *two-component* system of ions and conduction electrons is reduced to a *single-component* system consisting of pseudo-ions, and the statistical mechanics of the liquid metal with surface is described by the classical statistical mechanics of the pseudo-ions. For example, instead of the integro-differential equation (6.23) the following relation holds between $\varrho_+^{(1)}(z)$, $\varrho_+^{(2)}(z_1, \boldsymbol{r}_{12}) \equiv \varrho_+^{(2)}(z_1, z_2, r_{12})$ and $v(r_{12}, n(z))$

$$-k_B T \frac{\partial \varrho_+^{(1)}(z_1)}{\partial z_1} = \varrho_+^{(1)}(z_1) \frac{d}{dz_1} u(n(z_1))$$

$$+ \frac{1}{2} \int d\boldsymbol{r}_{12} \frac{\partial}{\partial z_1} v(r_{12}, n(z_1)) \varrho_+^{(2)}(z_1, \boldsymbol{r}_{12})$$

$$- \frac{1}{2} \int d\boldsymbol{r}_{12} \frac{z_{12}}{r_{12}} \left[ \frac{\partial}{\partial r_{12}} v(r_{12}, n(z_1)) \right.$$

$$\left. + \frac{\partial}{\partial r_{12}} v(r_{12}, n(z_1 + z_{12})) \right] \varrho_+^{(2)}(z_1, \boldsymbol{r}_{12}) \ . \tag{6.58}$$

In other words, for the determination of the pseudo-ion density profile $\varrho_+^{(1)}(z)$ we need the conduction electron density profile n(z), the pair interaction $v(r, n(z))$ and a model for $\varrho_+^{(2)}$. No numerical results based directly on (6.58) seem as yet to be available. Numerical values of the surface tension calculated for liquid Na, K and Al agree well with experimental data [6.54].

## 6.2 Dynamics of the Liquid Surface

In solid-state physics the *phonon concept* is of particular interest. In many cases the solid can be described in the *harmonic approximation*, and in this case the *dynamics* is given completely in terms of non-interacting phonons. The equation of motion for the $k^{th}$ particle with mass m of a monatomic system is expressed in the harmonic approximation by

$$\ddot{u}_i^k + \frac{1}{m} \sum_{\ell,j} \phi_{ij}^{k\ell} u_j^\ell = 0 \; ; \quad i = x, y, z \; , \tag{6.59}$$

where $\phi_{ij}^{k\ell}$ are the force constants and $u_i^k$ is the $i^{th}$ component of the time-dependent displacement from the mean position $r_0^k$. The solutions of (6.59) have the form of *three-dimensional* Bloch functions in the case of a system *without* surface:

$$u_i^k \propto \exp[i(q \cdot r_0^k - \omega t)] \; . \tag{6.60}$$

From the dispersion curves $\omega = \omega(q)$ the *phonon density of states* $g(\omega)$ can be extracted.

In the case of systems *with* free surfaces (semi-infinite crystal) there is only a *two-dimensional* invariance of the force constants and this implies that the normal-mode solutions, (6.60), have the form of *two-dimensional* Bloch functions (Chap. 2) [6.65].

The harmonic approximation breaks down in the case of liquids where the *anharmonicities* cannot be considered as small perturbations, and the phonon density of states $g(\omega)$ is no longer relevant. Thus, we have to find another proper function for the description of the dynamics of liquids. Such a function is the *velocity autocorrelation function*

$$\Psi(t) = \frac{\langle v(0) \cdot v(t) \rangle}{\langle v(0)^2 \rangle} \; , \tag{6.61}$$

where $v(t)$ is the velocity at time t for one atom of the ensemble and the brackets $\langle \ldots \rangle$ denote a statistical average [6.1]. The Fourier transform of $\Psi(t)$ yields a frequency spectrum $f(\omega)$ which is in the case of the harmonic solid just the phonon density of states

$$f(\omega) = g(\omega) \; . \tag{6.62}$$

The frequency spectrum is by definition

$$f(\omega) = \frac{2}{\pi} \int_0^\infty \Psi(t) \cos \omega t \, dt \; , \tag{6.63}$$

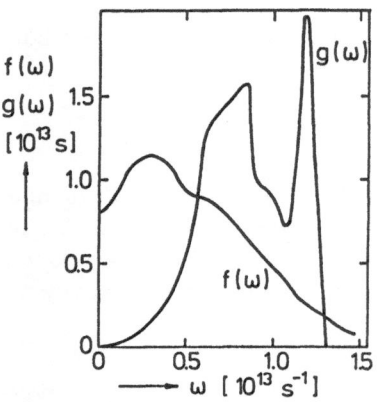

**Fig. 6.12.** Phonon density of states $g(\omega)$ for solid argon in the bulk [6.66] and the frequency spectrum $f(\omega)$ for liquid argon in the bulk [6.67]

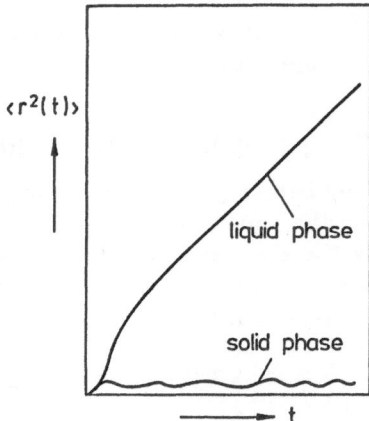

**Fig. 6.13.** The mean-square displacement as a function of time (schematic)

where $f(\omega)$ is normalized to unity:

$$\int_0^\infty f(\omega)d\omega = 1 \ . \tag{6.64}$$

In Fig. 6.12 the phonon density of states $g(\omega)$ for *solid* argon in the bulk [6.66] is compared with the frequency spectrum $f(\omega)$ for *liquid* argon in the bulk [6.67]. It can be seen from Fig. 6.12 that the anharmonicities in the liquid state shift the spectrum to lower frequencies. In particular, the value of $f(\omega)$ at $\omega = 0$ is not zero, and $f(\omega = 0)$ is related to the diffusion coefficient D as follows

$$D = \frac{\pi}{2}\frac{k_B T}{m}f(\omega = 0) \ . \tag{6.65}$$

The mean-square displacement $\langle r^2(t)\rangle$ can also be determined from $\Psi(t)$ and is given by [6.3]:

$$\langle r^2(t)\rangle = \frac{6k_B T}{m}\int_0^t (t-s)\Psi(s)ds \ . \tag{6.66}$$

For the asymptotic form of $\langle r^2(t)\rangle$ one obtains

$$\lim_{t\to\infty}\langle r^2(t)\rangle = 6Dt + \text{const}. \tag{6.67}$$

The mean-square displacement for both the solid and the liquid phase are shown schematically in Fig. 6.13.

The frequency spectrum $f(\omega)$ for liquids can be measured by inelastic neutron scattering experiments [6.3]:

$$f(\omega) = \lim_{q \to 0} \frac{2m}{k_B T} \frac{\omega^2}{q^2} S_s(q, \omega) \; , \tag{6.68}$$

where $S_s(q, \omega)$ is the incoherent scattering law and this can be directly measured.

The experimental determination of $f(\omega)$ at the surface should, in principle, be possible with the help of the inelastic scattering of atoms and electrons (Chap. 1). However, these techniques have not yet been applied in the study of systems with strong anharmonicities like liquids.

The *anisotropy* at the liquid surface suggests that the dynamical behaviour *normal* to the surface is different from that *parallel* to the liquid surface. Thus, at the surface two velocity autocorrelation functions may be defined

$$\Psi(t)_{\parallel} = \frac{\langle v_{\parallel}(0) \cdot v_{\parallel}(t) \rangle}{\langle v_{\parallel}(0)^2 \rangle} \; , \tag{6.69}$$

$$\Psi(t)_{\perp} = \frac{\langle v_{\perp}(0) \cdot v_{\perp}(t) \rangle}{\langle v_{\perp}(0)^2 \rangle} \; . \tag{6.70}$$

Less is known about the functions $\Psi(t)_{\parallel}$ and $\Psi(t)_{\perp}$, and we shall confine our remarks to the molecular-dynamics calculations on a *two-dimensional* liquid argon model at $T = 94.4\,K$, carried out by *Croxton* and *Ferrier* [6.28]. The single-particle distribution function $\varrho^{(1)}$ given in Fig. 6.5 was also calculated on the basis of this model system.

$\Psi(t)_{\perp}$ for this two-dimensional system is shown in Fig. 6.14, and it can be seen from Fig. 6.14 that $\Psi(t)_{\perp}$ oscillates strongly; these oscillations are much more pronounced than in the bulk of the liquid (broken curve in Fig. 6.14). *Croxton* and *Ferrier* discussed the pronounced oscillatory behaviour of $\Psi(t)_{\perp}$ in terms of the quasi-crystalline structure of the single-particle distribution function (Fig. 6.5). The frequency spectrum $f(\omega)_{\perp}$ perpendicular to the surface (the Fourier transform of $\Psi(t)_{\perp}$) is shown in Fig. 6.15. Instead of (6.64) the norm $f(0)_{\perp} = 1.0$ was used. The frequency spectrum for bulk argon is also plotted for comparison.

Figure 6.16 shows the mean-square displacements normal and parallel to the surface. In the same figure the result for the liquid in the bulk is plotted for comparison. From the linear parts of the curves ($\langle r^2(t) \rangle \sim t$, see (6.67)), Croxton and Ferrier estimated the diffusion coefficients, and they found

$$D_{\perp} = 1.05 \times 10^{-4} \frac{cm^2}{s} \; ,$$

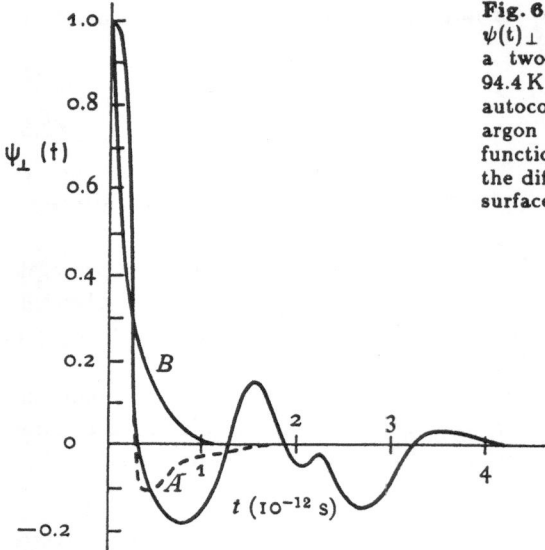

**Fig. 6.14.** Velocity autocorrelation function $\psi(t)_\perp$ perpendicular to the surface for a two-dimensional liquid argon model at 94.4 K [6.28]: Curve $A$ is the velocity autocorrelation function in the bulk of liquid argon [6.67]. Curve $B$ is the Langevin function $\exp(-k_B T/mD_\perp)t$, where $D_\perp$ is the diffusion coefficient perpendicular to the surface

$$D_\parallel = 5.22 \times 10^{-5} \frac{cm^2}{s} \ .$$

Both coefficients are larger than the diffusion coefficient in the bulk of liquid argon $(D = 2.43 \times 10^{-5} \, cm^2/s)$ [6.67].

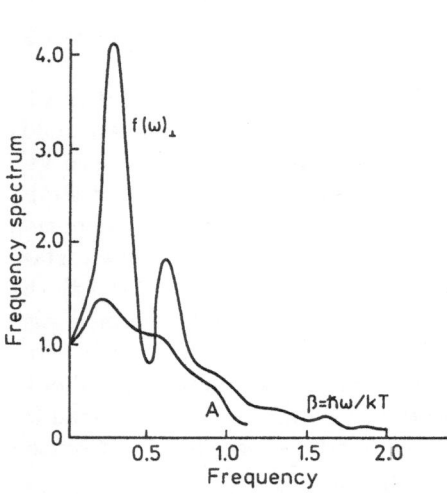

**Fig. 6.15.** The frequency spectrum $f(\omega)_\perp$ perpendicular to the surface (Fourier transform of $\psi(t)_\perp$ (Fig. 6.14)). Curve $A$ is the frequency spectrum in the bulk of liquid argon [6.67])

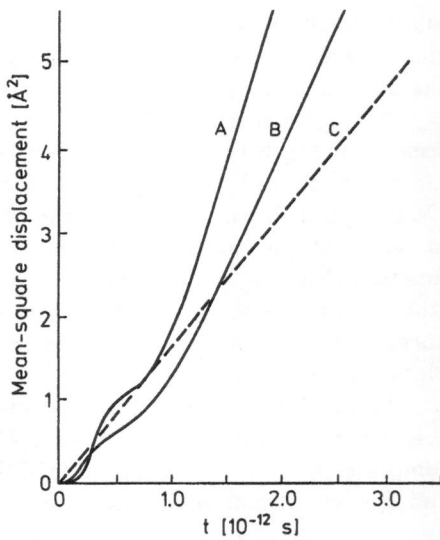

**Fig. 6.16.** Mean-square displacements for liquid argon (Curve $A$: perpendicular to the surface [6.28], Curve $B$: parallel to the surface [6.28], Curve $C$: in the bulk [6.67])

## 6.3 The Effect of Premelting. A Study for Krypton

In the preceding sections we have discussed the liquid with surface, i.e. we have investigated the liquid surface above the melting temperature. However, there is strong evidence that some layers near the surface are already in a *liquid-like* state at temperatures *below* the melting temperature. This phenomenon is the so-called *effect of premelting* and the occurrence of this effect can be understood empirically in terms of the *Lindemann* criterion [6.68]: a solid melts when the vibrational amplitudes of the particles reach a critical fraction ( $\sim 10\,\%$ ) of the nearest-neighbour distance. The vibrational amplitudes of the particles are significantly larger at the surface of crystals than in the bulk and, therefore, we expect that within the framework of *Lindemann*'s criterion the particles are already in a liquid-like state below the melting temperature. In this case the structure and dynamics of the surface particles have to be characterized by the quantities introduced in the preceding sections, i.e. the structure has to be described by distribution functions and the dynamics by the frequency spectrum $f(\omega)$ defined by (6.63) and not by the phonon density of states $g(\omega)$.

In this section we shall discuss the structural and dynamical properties of a noble-gas (krypton) system which shows the effect of surface premelting. The study is based on molecular-dynamics calculations, and it is a continuation of the discussion given in [6.1].

### 6.3.1 Model

The molecular-dynamics model used in the calculations of the surface dynamics is described extensively in [6.1,69,70]. Here its main characteristics will be summarized: The krypton atoms are arranged as a slab-shaped fcc crystal; the two free surfaces being (100) planes. For the (100) surface, which is an open, square lattice, the effect of premelting should be more pronounced than at the (111) close packed triangular surface. The slab consists of 11 layers, and its total number of atoms in the calculations is N = 550. For the interaction between the atoms we have used the potential of *Barker* et al. [6.71]; it is obviously more realistic than the Lennard-Jones potential [6.70]. For this system the classical Hamilton equations were solved by iteration.

It is discussed in [6.69] that effects due to three-body forces should be small. Periodical boundary conditions (pbc) were imposed with respect to translations parallel to the surface. A systematic analysis showed that effects due to the pbc can be ruled out. Our results should not be influenced by effects due to the finite size of the box: we varied the particle number from 108 to 864 and found no size effects in the case of N = 550. The calculations were done for temperatures of 7,70 and 102 K (the melting temperature is 116 K).

### 6.3.2 Results for the Distribution Functions

### a) Single-Particle Distribution Functions for Krypton

Again the coordinates x, y are parallel to the surface and the coordinate z is the direction perpendicular to the surface. Furthermore, we define m = 1 for

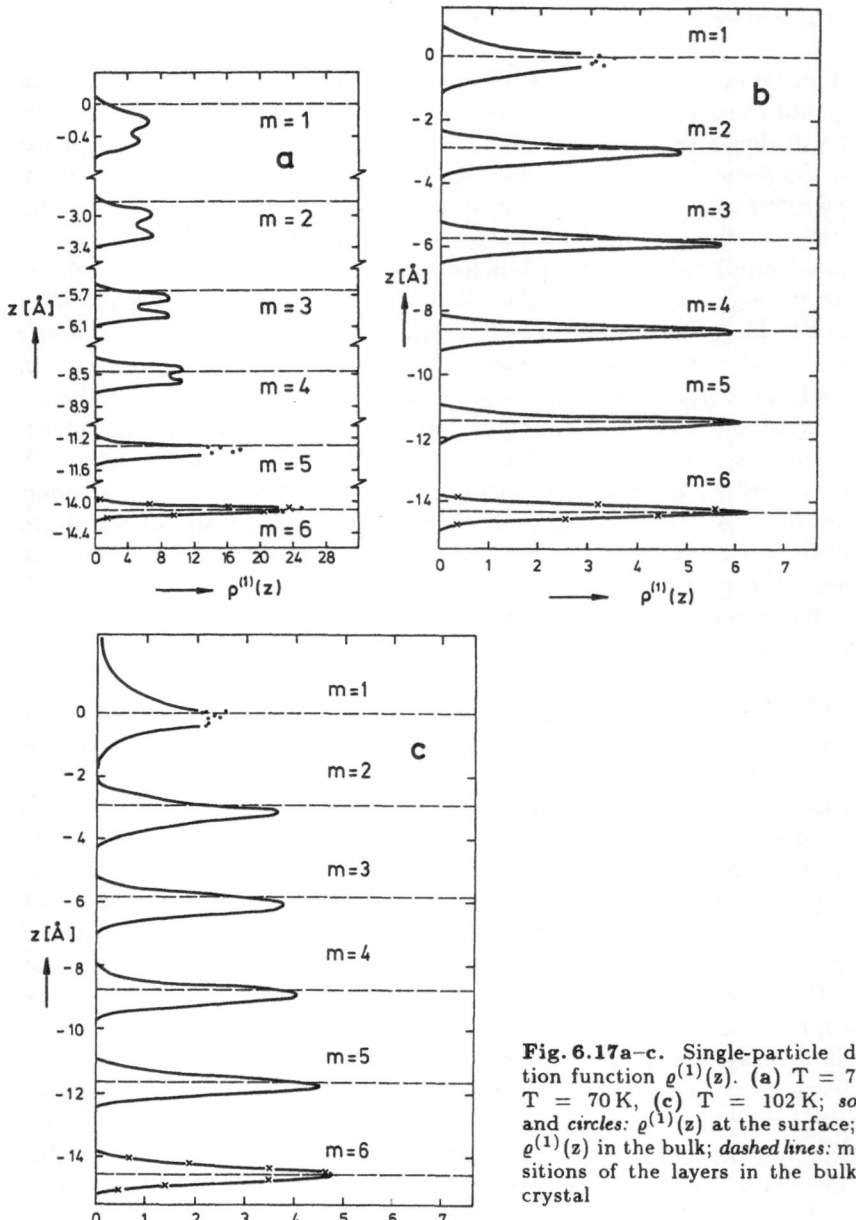

**Fig. 6.17a–c.** Single-particle distribution function $\varrho^{(1)}(z)$. (a) T = 7 K, (b) T = 70 K, (c) T = 102 K; *solid lines* and *circles:* $\varrho^{(1)}(z)$ at the surface; *crosses:* $\varrho^{(1)}(z)$ in the bulk; *dashed lines:* mean positions of the layers in the bulk of the crystal

the outermost layer, m = 2 for the plane just below the upper layer, and so on. In order to investigate how the particles are distributed in the z-direction, we have calculated the *single-particle distribution function* $\varrho^{(1)}(z)$. The results for $\varrho^{(1)}(z)$ for temperatures of 7 K, 70 K and 102 K are shown in Fig. 6.17. It can be seen from Fig. 6.17 that the single-particle distribution for m = 6 (innermost layer) agrees very well for all temperatures with the corresponding data of the bulk. There are, however, large effects in the outermost layer (m = 1): with decreasing m the heights of the peaks are getting smaller and their widths broader. This is consistent with the fact that the mean-square amplitudes of the particles are significantly larger at the surface than in the bulk of the crystal; the molecular-dynamics results of the mean-square amplitudes are shown in [6.1].

In the calculation for T = 7 K we observe a *double-peak structure* which is a typical feature of a classical oscillator; it is not due to double minima in the effective single particle potential but due to a dynamic effect which will be discussed in more detail in a forthcoming paper. For m = 5 and m = 6 this double-peak structure has not been resolved by our MD calculation and this is because the amplitudes of the oscillations are very small for m = 5 and m = 6 (0.07 Å for m = 6, Fig. 6.17). The double-peak structure has not been recognized in the calculations for 70 K and 102 K indicating that the layers are disturbed perpendicular to the surface; there must be a thermally induced *disorder*.

## b) Pair Correlation Function for Krypton

In the preceding subsection we have studied the structure perpendicular (z-direction) to the surface by means of the single-particle distribution function $\varrho^{(1)}(z)$. The points in the x,y-plane of layer m are equivalent and, therefore, the relative distances between the particles are relevant. Thus, it is instructive to study the structure parallel (x,y-directions) by means of the *pair correlation function* $g_2(r)$, i.e., the probability distribution for the *distances* (with respect to the x,y-coordinates) between *two* particles. $g_2(r)$ has been simply computed by means of the coordinates $x_i, y_i$ of all particles which belong to the layer m and it is given by

$$g_2(r) = \frac{1}{2\pi r \Delta r} \frac{n(r, \Delta r)}{\varrho} , \qquad (6.71)$$

where $\varrho$ is the two-dimensional macroscopic density of layer m, and $n(r, \Delta r)$ is the density in the two-dimensional spherical shell around a particle having the radii r and $r + \Delta r$. The z-coordinates of the particles are not considered in $g_2(r)$ but in $\varrho^{(1)}(z)$ (Fig. 6.17). Results for $g_2(r)$ for various layers m are represented in Figs. 6.18–20. It can be seen from Figs. 6.18–20 that $g_2(r)$ for the innermost (m = 6) layer agree very well for all temperatures (7 K, 70 K and 102 K) with that in the bulk of the crystal.

The results for $g_2(r)$ for the outermost (m = 1) layer at T = 7 K agrees well with those for the innermost (m = 6) layer. Thus, at T = 7 K there are no (or almost no) structural effects parallel to the surface. This is consistent with the fact that the mean-square displacements parallel to the surface are almost independent of m. In contrast to this behaviour we have found in Sect. 6.3.2a

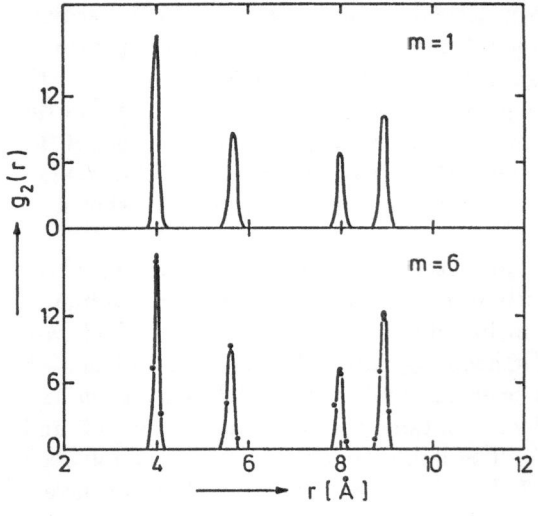

◄— **Fig. 6.18.** Pair correlation function $g_2(r)$ parallel to the surface (T = 7 K) : (—–) surface calculation; (···) bulk calculation

**Fig. 6.19.** Pair correlation function $g_2(r)$ parallel to the surface (T = 70 K) : (—–) surface calculation; (···) bulk calculation

**Fig. 6.20.** Pair correlation function $g_2(r)$ parallel to the surface (T = 102 K) : (—–) surface calculation; (···) bulk calculation

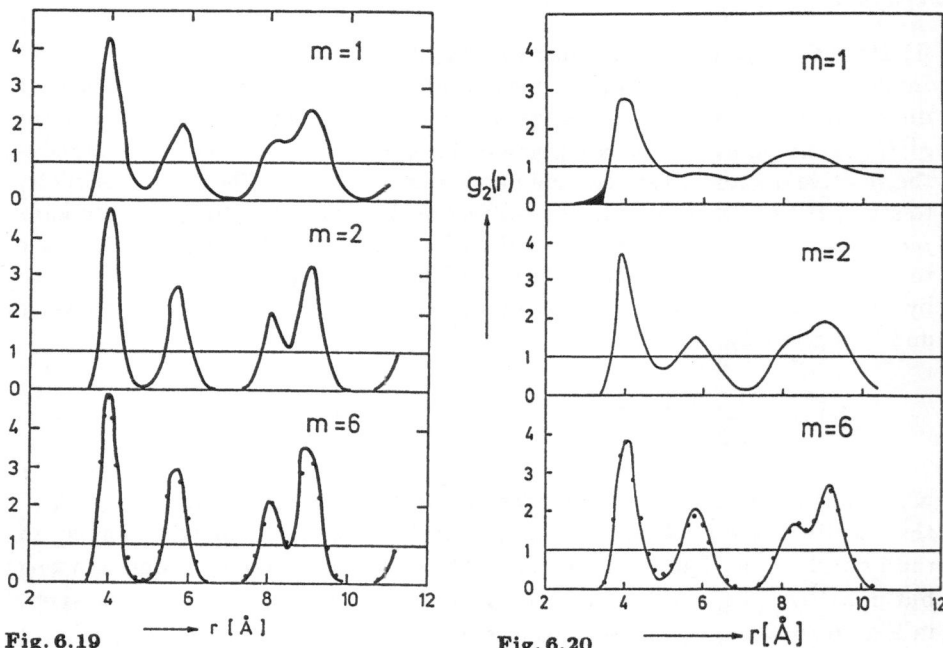

**Fig. 6.19**

**Fig. 6.20**

that in the calculation for 7 K the single-particle distribution function $\varrho^{(1)}$ varies strongly with m (Fig. 6.17).

In the calculation for T = 70 K we observe in $g_2(r)$ relatively large effects: the peaks of $g_2(r)$ of layer m = 1 are broadened and their heights are reduced in comparison to those of layer m = 6 (Fig. 6.19). The mean-square amplitude parallel to the surface is approximately 125 % larger for layer m = 1 than that for layer m = 6. $g_2(r)$ for layer m = 2 already agrees well with that of layer m = 6 (Fig. 6.19). Thus, at T = 70 K only the structure of the outermost layer is distinctly disturbed in the x,y-plane.

In the calculation for T = 102 K (Fig. 6.20) a periodic structure parallel to the surface for layer m = 1 can hardly be recognized; in [Ref. 6.1, Fig. 6.2] three-dimensional plots for the outermost layer at 7 K, 70 K and 102 K illustrate by comparison the degree of disorder at T = 102 K : Although the temperature is still 14 K below the melting point, the outermost layer is extensively disordered. In particular, it can be seen from [Ref. 6.1, Fig. 6.2] that not all the particles are arranged side by side but also one upon another leading to the "black" area in $g_2(r)$ (Fig. 6.20, m = 1). Due to the relatively small energy of the particles at T = 102 K this "black" area in $g_2(r)$ could not be occupied in the case of a *pure* two-dimensional layer, i.e. a layer without degree of freedom perpendicular to the surface.

### 6.3.3 Discussion of the Results for the Distribution Functions

#### a) Single-Particle Distribution Function for the Outermost Layer at T=7 K

The anharmonicities *perpendicular* to the surface are of considerable importance already for the temperature of T = 7 K. Although the mean vibrational amplitudes are 3.7 times larger than in the bulk, they are still small at 7 K and would not give rise to significant anharmonicities if the potential would be symmetrical as in the bulk. This is, however, not the case for the anharmonicities due to the *asymmetry* perpendicular to the surface: Let us consider particles of the outermost layer which are vibrating perpendicular to the surface (see also Fig. 6.21). Upon moving inward (in the direction 2 in Fig. 6.21), they collide with the particles of the second layer and the *repulsive* part of the potential is effective which causes the particles to reverse their direction of motion. However, when the particles are moving outward (in the direction of 1 in Fig. 6.21), they do not collide with any other particles, but the direction of

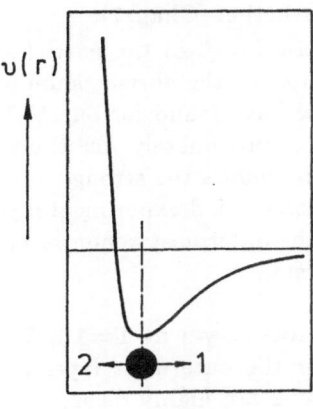

**Fig. 6.21.** Pair potential as a function of the distance r (schematic). The *vertical dashed line* denotes the equilibrium position of a particle at T = 0 K. The arrow towards *1* is the direction towards the vacuum; the arrow towards *2* is the direction towards the second layer of the crystal

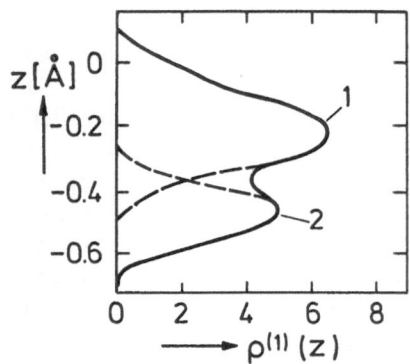

**Fig. 6.22.** Single-particle distribution function for the outermost layer (T = 7 K). The curve is an enlarged representation of Fig. 6.17a (m = 1)

motion is reversed by the *attractive* part of the potential in the crystal. Due to the *asymmetry* of the repulsive and the attractive parts of the pair potential, we should observe an *asymmetry* in the *single-particle distribution function* $\varrho^{(1)}(z)$. $\varrho^{(1)}(z)$ for the outermost layer at T = 7 K is shown in Fig. 6.22; it is an enlarged representation of Fig. 6.17a (m = 1). Within the *harmonic* approximation both peaks must have identical areas. This requirement is not fulfilled in Fig. 6.22: The area of peak 1 (it is the peak towards the vacuum) is approximately *two* times larger than that of peak 2 (it is the peak towards the crystal). The explanation of this effect is given by the picture outlined above: As the attractive part is weaker than the repulsive part of the potential, the probability to find a particle in the region where the potential is attractive is larger than to find it in the repulsive region.

The asymmetry-effect in $\varrho^{(1)}(z)$ (Fig. 6.22) reflects an anharmonicity in the motion of the particles and cannot be explained by the solutions of the dynamical matrix (i.e., in the harmonic approximation). In other words, we have to consider nonlinear interactions in the equation of motion. As the asymmetry effect in $\varrho^{(1)}(z)$ is very pronounced we may conclude that the anharmonicities *perpendicular* to the surface are large even at very low temperatures (7 K), and in our opinion they cannot be treated as small perturbation (Chap. 7).

The asymmetry in $\varrho^{(1)}(z)$ must lead to an *outward-shift* of the particles near the surface and, therefore, the *thermal expansion* at the surface should be larger than in the bulk of the crystal. In fact, we have found for our MD system that the thermal expansion at the surface is approximately *five* times larger than in the bulk of the krypton crystal. This resembles the strong thermal expansion at a xenon surface which has been determined experimentally [6.72]; it turned out that the thermal expansion of the outermost xenon layer is 4–5 times larger than in the bulk of the xenon crystal.

**b) The Dynamics of the Particles of the Outermost Layer at T=102 K**
From the shape of the pair correlation function for the outermost layer at T = 102 K (Fig. 6.20, m = 1) follows that the particles are highly disordered parallel to the surface, and the following question arises: Do the particles of

the outermost layer perform a *diffusive motion* at T = 102 K (the melting temperature is 116 K)?

In other words, is the outermost layer in a liquid-like state already 14 K below the melting temperature? In order to answer this question the mean-square amplitudes $\langle r^2(t) \rangle$ *parallel* to the surface were calculated as a function of time. We have determined $\langle r^2(t) \rangle$ from the molecular-dynamics data by [6.1]

$$\langle r^2(t) \rangle = \frac{1}{N_\tau} \frac{1}{N_L} \sum_{j=1}^{N_\tau} \sum_{i=1}^{N_L} [r_i(t + \tau_j) - r_i(\tau_j)]^2 \, , \tag{6.72}$$

where $r_i(t) = \{x_i(t), y_i(t)\}$ is the component of the position vector parallel to the surface of the $i^{th}$ particle at time t. $N_\tau$ is the number of time origins and $N_L$ is the number of particles in the outermost layer. $N_\tau$ was chosen to be 200. With $N_L = 50$ (Sect. 6.3.1) we obtain a statistical error of 1 %. The results for $\langle r^2(t) \rangle$ are represented in Fig. 6.23. From the slope of the linear part of $\langle r^2(t) \rangle$ we have estimated the diffusion coefficient D, and we obtained a value of $0.85 \times 10^{-5} \, cm^2/s$, which is also a typical value for *bulk liquids*.

Because D is so large and the structure of the layer is disordered at 102 K (see the Figs. 6.20 (m = 1) and 6.23) we may conclude that the outermost layer shows the effect of *surface premelting* already investigated for Lennard-Jones systems [6.73–75]. In [6.74] the diffusion process has not been studied, and in [6.73] the long-time gradient for $\langle r^2(t) \rangle$ is very small on the molecular-dynamics scale and, in our opinion, one is hardly able to recognize a diffusive motion. In [6.75] there is a diffusion process parallel and perpendicular to the surface; in our system there is no diffusion normal to the surface at T = 102 K indicating that the surface properties are very sensitive to small variations in the pair potential.

The diffuse motion parallel to the surface in the outermost layer indicates that *nonlinear* effects at T = 102 K should be very pronounced. In fact, due to the anharmonicities parallel to the surface the frequency spectrum $f(\omega)_{||}$ at T = 102 K (Fig. 6.24) is shifted to lower frequencies with respect to $f(\omega)_{||}$ at 7 K. In particular, it can be seen from Fig. 6.24 that $f(\omega = 0)_{||}$ at T = 102 K is not zero which is consistent with the fact that the particles perform a diffuse

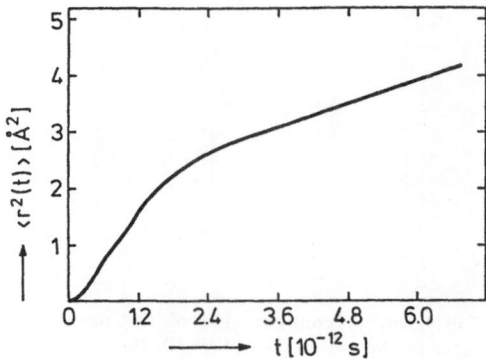

**Fig. 6.23.** Mean-square amplitudes parallel to the surface as a function of time (T = 102 K) for m = 1

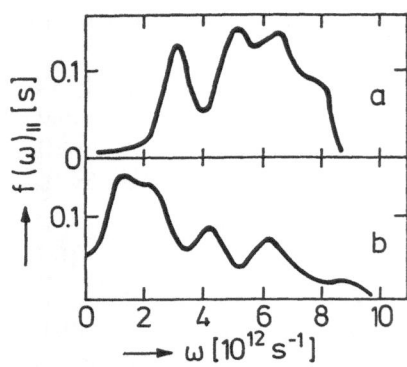

Fig. 6.24. Frequency spectrum parallel to the surface for m = 1. *a:* T = 7 K; *b:* T = 102 K

motion in the x,y-plane. The numerical value for the diffusion constant obtained from $f(\omega)_{\parallel}$ agrees well with that we have estimated from $\langle r^2(t)\rangle$. More details concerning the diffusion at the surface are given in Chap. 3.

### c) Melting Mechanism: Description at the Atomic Level

The diffusion process *parallel* to the surface is essentially influenced by the structure and dynamics *perpendicular* to the surface. From our analysis follows that the structure normal to the surface is disordered (Fig. 6.17c, m = 1) and, furthermore, that the particles vibrate in the z-direction almost independently from each other. Although there is no exchange of particles between different layers, the correlations in the motion perpendicular to the surface are almost zero. It can be recognized from the configuration given in Fig. 6.25 that the particles are randomly distributed in the z-direction.

Due to these effects the *density parallel to the surface* is lowered and the particles can perform a *diffusive motion* and the layer becomes liquid-like. It is probable that a diffusion process cannot take place without *configurational transition* normal to the surface, i.e. there would be no or almost no diffusion parallel to the surface if the particle motion took place strictly in two dimensions, i.e. in the x,y-plane.

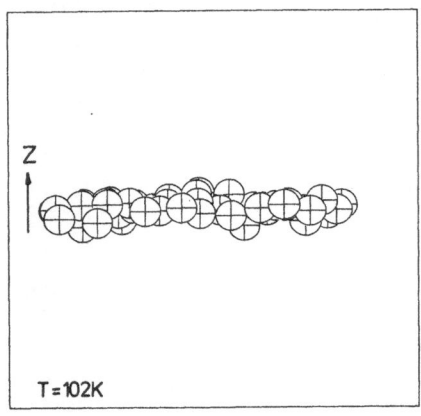

Fig. 6.25. A configuration of the outermost layer in the z-direction (T = 102 K)

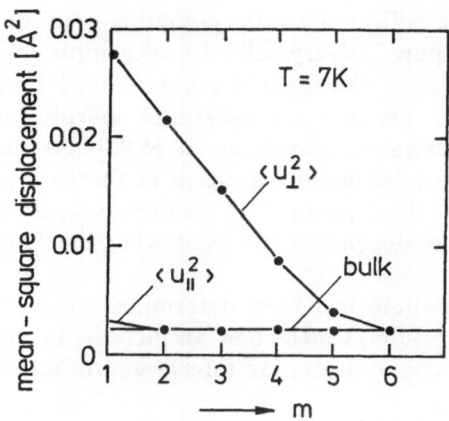

**Fig. 6.26.** Mean-square amplitudes for $T = 7$ K. m labels the layers starting with $m = 1$ for the outermost layer. $\langle u_\parallel^2 \rangle$ : parallel to the surface; $\langle u_\perp^2 \rangle$ : normal to the surface

In conclusion, the effect of premelting is mainly initiated by the disorder normal to the surface; this disorder is a thermally induced process. This disorder reflects an *anharmonicity*, and because of the *asymmetry* in the z-direction (Sect. 6.3.3a) the anharmonicity normal to the surface constitutes a large perturbation even at low temperatures ($T = 7$ K). This effect is also clearly reflected in the mean-square displacements: whereas $\langle u_\parallel^2 \rangle$ is at 7 K almost the same as in the bulk, $\langle u_\perp^2 \rangle$ is approximately 10 times larger than $\langle u_\parallel^2 \rangle$ (Fig. 6.26, $m = 1$).

The effect of premelting is of particular interest in connection with the *theory of melting*. One of the difficulties is that under normal conditions *undercooling* of liquids was observed [6.76–78] but not *superheating* of solids above their melting temperature. In the presence of a liquid outermost layer, superheating would be precluded because at the melting temperature the molten outermost layer acts as nucleation center for bulk melting.

*Other Explanations of Melting* [6.79,80]:

i) The absence of superheating can be explained by the fact that the sum of the liquid-vapour and the solid-liquid interfacial energies is less than the solid-vapour interfacial energy [6.81]. However, on the basis of thermodynamic arguments we can nothing learn about the mechanism at the atomic level.

ii) The liquid state is assumed to be a strongly distorted and perturbated form of the solid state [6.82]. This can be described by the introduction of *defects* (interstitial atoms [6.83], dislocations [6.84,85], vacancies [6.86]) into the perfect crystal lattice. The transition from the solid to the liquid phase is explained on the basis of defect generation.

iii) Melting is explained as an *instability* of the lattice and is due to the anharmonicity of the potential between the particles [6.87–91].

Recent *experimental* studies on lead have indicated a surface-melting-point depression. The electrical resistance of a pure polycrystalline lead sample was measured with high accuracy from 295 K to the melting point (600.7 K) by *Pokorny* and *Grimvall* [6.92]. It turned out that the resistance sharply increased above about 540 K. This behaviour has been explained in [6.92] in terms of a liquid-like surface structure, i.e. the solid-liquid transition at the surface of lead starts approximately 60 K below the bulk melting temperature. This observation is confirmed in a temperature-dependent ion shadowing/blocking study of a Pb(110) surface [6.93].

The specific heat of sodium and gallium has been determined close to the melting temperature [6.94,95]. The results, which show an increase in the isobaric specific heat below the melting temperature, are interpreted in terms of surface premelting.

The bending of the Arrhenius curve, which is observed for various materials (e.g., tungsten) for about $T/T_m > 0.75$, is interpreted in [6.96] as a change of surface transport from an atomic jump mechanism at low temperatures towards a viscous mechanism near the melting point; for $T/T_m > 0.75$ a molten surface region of thickness of about a few monolayers is assumed.

## 6.4 Final Remarks

In this chapter we have studied the structure and dynamics of simple liquid surfaces. The structure was described by *distribution* and *correlation* functions, and these functions have been introduced on the basis of statistical mechanics.

Models of the *single-particle distribution function* $\varrho^{(1)}$ have been studied and it turned out that the density profile perpendicular to the surface is *non-monotonic* but *oscillatory* in character. Furthermore, we have discussed in Sect. 6.3.3a that $\varrho^{(1)}$ is an important quantity in the study of *anharmonic* effects at the surface; it turned out that the anharmonicities perpendicular to the surface are of considerable importance even at low temperatures. From the shape of $\varrho^{(1)}$ below the melting point we have concluded that the outermost layer of a krypton system is in a liquid-like state *(effect of premelting)*. In conclusion, we believe that the knowledge of $\varrho^{(1)}$ is essential and its reliable experimental determination would be desirable. It was shown in [6.32–34] that the angular dependence of x-ray reflection at almost grazing angles can be used for the test of theoretical approaches for $\varrho^{(1)}$. This kind of experiment should be extended. In this connection the study of liquid surfaces by *synchrotron x-rays* (Chap. 5) is of considerable interest.

In the description of *thermodynamic functions* (pressure, energy, etc.) of the liquid surface also the *two-particle distribution function* $\varrho^{(2)}$ is of relevance. Approximations for $\varrho^{(2)}$ have been quoted. In connection with the effect of premelting we have demonstrated that also the pair *correlation function* $g_2$ is a proper quantity for characterising of the structure parallel to the surface. No experimental results for $g_2$ at the surface seem as yet to be available. The experimental determination of the Fourier transform of $g_2$ at the surface should, in principle, be possible with the help of scattering experiments [6.97,98], and such experiments should be performed.

Since the *harmonic* approximation breaks down in the case of liquids, the phonon density of states $g(\omega)$ is no longer relevant in the description of the dynamics of the liquid surface, and we have discussed in Sect. 6.2 that the Fourier transform of the *velocity autocorrelation function* becomes important. The experimental determination of $f(\omega)$ at the surface should, in principle, be possible by the inelastic scattering of atoms [6.99] and electrons (Chap. 1) [6.100,101]. However, these techniques have not yet been applied in the study of systems with strong anharmonicities such as liquids.

# References

6.1   W. Schommers: In *Structure and Dynamics of Surfaces I*, ed. by W. Schommers and P. v. Blanckenhagen, Topics Curr. Phys., Vol. 41 (Springer, Berlin, Heidelberg 1986)
6.2   N.H. March: In "Theory of Condensed Matter" (International Atomic Energy Agency, Vienna 1968)
6.3   P.A. Egelstaff: *An Introduction to the Liquid State*, (Academic, New York 1967)
      S.W. Lovesey, T. Springer (eds.): *Dynamics of Solids and Liquids*, Topics Curr. Phys., Vol. 3 (Springer, Berlin, Heidelberg 1976)
6.4   R. Block, W. Schommers: J. Phys. C8, 1997 (1975)
6.5   R. Abe: Prog. Theor. Phys. **21**, 421 (1959)
6.6   M.C. Abramo, H.P. Tosi: Nuovo Cim. **5**, 1044 (1972)
6.7   B.J. Alder: Phys. Rev. Lett. **12**, 317 (1964)
6.8   P.A. Egelstaff: Ann. Rev. Phys. Chem. **24**, 159 (1973)
6.9   E. Feenberg: *Theory of Quantum Fluids* (Academic, New York 1969)
6.10  P.A. Egelstaff, D.I. Page, C.R.T. Heard: J. Phys. C4, 1453 (1971)
6.11  J.S. Kirkwood: J. Chem. Phys. **56**, 2034 (1972)
6.12  J.A. Krumhansl, S. Wang: J. Chem. Phys. **56**, 2034 (1972)
6.13  H.J. Raveche, R.D. Mountain, W.B. Streett: J. Chem. Phys. **57**, 4999 (1972)
6.14  J.S. Rowlinson: Mod. Phys. **88**, 149 (1966)
6.15  W. Schommers: Phys. Rev. A**28**, 3599 (1983)
6.16  P.A. Egelstaff, J.J. Salacuse, W. Schommers, J. Ram: Phys. Rev. A**30**, 374 (1984)
6.17  W. Schommers: Phys. Rev. Lett. **38**, 1536 (1977)
6.18  W. Schommers: Phys. Rev. B**17**, 2057 (1978)
6.19  W. Schommers: Phys. Rev. A**22**, 2855 (1980)
6.20  R.H. Fowler: Proc. Roy. Soc. A**159**, 229 (1937)
6.21  J.S. Kirkwood, F.P. Buff: J. Chem. Phys. **17**, 338 (1949)
6.22  C.A. Croxton: *Liquid State Physics – A Statistical Mechanical Introduction* (Cambridge Univ. Press, London 1974)
6.23  C.A. Croxton: *Introduction to Liquid State Physics*, (Wiley, London 1975)
6.24  I.Z. Fisher: *Statistical Theory of Liquids* (Chicago Univ. Press, Chicago, Ill. 1964)
6.25  C.A. Croxton, R.P. Ferrier: J. Phys. C4, 1909 (1971)
6.26  C.A. Croxton: Adv. Phys. **22**, 385 (1973)
6.27  G.M. Nazarian: J. Chem. Phys. **56**, 1408 (1972)
6.28  C.A. Croxton, R.P. Ferrier: J. Phys. C4, 2447 (1971)
6.29  J. Gryko, S.A. Rice: J. Non-Crystalline Solids **61/62**, 703 (1984)
6.30  C.H. Woo, S. Wang, M. Matsuura: J. Phys. F5, 1836 (1975)
6.31  J. Gryko, S.A. Rice: J. Phys. F**12**, L245 (1982)
6.32  B.C. Lu, S.A. Rice: J. Chem. Phys. **68**, 5558 (1978)
6.33  D. Sluis, M.P. D'Evelyn, S.A. Rice: J. Chem. Phys. **78**, 1611 (1983)
6.34  L. Bosio, R. Cortes, A. Defrain, M. Omezine: J. Non-Crystalline Solids **61/62**, 697 (1984)

6.35  H.S. Green: In Handbuch Physik, Vol. 10, ed. by S. Flügge (Springer, Berlin, Heidelberg 1960) p. 79
6.36  M.V. Berry, S.R. Rezneck: J. Phys. A4, 77 (1971)
6.37  C.A. Croxton, R.P. Ferrier: J. Phys. C4, 1921 (1971)
6.38  C.A. Croxton, R.P. Ferrier: Phil. Mag. 24, 493 (1971)
6.39  R.C. Tolman: J. Chem. Phys. 17, 118 (1949)
6.40  T.L. Hill: J. Chem. Phys. 20, 141 (1952)
6.41  I.W. Plesner, O. Platz: J. Chem. Phys. 48, 5361 (1968)
6.42  S. Toxvaerd: J. Chem. Phys. 55, 3116 (1971)
6.43  J.W. Gibbs: Collected Works 1 (Lougman, Green and Co., New York 1928) p. 219
6.44  C. Herring: In Metal Interfaces (Am. Soc. Metals, Cleveland 1952)
6.45  S. Takeuchi (ed.): The Properties of Liquid Metals (Taylor and Francis, London 1973)
6.46  C.A. Croxton: Statistical Mechanics of the Liquid Surface (Wiley, Chichester 1980)
6.47  P.D. Shoemaker, G.W. Paul, L.E. Marc de Chazal: J. Chem. Phys. 52, 491 (1970)
6.48  F.B. Sprow, J.M. Prausnitz: Trans. Faraday Soc. 62, 1097 (1966)
6.49  N.D. Lang, W. Kohn: Phys. Rev. B1, 4555 (1970)
6.50  P. Hohenberg, W. Kohn: Phys. Rev. 136, B864 (1964)
6.51  W. Kohn, L.J. Sham: Phys. Rev. 140, A1133 (1965)
6.52  R. Evans: J. Phys. C7, 2808 (1974)
6.53  R. Kumaravadivel, R. Evans: J. Phys. C8, 793 (1975)
6.54  R. Evans, R. Kumaravadivel: J. Phys. C9, 1891 (1976)
6.55  K.K. Mon, D. Stroud: Phys. Rev. Lett. 45, 817 (1980)
6.56  S. Amokrane, J.P. Badiali, M.L. Rosinberg, J. Goodisman: J. Physique C8, 775 (1980)
6.57  S. Amokrane, J.P. Badiali, M.L. Rosinberg, J. Goodisman: J. Chem Phys. 75, 5543 (1981)
6.58  R. Evans, M. Hasegawa: J. Phys. C14, 5225 (1981)
6.59  M. Hasegawa, M. Watabe: J. Phys. C15, 353 (1982)
6.60  M. Hasegawa, M. Watabe: J. Non-Crystalline Solids 61/62, 707 (1984)
6.61  D.L. Price: Phys. Rev. A4, 358 (1971)
6.62  M. Hasegawa, M. Watabe: J. Phys. Soc. Japan 32, 14 (1972)
6.63  J.M. Ziman, Adv. Phys. 13, 89 (1964)
6.64  W.A. Harrison: Pseudopotentials in the Theory of Metals (Benjamin, New York 1966)
6.65  J.E. Black: In Structure and Dynamics of Surfaces I, ed. by W. Schommers and P. v. Blanckenhagen, Topics Curr. Phys., Vol. 41 (Springer, Berlin, Heidelberg 1986)
      G. Benedek, U. Valbusa (eds.): Dynamics of Gas-Surface Interaction, Springer Ser. Chem. Phys., Vol. 21 (Springer, Berlin, Heidelberg 1982)
6.66  J. Grindley, R. Howard: Lattice Dynamics, Proc. of Int'l. Conf. Copenhagen (1963)
6.67  A. Rahman: Phys. Rev. 136, A405 (1964)
6.68  F.A. Lindemann: Z. Phys. 14, 609 (1910)
6.69  W. Schommers: Phys. Rev. B32, 6845 (1985)
6.70  W. Schommers, P. v. Blanckenhagen: Vacuum 33, 733 (1983)
6.71  J.A. Barker, R.O. Watts, J.K. Lee, T.B. Schafer, Y.T. Lee: J. Chem. Phys. 61, 3081 (1974)
      M.L. Klein (ed.): Inert Gases, Springer Ser. Chem. Phys., Vol. 34 (Springer, Berlin, Heidelberg 1984)
6.72  A. Ignatiev, T.N. Rhodin: Phys. Rev. B8, 893 (1973)
6.73  J.Q. Broughton, L.V. Woodcock: J. Phys. C11, 2743 (1978)
6.74  R.M.J. Cotterill: Phil. Mag. 32, 1203 (1975)
6.75  J.Q. Broughton, G.H. Gilmer: J. Chem. Phys. 79, 5119 (1983)
6.76  L. Bosio, A. Defrain, I. Epelboin: J. Phys. (Paris) 21, 61 (1966)
6.77  L. Bosio, C.G. Windsor: Phys. Rev. Lett. 35, 1652 (1975)
6.78  W. Schommers: Solid State Commun. 21, 65 (1977)
6.79  R.M.J. Cotterill, J.K. Kristensen: Phil. Mag. 36, 453 (1977)
      J.K. Kristensen, R.M.J. Cotterill: Phil. Mag. 36, 437 (1977)
6.80  A.R. Ubbelohde: Melting and Crystal Structure (Clarendon, Oxford 1965)
6.81  L.D. Landau, E.M. Lifschitz: Statistical Physics (Pergamon, New York 1970)
6.82  F.C. Frank: Proc. Roy. Soc. A170, 182 (1939)
6.83  J.E. Lennard-Jones, A.F. Devonshire: Proc. Roy. Soc. A170, 464 (1939)
6.84  S.J. Mitzushima: Phys. Soc. Japan 15, 70 (1960)

6.85   D. Kuhlmann-Wilsdorf: Phys. Rev. **140**, A1599 (1965)
6.86   J. Frenkel: *Kinetic Theory of Liquids* (Dover, New York 1955)
6.87   Y. Ida: Phys. Rev. **187**, 951 (1969)
6.88   Y. Ida: Phys. Rev. B**1**, 2488 (1969)
6.89   T. Siklos: Acta Phys. Hung. **34**, 327 (1973)
6.90   L. Pietronero, E. Tosatti: Solid State Commun. **32**, 255 (1979)
6.91   C.S. Jayanthi, E. Tosatti, L. Pietronero: Phys. Rev. B**31**, 3456 (1985)
6.92   M. Pokorny, G. Grimvall: J. Phys. F**14**, 931 (1984)
6.93   J.W.M. Frenken, J.F. van der Veen: Phys. Rev. Lett. **54**, 134 (1985)
6.94   G. Fritsch, R.Lachner, H. Diletti, E. Lüscher: Phil. Mag. A**46**, 829 (1982)
6.95   G. Fritsch, H. Diletti, E. Lüscher: Phil. Mag. A**50**, 545 (1984)
6.96   Vu Thien Binh, P. Melinon: Surf. Sci. **161**, 234 (1985)
6.97   G. Comsa, B. Poelsema: Appl. Phys. A**38**, 153 (1985)
6.98   K.H. Rieder: In *Structure and Dynamics of Surfaces I*, ed. by W. Schommers and P. v. Blanckenhagen, Topics in Curr. Phys., Vol. 41 (Springer, Berlin, Heidelberg 1986)
6.99   J.P. Toennies: J. Vac. Sc. Techn. A**2**, 1055 (1984)
6.100  H. Ibach (ed.): *Electron Spectroscopy for Surface Analysis*, Topics Curr. Phys., Vol. 4 (Springer, Berlin, Heidelberg 1977)
6.101  M. Rocca, H. Ibach, S. Lehwald, T.S. Rahman: In *Structure and Dynamics of Surfaces I*, ed. by W. Schommers and P. v. Blanckenhagen, Topics Curr. Phys., Vol. 41 (Springer, Berlin, Heidelberg 1986)

## Additional References with Titles

Broughton, J.Q., Gilmer, G.H.: Thermodynamic Criteria for Grain-Boundary Melting: A Molecular-Dynamics Study. Phys. Rev. Lett. **56**, 2692 (1986)
Frenken, J.W.M., Marée, P.M.J., van der Veen, J.F.: Observation of Surface-Initiated Melting. Phys. Rev. B**34**, 7506 (1986)
Hayes, W.: Premelting. Contemp. Phys. **27**, 519 (1986)
Nenow, C.: Surface Premelting. Prog. Cryst. Growth and Charact. **9**, 185 (1984)
Nenow, D., Pavlovsky, A.: Crystal Morphology and Surface Melting, Morphology and Phase Equilibria of Minerals – IMA 1982, 181 (1986)
Van der Veen, J.F., Frenken, J.W.M.: Dynamics and Melting of Surfaces. Surface Science **178**, 382 (1986)

# 7. The Roughening Transition

H. van Beijeren and I. Nolden

With 24 Figures

The roughening transition of a crystal surface is characterized macroscopically by the disappearance of a facet of a given orientation from the equilibrium crystal shape. This corresponds to the disappearance of a cusp in the Wulff plot, a polar plot of surface tension vs. surface orientation. Microscopically the roughening transition is characterized by the free energy of a step on the facet becoming zero, or alternatively by the appearance of strong fluctuations in the location of the facet.

Both renormalization-group analyses and results for exactly solvable models show that the roughening transition generically is of Kosterlitz-Thouless type, i.e. it is an infinite-order transition with a very weak singularity in the free energy. This prediction is confirmed by experiments on helium crystals in contact with superfluid helium. Besides the rouhgening transition the relation between the Wulff plot and equilibrium crystal shape is discussed in general, and various possibilities for equilibrium crystal shapes are mentioned. A qualitative explanation is given why vicinal areas bordering to facets should exhibit a universal non-analytic shape. Finally, we discuss the aspects of surface melting vs. roughening and some possibilities for roughening transitions that are not of Kosterlitz-Thouless type.

## 7.1 Background

In 1951 in a pioneering paper *Burton* et al. [7.1] put forward the idea that a phase transition may occur in the equilibrium structure of crystal surfaces. Inspired by *Onsager*'s solution of the two-dimensional Ising model [7.2] they conjectured that singular crystal faces, i.e. faces with the orientation of stable facets on the crystal, in equilibrium with vapor, melt or a solution, would become rough above a certain transition temperature. Below this temperature perfect (that is, dislocation free) faces would grow by means of a nucleation mechanism requiring the surmounting of a large free energy barrier for the formation of a stable growth nucleus. A growing layer should be describable, at least approximately, by a two-dimensional Ising model, or rather lattice gas [7.3] in a state of coexisting "liquid" (the completed part of the growing layer) and "gas" (the empty part of the growth layer). Above the transition temperature no resistance against the formation of large clusters on the surface would exist and growth at all supersaturations would occur through a homogeneous deposition of matter on the growing surface. On the basis of this general picture the phase transition thus described has later been called the roughening transition.

*Burton* et al. made one important simplification in their description of the growth layer: they ignored the possibility of new layers starting to form before completion of the previous one. If the nucleation barrier is large this is fully justified since completion of a layer once a critical nucleus has been formed proceeds fast compared to the nucleation of the next layer. But close to the roughening transition growth occurs in several layers simultaneously. As a result this transition is not similar to that of the two-dimensional Ising model, as supposed in [7.1], but rather turns out to be of Kosterlitz-Thouless type [7.4–7]. This means it is a phase transition of infinite order; the singularity in $\gamma$, the surface tension or surface free energy per unit area, at the roughening temperature $T_R$ is extremely weak. All derivatives of $\gamma$ with respect to temperature exist and are smooth functions of temperature. Notably the surface specific heat does not exhibit a singularity, or even a maximum, at the transition point in contradiction to previous conjectures based on *Onsager*'s solution of the Ising model [7.1,8]. Furthermore, in a sense the surface is in a critical state for all temperatures $T \geq T_R$ : Though the only singularity in $\gamma$ due to the roughening transition occurs at $T_R$, the correlation length $\xi$ for fluctuations in the surface height diverges for all $T \geq T_R$ [7.4].

Macroscopically the roughening transition manifests itself in the equilibrium shape of crystals. In general, the surface of a crystal in equilibrium consists of facets, which are macroscopically flat, and/or rounded parts. The facets are crystal faces in a smooth state, i.e. below their roughening temperature, and the rounded parts may be considered as constructed from infinitesimal pieces of rough faces, i.e. faces above their roughening temperature. An example of an equilibrium shape containing both facets and rounded parts is shown in Fig. 7.1. If temperature is raised through $T_R(\hat{n})$, the roughening temperature

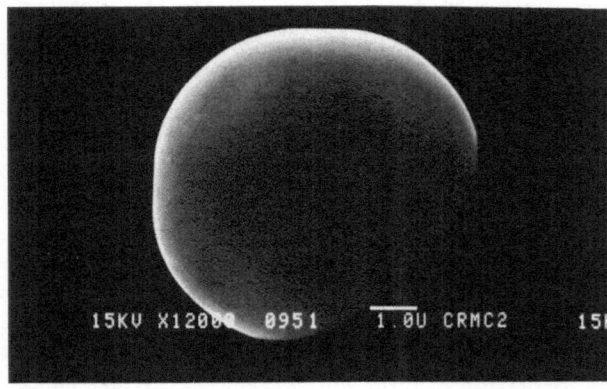

15KV X12000 0951 ‾1‾.‾0U CRMC2 15K

**Fig. 7.1.** Equilibrium shape of a lead crystallite at 520 K, as observed by *Heyraud* and *Métois* [7.10]. The diameter of the crystal is approximately 5 $\mu$m

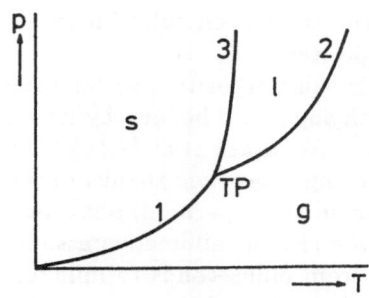

Fig. 7.2. Typical phase diagram of a simple substance. The solid, liquid and gaseous phase are denoted by s, $\ell$ and g, respectively. The triple point is denoted as TP

of a face of orientation n̂, the facet size of this face shrinks to zero and the orientation n̂ becomes part of a rounded area.

Experimental observation of the roughening transition is difficult for different reasons. First of all it is generally very difficult to produce crystals with an equilibrium shape. Shapes produced during growth generically are metastable and their relaxation rate increases rapidly with crystal size. In experiments by *Heyraud* and *Métois* [7.9,10] on metal crystallites of only a few micrometers diameter at temperatures of a few hundred degrees Kelvin equilibration times of a few days were observed. In addition, it appears that most observable crystal faces do not show a roughening transition within the physical range of their existence. Typically the surfaces of crystals in equilibrium with their melt are rough all along the melting line, down to the triple point, whereas the principal facets of crystals in equilibrium with their vapor tend to remain smooth along the sublimation line, up to the triple point (Fig. 7.2). For crystal vapor equilibrium this difficulty often can be circumvented by considering higher index crystal faces, which are less tightly packed, so they have lower roughening temperatures. But then the problem of the long relaxation times makes itself felt again, although relaxation times for the internal structure of existing facets are generally much shorter than the relaxation times for a full reconstruction of the crystal shape. Another experimental problem is the sensitivity of surface properties to impurities and dislocations, which may make it hard to decide whether an observed surface shape is really that of a pure crystal in equilibrium. Finally, even in cases where a roughening transition is observed the nature of the transition makes it very hard to pin down the roughening temperature. Typically facets disappear so gradually at the approach of the roughening temperature that they become practically unobservable already at temperatures distinctly below $T_R$.

In spite of these difficulties roughening transitions have been found experimentally in a number of systems. *Pavlovska* and *Nenow* studied the equilibrium shape of "negative crystals", i.e. vapor bubbles included within a crystal, in organic substances such as diphenyl [7.11], naphthalene [7.12], and tetrabrommethane [7.13], and observed the disappearance of certain facets within well-defined reproducible temperature ranges. Similar observations were reported by *Ohachi* and *Taniguchi* [7.14] on silver sulfide, and *Passerone* et al.

[7.15] observed roughening transitions on interfaces between solid Zn and a liquid Zn-Bi-In alloy, as a function of alloy composition.

However, the best observations of the roughening transition so far have been made for helium crystals in equilibrium with superfluid helium, by *Avron* et al. [7.16], by *Balibar* and *Castaing* [7.17], and by *Keshishev* et al. [7.18]. This system is ideally suited for observing the roughening transition: thanks to the extremely rapid transport of both heat and mass in the superfluid, relaxation to equilibrium is very fast. Crystals grown to the size of millimeters assume their equilibrium shape within periods ranging from milliseconds to minutes, depending on both external and internal parameters. In addition, these crystals are extremely pure because, with the exception of $^3$He, all impurities may be filtered out from the superfluid very efficiently. Thus far roughening transitions for three different types of facets have been observed, at temperatures of roughly 1.3 K [7.16–18], 0.9 K [7.16,18] and 0.35 K [7.19], respectively. In addition, *Balibar* and his group have exploited the exceptional properties of the helium system in some beautiful experiments [7.20], which confirm quantitatively the predicted Kosterlitz-Thouless character of the roughening transition. Most remarkably the most accurate determinations of the roughening temperature and other characteristics of the phase transition result from growth rate measurements.

Although roughening transitions are hard to observe in nature, they are not without relevance. We mentioned already the different growth modes at small supersaturation. For temperatures above $T_R$ growth is continuous, with a growth rate proportional to supersaturation, for a perfect crystal below $T_R$ one has nucleation growth with a growth rate roughly proportional to $\exp(-c/\Delta\mu)$ where $\Delta\mu$ is the supersaturation [7.21]. At temperatures far below $T_R$ nucleation growth in practice is dominated by spiral growth [7.1,22] driven by the presence of screw dislocations within the bulk of the crystal. It also has been observed that growth of entirely rough crystals usually becomes dendritic already for quite small supersaturation, whereas in the presence of stable facets growth shapes are faceted up till high supersaturation. This feature has been used by *Jackson* and *Miller* to demonstrate the occurrence of a roughening transition in hexachlorethane [7.23a] and it has also been observed by *Rolley* et al. in growth experiments on $^3$He [7.23b]. Further, it seems obvious that adsorptive and catalytic properties of surfaces will depend severely on whether the surface is rough or smooth.

The sequel of this chapter is organized as follows: In Sect. 7.2 we discuss the thermodynamics of crystal shapes, explain the Wulff construction relating the equilibrium crystal shape to the anisotropic surface tension and give a macroscopic description of the roughening transition. In Sect. 7.3 we provide a microscopic description of the surface structure and a microscopic characterization of the roughening transition in terms of the vanishing of the free energy of steps or alternatively the delocalization of the position of the crystal surface. Section 7.4 is devoted to the statistical mechanics of the roughening transition. We sketch the renormalization group treatment and quote its main results; we describe and discuss some exactly solvable models and give a semiquantitative explanation for the non-analytic shape of vicinal areas joining facets. In Sect. 7.5 we juxtapose roughening and surface premelting, we address the question whether all roughening transitions are of the same type, we discuss the relation between roughening and multilayer adsorption and we present some concluding remarks.

Finally, let us mention that we have not aimed at giving a complete description of all recent developments concerned with the roughening transition.

Our main goal has been to present a rather elementary introduction to this field, adding in a few rather specialistic comments at places where the extensive literature to our point of view contained some niches.

Further discussions of the roughening transition and other properties of equilibrium crystal surfaces can be found in reviews by *Weeks* [7.24], *Zia* [7.25], *Rottman* and *Wortis* [7.26], *Balibar* and *Castaing* [7.27] and *Balibar* and *Saam* [7.28] and in extensive lecture notes by *Nozières* [7.29].

# 7.2 Macroscopic Description of Equilibrium Crystal Shapes

### 7.2.1 The Wulff Construction

Let us suppose the surface tension $\gamma(\hat{n})$ of a macroscopically flat interface is given as a function of its orientation $\hat{n}$ (see, e.g., [7.29] on the problem of how to determine this function). Then the equilibrium crystal shape is the shape that minimizes the total free energy of the surface at a given fixed crystal volume. The solution of this minimization problem is obtained through the famous Wulff construction [7.26,30,31], as illustrated in Fig. 7.3: for each direction $\hat{n}$ construct a plane perpendicular to $\hat{n}$ at a distance $\lambda\gamma(\hat{n})$ from a given origin. Then the inner envelope of all these planes is the equilibrium crystal shape. To

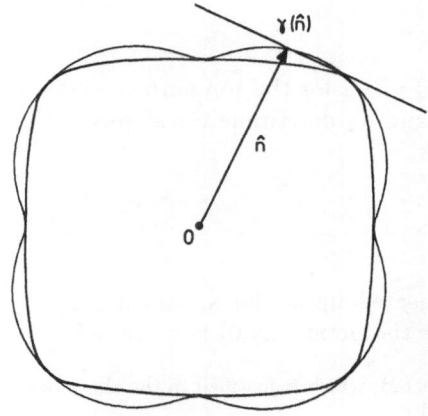

Fig. 7.3. Illustration of the Wulff construction by a two-dimensional example. The distance of the *thin curve* from the origin in the direction $\hat{n}$ equals $\lambda\gamma(\hat{n})$ and the *solid curve* depicts the equilibrium crystal shape

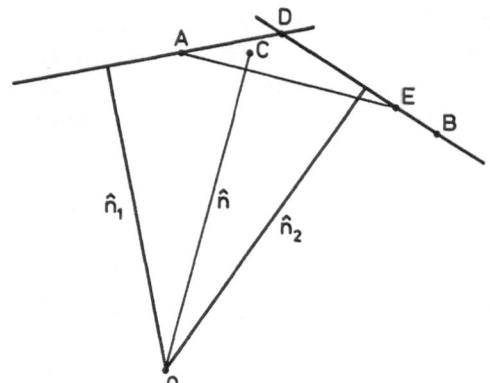

Fig. 7.4. Convexity property of the Wulff construction, explained in the text

understand this construction and obtain at the same time an analytic representation for it, we follow *Landau* and *Lifshitz*'s derivation [7.32]. It is not the most general one, e.g. it relies on $\gamma(\hat{n})$ being piecewise differentiable, but it has the advantage of being straightforward and clear.

Anticipating on the result we notice that the Wulff construction always produces a convex crystal shape. This is illustrated in Fig. 7.4: Suppose the orientations $\hat{n}_1$ and $\hat{n}_2$ are present on a crystal shape in the points A and B, respectively, and the orientation $\hat{n} \sim \mu \hat{n}_1 + (1 - \mu)\hat{n}_2$ with $0 < \mu < 1$ occurs at the point C. Then C has to be within the prismatic volume orthogonal to the triangle ADE, (the lines AD, BD and AE are perpendicular to $\hat{n}_1$, $\hat{n}_2$ and $\hat{n}$, respectively) otherwise either C would be outside the planes orthogonal to $\hat{n}_1$ and $\hat{n}_2$ or A would be outside the plane orthogonal to $\hat{n}$. A convex shape can always be divided into a top and a bottom surface, definable as those parts of the surface that are visible from the + and −z direction, respectively. Let us suppose the top and bottom surfaces can be described analytically in a Cartesian coordinate system by functions $z_+(x, y)$ and $z_-(x, y)$, respectively. The orientation of the crystal surface on top and bottom can be described as

$$\hat{n}_\alpha = \text{sign}\{\alpha\}(p_\alpha, q_\alpha, 1)/\sqrt{1 + p_\alpha^2 + q_\alpha^2} \ , \tag{7.1}$$

where $\alpha$ refers to top or bottom and $\text{sign}\{\alpha\} = +1$ for the top surface and $-1$ for the bottom surface. The parameters $p_\alpha$ and $q_\alpha$ determine the slopes of the crystal surface in the x and y directions, viz.

$$\frac{\partial z_\alpha}{\partial x} = p_\alpha \qquad \frac{\partial z_\alpha}{\partial y} = q_\alpha \ . \tag{7.2}$$

The surface free energy per unit area projected upon the x,y-surface of an infinitesimal surface element above or below the point (x,y,0) is given as[1]

---

[1] Here contributions due to curvature are neglected, which is justified in the thermodynamic limit, where the crystal volume goes to infinity.

$$f_\alpha(\hat{n}_\alpha) = \gamma(\hat{n}_\alpha)\sqrt{1 + p_\alpha^2 + q_\alpha^2} \ . \tag{7.3}$$

Now the problem of minimizing the total surface free energy under the constraint of given total volume can be attacked by solving the variational equation

$$\delta \int [f_+ + f_- - 2\lambda(z_+ - z_-)]dx\,dy = 0 \ , \tag{7.4}$$

where $\lambda$ is a Lagrange multiplier. Equation (7.4) can be transformed according to

$$\sum_\alpha \int \left( \frac{\partial f_\alpha}{\partial p_\alpha}\frac{\partial \delta z_\alpha}{\partial x} + \frac{\partial f_\alpha}{\partial q_\alpha}\frac{\partial \delta z_\alpha}{\partial y} - 2\lambda\,\text{sign}\{\alpha\}\delta z_\alpha \right)dx\,dy = 0 \ ,$$

where (7.2) was used. Setting $f_\alpha = z_\alpha = 0$ outside the crystal one may apply a partial integration, yielding

$$-\sum_\alpha \int \left( \frac{\partial}{\partial x}\frac{\partial f_\alpha}{\partial p_\alpha} + \frac{\partial}{\partial y}\frac{\partial f_\alpha}{\partial q_\alpha} + 2\lambda\,\text{sign}\{\alpha\} \right)\delta z_\alpha dx\,dy = 0 \ .$$

Now $\delta z_+(x,y)$ and $\delta z_-(x,y)$ may be treated as independent variations. Hence one obtains the equation

$$\frac{\partial}{\partial x}\frac{\partial f_\alpha}{\partial p_\alpha} + \frac{\partial}{\partial y}\frac{\partial f_\alpha}{\partial q_\alpha} + 2\lambda\,\text{sign}\{\alpha\} = 0 \ . \tag{7.5}$$

Following *Landau* and *Lifshitz* we find a solution

$$f_\alpha = \lambda\,\text{sign}\{\alpha\}(z_\alpha - p_\alpha x - q_\alpha y) \tag{7.6}$$

as can be checked by inserting (7.6) into (7.5) and using (7.2).

Considering (7.6) as an equation for a plane at fixed $p_\alpha$ and $q_\alpha$ and variable x,y and $z_\alpha$, one readily sees that this is just the Wulff plane perpendicular to $\hat{n}_\alpha$. However, interpreting (7.6) as defining the crystal shape, one must assume that $z_\alpha$, x and y are functions of $p_\alpha$ and $q_\alpha$, determined by (7.2 and 6) together. The explicit form of these functions is obtained by taking the partial derivatives of (7.6) with respect to $p_\alpha$ and $q_\alpha$ and using (7.2). The resulting representation of the crystal shape has the form

$$x = -\frac{1}{\lambda}\text{sign}\{\alpha\}\frac{\partial f}{\partial p_\alpha} \tag{7.7a}$$

$$y = -\frac{1}{\lambda}\text{sign}\{\alpha\}\frac{\partial f}{\partial q_\alpha} \tag{7.7b}$$

265

$$z_\alpha = \frac{1}{\lambda}\text{sign}\{\alpha\} \left( f_\alpha - p_\alpha \frac{\partial f}{\partial p_\alpha} - q_\alpha \frac{\partial f}{\partial q_\alpha} \right) . \tag{7.7c}$$

For each direction $\hat{n}_\alpha$, indeed, the surface defined by (7.7) is tangent to the Wulff plane defined by (7.6), hence the conclusion that (7.7) yields an analytic representation of the Wulff construction. The uniqueness of the solution (7.7) to eq. (7.5) was demonstrated in [7.33,34].

From (7.7c) one sees that the equilibrium crystal shape $z(x,y)$ is obtained as a Legendre transform of the surface tension per unit projected area [7.35]. Similarly f is obtained from the surface shape through the Legendre transform

$$f = z - x\frac{\partial z}{\partial x} - y\frac{\partial z}{\partial y} . \tag{7.8}$$

This relation enables one to reconstruct the Wulff plot (or the function $\gamma(\hat{n})$) from the equilibrium crystal shape. *Heyraud* and *Métois* have actually done this for the case of lead crystals in equilibrium with their vapor [7.10].

The Lagrange multiplier $\lambda$ can be identified from (7.5), with the aid of the thermodynamics of the interface [7.29] as

$$\lambda = \frac{1}{2}\frac{\varrho_s - \varrho_\ell}{\varrho_\ell}\delta p , \tag{7.9}$$

where $\varrho_s$ and $\varrho_\ell$, respectively, are the mass densities of the crystal and the fluid with which it is in equilibrium, while $\delta p$ is the pressure difference between fluid and crystal or, more precisely, the pressure difference between the fluid in equilibrium with the actual crystal and a similar fluid that would be in equilibrium with an infinite crystal at the same temperature [7.29].

Note that the applicability of the Wulff construction is not restricted to crystals in equilibrium with fluids. It applies equally well to other phase equilibria involving anisotropic materials, e.g. liquid crystals.

### 7.2.2 Facets, Edges and Corners

Facets appear in the equilibrium crystal shape if the Wulff plot exhibits cusps, as is illustrated in Fig. 7.5. If facets are present Landau's method must be

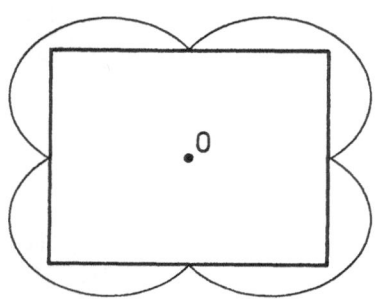

**Fig. 7.5.** Cusps in the Wulff plot lead to facets in the equilibrium crystal shape

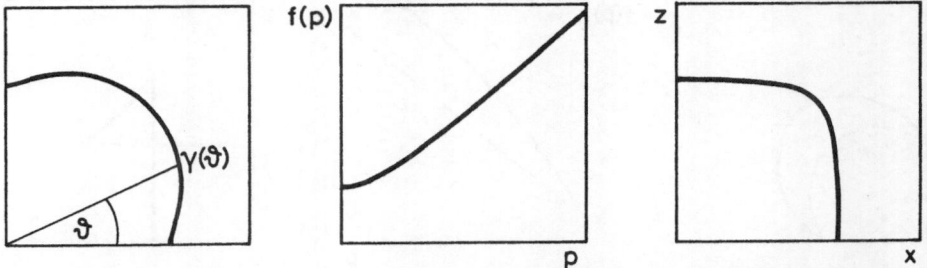

**Fig. 7.6.** The polar plot of **(a)** $\gamma$ as a function of $\hat{n} = (\sin\vartheta, 0, \cos\vartheta)$, **(b)** the surface free energy per unit area $f(p, 0)$ plotted as a function of $p$ and **(c)** quadrant of the section of the equilibrium crystal shape with the x,z-plane, for a fully rounded crystal

slightly generalized: Although (7.6) is satisfied even on the facets, (7.5) is not, since on the facet $p$ and $q$ do not depend on $x$ and $y$. However, the Wulff construction including facets does yield the correct equilibrium shape [7.33]. *Landau* and *Lifshitz*'s representation can be considered as the limiting case of a series of Wulff constructions in which the cusps are replaced by rounded tips that approach the cusps more and more.

In Figs. 7.6–10 we give some examples of the types of surface shapes that may arise from various Wulff plots. For simplicity, we always show the section of the x,z-plane with the Wulff plot, the resulting $f(p, 0)$ and the corresponding section of the equilibrium shape. In addition, we assume cubic symmetry.

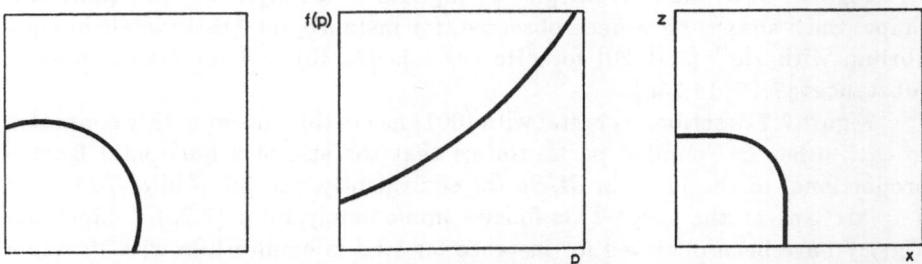

**Fig. 7.7.** The same as in Fig. 7.6 for a crystal with facets and rounded parts connected smoothly to each other

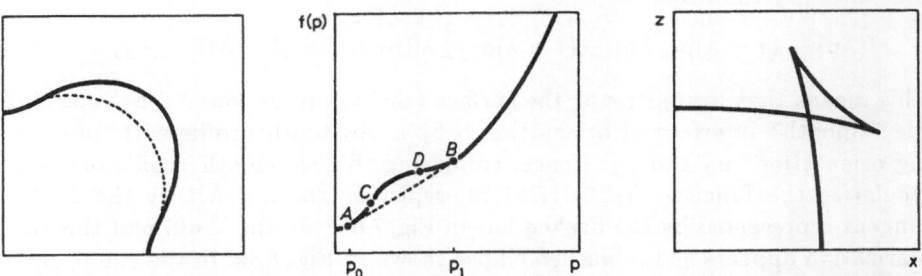

**Fig. 7.8.** The same as in Fig. 7.6 for a crystal with sharp edges between rounded parts. The regularized plots of $\gamma(\hat{n})$ and $f(p)$ are denoted by the *dashed circle* and line, respectively

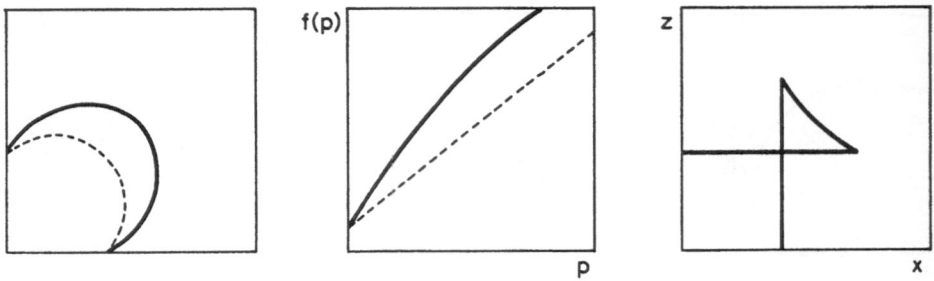

**Fig. 7.9.** The same as in Fig. 7.6 for a crystal of purely cubic equilibrium shape

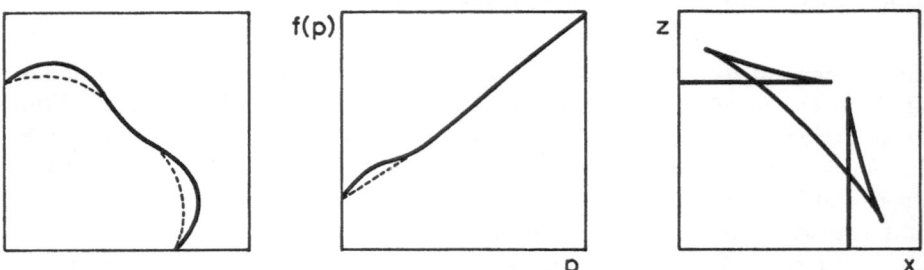

**Fig. 7.10.** The same as in Fig. 7.6 for a crystal with sharp edges between facets and rounded parts

Figure 7.6 shows a Wulff plot giving rise to a fully rounded equilibrium shape. Such shapes have been observed, for instance, for $^4$He crystals in equilibrium with He$^{II}$ [7.16,20] for $^3$He crystals [7.23b] and for several organic substances [7.12–14,23a].

Figure 7.7 describes a crystal with (001) facets that are smoothly connected to each other by rounded parts. Notice that the size of a horizontal facet is proportional to the jump in $\partial f/\partial p$ (or equivalently, the jump in $\partial\gamma/\partial\vartheta$, with $\vartheta = \text{arctg } p$) at the cusp! This follows immediately from (7.7a). Shapes like Fig. 7.7 have been observed for instance for lead in equilibrium with its vapor [7.10].

In Fig. 7.8 the Wulff plot gives rise to a function $f(p,0)$ that does not satisfy the convexity condition

$$f(\lambda p_1 + (1 - \lambda)p_2, \lambda q_1 + (1 - \lambda)q_2) \leq \lambda f(p_1,q_1) + (1 - \lambda)f(p_2,q_2) \ . \quad (7.10)$$

This means that for $p_0 < p < p_1$ the surface tension can be lowered by replacing the "smooth" interface of orientation $p$ by a saw-tooth profile with alternating orientations $p_0$ and $p_1$. Hence, employing this saw-tooth profile one may *regularize* the function $f(p,0)$ [7.31,38] replacing the arc **AB** by the double tangent represented by the dashed line in Fig. 7.8b[2]. In the Wulff plot this regularization appears as the dashed ellipse shown in Fig. 7.8a. In the equilibrium

---

[2] Notice the analogy to Maxwell's construction for the Van der Waals fluid [7.39].

crystal shape the saw-tooth configurations usually do not show up on three-dimensional crystals because they require an extra free energy proportional to the total length of the edges bordering the teeth. Therefore it is more favorable having just two faces, of slope $p_0$ and $p_1$, respectively, cutting each other under a sharp edge. Such edges between rounded parts of a crystal have been observed on colloidal liquid crystals, or tactoids, [7.40] and on grain boundaries of metals [7.41] and on metal-liquid interfaces [7.15]. In two-dimensional crystals at non-zero temperature, on the other hand, no sharp corners will occur in the boundary (at least in the case of short-range interactions). Due to the gain in entropy there are always saw-tooth configurations (possibly with very large teeth, but in the thermodynamic limit this does not matter) appearing as stable orientations in the Wulff plot. For three-dimensional crystals in a gravitational field saw-tooth configurations orthogonal to the field may also occur, [7.38] as the loss in edge free energy is offset by a gain in gravitational energy.

As to the arc AB appearing in Fig. 7.8b: like in Van der Waals theory the convex pieces AC and DB may be interpreted as representing metastable surface states, whereas the piece CD is instable and does not have an unambiguous physical interpretation. In Fig. 7.8a the arc ACDB corresponds to a wing, which has to be discarded from the equilibrium shape. Again, the arc pieces AC and DB may show up in metastable shapes. In the equilibrium shape deriving from the regularized Wulff plot the wing is reduced to a single point.

Figure 7.9 shows the Wulff plot $f(p,0)$ and the equilibrium shape of a purely cubic crystal. Again the regularized Wulff plot (consisting in this case of 8 spheres, each passing through the origin and one of the cube corners) and the regularized form of $f(p)$ are indicated by dashed lines and the wing in the equilibrium shape resulting from the non-regularized Wulff plot reduces to a point in the regularized equilibrium shape. The saw-tooth configurations building the regularized planes now are made up of strips of (100) and (001) planes. Regularized planes in arbitrary directions are built out of rectangular polygons in the (100), (001) and (010) planes. Again, if one assumes the edges between facets to have a positive free energy, the formation of saw-tooth configurations is unfavorable compared to the combination of two (or three) single facets. However, the stability of corners against rounding through the formation of hills and valleys made up of small stable facets, seems much harder to assess. The facet size now is determined by the jump in $\partial f^{\mathrm{reg}}/\partial p$, the slope of the regularized surface tension.

The Wulff plot in Fig. 7.10 gives rise to an equilibrium shape with sharp edges between facets and rounded parts. Again the dashed curves denote the regularized Wulff plot and surface tension per unit projected area.

We may conclude that equilibrium crystal shapes, in general, consist of facets and rounded areas, which may join each other either at sharp edges (discontinuous change of slope across the edge) or at smooth boundaries (continuous change of slope across the boundary). However, some additional features may appear:

*Grooves* on the Wulff plot [7.42] may give rise to cylindrical or conical faces in the equilibrium shape. Such surfaces, having one finite and one infinite radius of curvature, would be intermediate between facets and rounded faces. One could imagine this phenomenon to occur for surfaces of nematic polymer bundles, but we do not know of any actual realization. Wings in the crystal shape with more or less cylindrical symmetry may give rise to *conical points*. Such points may have been observed in tactoid shapes [7.40]. Finally, wings occurring on a rounded part of the crystal shape may reduce to edges with conical end points.

Again, we do not know of any physical example for this.

### 7.2.3 The Roughening Transition and Other Phase Transitions

The roughening transition can now be characterized as the disappearance of a cusp from the Wulff plot. Above $T_R$ the surface tension $\gamma(\hat{n})$ is a smoothly differentiable function of $\hat{n}$ for $\hat{n}$ in the direction of the cusp below $T_R$.

Other phase transitions may occur connected with the disappearance of sharp edges or conical points. These have been much less investigated than the roughening transition so far, although some theoretical work has been reported [7.43].

## 7.3 Microscopic Description of Crystal Surfaces

### 7.3.1 The Kossel Crystal, Lattice-Gas Models, Solid-on-Solid Models

The crudest and simplest microscopic model of a crystal is that of a compact structure packed together out of rigid elementary building blocks, which may, for instance, be of cubic shape, corresponding to lattice cells or parts thereoff. This model is known as the Kossel crystal [7.44,45]. It completely ignores lattice vibrations, electronic structure, dislocations and other essential features of realistic crystals, but in spite of this in most cases it seems to yield a good qualitative picture of realistic crystal surfaces. A slightly more refined description of a crystal in equilibrium with its vapor is provided by the lattice gas [7.3] or three-dimensional Ising model; in this model the unit building blocks are replaced by lattice cells which may be either empty or occupied by a single particle. In a typical two-phase equilibrium state there is a dense component, which can be identified as the crystal phase with a small concentration of vacancies, and a dilute phase, which can be identified as the vapor phase. It consists predominantly of empty sites with a small concentration of vapor particles, mostly monomers but some united in small clusters. The crystal surface can be defined microscopically as the contour separating the crystalline phase from the vapor, as illustrated in Fig. 7.11. Of course, the assumption of a lattice structure for the vapor is unrealistic, but for the description of the crystal-vapor interface this should be immaterial. In fact, for describing just this interface both the vacancies in the bulk of the crystal and the free vapor particles should

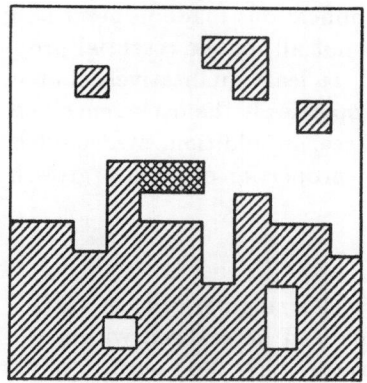

Fig. 7.11. A typical equilibrium state of the two-dimensional lattice gas at low temperature. The *dashed sites* are occupied, the remaining ones empty. The fat separation line is the microscopic phase boundary between the solid and the gaseous phase. The *cross-hatched area* indicates an overhang

not be of real importance, at least not at sufficiently low temperatures. If, in addition, one ignores the possibility of having overhangs in the surface shape (Fig. 7.11), which also is a very good approximation, at least at low temperatures, one arrives at the solid-on-solid or SOS model. It describes the interface in terms of a set of discrete-valued variables $h_{ij}$ defined on a two-dimensional lattice, describing up to which height the interface is built up at each lattice site [ij] (see Fig. 7.12 for an example).

One obtains a statistical mechanical model by assigning to each interface configuration $\{h_{ij}\}$ an energy $E\{h_{ij}\}$. The specific choice of the SOS model that has been used most is that of a simple cubic model, where $h_{ij}$ assumes integer values on a quadratic lattice and the energy of a configuration is proportional to the total number of unsaturated nearest-neighbour bonds, i.e. [3]

$$E\{h_{ij}\} = J \sum_{ij}(|h_{ij} - h_{i+1j}| + |h_{ij} - h_{ij+1}|) \ . \tag{7.11}$$

---

[3] The implicit assumption of having to do with a dislocation free crystal surface may be of importance (Sect. 7.5.2b).

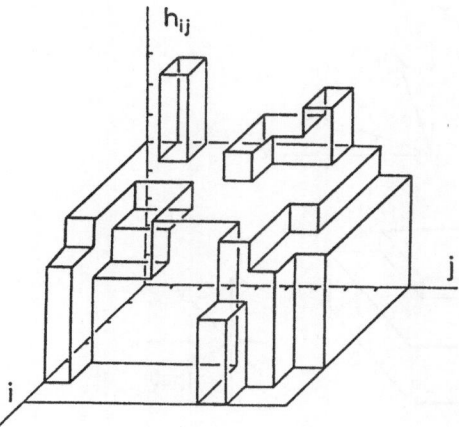

Fig. 7.12. An example configuration of the three dimensional solid-on-solid model

It is believed that, in spite of the enormous simplifications made in describing crystal interfaces by the SOS model, many (but not all) of the essential properties of crystal surfaces are still well described, at least qualitatively, but in many cases even quantitatively. Among these properties is the occurrence and character of roughening transitions, as we shall see. In addition, SOS models have been used with success for understanding properties of crystal growth (Chap. 3) [7.21,46].

### 7.3.2 Steps and Step Free Energies

Vicinal surfaces, i.e. surfaces under a tiny angle with a singular or facet surface (e.g., the (001) surface of a simple cubic lattice), can be realized within the SOS model as *stepped surfaces*. At zero temperature the (001) surface of an SOS model defined by (7.11) is perfectly smooth, as shown in Fig. 7.13a. The ground state of a (p01) surface consists of a collection of perfectly straight steps in the x direction[4] (Fig. 7.13b). The density of these steps per unit length in the x direction equals p, but their precise positions are arbitrary otherwise, hence the ground state is degenerate[5]. In the ground state of a (pq1) surface the steps are not straight any more, but instead exhibit a zig-zag structure such that their average direction is the (pq0) direction, as illustrated in Fig. 7.13c.

At non-zero temperature the straight steps of Fig. 7.13b develop kinks (90° bends) as well, as shown in Fig. 7.14 and the zig-zag steps of Fig. 7.13c develop

---

[4] The direction of a step is taken to be the direction of the height gradient across the step.
[5] This degeneracy is removed if there exist long-range interactions between steps, as is the case in practice (Sect. 7.4.4).

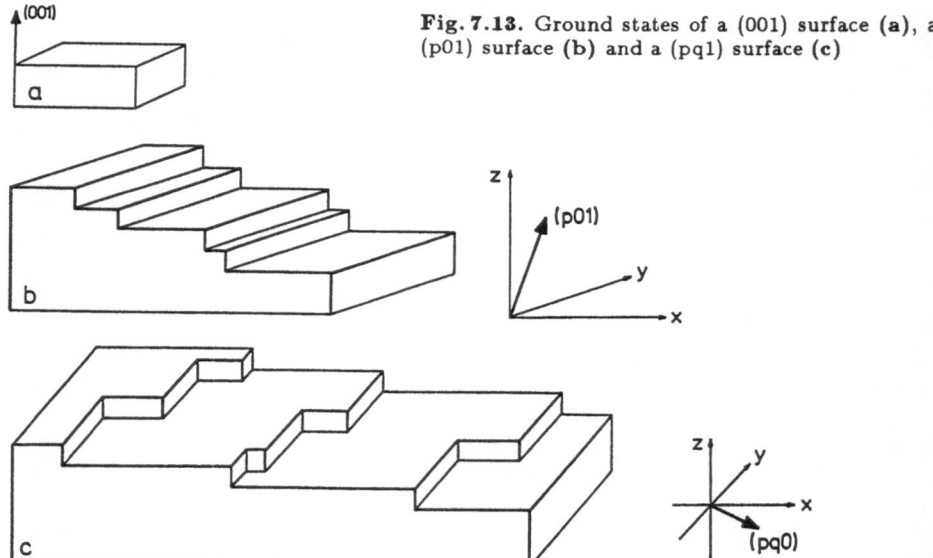

**Fig. 7.13.** Ground states of a (001) surface (**a**), a (p01) surface (**b**) and a (pq1) surface (**c**)

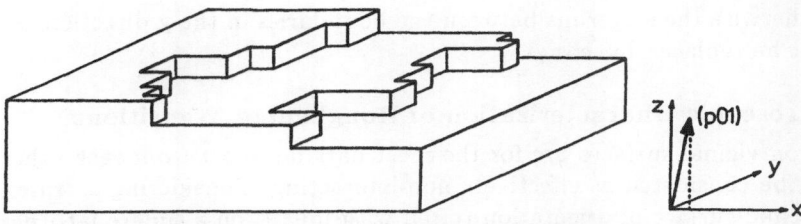

**Fig. 7.14.** Steps of Fig. (7.13b) with thermally induced kinks

additional kinks beyond the ones already present at $T = 0$. Furthermore the terraces between the steps develop thermal excitations in the form of small bumps (ad-atoms) and pits (vacancies). Still, the (pq1) surface may be considered as originating through the formation of steps in the (pq0) direction on the (001) surface.

A surface with a single step can be constructed by imposing appropriate boundary conditions on the set $\{h_{ij}\}$. E.g., a step in the x direction is obtained by requiring

$$\lim_{i \to \pm\infty} h_{ij} = n + \tfrac{1}{2}(1 + \text{sign}\{i\}) \tag{7.12}$$

with n some fixed integer. The *step free energy* of such a single step is defined as the difference of the free energy of an interface containing the step and that of a similar interface containing no steps at all (notice that this definition of the step free energy is quite general and not restricted to the SOS model!). To be more specific, let us consider a step under an angle $\varphi$ with the y direction on a quadratic surface with boundary conditions on $h_{ij}$ as indicated in Fig. 7.15a and compare to a flat surface with boundary conditions, as given in Fig. 7.15b. Then the step free energy per unit length and per unit step height is defined as

$$f^s(\varphi) = -k_B T \lim_{N \to \infty} \frac{|\sin \varphi|}{Na_y a_z} \log \frac{Z_N^{st}(\varphi)}{Z_N} \,, \tag{7.13}$$

where $k_B$ is Boltzmann's constant, and $Z_N^{st}$ and $Z_N$ are the partition functions for the interfaces with the two sets of boundary conditions, whereas $a_y$ and $a_z$ are the lattice constants in the y and z direction, respectively. We assumed

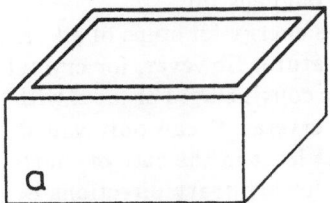 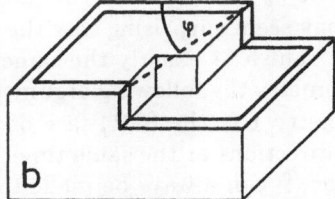

**Fig. 7.15.** Boundary conditions (——) giving rise to **(a)** a (001) surface without steps and **(b)** a (001) surface with a single step of orientation $(\cos \varphi, \sin \varphi)$

273

$\varphi \geq \pi/4$, otherwise the step runs between the boundaries in the y direction and $\sin \varphi$ must be replaced by $\cos \varphi$.

### 7.3.3 Microscopic Characterization of Roughening Transitions

The steps on vicinal surfaces are for the most part far apart from each other, hence may be considered as effectively non-interacting. Considering as an example a vicinal surface of orientation (p01), with $|p| \ll 1$, on a square lattice of size N one finds for this a surface tension of the form

$$f(p, 0, 1) = f(0, 0, 1) + |p| f^s(0) - \frac{k_B T}{N^2} \log \binom{N}{\varrho_s N}$$

$$+ \text{ interaction terms } + \text{ boundary terms} \tag{7.14}$$

with $\varrho_s = (a_x/a_z)|p|$. The second term is the sum of the step free energies of $\varrho_s N$ isolated steps, the third term estimates the entropy contribution due to the number of ways the steps can be distributed on the surface. The third term is of order $N^{-1}|p| \log |p|$, hence may be ignored in the thermodynamic limit, just as the boundary terms, which likewise are of order $N^{-1}$. The interaction terms could at most be of order $|p|$ and therefore can also be ignored for $p \to 0$. A rigorous derivation of (7.14) for the SOS model has been given in [7.47]. In the sequel of this subsection we restrict ourselves to the case that the step free energy has two-dimensional inversion symmetry, i.e. $f(p, q) = f(-p, -q)$. The general case will be discussed in the next subsection.

If $f^s(0) = f^s(\pi) > 0$, one sees from (7.14) that $f(p, 0, 1)$ exhibits a cusp as a function of p at $p = 0$. According to the discussion of Sect. 7.2 this implies that the equilibrium crystal shape exhibits a facet in the (001)-direction, and the diameter of this facet in the y direction is proportional to the step free energy for a step in the x direction. Generally $f(\lambda \cos \varphi, \lambda \sin \varphi, 1)$ as a function of $\lambda$ exhibits a cusp at $\lambda = 0$ under a slope given by

$$\left| \frac{\partial f}{\partial \lambda} \right|_{\lambda=0} = f^s(\varphi) \ . \tag{7.15}$$

In the previous section we have seen that the roughening transition corresponds to the disappearance of a cusp in the Wulff plot, hence we may conclude that the *roughening transition* can be characterized *microscopically* by the *vanishing of the step free energy*[6] for steps on the facet that roughens up.

At first it may seem surprising why the step free energy for steps of all orientations should vanish at exactly the same temperature. However, for crystal faces of high symmetry the following argument seems convincing: For faces with tetragonal symmetry, e.g. the (001) face of a cubic crystal, $f^s$ can only vanish for at least four directions at the same time. Then, at least in the case of short-range interactions, $f^s$ can always be made to vanish for arbitrary directions by

---

[6] Through (7.14) $f^s$ can be given a macroscopic meaning as the jump in slope of $\gamma$ at the cusp, but intrinsically a step is a microscopic object.

piecing together a step in the required direction out of line pieces in directions with vanishing step free energy[7] (this is the same argument explaining why at non-zero temperature equilibrium shapes in two dimensions show no sharp angles between rounded pieces). This argument generalizes immediately to the cases of triangular and hexagonal symmetry. But in the absence of inversion symmetry step free energies may become negative over certain ranges, without leading to unstable facets. This will be discussed in the next subsection. Once the free energy of infinitely long straight steps on a given crystal face equals zero for all directions, steps of arbitrary length will form spontaneously on this face. Let us consider again an interface on an $N \times N$ square with fixed boundary height n. For T above $T_R$ long closed steps will form on the surface and as N is increased several of these may become nested within each other. Hence the height $h_{00}$ of a point in the middle of the interface will fluctuate more and more about its average value n. This property provides an alternative way of characterizing rough states of the interface. One obtains

$$\lim_{N \to 0} \langle (h_{00} - \langle h_{00} \rangle)^2 \rangle = \infty \quad \text{in a rough phase}$$

$$= \text{finite in a smooth phase,} \tag{7.16}$$

where the brackets denote an average over an equilibrium distribution of interface configurations. Closely related to this is the height-height correlation function, describing the mean-square height difference between two points on the interface. Depending on whether one is above or below the roughening transition one finds for this

$$\lim_{R_{ij,k\ell} \to \infty} \langle (h_{ij} - h_{k\ell})^2 \rangle = \infty \ , \quad T > T_R \tag{7.17a}$$

$$= 2 \langle (h_{00} - \langle h_{00} \rangle)^2 \rangle \ , \quad T < T_R \tag{7.17b}$$

where $R_{ij,k\ell}$ denotes the distance between the lattice points (i,j) and (k,$\ell$). In writing (7.17b) we made the assumption that below $T_R$ interface fluctuations become uncorrelated at large distances from each other. In fact, one expects a decay that can be expressed in terms of a correlation length $\xi$, i.e.

$$\log \left[ \langle (h_{ij} - h_{k\ell})^2 \rangle - 2 \langle (h_{00} - \langle h_{00} \rangle)^2 \rangle \right] \sim - \frac{R_{ij,k\ell}}{\xi} \ , \quad R_{ij,k\ell} \to \infty \ . \tag{7.18}$$

Physically the vanishing of the step free energy at the roughening transition means that the steps get lost among the thermal fluctuations of the interface. This can be seen beautifully from simulation results on interfaces on

---

[7] A simple calculation, minimizing $f^s$ as a function of the number of line pieces, shows that the average length of the latter can be estimated as $\exp(E_c/k_B T)$, with $E_c$ the energy of a corner between two line pieces. Hence the line pieces remain of microscopic length, though they may become fairly long at low temperatures.

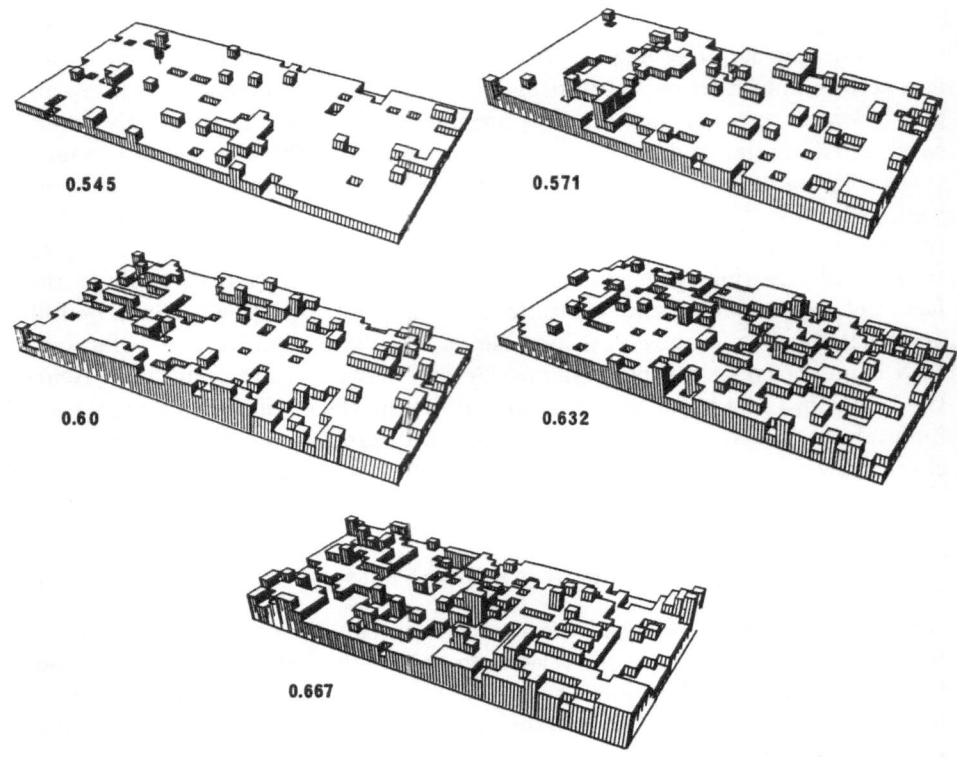

**Fig. 7.16.** Typical equilibrium configurations of (001) SOS surfaces at temperatures of 0.91 T$_R$ (0.545), 0.95 T$_R$ (0.571), 1.0 T$_R$ (0.60), 1.05 T$_R$ (0.632), 1.11 T$_R$ (0.667) [7.24]

the SOS model by *Weeks* and co-workers [7.21,24,48]. Figure 7.16 shows typical interfacial configurations with steps for some different temperatures. At the lowest temperatures, which are below T$_R$, the steps are clearly recognizable although they become more and more fuzzy as temperature increases. The difference with step-free surfaces at equal temperatures, shown in Fig. 7.17, is obvious. For T>T$_R$, however, one does not recognize the steps any more and the difference with a step-free surface gets lost.

### 7.3.4 Equilibrium Shapes of Facets. General Criterium for Roughening

The equilibrium shape of facets below their roughening temperature follows directly from the Wulff construction. To demonstrate this let us orient the crystal with the facet of interest orthogonal to the z direction. Then a vicinal face of orientation ($\varepsilon \cos \varphi$, $\varepsilon \sin \varphi$, 1) intersects the facet in a line under an angle $\varphi$ with the y axis and at a distance $\lambda f^s(\varphi)$ from the z axis. This is illustrated in Fig. 7.18 for the case $\varphi = 0$. The equilibrium shape of the facet is the inner envelope of the lines of intersection between the facet and all the vicinal surfaces. This corresponds to a two-dimensional Wulff construction

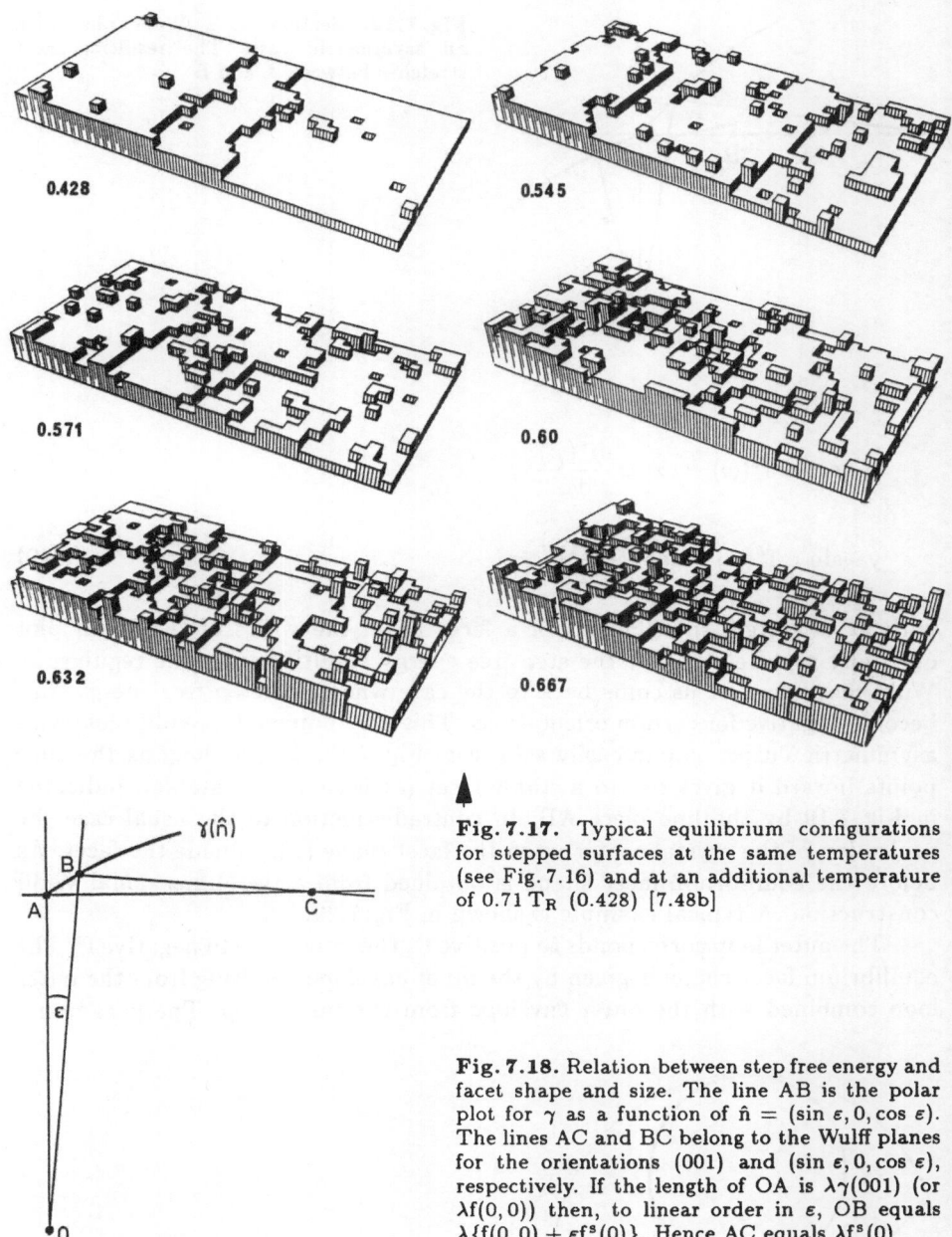

**Fig. 7.17.** Typical equilibrium configurations for stepped surfaces at the same temperatures (see Fig. 7.16) and at an additional temperature of $0.71\,T_R$ (0.428) [7.48b]

**Fig. 7.18.** Relation between step free energy and facet shape and size. The line AB is the polar plot for $\gamma$ as a function of $\hat{n} = (\sin\varepsilon, 0, \cos\varepsilon)$. The lines AC and BC belong to the Wulff planes for the orientations (001) and $(\sin\varepsilon, 0, \cos\varepsilon)$, respectively. If the length of OA is $\lambda\gamma(001)$ (or $\lambda f(0,0)$) then, to linear order in $\varepsilon$, OB equals $\lambda\{f(0,0) + \varepsilon f^s(0)\}$. Hence AC equals $\lambda f^s(0)$

yielding the equilibrium facet shape from the Wulff plot of the step free energy $f^s(p,q)$. From (7.7) in the limit $p, q \to 0$ one obtains, using the identity $f(p,q) = f(0,0) + f^s(p,q)\sqrt{p^2 + q^2}$ for $p, q \to 0$, the following parametrized form of the facet shape [7.1, Appendix D]

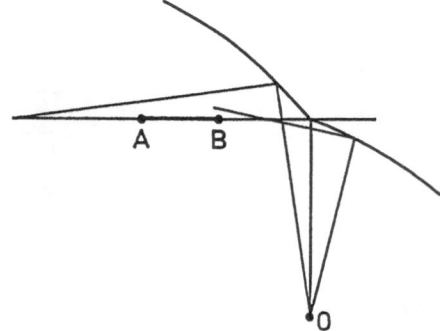

**Fig. 7.19.** Section of a Wulff plot with an asymmetric cusp. The resulting facet stretches between A and B

$$x \sim \cos \varphi \, f^s(\varphi) - \sin \varphi \frac{df^s(\varphi)}{d\varphi} \quad ,$$

$$y \sim \sin \varphi \, f^s(\varphi) + \cos \varphi \frac{df^s(\varphi)}{d\varphi} \qquad (7.19)$$

with $\varphi = \arctan(q/p)$. If parts of a facet lie in the wings of the Wulff plot one must replace $f^s(\varphi)$ by the step free energy resulting from the regularized Wulff plot. Now let us come back to the case where the step free energy may become negative for certain orientations. This may happen for Wulff plots with asymmetric cusps, as one easily sees from Fig. 7.19. Yet, as long as the cusp points inward it gives rise to a stable facet (at least locally stable), indicated in Fig. 7.19 by the line piece AB. In contradistinction to the usual case the projection of the crystal center upon the facet plane falls outside the facet. As before the equilibrium facet shape is obtained from a two-dimensional Wulff construction. A typical example is shown in Fig. 7.20.

The outer loop corresponds to positive $f^s$, the inner loop to negative $f^s$. The equilibrium facet shape is given by the inner envelope resulting from the outer loop combined with the outer envelope from the inner loop. The parametric

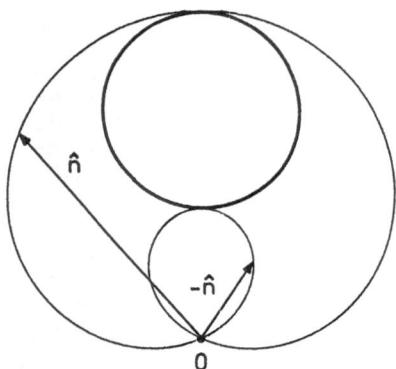

**Fig. 7.20.** Equilibrium facet shape for a Wulff plot including negative step free energies. The larger loop of the selfintersecting curve is the polar plot for positive $f^s$ as a function of orientation n̂, the smaller loop plots $-f^s$ in the direction $-$ n̂ for negative $f^s$. The fat curve describes the equilibrium facet shape

278

representation (7.19) remains valid. Disappearance of the cusp in the three-dimensional Wulff plot in favour of a smooth surface yields $f^s(\varphi) = -f^s(-\varphi)$, i.e. inner and outer loop become identical and indeed the facet disappears. The inverse statement, i.e. disappearance of the facet in the two-dimensional Wulff plot for $f^s$ requires the disappearance of the cusp in the three-dimensional Wulff plot, hence $f^s(\varphi) = -f^s(-\varphi)$ for all $\varphi$, is much harder to prove. If the cusp relaxes to a groove instead of a smooth surface, one would have a transition to a locally cylindrical crystal shape. Whether or not this is actually possible is not known, to our knowledge.

Physically the fact that one can have stable facets in spite of the fact that the step free energy is negative for a range of step directions, comes about because steps cannot appear independently of each other. A facet becomes de-localized through the formation of large terraces and this requires the vanishing of the free energy of *closed* steps. In the case of inversion symmetry indeed this only happens if $f^s(\varphi)$ becomes zero for arbitrary $\varphi$, but in general this condition must be relaxed to the requirement that $f^s(\varphi) + f^s(-\varphi)$ vanishes. An example where one actually has to use the weaker requirement is provided by the terrace-step-kink-model (TSK), which will be discussed in Sect. 7.4.3.

To summarize the preceding results: roughening of a crystal face is characterized microscopically by the vanishing of the free energy of a step on this face, or more precisely the mean of the free energy of a step and that of a step in the opposite direction. Alternatively it may be characterized by infinite fluctuations in the interface height.

## 7.4 Statistical Mechanics of Interface Models

Our knowledge of the precise nature of the roughening transition derives almost entirely from calculations on statistical mechanical models of the crystal surface, such as the SOS-model. These calculations mostly are either renormalization group calculations or exact calculations on solvable models. The results of both support each other completely.

As mentioned already in Sect. 7.1 the first model calculations were done by *Burton* et al. [7.1], who mapped a growth layer at the crystal surface on the two-dimensional Ising model. Although good over a large temperature range below the roughening transition the model breaks down near the roughening transition itself because it ignores the multilayered character of the surface in a rough or nearly rough state.

To our knowledge *Gallavotti* [7.49] was the first one to propose that in the three-dimensional Ising model a phase transition delocalizing the interface between phases of positive and negative magnetization might occur at a temperature below the bulk critical temperature. *Weeks* et al. found strong evidence for such a transition from low temperature expansions for moments of displacement of the interface [7.50]. These expansions were found to become divergent at temperatures, roughly 10 % above the critical temperature of the two-dimensional Ising model corresponding to a single layer, well below the critical temperature of the three-dimensional Ising model. These results contradict a conjecture by *Burton* et al. [Ref. 7.1, p. 344] that the actual roughening temperature of a crystal surface would be below that of the corresponding Ising model. In fact, for a large class of Ising models the latter could be shown to be a rigorous lower bound to the roughening temperature [7.51,52].

A major advance in the theory of the roughening transition was made by *Chui* and *Weeks* [7.4], who showed that the discrete Gaussian model, an SOS model with quadratic interactions, could be mapped exactly upon a two-dimensional Coulomb gas on a lattice. For the latter system *Berezinskii* [7.5], and *Kosterlitz* and *Thouless* [7.6] had found a phase transition, mapping to a roughening transition in the discrete Gaussian model. For quite some time, however, both the existence and the precise nature of this Kosterlitz-Thouless transition have been much disputed, hence the discovery of the body centered solid-on-solid model (BCSOS model), an exactly solvable model [7.53] supporting completely the Kosterlitz-Thouless picture of the phase transition, turned out quite helpful. For the ordinary SOS model *Fröhlich* and *Spencer* [7.54] proved the existence of a Kosterlitz-Thouless type roughening transition from the behaviour of the height-height correlation function at high and at low temperatures.

Several further calculations, both for the Kosterlitz-Thouless renormalization group theory and for the BCSOS model, have confirmed the picture sketched above and revealed additional properties, both of the roughening transition and of crystal surfaces in general. For instance, *Jayaprakash* et al. [7.55] found that at the roughening transition the curvature of the roughening facet jumps from zero to a universal value, and they [7.55] as well as independently *Blöte* and *Hilhorst* [7.56], and *Rottman* and *Wortis* [7.57] discovered a universal non-analytic shape for the rounding of a surface area bordering smoothly to a facet.

Some of these properties have been verified experimentally: the universal jump in curvature by *Wolf* et al. [7.20] and the non-analytic shape of the surface area by *Rottman* et al. [7.58]. Further, some more exactly solvable models were found. *Blöte* and *Hilhorst* [7.56,59] constructed a model that is equivalent to a solvable triangular dimer model, and which is especially suited for describing the neighbourhoods of (111) directions on cubic crystal surfaces. Without further extensions this model does not exhibit the non-analytic structure of areas bordering to facets, we mentioned above. Another exactly solvable model is the terrace-step-kink model (TSK model), describing surfaces of finite slope [7.60–62]. This model does exhibit roughening transitions, which are of Kosterlitz-Thouless type again, and it also confirms the results on the rounding of vicinal areas.

In Sect. 7.4.1 we discuss the results of the renormalization group theories, in Sect. 7.4.2 we introduce and discuss the BCSOS model, in Sect. 7.4.3 we discuss the TSK model and in Sect. 7.4.4 we present a simple argument due to *Gruber* and *Mullins* [7.60], explaining the non-analytic curvature of vicinal areas.

### 7.4.1 Renormalization-Group Results

Renormalization-group calculations usually start from the partition function of the SOS model

$$Z = \sum_{z_p=0,\pm a,\pm 2a \ldots} \exp\left[-\beta \sum_{p,q} V(z_p - z_q)\right] , \qquad (7.20)$$

where the summation over p and q runs over nearest-neighbour pairs, $\beta = 1/k_B T$ and a is the lattice unit in the z direction. Using the Fourier series expansion

$$\sum_{n=-\infty}^{\infty} \delta(x - na) = \frac{1}{a} \sum_{m=-\infty}^{\infty} e^{2\pi i x m/a} \qquad (7.21)$$

one may rewrite (7.20) as

$$Z = \left(\prod_p \int_{-\infty}^{\infty} dz_p\right) \sum_{m_p=-\infty}^{\infty} \exp\left[-\beta \sum_{p,q} V(z_p - z_q) + \sum_p 2\pi i z_p m_p/a\right] , \qquad (7.22)$$

where factors a were absorbed in the normalization. Next the summations over $m_p$ are restricted to $m_p = 0, \pm 1$ and the potential $V(z)$ is approximated by a quadratic potential with the result

$$Z = \sum_{m_p=0,\pm 1} (\prod_p \int_{-\infty}^{\infty} dz_p\, K^{|m_p|}) \exp\left[-\frac{\alpha}{2} \sum_{p,q} (z_p - z_q)^2\right.$$
$$\left. + \sum_p 2\pi i z_p m_p/a\right] . \qquad (7.23)$$

Here K has been substituted instead of unity, because in the renormalization procedure this prefactor changes its value. At least for large $\beta$ the prefactor $\alpha$ may be identified with $2\beta[V(1) - V(0)]$, so that for configurations in which nearest-neighbour height differences of 0 and $\pm 1$ are predominant, as will be the case for small T, the actual potential is approximated well by the quadratic one. The physical justification behind the approximations made in (7.23) is that the dominant characteristic properties of the Hamiltonian remain conserved, i.e. the favoring of integer values for $z_p$ and the presence of short-range attractive forces that can be expressed in these variables. Therefore one expects the nature of the phase transition to remain unchanged. The model (7.23) has been introduced by *Villain* [7.63]. Its relatively simple structure simplifies the further analysis.

There are two equivalent ways of deriving renormalization group equations from (7.23). *Kosterlitz* [7.6b] proceeds by performing the Gaussian integrals over $z_p$, arriving approximately at

$$Z = Z_G \sum_{m_p=0,\pm 1}' K^{\Sigma |m_p|} \exp\left[-\frac{\pi}{\alpha} \sum_{p,q} -\log\,(r_{pq}/a) m_p m_q\right]\,, \qquad (7.24)$$

where $Z_G$ is the partition function of the continuous Gaussian model, obtained by setting all $m_p = 0$ in (7.23) and $r_{pq}$ is the distance between sites p and q. The remaining term is the grand partition function of a two-dimensional Coulomb gas with charges $\pm 1$ at activity K and temperature $T = \alpha/\pi k_B$. The prime on the summation over $m_p$ denotes the neutrality condition $\sum_p m_p = 0$. *Kosterlitz* applied a renormalization-group transformation by integrating out short-range interactions and rescaling the length variables. He found that under this procedure the Hamiltonian approximately maintains its form, but the parameters $\alpha$ and K are transformed to different values. By looking for fixed points of this transformation he obtained the roughening transition (in our interpretation!) and could determine its critical properties.

In an alternative but equivalent procedure (7.23) is approximated for small K as [7.20,29,64]

$$Z = \left(\prod_P \int_{-\infty}^{\infty} dz_p\right) \exp\left(-\frac{\alpha}{2} \sum_{p,q} (z_p - z_q)^2 + 2K \sum_p \cos 2\pi z_p/a\right)\,. \quad (7.25)$$

Replacing the discrete sums over p and q by an integral over $\varrho = (x,y)$ one obtains the partition function of the sine-Gordon field theory [7.65]

$$Z = \int d\varrho \int_{-\infty}^{\infty} dz(\varrho) \exp\left\{-\frac{\alpha}{2} \int d\varrho \int d\varrho' [z(\varrho) - z(\varrho')]^2\right.$$
$$\left. + 2K \int d\varrho \cos 2\pi z(\varrho)/a\right\}\,. \qquad (7.26)$$

As a prize for giving up the discrete lattice structure one has to regularize (7.26) by imposing a momentum cut-off on the Fourier transform $\hat{z}(k)$ of $z(\varrho)$ at $K = \Lambda = 2\pi/a_h$ where $a_h$ is the lattice spacing in the horizontal plane. Renormalization now is performed by integrating out $\hat{z}(k)$ over an outer momentum shell $\Lambda - \delta\Lambda < k < \Lambda$ and renormalizing the length scale such that the new cut-off $\Lambda - \delta\Lambda$ is rescaled to $\Lambda$. Again, one recovers (7.26) with renormalized values for $\alpha$ and K. The lines of flow in the $\alpha,k$–plane along which the renormalization proceeds are precisely the same ones as obtained by *Kosterlitz*.

For the details of these renormalization-group calculations we refer to the original publications, here we just mention the main results:

a) **Roughening Temperature**
Although the roughening temperature obviously is not a universal constant, *Kosterlitz'* renormalization-group method allows for its determination with remarkable accuracy [7.6b,25]. The predicted value for the discrete Gaussian model is of the order [7.25]

$$T_R \approx 1.45 J/k_B$$

which is quite close to the estimates from computer simulations [7.66].

## b) Surface Tension

The surface tension exhibits a very weak singularity at the roughening transition, which is of the form [7.6b]

$$\gamma_{\text{sing}} \approx B \exp\left(-\frac{C}{|T - T_R|^{1/2}}\right) \tag{7.27}$$

for T in the neighbourhood of $T_R$. One sees that all derivatives of $\gamma_{\text{sing}}$ with respect to temperature are smooth functions of T and vanish at $T = T_R$. The constants B and C are non-universal.

## c) Step Free Energies

The free energy of a step is found to behave as

$$
\begin{aligned}
f^s &\approx f^0 \exp[-A/(T_R - T)^{1/2}] \ , \quad T < T_R \\
&= 0 \ , \quad T > T_R
\end{aligned}
\tag{7.28}
$$

where $f^0$ and A are non-universal constants again. This result is obtained by observing that $f^s$ is the dual conjugate of the inverse correlation length in the XY model [7.67,68] (see below) and using *Kosterlitz'* results [7.6b] for the latter. From (7.28) and the proportionality between step free energy and facet size discussed in Sect. 7.3 one sees that indeed the vanishing of a facet as $T_R$ is approached from below is extremely smooth, hence hard to observe experimentally.

## d) Correlation Length

For temperatures below, but close to the roughening temperature the correlation length, which is the characteristic length for correlations between thermal excitations of the crystal surface behaves as[8] [7.25,69]

$$
\begin{aligned}
\xi &\approx \xi^0 \exp[A/(T_R - T)^{1/2}] \ , \quad T < T_R \\
&= \infty \ , \quad T \geq T_R
\end{aligned}
\tag{7.29}
$$

with $\xi^0$ again a non-universal constant and A the same as in (7.28).

---

[8] The arguments leading to (7.28,29) are somewhat subtle: The step free energy by dual conjugation is identified with the logarithm of the spin-spin correlation function in the XY model at the dual temperature [7.64,65]. The correlation length in (7.29), on the other hand, is the dual conjugate of the correlation length for correlations between vortices in the XY model [7.4]. The identity of the spin-spin and the vortex-vortex correlation length follows from the analysis by *Kosterlitz* [7.66].

### e) Height-Height Correlation Function

The height-height correlation function, introduced in (7.17), behaves as [7.4,69]

$$G(r) = \frac{1}{2\pi} K(T) a_v^2 \log \, [a_h^2 (r^{-2} + \xi^{-2})]^{-1} \tag{7.30}$$

with the following identifications: $G(r)$ equals $\langle (h_{ij} - h_{k\ell})^2 \rangle$ with $r = R_{ij,k\ell}$; $a_v$ and $a_h$, respectively, are the lattice constants in the directions orthogonal and parallel to the plane of the surface under consideration (where we assumed implicitly $a_h$ is the same for different principal directions in this plane); $\xi$ is the correlation length introduced in (7.29), hence for $T > T_R$ one has $\xi^{-1} = 0$. The coefficient $K(T)$ is an increasing function of temperature. Renormalization-group theory produces specific predictions about its behavior. At the roughening temperature $K$ assumes the universal value [7.69]

$$K(T_R) = \frac{2}{\pi} \, . \tag{7.31}$$

This value is approached from above through a square root cusp [7.25], i.e.

$$K(T) = \frac{2}{\pi} + C(T - T_R)^{1/2} \tag{7.32}$$

for $T > T_R$, but $(T - T_R)/T_R \ll 1$. The constant $C$ is non-universal again. The universal value (7.31) of $K(T_R)$ has been used by *Shugard* et al. [7.66] to estimate roughening temperatures of model systems from Monte-Carlo simulation results. The square root cusp is very little pronounced for these systems and has not been observed. For temperatures below $T_R$ the height-height correlation function saturates for large $r$ at the value $K(T) a_v^2/\pi \log \, (\xi/a_h)$. Hence, from (7.17b) and (7.29) it follows that the interface width diverges as $(T_R - T)^{-1/4}$ as $T$ approaches $T_R$ from below [7.69]. Finally it should be remarked that (7.30) holds for anisotropic surfaces as well, at least in the limit of large $r$ [7.62].

### f) Jump in Curvature and Surface Stiffness

One of the most striking predictions of renormalization-group theory is that at $T = T_R$ the radius of curvature for the surface element with the orientation of the roughening facet jumps from $\infty$ (for $T < T_R$) to the universal value [7.55]

$$R_c = \frac{z_0 (k_B T_R) \pi}{2 \gamma_0 a_v^2} \, . \tag{7.33}$$

Here $z_0$ is the distance from the tangent plane at the surface to the crystal center and $\gamma_0$ is the surface tension per unit area (Sect. 7.2.1). For an anisotropic surface $R_c$ in (7.33) has to be replaced by [7.70,43] $(R_c^{(1)} R_c^{(2)})^{1/2}$, where $R_c^{(1)}$ and $R_c^{(2)}$ are the principal radii of curvature.

An alternative way of expressing the above relations is in terms of the *surface stiffness,* which measures the resistance of the surface against bending. If $\gamma(p,q)$ is the surface tension of the (pq1) direction then the surface stiffness against bending around the $(-qp0)$-axis is defined as

$$\alpha^{(0)}(p,q) = \gamma(p,q) + \partial^2\gamma(p,q)/\partial\Theta^2 \ , \tag{7.34}$$

where $\Theta = \mathrm{arctg}(p^2 + q^2)^{1/2}$ and $p/q$ is kept constant. The universal jump in curvature at $T = T_R$ now translates to [7.20,70][9]

$$\left\{ \frac{\partial^2 f}{\partial p^2} \frac{\partial^2 f}{\partial q^2} - \left( \frac{\partial^2 f}{\partial p\,\partial q} \right)^2 \right\}^{1/2}_{p=q=0} = \frac{k_B T_R \pi}{2a_v^2} \tag{7.35}$$

for the general anisotropic case. In Appendix 7.A we demonstrate how to derive (7.35) from (7.33).

For $T>T_R$ but $(T - T_R)/T_R \ll 1$ the surface stiffness as a function of temperature behaves as [7.20]

$$\alpha(T) = \alpha(T_R)[1 - C(T - T_R)^{1/2}] \ , \tag{7.36}$$

where C is again a non-universal constant, which, according to *Wolf* et al. [7.20], takes the value $C = \pi/2A$, with A the constant appearing in (7.28,29). Finally, at the roughening temperature the dependence of surface stiffness on p and q is very strong and can be obtained in quantitative form from renormalization group again [7.20].

Experimental confirmation of (7.33,35) has been obtained by *Wolf* et al. [7.20], whereas *Babkin* et al. [7.37] claim their experiments contradict these predictions. It should be noted, however, that *Wolf* et al. had to apply appreciable corrections to account for angle dependence of $\alpha$ (the measured value of this quantity represents some angular average), even though the range of angles they observe was very small.

### 7.4.2 The BCSOS Model

The BCSOS (body centered solid-on-solid) model [7.53] is a solid-on-solid model defined on a quadratic lattice that is divided into two sublattices (Fig. 7.21). On one of the sublattices the height variables may assume only integer values $(0, \pm1, \pm2, \ldots)$ and on the other sublattice only half-integer values $(\pm1/2, \pm3/2, \ldots)$. In addition, the height differences between neighbouring sites are constrained to be $\pm1/2$. This model is isomorphic to Lieb's six vertex model: drawing arrows on the bonds of the dual lattice, such that always the higher of the two neighbouring height variables is to the right of the arrow, one generates configurations of the six vertex model, which correspond one-to-one (up to an overall vertical shift of the height variables, a symmetry which can be

---

[9] According to *Wolf* et al. [7.20] the universal value of the surface stiffness at $T_R$ is directly related to the universal prefactor of the height-height correlation function at the roughening temperature. We do not understand this connection, especially since in the anisotropic case the two quantities seem to behave differently at $T_R$.

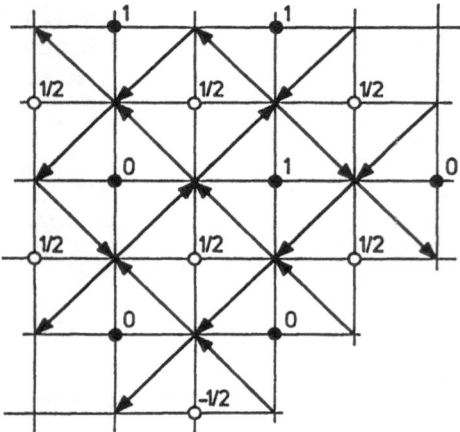

**Fig. 7.21.** Construction of a 6-vertex configuration from a BCSOS-model configuration

removed by fixing the value of one single height variable) to the allowed configurations of the BCSOS model. The ice-rule (at each node of the dual lattice two arrows point inward and two point outward) is an immediate consequence of the uniqueness of the height variables [7.53] (in the presence of screw dislocations in the crystal lattice this has to be mitigated, see Sect. 7.5). An example for the mapping from BCSOS to six vertex configurations is shown in Fig. 7.21. By assigning energies $\varepsilon_1 \ldots \varepsilon_6$, as indicated in Fig. 7.22, to the six possible vertices one introduces interactions between the height variables. In [7.53] it is shown that these interactions obtain, in a most natural way, from a lattice gas on a body centered cubic crystal lattice with strong nearest-neighbour attraction and much weaker next-nearest neighbour interaction. If the latter is attractive the corresponding 6-vertex model is the F model, for which Lieb has obtained an exact solution [7.71]. From this solution it follows that the surface tension of the (001) plane exhibits a phase transition of Kosterlitz-Thouless type. All the renormalization-group predictions discussed in Sects. 7.4.1b and c [7.53], in Sect. 7.4.1d [7.72], in Sect. 7.4.1e (for technical reasons this has only been shown at the single temperature $T = 2T_R$ [7.55]) are confirmed.

*Sutherland* et al. [7.73b] have extended Lieb's result to a solution for an arbitrary choice of vertex weights (which is expressed, however, in terms of the solution of a rather intricate nonlinear integral equation). From this so-

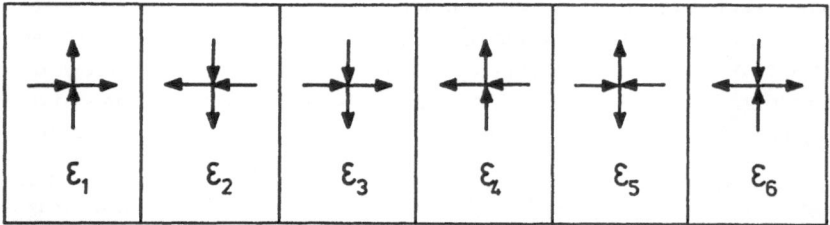

**Fig. 7.22.** Allowed vertex configurations with the assigned energies

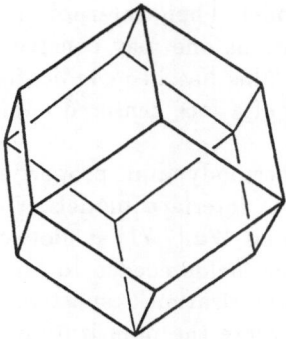

**Fig. 7.23.** The rhombododecahedron

lution one may extract the surface tension for all temperatures and p and q values satisfying $|p|, |q| \leq 1$. The latter restriction should be interpreted as follows: The equilibrium shape of the Ising model on a bcc lattice with infinitely strong nearest-neighbour attraction is a rhombododecahedron, bounded by facets of (110)-type, as one finds straightforwardly from the Wulff construction (Fig. 7.23). If the strength of the nearest-neighbour coupling is relaxed to finite values, which are still $\gg k_B T$, this shape persists but the edges and corners are rounded. Then the BCSOS model describes the equilibrium shape of the six corners formed by the intersection of four (110)-type facets. In the case of the F model (next-nearest neighbour attraction) these rounded corners at low temperatures develop a facet of (001)-type, which disappears at the roughening temperature of the F model. At still higher temperatures one may neglect the nearest-neighbour interactions completely and approximate the (110) ori-

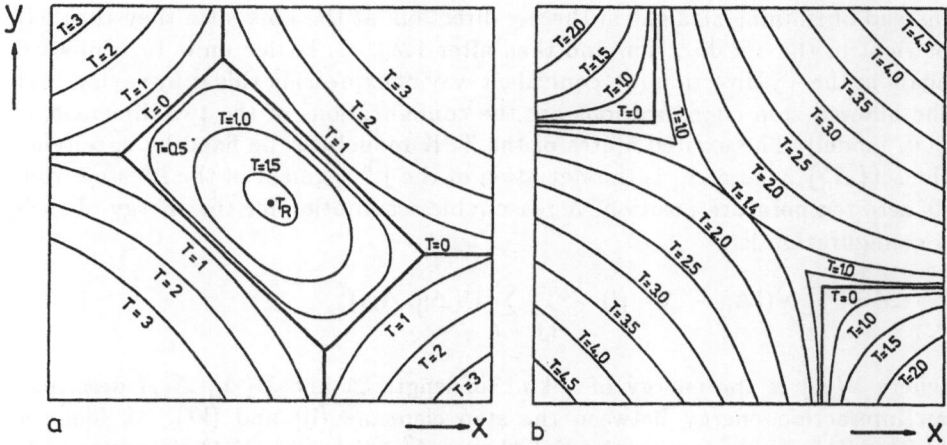

**Fig. 7.24.** (a) Facet boundaries of an anisotropic BCSOS model with next nearest-neighbour attraction for a set of different temperatures. This is a reinterpretation of a phase diagram for the F model of antiferroelectricity by *Lieb* and *Wu* [7.71]. In the units chosen $T_R = 2.08$. (b) Facet boundaries of an anisotropic BCSOS model with next nearest-neighbour repulsions for a set of different temperatures. This is a reinterpretation of a phase diagram for the KDP model of ferroelectricity by *Lieb* and *Wu* [7.71]

entations and their neighbourhoods by another F model. Then, interpolating between these results for different (110)-type orientations one may construct approximate equilibrium shapes for the full crystal. This has been indicated in [7.55] and worked out in more detail for the case of a face centered cubic crystal in [7.75]. In this connection see also [7.26].

It is noteworthy that many earlier results on the thermodynamic properties of six vertex models can be reinterpreted in terms of interface models. For example, Figs. (7.24a,b) have been presented by *Lieb* and *Wu* [7.71] as plots of critical field strengths at different temperatures for the fields needed to force the antiferroelectric F model into a state of non-zero polarization, respectively, to force the ferromagnetic KDP model into a state where the polarization is not frozen in at its maximal value. They may be interpreted alternatively as the plots of equilibrium shapes and sizes of facets in the BCSOS model.

Finally, let us mention that the BCSOS model also confirms the universal non-analytic shape of vicinal areas [7.55] that will be discussed in the Sect. 7.4.4.

### 7.4.3 The Terrace-Step-Kink Model

The terrace-step-kink model, or TSK model, describes surfaces with a finite density of steps, more specifically (p01) surfaces of cubic crystals. Assuming an elastic and/or dipolar repulsion between steps (Sect. 7.4.4b) one finds the ground state for such a surface as a regular array of perfectly straight steps with a spacing 1/p between neighbouring steps (i.e., if 1/p is an integer, or a half-integer in the case of an fcc or bcc lattice, otherwise the ground state is more complex [7.69]). For non-zero temperature the only allowed excitations of the interface exist in the formation of kinks in the step edges (Fig. 7.14), i.e. instead of running straight in the +y direction, at the kink sites they turn left or right in the ±x direction and then after 1,2,3,... lattice units turn upward again in the +y direction (an equivalent way of expressing this is by saying that the allowed step configurations are the configurations of the two-dimensional SOS model). The excited states of the TSK model can be fixed by specifying the set$\{\Delta_{ij}\}$, where $\Delta_{ij}$ is the deviation of the $j^{th}$ segment of the $i^{th}$ step from its zero temperature position. A reasonable assumption for the energy of such a configuration is

$$E = \sum_{i,j} V(|\Delta_{ij} - \Delta_{ij+1}|) + \sum_{i,j} \sum_{k,\ell} U(\Delta_{ij}, \Delta_{k\ell}) \ , \qquad (7.37)$$

where $V(\Delta)$ is the energy of a kink of length $\Delta$ and $U(\Delta_{ij}, \Delta_{k\ell})$ describes an interaction energy between the step elements (ij) and (k$\ell$). As long as $k_BT \ll V(1)$ the approximation $V(\Delta) = \varepsilon\Delta^2$, with $\varepsilon = V(1)$, is very good, since kinks of length $>1$ basically do not occur. If, in addition, one sets $U(\Delta_{ij}, \Delta_{k\ell}) = \delta_{ik-1}\delta_{j\ell}\alpha(\Delta_{ij} - \Delta_{kj})^2$, (7.37) becomes the energy function of an anisotropic discrete Gaussian model. Hence the resulting phase transition is a Kosterlitz-Thouless-type roughening transition of the (p01) face. It is characterized by a delocalization of the steps with respect to their zero-temperature positions and by fluctuations $(\Delta_{ij} - \Delta_{ik})^2$ along a single step that grow as $\log |j - k|$ with increasing distance between j and k. Due to the anisotropy the fluctuations in the step positions are much larger than the fluctuations along a step.

Another choice of the potentials yields an exactly solvable model[10] [7.61,74]

$$V(\Delta) = \delta_{\Delta 0} + \varepsilon \delta_{\Delta 1} \quad \text{(with } \Delta = 0 \text{ or } 1 \text{ only)} \tag{7.38a}$$

$$U(\Delta_{ij}, \Delta_{k\ell}) = \delta_{ik-1} \delta_{j\ell} g(\Delta_{ij} - \Delta_{kj}) \tag{7.38b}$$

with

$$g(n) = \infty \quad , \quad n < -1$$
$$g(-1) = \alpha$$
$$g(n) = 0 \quad , \quad n \geq 0 \ . \tag{7.38c}$$

The model can be mapped upon the XXZ chain in a field [7.61,75] which has been solved by *Yang* and *Yang* [7.76]. It exhibits a roughening transition of Kosterlitz-Thouless type as well. It may be expected to yield a reasonable description of real crystal surfaces under the following conditions:

i)    The kink energy $\varepsilon$ must be $\gg k_B T$ so that kinks of more than unit length are rare (they are forbidden by (7.38a)).

ii)   The interaction between neighbouring steps must be a rapidly decreasing function of their distance; a more realistic choice for g(n) than (7.38c) should satisfy $g(-2) \gg k_B T$ and $g(n) \ll k_B T$ for $n > 0$.

Obviously there will at best be only a restricted range of temperatures where these conditions are met and for each rational value of p this characteristic range will be different. As a consequence of this *Burkov* [7.77] and *Schulz* [7.74] predicted an infinite sequence of roughening temperatures for (m/n01)-surfaces, with $T_R$ depending only weakly on n, but equilibrium facet sizes that decrease rapidly with growing n. The models (7.37 and 38) obviously may have a negative "step free energy", as a possibility mentioned already in Sect. 7.3.3. Without interaction between steps f(p,0,1) would be just a linear function of |p|. At low temperature the interactions produce cusps at rational p-values, e.g. $p = 1/n$, because reduction and increase by one unit of the distance between neighbouring steps does not change the interaction energy by opposite amounts. If the linear term in |p| does not vanish, the resulting cusp in f(p,0,1) is of the type shown in Fig. 7.19. It is somewhat confusing in this case to relate the jump in $\partial f / \partial p$ to a step free energy, as was done in Sect. 7.3.2, because it is not the free energy of the steps promoting the (001) face to a (p01) face. However, if one interprets the variables $\Delta_{ij}$ as variables of a SOS model, with

---

[10] In [7.62], where this model has also been treated, the suggestion was raised that $\Delta$ may assume values $> 1$, which is not the case.

an energy function of the form (7.37), the cusp in $f(p,0,1)$ can be attributed to the appearance of a step in this related SOS model. In the real model this quasi step corresponds to a unit shift to the left or right of all real steps beyond a certain value of x.

*Selke* and *Szpilka* [7.78] have studied the model (7.37) in computer simulations and confirmed the Kosterlitz-Thouless character of its roughening transition. The aim of this work was to make a comparison to experimental results on the scattering from stepped surfaces. There seems to be good qualitative agreement with experiments for Cu by *Lapujoulade* et al. [7.62], but the interpretation of these experiments is still controversial; *Conrad* et al. [7.79] on the basis of very similar experiments for Ni concluded to a much lower range of temperatures for the occurrence of roughening on stepped faces. Obviously more experimental work is called for in this area. Finally, if the interaction between steps is attractive (this may be the result of dipolar forces) a first-order phase transition may occur, i.e. at low temperature the steps on a (p01) face may all collapse together, leaving a (001) facet together with e.g. a (100) facet, separated by a sharp edge [7.74]. On raising the temperature the entropic repulsion will be increased and eventually stabilize the (p01) direction.

### 7.4.4 Universal Shape of Vicinal Areas

Consider again the free energy per unit projected area $f(p,q,1)$ for vicinal faces near a faceted (001) face, as a function of p for q = 0. For small p this function can be expanded as [7.27,55]

$$f(p, 0, 1) = f^{(0)} + |p|f^s + |p|^3 C + \dots . \tag{7.39}$$

The absence of a quadratic term in this expansion is most remarkable. It will be explained below. Given (7.39) the crystal shape in the y,z-plane follows from (7.7) as

$$z = z_0 \left( 1 - \frac{2C}{f^{(0)}} |p|^3 + \dots \right) , \tag{7.40a}$$

$$x = \pm x_0 \left( 1 + \frac{3C}{f^s} p^2 \right) \tag{7.40b}$$

with $z_0 = f^{(0)}/\lambda$ and $x_0 = f^s/\lambda$. Hence,

$$(z - z_0) = -2/3\,c^{-1/2}|x - x_0|^{3/2} . \tag{7.41}$$

This 3/2 power for $z - z_0$ as a function of $|x - x_0|$ establishes the non-analytic shape of vicinal areas we referred to above. As a consequence the radius of curvature at the facet boundary is given as

$$R_y = \left( \frac{\partial^2 z}{\partial x^2} \right)_{z_0^+}^{-1} = 0 . \tag{7.42}$$

Notice that a non-zero quadratic term in (7.39) would give rise to a parabolic profile and to a finite radius of curvature.

The basic equation (7.39) may be interpreted as follows: the linear term in $|p|$ sums up the free energy contributions of single non-interacting steps and the $|p|^3$ term contains the dominant contribution resulting from interactions between steps. Naively one might expect this interaction free energy to be proportional to the square of the step density $|p|$, so why is it actually proportional to $|p|^3$? The explanation requires both entropic and energetic arguments. The entropic interactions can be understood qualitatively from a calculation, given by *Gruber* and *Mullins* [7.60], who consider the effect of the non-crossing condition for neighbouring steps on the surface tension. To simplify the analysis they investigate a single TSK-type step, constrained between straight walls rather than between its neighbour steps. These boundary conditions produce an entropic repulsion between step and walls. If the step is in the x direction and the walls are located at $x = \pm L$, the probability density $P(n)$ for the step passing through $x = n$ at some given y coordinate is given as [7.60]

$$P(n) = (1/L)\cos^2[\pi n/2L] \tag{7.43}$$

and the free energy of the step is found to be of the form

$$f^s(L) = f^s + CL^{-2} \ . \tag{7.44}$$

By staying away from the walls the step reduces its decrease of entropy due to the non-crossing restriction and thereby reduces the increase of free energy from a term proportional to $L^{-1}$ (which would be the result if $P(n)$ were constant) to one proportional to $L^{-2}$. If the fixed boundaries are replaced by the neighbouring steps the effective distance L becomes proportional to $|p|^{-1}$, hence the free energy per step behaves as $f^s + Cp^2$ and (7.39) results.

In addition to the entropic repulsion, one usually has elastic interactions between steps mediated through the lattice, and dipole-dipole interactions due to local charge distributions. Both types of interactions for large distances between steps assume the form [7.62]

$$v = \frac{C}{A}\int_0^A dy |x_1(y) - x_2(y)|^{-2} \ , \tag{7.45}$$

where $x_i(y)$ denotes the x-coordinate of the $i^{th}$ step at position y, v is the interaction energy per unit step length and the integral runs between the boundaries of the facet on which the steps appear. For small p the separation between steps typically is large, so the step interaction will hardly influence the step profiles[11]. Hence, using (7.43) one may obtain the estimate

---

[11] That is, provided the steps do not attract each other so strongly so as to make the vicinal faces unstable (Sect. 7.4.3).

$$\mathbf{v} \approx \int_{-\frac{1}{2}|p|^{-1}}^{\frac{1}{2}|p|^{-1}} 2|p|dx \cos^2(\pi px)C(x + \tfrac{1}{2}|p|^{-1})^{-2} \sim p^3 \ . \tag{7.46}$$

## 7.5 Discussion

Since *Burton* et al. introduced the concept of a roughening transition [7.1], our understanding of this phenomenon has deepened enormously. As discussed in the preceding sections, the roughening transition may be characterized on the one hand by a very gradual disappearance of a facet of given orientation on a crystal, on the other hand by a divergence in the width of a crystal surface with the orientation of this facet; this divergence is such that the interface width increases as the square root of the logarithm of the interface diameter. This general picture emerges from both renormalization-group calculations and results for exactly solvable models of crystal surfaces. Experimentally, roughening transitions, occurring mostly as a function of temperature, have been observed on plastic crystals in contact with their vapor, on $^4$He in contact with the superfluid phase and on a few other systems. For $^4$He the theoretical predictions about the nature of the transition could be confirmed quantitatively. On liquid-solid interfaces in metal alloys roughening transitions also have been observed in dependence of composition.

In spite of all the progress made, some controversies and open questions remain. We address two of them below:

### 7.5.1 Roughening Versus Surface Melting

*Burton* et al. [7.1] instead of using the term roughening rather talk of surface melting. The same identification of roughening transition and surface melting, or surface premelting in more modern terminology, can be found in more recent papers, e.g. in a recent review by *Nenow* [7.80].

Are the two in fact the same? Roughening, as remarked above, is characterized by a divergence of the surface width or, more precisely, by a delocalization of its position in the orthogonal direction. Surface premelting, on the other hand, may be characterized by the following two features (Chap. 6), the first of which is of static character and the second one dynamic:

i) The outer layer or layers of the melting surface lose their two-dimensional lattice structure, i.e. the average density of molecules within the plane of these layers becomes roughly uniform.

ii) The molecules, which are basically frozen into fixed positions below the transition, become mobile at the surface premelting. At the same time the surface layer loses its resistance against two-dimensional shearing.

In actual fact one should expect this characterization of the differences between the solid and the surface molten state to be exaggerated. The mobility in the solid state would not be exactly zero, because hopping diffusion through exchange of particles and vacancies in the surface layer may occur. On the other hand, due to the influence of the underlying crystal one cannot expect the density in the surface layer to be entirely translationally invariant; some periodic modulation, however slight, must remain. Hence, surface premelting must be a first-order transition between two states with the same basic symmetry, but with the solid state having strong modulation and low mobility, the surface molten state having weak modulation and high mobility.

There seems to be no compelling reason why a roughening transition should be accompanied by a sudden decrease in lattice ordering within the surface layer. In fact, both in computer simulations and in real experiments surface premelting is usually found in somewhat loosely packed crystal faces at temperatures near the bulk melting temperature, i.e. far above the roughening temperature. Recently *Frenken* and *Van der Veen* observed a premelting transition in lead [7.81] at a temperature for which the crystal face they considered, according to *Heyraud* and *Métois* [7.10], is well within the rough phase. On the other hand, one could also imagine surface premelting occurring in a narrow surface area without a delocalization in the orthogonal direction. *Jayanthi* et al. [7.82] conjectured this might happen for low-index faces of Cu, though at temperatures appearing to be very close to or even above the bulk melting temperature. It seems safe to conclude that surface premelting and surface roughening are completely different phenomena, although, of course, they may have a mutual influence. The subject of surface premelting seems to offer no less open questions than the roughening transition. E.g., it is not obvious to us whether one is certain that surface premelting is a sharp phase transition and not a continuous transition with a rapid but smooth change in mobility and surface modulation over a small temperature range. Obviously, about the mutual effects of roughening on surface premelting and vice versa even less is known. Further we want to point out that, in principle, a still different type of surface premelting may exist, related to the original model of *Burton* et al. [7.1]. Let us recall that in this model a growth layer below the transition temperature may exist in two phases, a dilute one describing the empty parts of the layer and a dense phase describing the grown-on parts. These two phases may coexist, separated by a well-defined phase boundary. Above the transition temperature there exists only one equilibrium phase, which implies that growth must proceed homogeneously throughout the growth layer. The character of the phase transition is that of the two-dimensional Ising model or lattice gas.

Transitions of this type may in fact occur for crystal face orientations where inequivalent layers are stacked upon each other. This has been demonstrated by *Knops* on the example of a BCSOS model with inequivalent sublattices [7.83], i.e. the next-nearest-neighbour bonds on one simple cubic sublattice are stronger than those on the other one. In this case no roughening transition occurs at finite temperature, but the loosely packed sublattice undergoes an Ising-type phase transition. Above the transition temperature the upmost layer of the tightly bound sublattice remains smooth, but it is covered halfways and in an irregular fashion by particles of the loosely bound phase. The disorder in the distribution of the latter prohibits the formation of large terraces of the tightly bound phase on top, thus suppressing the occurrence of roughening at finite temperatures. There is no reason for two-dimensional demodulation or for a strong increase of mobility within the loosely bound top layer at the transition temperature, hence the transition cannot be called surface melting in the sense described above.

The Blöte-Hilhorst model also may describe crystal faces built of inequivalent layers, in this case of three different types [7.59]. Again, at finite temperatures no roughening transition occurs, and within the subclass of exactly solvable models no Ising-type transitions are found either. It should be possible to recover the latter by adding interactions between particles in the same layer, although the model would not be exactly solvable any more by known means.

An intriguing question is, whether Ising-type transitions would also be possible without a stacking of inequivalent layers. For instance, one could ask oneself whether the quantum roughening of $^4$He surfaces at decreasing temperature advocated by *Andreev* and *Parshin* [7.84], but refuted as a roughening transition by *Fisher* and *Weeks* [7.70], could perhaps be an Ising-type transition in reality. In this connection it should also be noted that *Castaing* et al. have observed phenomena on $^4$He at 0.21 K with the appearance of a critical slowing down [7.17,85], for which apparently no satisfactory explanation has been found yet.

### 7.5.2 Are All Roughening Transitions of Kosterlitz-Thouless Type?

So far all explicit calculations giving rise to roughening transitions led to phase transitions of Kosterlitz-Thouless type. Does this mean all roughening transitions actually are of this type? In fact this is not the case and below we discuss three possible causes for roughening transitions of different character.

#### a) Inequivalent Layers

In Sect. 7.5.1 we remarked that BCSOS models and Blöte-Hilhorst models with inequivalent sublattices do not show roughening transitions at finite temperatures. Roughening does occur, however, in the limit $T \to \infty$. As a function of temperature these transitions cannot be of Kosterlitz-Thouless type, since this would require a finite roughening temperature, but even as a function of inverse temperature the character of this roughening transition is not of the same form as that of a Kosterlitz-Thouless transition [7.83,59]. In a way the infinite roughening temperature is an artefact of these types of model, due to the imposition of a solid-on-solid condition at all temperatures. On relaxing this condition one would presumably recover transitions of Kosterlitz-Thouless type at finite temperatures.

#### b) Impurities and Lattice Faults

The presence of lattice faults or impurities may change the character of the roughening transitions. Explicitly we want to discuss two examples:

**Screw Dislocation.** End points of screw dislocations sticking out of a crystal surface give rise to steps running between screw dislocations of opposite helicity. These steps are the well-known cause for spiral growth on smooth surfaces [7.1]. In the BCSOS model screw dislocations of Burgers vector ±4 can be incorporated by extending the corresponding six-vertex model to a symmetric eight-vertex model [7.53]. From *Baxter*'s solution of the latter [7.86] it is well known that the critical behavior is modified consequently, and, in fact, the critical exponents may assume basically all possible consistent sets of values. However, these results cannot be applied right away to describe the properties of realistic crystal surfaces: in the eight-vertex models the end points of screw dislocations are assigned an energy and next all thermodynamic properties are determined by applying equilibrium ensemble averages. This means that densities of screw dislocations and correlations between their positions depend on temperature and may be expected to change drastically on passing through the critical point. In real crystals the end points of screw dislocations are fixed by the structure of the bulk, which will hardly be influenced by a roughening transition somewhere at the surface. Hence, the way from the eight-vertex model to realistic crystal surfaces is a tricky one, which has to be laid out yet.

**The Influence of Impurities.** Recently, *Huse* and *Henley* have demonstrated that in a two-dimensional crystal the presence of fixed random impurities (specifically: random bonds in a two-dimensional Ising model) severely influences the way in which an interface roughens (the roughening temperature is zero in their specific case) [7.87]. Their renormalization-group analysis indicates that also in three dimensions random impurities will change the character of the roughening transition. It is our feeling that this subject deserves further investigation, experimentally as well as theoretically.

### c) First-Order Transitions

If a crystal surface may organize itself into two distinctly different states, one of which, at a given temperature, happens to be smooth and the other one to be rough, one could imagine a first-order transition between these two states happening. This then may be characterized as a first-order roughening transition [7.88]. A somewhat trivial way in which this could happen is as a secondary effect of a first-order bulk transition. For instance, at the triple point a solid-gas interface is replaced by a solid-liquid interface, and in typical cases the former is smooth (at least for low-index faces) and the latter rough. Recently, *Dam* discussed another example [7.89], where a structural transition in the bulk of a crystal also gave rise to the change from a smooth to a rough interface. Clear cut examples of first-order roughening transitions that are not triggered by a bulk phase transition are not known to us. But *Eusthatopoulos* in [7.41] showed some experimental results for an interface between solid Zn and liquid In, provoking the impression that a first-order transition may be at hand [7.88].

In conclusion: our knowledge of the theory of the roughening transition has increased enormously over the past years, and at the same time ample experimental confirmation has started to build up. Yet there are several challenging problems left regarding, e.g., the possibility of having roughening transitions of different type than the ones most studied so far, the importance of impurities and lattice faults and the influence of growth conditions on the roughening transition [7.90]. Experimentally the roughening transition seems to be one of the hardest phase transitions to investigate, yet in view of the large advances made in the past few years it does not seem overly optimistic to expect several new results in the years to come.

### 7.5.3 Multilayer Adsorption and Roughening

Finally we discuss the relation between roughening and multilayer adsorption. When adsorbing layers of crystalline material on an attracting substrate, one may observe phase transitions connected with the adsorption of the $n^{th}$ layer. At sufficiently low temperature and an appropriate value $p_n(T)$ of the pressure (or, equivalently, the density of the adsorbate in the dilute phase) both the state with n adsorbed layers and that with n − 1 layers will be in equilibrium. This leads to a first-order transition in the adsorbate isotherms: at $p = p_n(T)$ the amount of adsorbed material jumps discontinuously from the value for the (n − 1)-layer adsorbate to that for the n-layer adsorbate. Both theoretical calculations [7.91], computer simulations [7.92] and experiments by *Miranda* et al. [7.93] indicate that at fixed temperature $p_n(T)$ is an increasing function of n. For each layer n there appears to be a critical temperature $T_n^c$ above which the growth of the $n^{th}$ layer does not proceed through a nucleation mechanism but through continuous growth. For each finite n the phase transition is expected to be in the universality class of the two-dimensional Ising model. In the limit $n \to \infty$ $T_n^c$ should approach the roughening temperature of the corresponding surface of a pure adsorbate crystal: the vanishing of the nucleation barrier was seen to be one of the characteristics of the roughening transition.

As temperature is raised towards the roughening temperature, the jumps in the amount of adsorbed material remain for increasingly larger values of n only, and at and beyond the roughening temperature the isotherms are continuous, approaching infinity as p approaches the bulk vapor pressure. The general picture given above is confirmed by the experiments of *Miranda* et al. [7.93] for Xe adsorbed in Pd, although no independent measurements of the roughening temperature of the relevant Xe surface are available.

## Appendix 7.A:
## Surface Stiffness and Universal Jump in Curvature

Consider first a bending around the y-axis. In (7.7) one may expand f and $\partial f / \partial p$ around $(p, q) = (0, 0)$ with the result

$$z(p) = \frac{1}{\lambda} \left( f^{(0)} - \frac{1}{2} p^2 \frac{\partial^2 f}{\partial p^2}^{(0)} + \cdots \right) \tag{7.A.1}$$

with the superscript (0) denoting the value at $(p, q) = (0, 0)$. On the other hand, if $R_x$ is the radius of curvature in the x,z-plane, one also has

$$z(p) = z_0 - \tfrac{1}{2}p^2 R_x \ . \tag{7.A.2}$$

From (7.3) one sees that for small p the surface tension $\gamma$ and the surface tension per unit projected area f are related by

$$\gamma(p, 0) = f^{(0)}(1 - \tfrac{1}{2}p^2 + \ldots) \ . \tag{7.A.3}$$

In the limit $p \to 0$ the partial derivative $\partial^2/\partial^2\Theta^2$ becomes identical to $\partial^2/\partial p^2$, hence from (7.34) and (7.A.3) one obtains

$$\alpha_y^{(0)} = (\partial^2 f/\partial p^2)^{(0)} \tag{7.A.4}$$

where the subscript y denotes the bending axis. Hence from (7.A.1,2) one obtains

$$\alpha_y^{(0)} = \lambda R_x \ . \tag{7.A.5}$$

The constant $\lambda$ may be expressed as

$$\lambda = \gamma_0/z_0 \tag{7.A.6}$$

according to (7.A.1,3). Inserting this into (7.A.5) and substituting (7.33) for $R_x$ one finds

$$\alpha_y^{(0)} = \pi k_B T_R/2a_v^2 \ . \tag{7.A.7}$$

In the general case $\alpha_y^{(0)}$ must be replaced by

$$(\alpha_y^{(0)}\alpha_x^{(0)})^{1/2} = [(\partial^2 f/\partial p^2)(\partial^2 f/\partial q^2)]^{1/2}$$

if the principal radii of curvature lie in the x,z and the y,z-plane. If the principal axes lie in arbitrary planes (7.A.7) generalizes to (7.35).

*Acknowledgements* Discussions with Drs. Balibar, Gallet, Nozières, Schommers, Selke, Van der Veen and Villain have been very helpful to us. Figure 7.1 was courteously made available to us by Dr. Métois, Figs. 7.16 and 17 by Dr. Weeks and Figs. 7.24a,b were taken from [7.71] with the consent of Dr. Lieb.

# References

7.1 W.K. Burton, N. Cabrera, F.C. Frank: Phil. Trans. Roy. Soc. (London) **243A**, 299 (1951)
7.2 L. Onsager: Phys. Rev. **65**, 117 (1944)
7.3 C.N. Yang, T.D. Lee: Phys. Rev. **87**, 404, 410 (1952)
7.4 S.T. Chui, J.D. Weeks: Phys. Rev. B**14**, 4978 (1976)
7.5 V.L. Berezinskii: Zh. Eksp. Teor. Fiz. **59**, 907 (1970); **61**, 1144 (1972) [Sov. Phys. JETP **32**, 493 (1971); **34**, 610 (1972)]
7.6 J.M. Kosterlitz, D.J. Thouless: J. Phys. C**6**, 1181 (1973)
    J.M. Kosterlitz: J. Phys. C**7**, 1046 (1974)
7.7 J.V. José, L.P. Kadanoff, S. Kirkpatrick, D.R. Nelson: Phys. Rev. B**16**, 1217 (1977)
7.8 R.A. Hunt, B. Gale: J. Phys. C**6**, 3571 (1973)
7.9 J.C. Heyraud, J.J. Métois: J. Cryst. Growth **50**, 571 (1980)
7.10 J.C. Heyraud, J.J. Métois: Surf. Sci. **128**, 334 (1983)
7.11 A. Pavlovska, D. Nenow: Surf. Sci. **27**, 211 (1971)
7.12 A. Pavlovska, D. Nenow: J. Cryst. Growth **12**, 9 (1972)
7.13 A. Pavlovska, D. Nenow: J. Cryst. Growth **39**, 346 (1977); Kristall und Technik **12**, 473 (1977)
7.14 T. Ohachi, I. Taniguchi: J. Cryst. Growth **65**, 84 (1983)
7.15 A. Passerone, S. Sangiorgi, N. Eustathopoulos: Scripta Meta **14**, 1089 (1980)
7.16 J.E. Avron, L.S. Balfour, C.G. Kuper, J. Landau, S.G. Lipson, L.S. Schulman: Phys. Rev. Lett. **45**, 814 (1980)
7.17 S. Balibar, E. Castaing: J. Phys. Lett. (Paris) **41**, L329 (1980)
7.18 K.O. Keshishev, A. Ya Parshin, A.V. Babkin: Zh. Exp. Theor. Fiz. **80**, 716 (1981) [Sov. Phys. JETP **53**, 362 (1981)]
7.19 P.E. Wolf, S. Balibar, F. Gallet: Phys. Rev. Lett. **51**, 1366 (1983)
7.20 P.E. Wolf, F. Gallet, S. Balibar, E. Rolley, P. Nozières: J. Phys. (Paris) **46**, 1987 (1985)
    F. Gallet, P. Nozières, S. Balibar, E. Rolley: Dynamic broadening of the roughening transition, submitted to Europhys. Lett.
7.21 J.D. Weeks, G.H. Gilmer: In Adv. Chem. Phys. **40**, 157 (Wiley, New York 1979)
7.22 F.C. Frank: Disc. Faraday Soc. **5**, 132 (1949)
7.23 K.A. Jackson, C.E. Miller: J. Cryst. Growth **40**, 169 (1977)
    E. Rolley, S. Balibar, F. Gallet: The first roughening transition of $^3$He Crystals, submitted to Europhys. Lett.
7.24 J.D. Weeks: The roughening transition, in *Ordering in Strongly Fluctuating Condensed Matter Systems*, ed. by T. Riste (Plenum, New York 1980) p. 293
7.25 R.K.P. Zia: Interfacial problems in statistical physics, in: Statistical and particle physics, common problems and techniques, Proc. Scottish Universities Summer School in Physics, ed. by K. Bowler and A. McKane (Scottish Univ. Summer School in Physics, Edinburgh 1984)
7.26 C. Rottman, M. Wortis: Phys. Rpt. **103**, 59 (1984)
7.27 S. Balibar, B. Castaing: Surf. Sci. Rpt. **5**, 87 (1985)
7.28 S. Balibar, W. Saam: In preparation
7.29 P. Nozières: Lectures at the Collège de France, 1984, to be published
7.30 G. Wulff: Z. Krist. Mineral. **34**, 449 (1901)
7.31 C. Herring: Phys. Rev. **82**, 87 (1951)
7.32 L.D. Landau, E.M. Lifshitz: *Statistical Physics*, Vol. 1 (Pergamon, Oxford 1980) p. 155
7.33 J.E. Taylor: Symp. Mathem. **14**, 499 (1974)
7.34 A. Dinghas: Z. Krist. **105**, 304 (1944)
7.35 A.F. Andreev: Zh. Eksp. Teor. Fiz. **80**, 2042 (1981) [Sov. Phys. JETP **53**, 1063 (1981)]
7.36 P. Pieranski, R. Barbet-Massin, P.E. Cladis: Phys. Rev. A**31**, 3912 (1985)
7.37 A.V. Babkin, K.O. Keshishev, D.B. Kopeliovitch, A. Ya Parshin: Pisma Zh. Eksp. Teor. Fiz. **39**, 519 (1984) [Sov. Phys. JETP Lett. **39**, 633 (1984)]
7.38 J.E. Avron, J.E. Taylor, R.K.P. Zia: J. Stat. Phys. **33**, 493 (1983)
7.39 See, e.g., M.E. Fisher: J. Math. Phys. **5**, 944 (1964)
7.40 J.H.L. Watson, W. Heller, W. Wojtowicz: J. Chem. Phys. **16**, 997 (1948)
7.41 N. Eustathopoulos: Int. Met. Rev. **28**, 189 (1983)

7.42  J.W. Cahn, D.W. Hoffman: Acta Met. **22**, 1205 (1974)
7.43  C. Jayaprakash, W.F. Saam: Phys. Rev. B**30**, 3916 (1984)
7.44  W. Kossel: Nachr. Ges. Wiss. Göttingen, Mathemat./Physikal. Klasse S. 135 (1927)
7.45  J.N. Stranski: Z. Phys. Chemie **136**, 259 (1928)
7.46  S.T. Chui, J.D. Weeks: Phys. Rev. Lett. **40**, 733 (1978)
7.47  J. Bricmont, A. El Mellouki, J. Fröhlich: J. Stat. Phys. **42**, 743 (1986)
7.48  H.J. Leamy, G.H. Gilmer: J. Cryst. Growth **24/25**, 499 (1974)
7.49  H. van Beijeren, G. Gallavotti: Lett. Nuovo Cim. **4**, 699 (1972)
7.50  J.D. Weeks, G.H. Gilmer, H.J. Leamy: Phys. Rev. Lett. **31**, 549 (1973)
7.51  H. van Beijeren: Commun. Math. Phys. **40**, 1 (1975)
7.52  J. Bricmont, J.L. Lebowitz: private communication
7.53  H. van Beijeren: Phys. Rev. Lett. **38**, 993 (1977)
7.54  J. Fröhlich, T. Spencer: Commun. Math. Phys. **81**, 527 (1981)
7.55  C. Jayaprakash, W.F. Saam, S. Teitel: Phys. Rev. Lett. **50**, 2017 (1983)
7.56  H.W.J. Blöte, H.J. Hilhorst: J. Phys. A**15**, L631 (1982)
7.57  C. Rottman, M. Wortis: Phys. Rev. B**29**, 328 (1984)
7.58  C. Rottman, M. Wortis, J.C. Heyraud, J.J. Métois: Phys. Rev. Lett. **52**, 1009 (1984)
7.59  B. Nienhuis, H.J. Hilhorst, H.W.J. Blöte: J. Phys. A**17**, 3559 (1984)
7.60  E.E. Gruber, W.W. Mullins: J. Phys. Chem. Solids **28**, 875 (1967)
7.61  C. Jayaprakash, C. Rottman, W.F. Saam: Phys. Rev. B**30**, 6549 (1984)
7.62  J. Villain, D.R. Grempel, J. Lapujoulade: J. Phys. F**15**, 809 (1985)
7.63  J. Villain: J. Phys. (Paris) **36**, 581 (1975)
7.64  T. Ohta, D. Jasnov: Phys. Rev. B**20**, 139 (1979)
7.65  P.B. Wiegmann: J. Phys. C**11**, 1583 (1978)
7.66  W.J. Shugard, J.D. Weeks, G.H. Gilmer: Phys. Rev. Lett. **41**, 1399 (1978)
7.67  J.P. van der Eerden, H.J.F. Knops: Phys. Lett. **66A**, 334 (1978)
7.68  R.H. Swendsen: Phys. Rev. B**17**, 3710 (1978)
7.69  T. Ohta, K. Kawasaki: Prog. Theor. Phys. **60**, 365 (1978)
7.70  D.S. Fisher, J.D. Weeks: Phys. Rev. Lett. **50**, 1077 (1983)
7.71  E.H. Lieb, F.Y. Wu: Two-dimensional ferroelectric models, in *Phase Transitions and Critical Phenomena*, Vol. I, ed. by C. Domb and M.S. Green (Academic, New York 1972)
7.72  P.J. Forrester: J. Phys. A**19**, L143 (1986)
7.73  C.P. Yang: Phys. Rev. Lett. **19**, 586 (1967)
      B. Sutherland, C.N. Yang, C.P. Yang: ibid. p. 588
7.74  H.J. Schulz: J. Phys. (Paris) **46**, 257 (1985)
7.75  C. Jayaprakash, W.F. Saam: Phys. Rev. B**30**, 3917 (1984)
7.76  C.N. Yang, C.P. Yang: Phys. Rev. **150**, 321, 327 (1966)
7.77  S.E. Burkov: J. Phys. (Paris) **46**, 317 (1985)
7.78  W. Selke, A.M. Szpilka: Z. Physik B**62**, 381 (1986)
7.79  E.H. Conrad, R.M. Aten, D.S. Kaufman, L.R. Allen, T. Engel, M. den Nijs, E.K. Riedel: J. Chem. Phys. **84**, 1015 (1986)
7.80  D. Nenow: Prog. Cryst. Growth and Charact. **9**, 185 (1984)
7.81  J.W.M. Frenken, J.F. van der Veen: Phys. Rev. Lett. **54**, 134 (1985)
7.82  C.S. Jayanthi, E. Tosatti, A. Fasolino, L. Pietronero: Surf. Sci. **152/153**, 155 (1984)
7.83  H.J.F. Knops: Phys. Rev. B**20**, 4670 (1979)
7.84  A.F. Andreev, A. Ya Parshin: Zh. Eksp. Teor. Fiz. **75**, 1511 (1978) [Sov. Phys. JETP **48**, 763 (1978)]
7.85  L. Puech, B. Hebral, D. Thoulouze, B. Castaing: J. Phys. Lett. (Paris) **44**, L159 (1983)
7.86  R.J. Baxter: *Exactly Solved Models in Statistical Mechanics* (Academic, London 1982)
7.87  D.A. Huse, C.L. Henley: Phys. Rev. Lett. **54**, 2708 (1985)
7.88  P. Nozières: private communication
7.89  B. Dam: Phys. Rev. Lett. **55**, 2806 (1985)
7.90  W. van Saarloos, G.H. Gilmer: Dynamical properties of long wavelength interface fluctuations during nucleation dominated crystal growth, submitted to Phys.Rev. B.
7.91  M.J. de Oliveira, R.B. Griffiths: Surf. Sci. **71**, 687 (1978)
      J.D. Weeks: Phys. Rev. B**26**, 3998 (1982)
      R. Pandit, M. Schick, M. Wortis: Phys. Rev. B**26**, 5112 (1982)
      W.F. Saam: Surf. Sci. **125**, 253 (1983)

7.92 J.M. Kim, D.P. Landau: Surf. Sci. **110**, 415 (1981)
7.93 R. Miranda, E.V. Albano, S. Daiser, G. Ertl, K. Wandelt: Phys. Rev. Lett. **51**, 782 (1983)
R. Miranda, E.V. Albano, S. Daiser, G. Ertl, K. Wandelt: J. Chem. Phys. **80**, 2931 (1984)


# Additional References with Titles

Abraham, D.B.: Surface structures and phase transitions – exact results, in "Phase Transitions and Critical Phenomena", Vol. 10, ed. by C. Domb and J.L. Lebowitz (Academic, London 1986)

Abraham, D.B.: Surface Reconstruction in Crystals. Phys. Rev. Lett. **51**, 1279 (1983)

Andrews, G.F., Forrester, P.J.: Height probabilities in solid-on-solid models: J. Phys. **A19**, L923 (1986)

Bonzel, H.P., Preuss, E., Steffen, B.: The Dynamical Behavior of Periodic surface Profiles on Metals Under the Influence of Anisotropic Surface Energy. Appl. Phys. **A35**, 1 (1984)

Carmi, Y., Lipson, S.G., Polturak, E.: The Critical Behaviour of Vicinal Surfaces of $^4$He, preprint

Conrad, E.H., Aten, R.M., Kaufman, D.S., Allen, L.R., Engel, T., den Nijs, M., Riedel, E.: Erratum: Observations of Surface Roughening on Ni(115) [J. Chem. Phys. **84**, 1015 (1986)], J. Chem. Phys. **85**, 4756 (1986)

Fabre, F., Gorse, D., Lapujoulade, J., Salonon, B.: An Experimental Study of the Roughening Transition on a Stepped Surface: Cu(115). Europhys. Lett. **3**, 737 (1987)

Forrester, P.J.: Exact calculation of the local height probabilities in the body-centred SOS model. J. Phys. **A19**, L143 (1986)

Gorse, D., Lapupjoulade, J., Pontikis, V.: A Molecular Dynamics Study of the Thermal Behaviour of Cu(115). Surf. Sci. **178**, 434 (1986)

Heyraud, J.C., Métois, J.J.: Surface Free Energy Anisotropy Measurement of Indium, Surf. Sci. **177**, 213 (1986)

Heyraud, J.C., Métois, J.J.: Growth Shapes of Metallic Crystals and Roughening Transition. J. Crystal Growth **75**, 173 (1986)

Levi, A.C., Tosatti, E.: On the Roughening of Molten Surfaces, Surf. Sci. **178**, 425 (1986)

Lapujoulade, J.: Molecular Beam Study of Surface Roughening Transitions. Surf. Sci. **178**, 406 (1986)

Métois, J.J., Heyraud, J.C.: Analysis of the Critical Behaviour of Curved Regions in Equilibrium Shapes of In Crystals. Surf. Sci. **178**, 406 (1986)

den Nijs, M.: Antiferromagnetic restricted solid-on-solid model: Ising models on rough surfaces. Phys. Rev. **B32**, 4785 (1985)

den Nijs, M.: Corrections to scaling and self-duality in the restricted solid-on-solid model. J. Phys. **A18**, L549 (1985)

Rys, F.S.: Critical and Noncritical Roughening of Surfaces. Phys. Rev. Lett. **56**, 624 (1986)

Rys, F.S.: On the Roughening of Solid Surfaces. Surf. Sci. **178**, 382 (1986)

Saito, Y., Müller-Krumbhaar, H.: Two-dimensional Coulomb gas: A Monte-Carlo Study. Phys. Rev. **B23**, 308 (1981)

Selke, W.: Kinetics of Roughening in a Model of a Stepped Surface, preprint

Swendsen, R.H.: Monte Carlo study of the Coulomb gas and the Villain XY model in the discrete Gaussian roughening representation. Phys. Rev. **B18**, 492 (1978)

Trayanov, A., Nenow, D.: Surface Roughening and Quasi-liquid Layer, J. Crystal Growth **74**, 375 (1986)

Veen, J.F.v.d., Frenken, J.W.M.: Dynamics and Melting of Surfaces. Surf. Sci. **178**, 382 (1986)

Villain, J.: Healing of a Rough Surface at Low Temperature. Europhys. Lett. **2**, 531, (1986)

Zia, R.K.P., Gittis, A.: Effects of Gravity on Equilibrium Crystal Shapes: Droplets hung on a Wall. Phys. Rev. **B35**, 5907 (1987)

# 8. Structural and Dynamical Aspects of Adsorption and Desorption

G. Doyen

With 4 Figures

Experimental and theoretical aspects of adsorption and desorption are reviewed. The emphasis lies on general ideas and interconnections between various approaches. Some recent promising developments are presented as selected topics.

On the experimental side a large data base of sticking coefficients, flash desorption spectra and desorption orders has been accumulated during the last two decades. Details of the dynamics are revealed with the help of the more recent molecular-beam experiments and state-specific detection techniques.

On the theoretical side the classical-stochastic approaches are developed furthest. Semi-classical approximations became quite popular during the last years. Within the context discussed here they have been mostly applied to study non-adiabatic adsorption. Purely quantum-mechanical approaches are considered to be more complicated and are therefore most frequently applied in low-order perturbation theory. Recently developed more general quantum techniques are discussed separately as a selected problem.

The other special subsections are devoted to the sticking problem at zero temperature and zero kinetic energy, to non-adiabatic adsorption and desorption, to the importance of constraints and their efficient treatment in information theoretical approaches, and to the long-established but still poorly understood compensation effect for desorption.

## 8.1 General

The interaction of gas particles with solid surfaces is routinely used as an experimental tool for studying the structure and the dynamical behaviour of surfaces [Ref. 8.1, Chaps. 2 and 7]. Molecular-beam techniques are also applied to study the gas-solid interaction itself. Excellent reviews covering this field exist ([Ref. 8.1, Chap. 2] and [8.2,3]).

The diagram below indicates the possible channels of the scattering process. Full lines connect those channels which will be considered in detail in this review.

The incoming particle has two possibilities: either it is scattered back directly, or it will be adsorbed and stays for some time on the surface. Direct backscattering can either be elastic or inelastic; elastic reflection occurs specularly or, if the potential is periodically corrugated, diffraction can be observed.

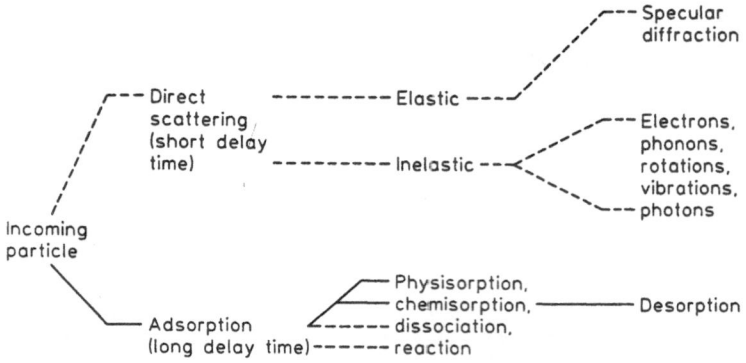

Inelasticity is due to the fact that the surface provides a continuum of low-energy excitations to which the incoming particle can couple. Lattice vibrations or phonons are of obvious interest in this respect. For metal surfaces, electron-hole pair excitations have to be considered as well. The relative importance of these two kinds of excitations is a frequently posed question which has not been settled yet. This problem will be addressed in Sect. 8.4.5.

An adsorbing particle has the possibilities of physisorption, chemisorption, dissociation, or it might chemically react with other atoms or molecules present on the surface. After a certain time depending on the substrate temperature $T_s$ the particles and products will desorb.

Adsorption and desorption are the two most elementary steps concerning the chemical interaction. A thorough understanding of these two processes is a necessary prerequisite for gaining any insight into the secrets of heterogeneous catalysis. Consequently much effort has been invested into the experimental investigation of these processes. A vast literature exists, but comprehensive review articles are available [8.5]. It is clearly outside the scope of this chapter to provide a comprehensive discussion of all experimental approaches, but a few methods and relevant results will be considered.

A general theoretical description of adsorption and desorption processes encompasses three main steps:

i)   The calculation of adiabatic potential energy surfaces,
ii)  the investigation of the dynamical movement on these hypersurfaces, and
iii) inclusion of non-adiabatic transitions between different energy surfaces.

The first fundamental task, which is not an issue of this review, has only been solved (with satisfying results) for very few and simple cases. One example is the evaluation of adiabatic rare gas-metal potentials. In this case potentials can be fitted to the experimental scattering data [8.6–8]. First principles evaluations have been tried using various approaches: Effective medium approximation [8.9–11], expansion in powers of overlap [8.12], model Hamiltonian approach [8.13], S-matrix formalism for evanescent waves [8.14]. For more complicated systems like diatomic molecules interacting with transition metal surfaces, theoretical results are more scant and presumably of lower quality [8.15,16]. Theoretical potentials can be tested solving Item (ii) for them and comparing to the experimental data. For attacking Item (i) for reactive systems, theoreticians have to rely on empirical methods embodying the available experimental data into an anticipated shape of the potential [8.17].

The dynamics on a given potential hypersurface can be studied from a classical, semiclassical or quantum-mechanical point of view. All three approaches will be briefly considered in the following. In these cases rather exhaustive reviews exist as well: The classical approach (combined with stochastic elements for describing the heat bath) is developed furthest and has been summarized several times ([8.18] and references therein). Semi-classical methods have gained much popularity during the last years. A listing of the more recent contributions can be found in [8.4]. Quantum treatments have been mostly restricted to elastic scattering. Inelastic processes were included in Distorted-Wave Born Approximation (DWBA). The state of this kind of theory up to 1970 has been described by Goodman [8.19]. Later developments have been discussed in [8.20].

A large variety of theoretical approaches have been tried in order to understand sticking and desorption. Prominent examples are the statistical theories and the purely classical theories. The statistical approaches started with Eyring's Absolute Rate Theory (ART) (also called transition state theory (TST)) have later been extended by theories which pay more attention to conformational and dynamical aspects. In the order of increasing sophistication they are ART, Kramers' theory, information theory and Generalized Langevin Equation (GLE). The quantum formulation by Lin and Adelman – the so-called Mixed Quantum Model (MQM) (Sect. 8.4.4) – is an example of a quantum partitioning model. These models appear promising with respect to a more detailed quantum mechanical description of sticking and desorption.

Classical theories have been improved to account for quantum effects. This led to the semiclassical trajectory methods, which found much attention during the recent years. They may be considered to be approximations to the exact quantum theory, if the latter is formulated in terms of path integrals.

The interconnections between the various methods are sketched in the diagram below.

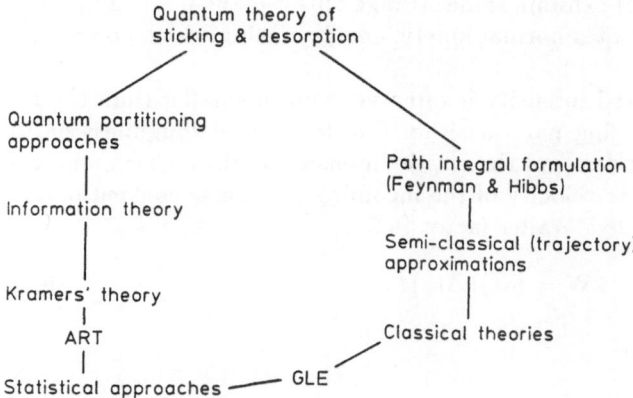

303

As review articles giving a rather complete account of all existing work, theoretical as well as experimental, are numerous, the philosophy of this chapter is to emphazise the major ideas and to put them into relative perspective. This will make obvious some shortcomings in our present understanding of the subject and display the challenges faced by the scientists working in the field. Section 8.4.3 is rather detailed and extensive in this respect. Besides reflecting a special taste of the author, the summary of the sticking problem given there is definitely novel and not to be found anywhere else in the literature. It might be skipped by less specialized readers, because some aspects of the problem are mentioned in the more general Sects. 8.2 and 3.

If not explicitly mentioned otherwise, atomic units are used throughout ($\hbar = 1$).

## 8.2 Basic Experimental Facts

In this section some basic experimental facts are discussed, which serve to illustrate the realm of phenomena to be considered. This is a rather limited and somewhat arbitrary choice which should not be misinterpreted as a general and complete overview of the present state of the experimental art in this field. For the latter purpose the reader should consult one of the review articles mentioned in the general section.

### 8.2.1 Elastic Scattering at Surfaces

An important but sometimes not sufficiently appreciated point is that elastic scattering from a vibrating surface constitutes a quantum mechanical phenomenon [8.21] which is not present in classical theories. Experimentally, elastic reflection can be studied using the molecular beam technique. Rare-gas atoms are elastically reflected from metal surfaces with diffraction peaks being observed for the more open faces exhibiting larger corrugations (see [8.13] for references). Due to improved experimental resolution diffraction has recently also been observed for the heavier rare gases (neon and argon) and for diatomic molecules ($H_2$ and, possibly, $N_2$) [8.22]. This is especially remarkable for the heavier species, because it demonstrates that contrary to a widespread believe Ar and $N_2$ scattering cannot be completely understood in purely classical terms. $N_2$ scattering from Cu(110) exhibits some strange out-of-plane intensity [8.22]; $N_2$ sticking on W(110) violates normal kinetic energy scaling which commonly holds approximately.

The elastically scattered intensity is often very much smaller than the incoming intensity. The missing part is either due to surface roughness or to inelastic events. The dependence of the elastic intensity on the surface temperature and the perpendicular velocity of the incoming particle is conventionally described in terms of a Debye-Waller factor [8.23]:

$$I_{el}/I_0 = \exp(-2W) \; ; \quad 2W = \langle u^2 \rangle (\Delta k_z)^2 \tag{8.1}$$

$\Delta k_z$ is the perpendicular momentum transfer during the scattering event and $\langle u^2 \rangle$ is a mean-quadratic displacement of the surface particle from its equilibrium position. Various theories [8.21,24–27] offer ways of calculating these quantities. $[1 - \exp(-2W)]$ is the total portion removed from the elastic channel and therefore contains trapping events as well. If the latter are important, a linear dependence of W on the surface temperature and the perpendicular kinetic energy can no longer be expected. For low surface temperatures and small kinetic energies significant deviation from the linear behaviour has been observed in the sense that inelasticity increases considerably. Among the investigated systems for which this has been found are He/Cu(117) [8.28], $H_2$/Cu(100) [8.29], He,Ne/Cu(100) [8.30], H/NaF(001) [8.31].

In this connection the very low electronically elastic scattering probabilities for metastable He atoms seem to provide a challenge to theorists. *Craig* and *Dickinson* [8.32] performed these experiments for a single crystalline germanium(100) surface. For their given sensitivity they found no reflected metastable (i.e., electronically excited) particles. After oxidation the reflection coefficient was $10^{-2}$. The main portion of the reflected intensity showed then a cosine angular distribution which in this case is indicative of a rough surface. *Conrad* et al. [8.33] performed similar experiments for a Pd(110) surface covered by a densely packed CO-layer. They detected $10^{-4}$ of the incoming intensity in a quasi-specular angular distribution and $10^{-3}$ in a cosine distribution. This is especially remarkable, because in gas-phase experiments the portion of He atoms surviving the scattering process in their metastable state is larger than 50%. The relevance of these experiments for adsorption – desorption phenomena is that they might give a clue for understanding the large observed inelasticity for weak interaction which is difficult to understand using standard text book quantum scattering theory. These problems will be outlined in detail below (Sect. 8.4.3).

### 8.2.2 Trapping and Sticking

The probability s that a particle hitting a solid surface will not be reflected back into the vacuum is called adsorption probability, sticking probability or sticking coefficient. The adsorption probability for a clean surface (zero coverage sticking probability) is commonly referred to as initial sticking probability. During the course of an experiment often an adsorption layer will develop and the adsorption probability will change, because the sticking probability is different for an adsorbed layer. These semantic definitions are rather ambiguous, because every particle will finally escape back into the vacuum (via a desorption process), if enough time elapses. If the adsorption lifetime is small compared with the characteristic experimental time period, the term "trapping probability" is commonly used. This situation prevails for weak adsorption (so-called physisorption). In the case of diatomics this corresponds usually either to trapping into a molecularly adsorbed state on the clean surface (intrinsic precursor) or to trapping into a physisorbed state above a preadsorbed layer (extrinsic precursor).

The common definition of the sticking coefficient is:

$$s = \frac{\text{rate of adsorption}}{\text{rate of bombardment}} = \frac{F_0 - F_{\text{refl}}}{F_0} = \frac{F_{\text{stick}}}{F_0} \; . \tag{8.2}$$

Here $F_0$ and $F_{\text{refl}}$ are the incident and reflected fluxes, respectively. Equation (8.2) contains, of course, the ambiguities mentioned above, because the measured rate of adsorption (or the measured reflected flux, respectively) depends on the adsorption lifetime and the experimental time scale. The latter can be defined precisely in a molecular beam experiment by modulating the incident beam with a known frequency and by recording the resulting modulated signal by means of a lock-in amplifier. If the adsorption lifetime is long compared to the inverse frequency, desorbing particles do not contribute to the modulated signal and an accurate determination of $F_{\text{refl}}$ is possible. With this method a time resolution of ca. $10^{-5}$ s can be achieved. For shorter adsorption lifetimes the angular distribution of the scattered intensity might give some indication, if a clear distinction into lobular and diffuse (power-of-cosine law) scattering is possible. The fraction of trapped particles can then be identified with the fraction of diffusely scattered intensity. If the adsorption lifetime approaches the time for a single collision, even this method will fail. The most elegant way of separating the trapping-desorption channel is by means of the time-of-flight technique [8.34a]. Xe scattering from Pt(111) has been analyzed using this method [8.34b].

*Morris* et al. [8.5] collected a rather complete database of measured sticking probabilities at metal surfaces. As an example of a typical experiment we briefly review the investigations by *Wang* and *Gomer* [8.35], who used an elaborate reflection detector technique. For the adsorption of $O_2$, CO and Xe on tungsten they found that for decreasing surface and gas temperatures the sticking coefficient tends to unity. This is an observation which is supported by intuition and experimental experience. In the mentioned experiment the whole vacuum chamber was immersed into liquid helium in order to secure that all particles hitting the walls would be adsorbed and hence would not interfere with the measured signal. This rather obvious fact has, however, posed severe problems to the quantum theory of adsorption (Sects. 8.3.3 and 8.4.3).

In this context an experiment deserves special attention which has unambiguously demonstrated that the sticking probability approaches unity at very low surface and gas temperatures [8.36]. The data reveals that s is between 2/3 and 1 for $^4$He incident with energies between 10 and 20 K on a metallic or dielectric surface at low temperature ( <4 K). It is also found that this result does not depend on whether the metallic surface is very nearly bare, or is covered instead with close to a monolayer of preadsorbed $^4$He. At higher surface temperature (19 K) the data could be interpreted to give s = 0. They are thus consistent with earlier accommodation coefficient measurements [8.37]. This latter result arises from a disappearance of a distinction between direct reflection and redesorption, phenomena that may be separated in time only at lower temperature.

For CO/W Wang and Gomer investigated the variation of the sticking coefficient with CO coverage. At cold surfaces s stays near unity up to very large coverages. This implies, of course, that s has to be near unity on the adsorbed layer, although the binding energy of CO on a preadsorbed CO layer is much smaller than on a clean tungsten surface. These findings tie in with the observation of Sivani et al. that sticking of He on a preadsorbed He layer is nearly complete as well.

The high efficiency of CO sticking on a CO layer is not restricted to small kinetic energies. In molecular beam experiments using CO molecules with 80 meV kinetic energy incident under a small polar angle on a CO covered Pd surface the same behaviour is found [8.38].

Structural effects on the adsorption probabilities are found in the dependence on the various crystal planes of transition metal surfaces. This appears to be more important for $H_2$ and $N_2$ than for CO and $CO_2$. Generally, the more close-packed faces show a smaller sticking coefficient than the more open planes [8.5].

CO has a large sticking coefficient on nearly all investigated transition metal surfaces [8.5]. Contrary to this the adsorption probability on semiconductor surfaces is very small. Accurate values are difficult to find in the literature. *Castro* estimated the sticking coefficient for CO on Si(111) to be smaller than $10^{-5}$ [8.39]. Detailed understanding of the interaction of gases with semiconductor surfaces is limited to two or three gases (hydrogen, oxygen and possibly chlorine) and to two semiconductors (Si and GaAs) [8.40]. $H_2$ adsorbs on silicon with very low probability [8.41]. In atomic form, however, hydrogen sticks readily on Si with sticking coefficient close to unity [8.42]. The initial sticking probability for $O_2$ on germanium and silicon is in the range of $10^{-2}$ to $10^{-3}$ [8.43–47]. It is a strong function of the surface step density [8.48] and can vary by more than two decades with changing steps density.

Small sticking coefficients do not necessarily imply increased elastic reflection. Even for He scattering from a Si(100) surface, an elastic intensity of only roughly $10^{-3}$ has been reported [8.49]. The dominating process appears to be inelastic reflection, also called direct inelastic scattering.

### 8.2.3 Direct Inelastic Scattering

Elastic scattering from surfaces appears to be a comparatively rare event, which is only observed for few gases, preferably He, Ne and $H_2$. If $H_2$ is replaced by HD or $D_2$, the elastic intensity is decreased considerably [8.50–53]. The angular distribution is then dominated by a broad, lobular but not exactly specular background intensity, which is referred to as direct inelastic scattering. The isotope effect for hydrogen is commonly ascribed to varying rotational energies and to the increased mass [8.53]. For most of the other chemically interesting gases (CO, $O_2$, $CO_2$, NO, $N_2O$) no elastic scattering has been detected at all. Interesting intermediate cases appear to be Ar and $N_2$.

Direct inelastic scattering exhibits a variety of phenomena and yields information about the coupling to the surface excitations (cf. Sect. 8.4.5 about the relative role of phonons and electron-hole pairs), but it is not a direct issue of this review.

### 8.2.4 Desorption

Molecular-beam experiments show besides the elastic peaks and the lobular direct inelastic intensity an additional contribution to the scattered flux which is symmetric around the surface normal. This intensity originates from desorbing particles, which after the sticking process spent enough time near the solid surface in order to achieve energy accommodation with the heat bath [8.2]. The angular dependence obeys often the cosine law. This is to be expected, if the sticking coefficient is independent of the angle of incidence and if gas phase and solid are in equilibrium [8.54]. Typical desorption experiments are, however, not performed under equilibrium conditions and deviations from the simple cosine law are observed rather frequently.

Another frequently observed but still poorly understood phenomenon is the so-called compensation effect. It stipulates that for the same adsorbed species the desorption rate is (within certain limits) independent of the adsorption energy. Because the desorption rate may be written as a product of a frequency factor $v_0$ and an exponential term $\exp(-E_{ad}/k_B T_S)$ ($E_{ad}$: adsorption energy; $k_B$: Boltzmann's constant; $T_S$: substrate temperature), this means that an increased adsorption energy implies an increased frequency factor as well [8.55,56]. This phenomenon will be studied in more detail in Sect. 8.4.2.

The most widely employed technique for studying desorption is temperature-programmed desorption (TPD) [8.5]. The molecular-beam technique has the advantage that the crystal can be maintained at constant temperature and that by chopping the incident beam the lifetimes of the adsorbed species can be monitored directly. More rarely performed are isothermal desorption experiments, where the substrate temperature is kept constant. At the time $t = 0$ the equilibrium pressure is turned off and the time dependence of the coverage is monitored. The advantage of this method is that diffusion or changing mobility does not influence the results.

Another method has been developed by *Comsa* and *David* [8.57,58]. Here the gas diffuses from the back side of the crystal to the surface from which it finally desorbs. In this case, however, the pathway of desorption could be considerably different from the one monitored by the previously described methods.

The quantities measured in desorption experiments are the adsorption energy, the frequency factor, angular and, more recently, velocity and internal state distributions of the desorbing particles. Also obtained is the order of the desorption process, which contains information about the mechanism of the desorption reaction.

The position of the maximum of a particular thermal desorption curve on the temperature scale in a TPD-experiment depends both on the adsorption energy and on the frequency factor. An error in the adsorption energy will usually lead to a considerable error in the estimated frequency factor. For the well investigated system hydrogen on Pt(111), for example, the reported frequency factors differ by several orders of magnitude [8.59].

The geometric structure of the solid surface is found to have a large influence on the desorption behaviour. This applies especially for imperfections like steps and kinks and many controversial results reported in the literature might, in fact, be due to disparate surface conditions. This point was carefully examined for NO desorbing from Pt(111) [8.60]. Adsorbate induced structural changes can lead to drastic variations of the frequency factor as well [8.61].

Effects on the internal energy (electronic, vibrational and rotational degrees of freedom) of the scattering molecule can be measured by applying state specific detection techniques: Laser induced fluorescence (LIF), multiphoton ionization, infrared excitation with bolometric detection and infrared emission techniques.

For NO/Ag(111) the trapping-desorption channel was studied in a molecular-beam experiment under glancing-incidence conditions [8.4]. The angular distribution was then almost entirely cosine. At substrate temperature $T_S = 100\,\text{K}$ a mean-residence time of $2.1 \times 10^{-3}\,\text{s}$ was observed, which, with an assumed pre-exponential factor for desorption of $10^{13}\,\text{s}^{-1}$, gave an estimated adsorption energy of $0.2\,\text{eV}$. The rotational state distribution was Boltzmann and for low surface temperature $T_S$ ($T_S < 250\,\text{K}$) the rotational temperature, $T_R$, was equal to $T_S$ while for higher substrate temperature it was less.

There are many similar systems for which the rotational state distribution is found to be of Boltzmann form but with $T_R < T_S$. Among them are NO/Ge [8.62], NO/Pt(111), NO/graphite, CO/Pt, NO/Ru(001). The chemical interaction of NO and CO with Pt and Ru is stronger than with Ag which implies a longer residence time. *Segner* et al. [8.63] estimate this in the case of NO/Pt(111) to be 3s for $T_S = 400\,\text{K}$ and $5 \times 10^{-7}\,\text{s}$ for $T_S = 800\,\text{K}$.

For NO/Pt(111) and NO/Ru(001) [8.64] vibrational state distributions have been measured as well. They appear to be Boltzmann with the vibrational temperature being equal to the surface temperature. Quite recently the rotational and vibrational distributions of $H_2$ and $D_2$ recombinatively desorbing from clean Cu(110) and Cu(111) surfaces following atomic permeation through the crystal have been studied using multiphoton ionization combined with time-of-flight mass spectrometry [8.65]. Rotational distributions are found to be non-Boltzmann and to possess mean-rotational energies which are 0.8–0.9 of the surface temperature $T_S$. The population of the first excited vibrational state was found to be 50–90 times greater than predicted by a Boltzmann distribution.

## 8.3 Overview of the Different Theoretical Approaches

Kinetic theories which try to understand the development of the interacting gas/solid system on a time scale which is long compared to typical molecular interaction times ($10^{-10}$ to $10^{-15}\,\text{s}$) are not considered in detail in this chapter. Also the more phenomenological theories of Eyring (so-called "transition state theory" (TST), see [8.66]), and *Kramers* [8.67] will only be mentioned as limiting cases. *Kramers* described the reaction as the Brownian motion of the corresponding phase point over the activation barrier. TST emerges as a special case of Kramers' more general treatment. This approach is frequently applied to desorption phenomena and has been reviewed several times [8.5]. It is strictly applicable only in the zero coverage limit, but has been extended to account for lateral interactions between adsorbed molecules [8.68,69].

In the framework of this type of theories it has often been pointed out, that there is a close connection between the rate of desorption and the sticking coefficient s multiplied by the Boltzmann factor [8.68]:

$$k_{desorp} = const\, s(E_i) exp(-E_i/(k_B T_S))$$

Here desorption into the state $|i\rangle$ with energy $E_i$ is considered and the sticking coefficient applies for adsorption from that initial state. From the kinetic theories it appears, as if this relationship is based on some kind of equilibrium condition between the gas phase and the adsorbed phase. In fact it might be more general, because it already follows directly from the fundamental equations (8.9) and (8.36) below.

### 8.3.1 Classical and Classical-Stochastic Theories

Classical theories have been among the first to be applied in order to understand the dynamics of the gas-surface interaction. Famous are the simple cube models [8.70–72], which still are often valuable for a first qualitative interpretation of the experiments. During the years they have been generalized to include an attractive square well [8.73] (which allows the evaluation of sticking coefficients [8.74]) and rotational energy exchange [8.75]. A modification into "hard spherical caps" has been tried as well [8.76].

The mentioned models are simple to solve (e.g., the dynamics of the hard-cube model follows as a direct consequence of conservation of total perpendicular momentum and energy), because of the special potential assumed, which leads only to impulsive scattering, and because they involve only a two-body problem (Sect. 8.4.1). For a general potential and a many particle problem one has to solve the Newtonian equations of motion on a computer. This leads to the method of molecular-dynamics [Ref. 8.1, Chap. 6]. Classical model calculations of this character have been first performed by *McClure* [8.77] and led to the discovery of the surface rainbow effect.

The molecular-dynamics method is not very much suited for the processes we are interested in, namely sticking and desorption, because these are slow and therefore it would require excessive computer time to simulate them. In order to mitigate this problem the Generalized Langevin Equation (GLE) method was developed [8.78,79]. The idea is to choose a small number of substrate atoms which directly interact with the gas particle and to incorporate the remaining part of the solid (the "heat bath") into a friction term and a fluctuating force which are added to the equation of motion. The GLE procedure was applied to sticking and thermal desorption of Ar and Xe from Pt(111). There now arises the problem of which trajectories to identify with trapped species. *Tully* [8.79] decided to call trapped atoms those which had total energy (kinetic energy plus energy of interaction with the surface) more negative than $-3k_BT_S$. These atoms are considered to be "effectively equilibrated". The time reversed trajectories were then used to determine desorption rates, which were obtained by multiplying a "transition state theory factor" by the sticking coefficient.

A similar simulation for NO interacting with Ag(111) gave the result that the sticking probability decreases with increasing initial rotational energy [8.80]. This ties in with detailed balancing arguments, because, at equilibrium, scattered molecules were shown to have higher than average rotational energy so that desorbing molecules must have lower than average rotational energy.

While the definition of the sticking probability is problematic, in general, for classical theories, it becomes obvious and unique at zero temperature and zero incoming kinetic energy, because then sticking is the only possible inelasticity. In this limiting case all classical theories give unit probability for trapping. This result is already obtained with a simple model, which was investigated 25 years ago by *Zwanzig* [8.81]. It consists of a one-dimensional chain of substrate atoms which are connected by springs. The incoming gas atom interacts with the surface atom via a cut-off harmonic oscillator potential. Solving the classical equations of motion one finds that below a critical kinetic energy (which depends on the mass ratio and the depth of the adsorption well) sticking occurs with unit probability, whereas above the critical energy the particle will be reflected with certainty. More realistic potentials and three-dimensional models give analogous results, but one finds a continuous decrease of the adsorption probability from unity with increasing kinetic energy [8.82,83]. The physical picture behind it is the following: The collision of the gas atom with the surface leads to a local deformation of the lattice, which is transmitted to the neighbouring substrate atoms so that lattice waves are set up propagating away from the point of impact and taking away irreversibly the kinetic energy of the incident particle.

Dissociative adsorption has been theoretically studied by *Gelb* and *Cardillo* [8.84], who tried to simulate the experimental results of *Balooch* et al. [8.85] for the activated chemisorption of hydrogen on copper. They performed classical trajectory calculations using empirical potentials of the London-Eyring-Polanyi-Sato (LEPS) form, but were unable to obtain good agreement with experiment.

Several classical studies of desorption using approximate models have been published. In these theories it is generally assumed that the adparticle desorbs, if its potential energy becomes equal to its adsorption energy. Applying a linearized theory where the crystal is modelled by a one dimensional chain of harmonic oscillators, *Goodman* [8.86] confirmed the early result of *Frenkel* [8.87] derived from thermodynamic arguments. *Armand* [8.88] supported these findings by treating the same model in the framework of collision theory assuming thermal equilibrium. This line of approach was extended to three-dimensional models by Beeby and collaborators who investigated various kinds of adatom surface interaction: a potential depending only on the relative distance between the adatom and one surface atom along the surface normal [8.89]; a dangling adatom linked to a surface atom by a three-dimensional harmonic potential [8.90]; a Morse interaction potential [8.91].

Concluding we may state the following difficulties with classical theories:

i)    Elastic scattering and diffraction cannot be described,
ii)   difficulties (or ambiguities) with the definition of sticking and desorption,
iii)  computational expense for calculating many trajectories and slow processes like trapping and desorption, and
iv)   non-adiabatic effects (i.e., transition between different adiabatic hypersurfaces) and other quantum effects (like tunneling and zero-point motion) can be taken into account only in a rather artificial way [8.92].

The advantages are:

i)    A clear picture of the physics involved,
ii)   the correct (experimentally established) behaviour of the sticking coefficient at zero temperature and zero kinetic energy (although this is just the regime where classical theories should fail), and
iii)  elegant inclusion of heat bath and continuum effects via the GLE procedure.

### 8.3.2 Semiclassical Theories

Semiclassical theories represent a compromise in treating the quantum mechanical many particle problem. Compared to the classical theories they are better suited for handling elastic scattering, i.e., for calculating Debye-Waller factors.

The work by *Beeby* [8.93] treats the substrate atoms classically, whereas most other investigators prefer to use the semiclassical approximation for the gas-atom motion [8.21,94–96]. *Bendow* and *Ying* [8.94] provided one of the few semiclassical approaches (WKB in this case) to desorption. It is a multi-phonon theory based on the Glauber formalism [8.97]. Calculations were only performed for one phonon processes.

Presently very popular is the trajectory approximation, where the scattering particle provides a localized time-dependent potential for the solid without being influenced itself by the reacting substrate, i.e., recoil effects are neglected completely. In the latter case the total energy is not conserved. The application of this approach to gas atom-phonon scattering was pioneered in [8.98]. It is often used to study non-adiabatic effects (i.e., electron-hole pair excitations) in gas-metal collisions [8.99–102].

A central quantity in these theories is the probability $P(E', E)$ that the scattering particle with kinetic energy $E'$ loses the energy $E$ on its way. The sticking probability is then defined as

$$s = \int\limits_{E'}^{\infty} P(E', E) dE \ . \tag{8.3}$$

If the particle has lost an amount of energy greater than its kinetic energy, it cannot leave the adsorption well. Nevertheless the semiclassical theory assumes that the gas particle will continue on its way out without being influenced by the surface. This is a somewhat disturbing aspect of the trajectory approximation. Early efforts [8.103] concentrated on the friction coefficient, which is related to the center of gravity of $P(E', E)$.

Although different approximations are employed in the various semiclassical theories [8.93–104], the results are very similar, especially concerning the Debye-Waller factor [8.5].

Recently semiclassical wave-packet approaches have been devised, which appear to be rather promising [8.105]. Very notable is also the self-consistent semiclassical approach to inelastic scattering developed by *Kumamoto* and *Silbey* [8.106]. Contrary to the simple trajectory approximation the effect of the reacting surface on the particle trajectory is calculated self-consistently.

Sebastian addressed the question of what is the best classical description of a particle in interaction with a bath of quantum oscillators. He used the Feynman path integral representation of quantum mechanics to derive the equation of motion for the classical subsystem [8.107].

Attempts with semiclassical methods are quite numerous. They need not all be listed here, because they have been mostly applied to surface diffraction problems or to inelastic scattering. Applications to adsorption and in particular desorption seem to be much more rare and less conclusive.

Depending on the various approximations, semiclassical theories might possess one or several of the following shortcomings:

i)     Violation of unitarity,
ii)    non-conservation of total energy,
iii)   neglect of recoil effects, and
iv)    the simple trajectory approximation is not suitable for desorption studies.

Their obvious advantage compared to the classical theories is that they contain quantum effects like elastic scattering, diffraction and, perhaps, tunneling.

### 8.3.3 Quantum Theories

Among the full quantum theories only those for elastic scattering from surfaces are in a highly developed state. Here excellent reviews exist [8.108,5]. The prevailing methods are integral equations (Lippmann-Schwinger equation) and close-coupling procedures. The latter can be extended to include rotational and vibrational states of the scattering molecule. In this case, however, the number of channels to be included will increase enormously, especially if one aims at convergence with respect to increasing the number of channels. The so-called "sudden approximation" has been tested agains clos-coupling calculations with reasonable success [8.109].

The described methods are suited to handle so-called "selective adsorption", where upon collision with a hard surface a part of the perpendicular translational gas particle energy is either transformed into parallel translational energy, rotational energy or molecular vibrational energy so that the gas particle resides for some time near the surface before being reflected out again by a similar process. These processes can be identified in the experimental diffraction data and they might indeed be important as "precursors" for the final sticking event where the energy will be transferred to the solid [8.110]. They will, however, not be considered further in this chapter, because they do not involve the dynamics of the surface motion.

A quantum theory of inelastic surface scattering was first developed in the non-unitary Distorted-Wave Born Approximation (DWBA) by *Lennard-Jones* and his collaborators [8.111]. It remained essentially the only quantum-mechanical approach for more than forty years, although some improvements – like including second-order effects [8.112] and constructing a unitarized version [8.113] – have been tried.

Recently the one-phonon DWBA has been applied to investigate, how rotationally mediated selective adsorption resonances affect sticking due to phonons [8.114]. The surface was modelled by a semi-infinite isotropic, elastic continuum. The effects were found to be weak for $H_2$, but strong for HD, because the asymmetry of the molecule causes stronger translational rotational coupling.

These theories predict that the trapping probability should be zero in the low-energy and low-surface-temperature limit. The reason for this is obvious and has been termed "transmission problem". As the gas atom cannot penetrate into the solid, the amplitude of the core wave functions (i.e., of the eigenfunctions of the static potential) has to tend to zero near the surface. At long wave-lengths of the gas atom ($k \longrightarrow 0$) the core wave functions therefore assume their maximum at macroscopic distances from the surface. Due to normalization the amplitude over the adsorption well is then practically zero and hence the transition matrix element for transitions into states bound in the adsorption well vanishes. This behaviour is often interpreted as meaning that the gas atom is reflected before penetrating into the surface region. For some time the theorists therefore hoped that the situation would be cured by using realistic long-range potentials [8.115–117]. *Brenig* and co-workers demonstrated, however, that this is not the case [8.118,119].

Another speculation was that the predicted paradoxal result might be a consequence of the perturbation (and hence one-phonon) character of the DWBA. Obviously these theories do not describe the lattice deformation near the point of impact and therefore the mechanism which leads to trapping probability unity in the classical models (Sect. 8.3.1) is not included. Attempts in this direction were made [8.120] but did not lead to the desired result. The present author suggested a dimensionality effect as a possible way out. This will be discussed in more detail below (Sect. 8.4.3).

*Kreuzer* and collaborators developed a quantum-statistical theory of desorption [8.121, 122]. They used a linear boson coupling Hamiltonian, (8.4) below, and evaluated the bound-state occupation numbers in second-order time-dependent perturbation theory. Since this is only valid for weak adsorbate-substrate coupling, application was for He desorbing from graphite or constantan. Further simple perturbative model calculations for desorption have been published by different authors [8.123–125].

Non-perturbative quantum theories which are applicable to adsorption-desorption phenomena appear to be rare. *Wolken* [8.126] gives a close-coupling version which tries to include phonon effects by discretization of the continuous phonon spectrum, but the details have only been worked out for one phonon exchange. *Micha* applied a many-body approach for polyatomic systems to energy transfer in gas atom-surface scattering problems [8.127]. *Newns* [8.128] improved the trajectory approximation using path-integral methods. He gained two additional terms: An optical potential coming from the response of the surface to the particle and a recoil term. Recently *Armand* and *Manson* [8.129] managed to evaluated the Debye-Waller factor exactly for a flat vibrating surface by summing the perturbation series completely.

The Hamiltonian which is most frequently studied in semiclassical and quantum theories has the following form

$$H = P^2/(2m) + V(R) + \sum_i \omega_i(b_i^\dagger b_i + 1/2) + \sum_i B_i F(R)(b_i + b_i^\dagger) , \quad (8.4)$$

where P and m are momentum and the mass of the gas particle, respectively. $b_i^\dagger$ and $b_i$ are the boson creation and annihilation operators for the $i^{th}$ mode, $\omega_i$ is the corresponding frequency. The first two terms in (8.4) describe the gas atom moving in a static potential. The second term represents the so-called bath of bosons. They are the relevant surface excitations, which might be phonons or tomonagons (i.e., electron-hole pairs, see Sect. 8.4.5). The coupling to these modes is linear in the boson displacement $(b_i + b_i^\dagger)$ and depends on the gas particle position R via the operator F(R). The coupling constants $B_i$ determine how this coupling depends on the boson mode i. The linearity in the boson displacement is only reasonable for weak interaction. It might lead to serious artefacts for reactive and heavy gas particles.

*Gadzuk* et al. [8.130] attempted a quantum-mechanical explanation of the general observation that the rotational temperature $T_R$ of desorbing molecules is smaller than the substrate temperature. They treat the model of a hindered rotor moving in an infinite cone well. They show that sudden unhindering can lead to high rotational energies due to the initial zero-point motion. They also gave classical arguments that "sudden unhindering" with simultaneous release from the surface can lead to partitioning of hindered rotational energy between rotation and translation, and hence to lowered apparent rotational temperatures. In a second publication [8.131] they investigated additional models of hindrance, among them a Kronig-Penney model for the azimuthal modulation of the potential.

Gadzuk et al. did not treat the dynamics of the desorption process. They postulate a "model dynamics" where the hindering potential is suddenly switched off and the probability of observing a certain final rotational state of the desorbing molecule is given by the corresponding Franck-Condon factor. The latter is the overlap between the particle wave function in the hindrance potential and the free particle state of the desorbed molecule. This explanation assumes that the scattering behaviour is dominated by the bounds on the motion (cf. the discussion of constraints in Sect. 8.4) and that the details of the dynamics are of minor importance and washed out by thermal averaging.

A more detailed quantum theoretical description of the dynamics is obtained within the "deformation resonance" approach [8.132], which will be described below (Sect. 8.4.5).

## 8.4 Selected Problems

### 8.4.1 Structural and Dynamical Constraints

There are several effects which in any experiment tend to mask the influence of details of the gas-surface interaction. These are: Averaging over a range of rotational and translational states in the incident beam and surface thermal motion. The importance of these effects can be appreciated from the success of the simple cube models in explaining low-resolution experimental data for Maxwellian beams. Nobody would seriously claim that the cube models give a realistic description of the interaction, but they include the two most important constraints: conservation of parallel momentum of the gas atom and conservation of the total pependicular momentum and the total perpendicular energy of the system (Fig. 8.1). In an experiment these constraints will only be obeyed approximately, but fine details are averaged out due to the thermal distributions in the beam and in the surface motion.

Whereas experimenters might try to reduce the rotational temperature of the incident beam, increase the velocity ratio and cool the surface more and more in order to bear out special details and quantum effects, it is the task of the theorists to study the influence on the predicted final channel intensities, if starting from an initial thermal distribution more and more constraints are

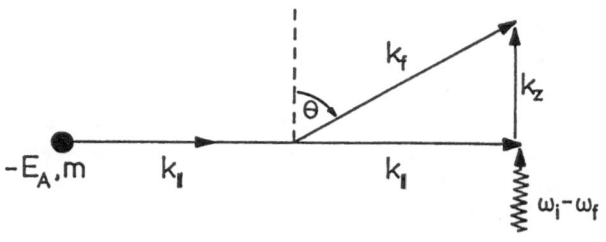

**Fig. 8.1** Schematic illustration of constraints used in the cube models and in the described desorption model

introduced. This is done in a systematic way by the information theoretical approach developed by *Levine* and collaborators [8.133,134]. Using surprisal analysis [8.135–137] they can determine in how much the experimental data deviates from a quasi-equilibrium situation as desribed by absolute rate theory (also called transition state theory). Applying reasonable constraints they can predict scattering distributions and sum rules without having to go into dynamical details [8.138,139].

*Meyer* and *Levine* analyzed (and generalized) the simple hard-cube model from an information theoretical point of view [8.140]. As constraints they assign definite values to the average energy $\langle E \rangle$ transferred to the solid, to the average value of the parallel momentum transfer $\langle q^2 \rangle$ and to the width of the energy distribution $\langle E^2 \rangle - \langle E \rangle^2$. The ansatz for the scattering distribution which is the least biased one under the given constraints (i.e., which has maximal entropy [8.141]), has the form:

$$P(k_i, k_f) = N \exp(-c_1 E - c_2 E^2 - c_3 q^2) . \tag{8.5}$$

N is the normalization constant. The Lagrange multipliers $c_1$, $c_2$ and $c_3$ are fixed by the condition that the expectation values of E, $E^2$ and $q^2$ calculated with the probability density $P(k_i, k_f)$ agree with the specified values for the constraints. *Meyer* and *Levine* chose

$$\langle E \rangle = |k_i - k_f|^2 / (2M) \tag{8.6}$$

which is the energy transfer in an impulsive, binary collision. $k_i$ and $k_f$ are the initial and final gas atom momenta, respectively. M denotes the mass of the target atom.

$$\langle E^2 \rangle - \langle E \rangle^2 = 2k_B T_S \langle E \rangle \tag{8.7}$$

is the high temperature Boltzmann limit, and

$$\langle q^2 \rangle = k_B T_S \langle E \rangle / c^2 + r^2 , \tag{8.8}$$

where c is the velocity of sound and $r^2$ denotes the width of the parallel momentum transfer distribution. These two quantities can be used as fitting parameters when analyzing experimental data. The mathematics is non-trivial, if one intends to calculate the normalization constant N and the Lagrange multipliers exactly.

The condition for $\langle E \rangle$ means just conservation of energy in the binary collision. If one specifies $\langle q^2 \rangle = 0$, i.e., conservation of parallel momentum for the gas atom, one has all the constraints defining the hard-cube model. *Meyer* and *Levine* demonstrated rigorously that in this case the maximal entropy procedure reproduces the well-known scattering distribution of the hard-cube model.

Performing the thermal averaging under given constraints can demonstrate interrelations in the experimental data, which might be very useful. For desorption phenomena relations between the average velocity in the direction $\theta$, the angular distribution and the angular dependence of the sticking rate have been derived in a conventional fashion neglecting the internal degrees of freedom of the particle [8.142]. The following constraints were applied: (i) The motions of the gas atom parallel and perpendicular to the surface are independent. (ii) Coupling to the solid excitations (phonons or electron-hole pairs) occurs only via the perpendicular motion. These two assumptions imply, of course, that the parallel momentum of the gas atom is conserved during the desorption process. The two constraints. which are also valid for the cube models, are illustrated in Fig. 8.1. In the following the condition of an impulsive binary gas-surface interaction is, however, not used and it will turn out that the character of this interaction influences the desorption flux in a characteristic way. The general formula for the desorption rate is

$$2\pi/Z(T_S) \sum_{f,i} |\langle f|T|i\rangle|^2 \exp(-E_i/(k_B T_S))\delta(E_f - E_i) \; , \tag{8.9}$$

where i represents the set of quantum numbers for the adsorbed situation, and f the set of quantum numbers after desorption. $\langle f|T|i\rangle$ is the transition matrix element, which contains the information about the dynamical details of the process. $Z(T_S)$ is the partition function of the adsorbate plus heat bath.

Due to the imposed constraints we have the following relations:

$$k_z^2 + k_\parallel^2 = k_z^2/\cos^2\theta \; , \tag{8.10}$$

$$\omega_i = \omega_f + E_A + k_z^2/(2m) \tag{8.11}$$

$k_\parallel$ is the conserved parallel momentum of the gas atom, $k_z$ is the perpendicular momentum transfer. $\omega_i$ and $\omega_f$ are the phonon energies in the initial and final states, respectively. $E_A$ is the energetic separation of the adsorbed gas atom state from the vacuum level and is positive. One introduces the "dynamic function" which contains the physical details (influence of the interaction potential, phonon spectrum) of the desorption process:

$$D(k_z, T_S) = \int_0^\infty d\omega_f |\langle f|T|i\rangle|^2 R_{ph}(\omega_f)\exp(-\omega_f/(k_B T_S)) \tag{8.12}$$

with $R_{ph}$ being the phonon density of states. Inserting this into (8.9) yields for the desorption rate into the direction $\theta$

$$k(\theta) = R(k_\parallel)\exp(-E_A/(k_B T_S))$$
$$\times \int_0^\infty dk_z D(k_z, T_S)\exp[-k_z^2/(2mk_B T_S)] \; . \tag{8.13}$$

$R(k_\parallel)$ is the density of states in k-space for the parallel gas particle movement. The angular distribution of the desorbing particles is now completely determined by the dynamical function $D(k_z, T_S)$. Assume that it depends via a power law on the momentum transfer

$$D(k_z, T_S) = F(T_S)k_z^n \; . \tag{8.14}$$

$F(T_S)$ is an unspecified function of the substrate temperature. In the framework of the information theoretical approach this would act as an additional constraint. The integration can be performed and yields

$$k(n, \theta) = Z^{-1}(T_S)F(T_S)R(k_\parallel)(mk_B T_S)^{(n+1)/2}$$
$$\Gamma((n+1)/2)\cos^{n+1}\theta \exp(-E_A/(k_B T_S)) \tag{8.15}$$

$\Gamma$ is the Gamma function, n is an integer. It is now easy to evaluate the expectation value of the kinetic energy *perpendicular* to the surface

$$\langle E_z \rangle = [(3+n)/2] k_B T_S \text{ for the angular distribution } \cos^n \theta \; . \tag{8.16}$$

Because of the independence of the perpendicular and parallel motion, the average energy in direction $\theta$ is

$$\langle E_0 \rangle = \langle E_z \rangle \cos^2 \theta + 2k_B T_S \sin^2 \theta \; . \tag{8.17}$$

$2k_B T_S$ is the average energy in a certain direction for a Maxwell distribution.

The results of this theory are in agreement with experiment. Especially satisfying is that correct velocity ratios are obtained and that the angular dependence of the average energy and the velocity ratio is described qualitatively correctly.

Assuming that the dynamic factor $F(T_S)$ shows only a weak dependence on the surfaces temperature, (8.15) also predicts the law of temperature variation for a given desorption angular distribution. It is $T_S^{n/2}$ for $\cos^n \theta$. This relationship has qualitatively been verified in the experimental investigation of $CO_2$ desorption from Pt(111) [8.143].

The transition matrix elements for desorption and for sticking can be related to each other by time reversal and applying the theorem of *Gell-Mann* and *Goldberger* [8.155], which yields

$$D(k_z, T_S) = k_{stick}(k_z, T_S)Z_{ph}(T_S)/R_{ad} \; . \tag{8.18}$$

$Z_{ph}$ is the phonon partition function and $R_{ad}$ is the density of gas atom states at the surface. This result is independent of any particular model for the particle-surface interaction. If the desorption rate varies according to $\cos^n \theta$ one concludes from (8.14 and 15) that D varies like $k_z^{n-1} \cos^{n-1} \theta$. Because the adparticle density of states is constant in k-space, $k_{stick}$ is then proportional to $\cos^{n-1} \theta$. This conclusion is also supported by the available experimental data [8.144].

### 8.4.2 The Compensation Effect

Though adsorption and desorption processes exhibit a rather large diversity of phenomena, there appears to be one general regularity: The compensation effect for desorption. It describes the observation that an increasing adsorption energy (or more precisely: activation energy for desorption) commonly implies an increasing frequency factor and vice versa so that the rate of desorption stays approximately constant. Sometimes it can be sharpened to the statement that for a series of similar desorption reactions the Arrhenius plots (i.e., logarithm of the rate versus reciprocal absolute temperature) pass through a common point corresponding to the so-called compensation temperature.

The compensation effect is not restricted to desorption phenomena. It was then discovered sixty years ago in heterogeneous catalysis [8.145] and since it has been found in various thermally activated rate processes. A detailed review article has been written by *Galway* [8.146] who summarized various explanations and maintains that no general understanding has been achieved. *Peacock-Lopez* and *Suhl* [8.147] tried to give such a general explanation applying to all observed cases. The main idea of their argument is repeated here for the example of a phonon driven desorption process.

Starting from (8.13) and introducing the energy transfer E from the phonons to the gas atom

$$E = E_A + k_z^2/(2m) \ , \tag{8.19}$$

the desorption rate in perpendicular direction can be written in the form

$$k(\theta = 0) = R(k_{||})/Z(T_S) \int_0^\infty dE\,\mathrm{sqrt}(2(E - E_A))D(E, T_S)\exp(-E/(k_B T_S)) \ . \tag{8.20}$$

The quantity which governs the dynamics and which has to be responsible for the compensation effect is $\mathrm{sqrt}(2(E - E_A))D(E, T_S)$. If it behaves like

$$\mathrm{const} \times (1/(k_B T_c) - 1/(k_B T_S)) \times \exp(E/(k_B T_S))$$

then the integration can be performed easily and yields

$$k(\theta) = R(k_{||})/Z(T_S) \times \mathrm{const} \times \exp(E_A/(k_B T_c) - E_A/(k_B T_S)) \tag{8.21}$$

for $T_S < T_c$. $T_c$ is obviously the compensation temperature, because for $T_S = T_c$ the rate becomes independent of the activation barrier $E_A$.

Crucial for this result is the exponential increase of the dynamic function $D(E, T_S)$ with the energy transfer E. Using the relation $\omega_i - E = \omega_f$ it has the form

$$D(E, T_S) = \int_{E_A}^\infty d\omega_i |T_1(\omega_i, E)|^2 R_{ph}(\omega_i - E)\exp[-(\omega_i - E)/(k_B T_S)] \ . \tag{8.22}$$

In the discussed work it is assumed that the density of accessible states of the solid $R_{ph}$ increases exponentially with E and that this dominates the behaviour of the dynamic function $D(E, T_S)$. The interpretation is then in terms of an entropy effect of the heat bath, i.e., the number of ways in which the heat bath can provide the necessary amount of energy increases exponentially with the energy transfer.

Mathematical difficulties are encountered with the integrations (e.g., obviously a cut-off energy has to be introduced for $T > T_c$) and the question whether the exponential increase of the dynamic function with energy exchange is a universal phenomenon needs further investigation.

For weak interaction, when one-phonon processes dominate, the dynamic function is, in fact, more likely to be dominated by the transition matrix element. If the latter behaves in a particular way as a function of the activation energy, this can give rise to a compensation effect as well. This is indeed the explanation given by *Sommer* and *Kreuzer* [8.148] for the compensation effect they predict for $^3$He desorption from graphite. Based on Hartree-Fock-like wave functions calculated from a mean-field theory they evaluate the desorption rate in first-order DWB approximation. As coverage builts up the He atoms are confined to a second layer where the effective potential is less attractive and hence the activation energy for desorption is smaller. The gradient of this potential, which is responsible for the phonon coupling and determines the magnitude of the transition matrix elements, is smaller as well in the second layer and therefore the dynamic function will decrease with decreasing $E_A$. The calculation shows that the pre-exponential drops by two orders of magnitude as the second layer is built up.

A third mechanism for producing a compensation effect might be an adsorbate induced modification of the phonon structure. If, for example, a phonon mode turns soft with increasing adsorption strength, the rate might increase due to the decreasing partition function $Z(T_S)$ in the denominator of (8.20) [8.149].

A similar influence of the partition function is to be expected, if the structure of the adsorbate layer changes from disordered free particle-like behaviour to an ordered structure. This was found to happen for CO/Ru near relative coverage 0.5 and led to extremely large frequency factors [8.56].

*McCoy* [8.150] proposed that the compensation effect for heterogeneous dissociation might be due to relaxation of the anharmonic molecular oscillator. He supposed that when the molecule is adsorbed at a catalytically active site, the bond can become more anharmonic. The activation energy and the pre-exponential factor will decrease then. *McCoy* predicted a very large compensation effect. It was, however, demonstrated [8.151] that there was a mistake in this analysis. The correct equations predict only a very weak compensation effect which cannot account for the experimental observations.

A too weak compensation effect as compared to the experimental results of *Bauer* et al. [8.55] was also found in the classical theory of desorption developed by *Armand* [8.152]. Perfect thermal equilibrium during the desorption process was assumed here and consequently the velocity distribution of the desorbing particles was Maxwellian with a cosine shaped angular distribution. The magnitude of the compensation effect is strongly dependent on the mean-thermal potential energy contained in the bond, which in *Armand*'s theory cannot become larger than $2k_B T_S$. The adatom desorbs, if the potential energy contained in the bond becomes equal to the adsorption energy. Armand speculated that a larger mean-thermal potential energy concentration in the bond and hence larger pre-exponential factors might result from a more sophisticated and more extending adatom-surface interaction.

### 8.4.3 The Sticking Coefficient –
### Definition and Zero Temperature/Zero Kinetic Energy Behaviour

As mentioned at several places in this chapter, the definition of the sticking coefficient is a difficult problem, in experiment as well as in theories of all kinds. This section elaborates on the theoretical definition paying special attention to quantum theories. In a classical description the difficulty is to decide at which time or after which energy loss the gas particle should be considered as stuck. *Beeby* and *Agrawal* [8.153] found that even for the simple one-dimensional model of *Zwanzig* (Sect. 8.3.2) the analysis becomes involved once multiple collisions are likely and the sticking coefficient depends significantly on the bulk properties.

The difficulty encountered in the context of quantum scattering theory is of a different kind. Standard text-book scattering theory [8.154] assumes that the scattering potential is ineffective in the remote past and in the distant future, because the particle is far from the scatterer in these limits. For a reflected particle this is true, and text-book scattering theory can be applied for suitable Hamiltonians (watch, however, gettering Hamiltonians and the dimensionality effect discussed below) to calculate the reflectivity (i.e., the Debye-Waller factor) as the absolute square of the forward scattering matrix element:

$$\exp(-2W) = |\langle i - |i + \rangle|^2 = |S_{ii}|^2 . \tag{8.23}$$

On the other hand, for a stuck particle the interaction with the solid is always effective. For this reason it is necessary to define the asymptotic scattering states corresponding to a stuck particle in a different way.

*Doyen* and *Grimley* [8.156] proposed the following definition:

Sticking corresponds to a transition from an initial product state $|k\rangle|m\rangle$ describing the state of an elastically scattered particle (see below) and a phonon state $|m\rangle$ to a final state $|s\rangle$ in which the atom is bound to the solid surface. For an experiment where sticking is detected by measuring the adsorbate coverage, the final state is an eigenstate of a Hamiltonian $H_{stick}$ which differs from the total Hamiltonian by an amount $V_{stick}$. An important point is that during the measurement the target is kept at constant temperature, which means that every created phonon (boson) with quantum numbers q is destroyed again.

This is equivalent to a measurement of the phonon q, thus switching off the interaction between the particle and this phonon. We therefore make the crucial observation that |s⟩ is a product state describing the wave function for the bound gas atom interacting with the non-emitted phonons and the n phonons $q_1, \ldots, q_n$ which have been emitted during the sticking process.

In their investigation of electron-hole pair excitations (Sect. 8.4.5) *Brivio* and *Grimley* [8.157] considered a "gettering" Hamiltonian of the kind

$$H_{gett} = |k0\rangle E_{k0} \langle k0| + \sum_m |m\rangle E_m \langle m| + \sum_m (|m\rangle \langle m|W|k0\rangle \langle k0|$$

$$+ |k0\rangle \langle k0|W|m\rangle \langle m|) , \qquad (8.24)$$

where |k0⟩ is the only state for a reflecting gas particle. The set of states |m⟩ has the gas particle adsorbed at the surface. Hence the Hamiltonian allows only for elastic scattering or sticking (Fig. 8.2). The Hamiltonian has the same structure as those considered by *Wolff* [8.158], *Anderson* [8.159], and *Newns* [8.160], and is therefore familiar and readily solved. The Green operator, e.g., has the matrix elements (for an orthogonal basis set):

$$\langle k0|G(E)|k0\rangle = [E - E_{k0} - q_{k0}(E)]^{-1} \qquad \text{with} \qquad (8.25)$$

$$q_{k0}(E) = \sum_m \frac{|\langle k0|W|m\rangle|^2}{E - E_m} \qquad (8.26)$$

$$\langle k0|G(E)|m\rangle = \langle k0|G(E)|k0\rangle \langle k0|W|m\rangle (E - E_m)^{-1} , \qquad (8.27)$$

$$\langle m|G(E)|n\rangle = (E - E_m)^{-1}[\delta_{mn} + \langle m|W|k0\rangle$$

$$\times \langle k0|G(E)|k0\rangle \langle k0|W|n\rangle (E - E_n)^{-1}] . \qquad (8.28)$$

If one now tries to construct the scattering state |k0 +⟩ by using the Lippmann-Schwinger equation one finds

$$\langle k0|k0 +\rangle = 1 + \sum_n \langle k0|G(E_{k0})|n\rangle \langle n|W|k0\rangle$$

$$= 1 + q_{k0}(E_{k0}) \langle k0|G(E_{k0})|k0\rangle$$

$$= (E_{k0} - E_{k0})/q_{k0}(E_{k0})$$

$$= 0 , \qquad (8.29)$$

$$\langle m|k0 +\rangle = \sum_n \langle m|G(E_{k0})|n\rangle \langle n|W|k0\rangle$$

$$= \langle m|W|k0\rangle (1 + q_{k0}(E_{k0}) \langle k0|G(E_{k0})|k0\rangle)/(E_{k0} - E_m)$$

$$= \langle k0|k0 +\rangle \langle m|W|k0\rangle/(E_{k0} - E_m)$$

$$= 0 . \qquad (8.30)$$

$\langle m|k0 + \rangle$ is zero also on the energy shell, because the limit $E_m \longrightarrow E_{k0}$ has to be performed last. Hence one has the remarkable and perhaps surprising result that the scattering state created from the Lippmann-Schwinger equation is identically zero. This implies that all transition matrix elements $\langle m|W|k0+ \rangle$ and $\langle k0|W|k0 + \rangle$ will vanish and hence all transition rates. The result is a consequence of the special structure of the gettering Hamiltonian which does not present a well-posed scattering problem. It describes a state $|k0\rangle$ immersed into a continuum and it is physically obvious that a particle which is initially in the state $|k0\rangle$ will finally disappear into the continuum. This is probably what the result means and this was anticipated by Brivio and Grimley. The result depends also on the fact that the matrix elements were evaluated *on the energy shell*, which means that the (total) energy has to be conserved during a scattering event and also during a decay process.

Brivio and Grimley interpret $|k0\rangle$ as a state where the gas particle reflects between two walls, one inert and the other, the solid surface, reactive. If one uses a Hamiltonian of the kind described by (8.24) to describe sticking at surfaces, one has to look for new reasonable definitions of scattering probabilities. For this purpose it is favourable that the self-energy $q_{k0}$ has a well-defined meaning. Its imaginary part on the energy shell can be connected with the decay rate and serve as a substitute for the sticking rate $k_{stick}$. If one defines then an "elastic scattering rate" as the collision number per unit time, the following definition for the sticking coefficient is tempting:

$$s_{gett} = k_{stick}/(k_{stick} + n_{coll}) . \tag{8.31}$$

This is the definition used by Brivio and Grimley. It has been earlier suggested by *Doyen* [8.161], who replaced $n_{coll}$ by the incoming particle flux. It has the property that it merges into the common definition used in the perturbation treatments [8.162], if the sticking probability is small, i.e., if $k_{stick} \ll n_{coll}$:

$$s_{pert} = k_{stick}/n_{coll} . \tag{8.32}$$

The dilemma with the definition (8.31) is that it is not compatible with the scattering theory definition for large sticking rates. To see this we start from the relation between the scattering matrix (S-matrix) and the transition matrix (T-matrix)

$$S_{fi} = \langle f - |i + \rangle = S_{fi} - 2\pi i \langle f|T|i \rangle \delta(E_f - E_i) . \tag{8.33}$$

This is the standard notation of time-independent scattering theory, where the S-matrix elements are distributions (i.e., linear functionals over a parameter Hilbert space). In order to convert this into complex numbers, one has to project onto well-behaved test functions. The $\delta$-function will then transform into the (many-particle) density of states

$$\delta(E_f - E_i) \longrightarrow R(E_f) . \tag{8.34}$$

The absolute square of the S-matrix element is now

$$|S_{fi}|^2 = \delta_{fi} + 4\pi^2 |\langle f|T|i\rangle|^2 R^2(E_f) \ . \tag{8.35}$$

The transition rate is only defined for $f \neq i$ and given by

$$k_{fi} = 2\pi |\langle f|T|i\rangle|^2 R(E_f) = |S_{fi}|^2 / (2\pi R(E_F)) \quad \text{for} \quad f \neq i \ . \tag{8.36}$$

Unitarity requires that

$$\sum_f |S_{fi}|^2 = 1 \ . \tag{8.37}$$

Summing the transition rates over all inelastic channels ($f \neq i$) and using (8.36 and 37) one obtains

$$\sum_{f \neq i} k_{fi} = (1 - |S_{ii}|^2) / [2\pi R(E_i)] \ . \tag{8.38}$$

Putting this back into (8.36) yields

$$|S_{fi}|^2 = 2\pi R(E_f) k_{fi} = (1 - |S_{ii}|^2) k_{fi} / \sum_{f \neq i} k_{fi} \ . \tag{8.39}$$

This is different from the ad hoc definition (8.31), because nothing like a collision number or a rate for elastic scattering appears. In fact, such a quantity is not defined within the context of quantum-scattering theory, because *specular elastic scattering at surfaces is forward scattering*. $2\pi |\langle i|T|i\rangle|^2 R(E_i)$ appears in the optical theorem but does not have the meaning of an elastic (forward) scattering rate.

Equation (8.39) might be called a branching ratio for the inelastic flux. It is applicable in all cases where information about the specular elastic reflectivity exists. For suitable Hamiltonians the latter can be identified with the Debye-Waller factor. Equation (8.39) yields always unitary results, even if the Debye-Waller factor and the inelastic transition rates are evaluated only approximately. It has the advantage of giving the exact results in those cases, where a Hamiltonian presenting a well-posed scattering problem is solved exactly by the methods of standard text-book scattering theory.

Besides its definition the other fundamental problem with the sticking coefficient is its behaviour at zero temperature and zero kinetic energy. In the following a simplified version of the way of reasoning developed in a series of publications [8.33,161,163,164] is presented. The discussion is based on box normalization, which displays explicitly the order of magnitude of the matrix elements and the influence of the dimensionality of the problem. To analyze the situation the following basis wave functions are used. They are normalized over a volume of dimensions $L \times L \times 2L$. Normal to the surface the box has the length $2L$ and the solid surface divides it into two sets: Delocalized functions where the gas atom is far from the surface and localized ones where the gas is in direct interaction with the solid (Sect. 8.4.4):

$$|k_z, k_{||}, q\rangle = L^{-3/2} 2^{1/2} \sin k_z z \exp(ik_{||}r) |S_q\rangle \quad , \tag{8.40}$$

where $k_z$ and $k_{||}$ are the perpendicular and parallel momentum of the gas atom, respectively, q stands for the phonon quantum numbers, z is the perpendicular distance of the gas atom and r indicates its parallel position. The normalization constant is only displayed for the gas atom part, for the phonon part it is absorbed in $|S_q\rangle$. The localized states are

$$|m\rangle = |k_{||}, q\rangle = L^{-1} a(z) \exp(ik_{||}r) |S_q\rangle \quad , \tag{8.41}$$

where $a(z)$ is localized in z-direction and normalized, which explains the different exponent of L as compared to the delocalized functions. The gas-atom potential is here chosen, for simplicity, as an infinite wall, for which the delocalized states are eigenfunctions. This special choice does not affect the conclusions. The gas atom-phonon interaction has the form

$$V_{A-ph} = \sum_{k_{||}} L^{-2} |a(z)\rangle |\exp(ik_{||}r)\rangle \langle \exp(ik_{||}r)| \langle a(z)|$$
$$\times L^{-3/2} \sum_{\ell_{||}, q} |\exp(ik_{||}r)\rangle g(k_{||}, \ell_{||}, q) \langle \exp(i\ell_{||}r)| (b_q^\dagger + b_q) \quad . \tag{8.42}$$

The factor $L^{-3/2}$ in front of the second sum normalizes the *phonons*. $V_{A-ph}$ is only effective, if the gas atom is near the surface (see [8.154] for motivation and comments on such a form of the phonon coupling). The matrix elements of $V_{A-ph}$ between basis states of different subsets are

$$\langle k_z, k_{||}, p | V_{A-ph} | \ell_{||}, q \rangle = L^{-2} \langle \sin k_z z | a(z) \rangle G(k_{||}, \ell_{||}, p, q) \quad . \tag{8.43}$$

Here *all* normalization factors involving L are displayed. The normalization factors for the parallel gas particle movement have been integrated out. The phonon normalization factor $L^{-3/2}$ survives as does the factor $L^{-1/2}$ from the perpendicular gas particle motion with momentum $k_z$. The quantities $G(k_{||}, \ell_{||}, p, q)$ are numbers depending on the phonon coupling. Going now to zero kinetic energy means for box normalization to take the lowest possible value of the gas atom momentum which is $k_z = 2\pi/L$. Near the surface, where $z \ll L$, $\sin k_z z$ can then be replaced by $2\pi z/L$. The matrix elements are in this limit

$$\langle k_z, k_{||}, p | V_{A-ph} | \ell_{||}, q \rangle = 2\pi L^{-3} \langle z | a(z) \rangle G(k_{||}, \ell_{||}, p, q) \quad . \tag{8.44}$$

Matrix elements of the operator $V_{A-ph}$ between two delocalized basis functions are in this limit of order $L^{-9/2}$, because in comparison to (8.44) one gains another factor $L^{-3/2}$ from the additional $\sin k_z z$ — part ($L^{-1/2}$ from normalization and $L^{-1}$ from $k_z \longrightarrow 0$). The corresponding matrix elements in the localized subset are of order $L^{-3/2}$ and are, of course, independent of the gas particle momentum, but they are not relevant for our discussion of the gas atom-phonon coupling. The indirect interaction between delocalized gas atom states via the phonon continuum can easily be estimated to be of the order $L^{-6}$.

Performing the limit to infinite normalization volume one observes that all coupling matrix elements vanish. But this does not imply that phonon excitation will not occur, because in this limit the density of (phonon) states tends to infinity. What matters is the mixing between delocalized and localized wave functions. Perturbation theory can be used to estimate this mixing, because low kinetic energy means weak coupling due to the transmission problem (the amplitude of the gas atom wave function near the surface is multiplied by an additional factor $L^{-1}$ at zero kinetic energy, see Sect. 8.3.3). The mixing coefficients between delocalized and localized states are

$$C_{km} = V_{km}/(E_k - E_m) \ . \tag{8.45}$$

The smallest energetic separation between two such energy levels is of order $L^{-3}$. $V_{km}$ is the matrix element (8.44), which also carries a factor $L^{-3}$. These two factors cancel and $C_{km}$ is hence of order unity also for infinite normalization volume. The mixing coefficient between two delocalized states is

$$C_{k\ell} = V_{k\ell}/(E_k - E_\ell) \ . \tag{8.46}$$

$V_{k\ell}$ is of order $L^{-9/2}$ so that for the smallest energetic separation $C_{k\ell}$ is of order $L^{-3/2}$ which vanishes in the limit $L \longrightarrow \infty$.

In the limit $L \longrightarrow \infty$ one therefore needs only to retain for the phonon coupling those terms which tend to zero most slowly. This means that inelastic reflection can be neglected, because both direct and indirect terms vanish in higher order due to the additional factors $L^{-3/2}$ and $L^{-3}$, respectively. Selecting the slowest asymptotic incoming gas atom state $|k0\rangle$ one now concludes that $|k0\rangle$ does not couple to any outgoing (i.e., reflecting) gas atom states, but only to stuck states. The situation is schematically illustrated in Fig. 8.2. This means that the interaction of this asymptotic state can be approximated by a Hamiltonian which has exactly the same structure as the gettering Hamiltonian described above. This Hamiltonian will yield the correct mixing of wave functions.

This is not the case at finite kinetic energies when $k_z$ is of order unity, although one still has that $V_{k\ell}$ (of order $L^{-5/2}$ then) tends faster to zero with $L \longrightarrow \infty$ than $V_{km}$ (of order $L^{-2}$ then). But perturbation theory yields then order of magnitude $L^{1/2}$ for $C_{k\ell}$ which diverges for $L \longrightarrow \infty$ as does $C_{km}$ (of order $L$ then). These findings mean that perturbation theory breaks down, but that mixing will be important in *both* cases and the gettering Hamiltonian does not describe the situation.

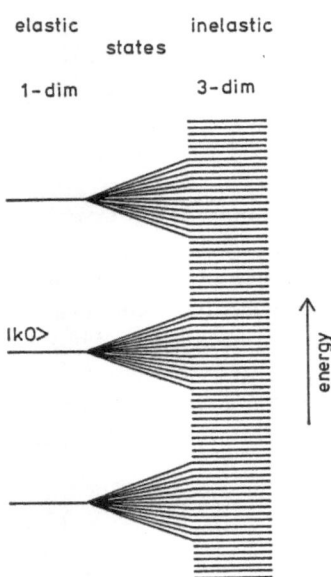

elastic        inelastic

states

1-dim        3-dim

|k0>

energy

**Fig. 8.2.** Level scheme for gas-surface interaction in the low kinetic energy regime, reducing to a sum of "gettering Hamiltonians" (see text)

The analysis of the zero kinetic energy behaviour can now be summarized as follows:

i) Evaluating the coupling coefficients in perturbation theory leads to the conclusion that the gettering Hamiltonian properly describes the zero kinetic energy behaviour.

ii) Standard text-book scattering theory must not be applied to the gettering Hamiltonian.

iii) The evaluation of the sticking rate is based on the relation

$$
\begin{aligned}
k_{stick} &= d/(dt)|\langle s|k0, eigen\rangle|^2 \\
&= 2\pi|\langle s|V_{stick}|k0, eigen\rangle|^2 \delta(E_{k0} - E_s) \ ,
\end{aligned}
\tag{8.47}
$$

Using the perturbation coefficients yields

$$
\begin{aligned}
k_{stick} &= \sum_m |C_{km}|^2 |\langle s|V_{stick}|m\rangle|^2 \delta(E_{k0} - E_s) \\
&= |C_{km}|^2 |\langle s|V_{stick}|m\rangle|^2 R_{ph} \ .
\end{aligned}
\tag{8.48}
$$

This is of order of magnitude unity. $|s\rangle$ and $V_{stick}$ are the final state for sticking and the transition inducing potential, respectively, as discussed above. Note that $V_{stick}$ is different from W in (8.24) and from $V_{A-ph}$ in (8.42).

iv) $n_{coll}$ is of order $L^{-2}$ at zero kinetic energy, because only the perpendicular flux is of importance. Applying the ad-hoc formula (8.31) yields then sticking coefficient unity [8.161]. (In [8.161] the *language* of standard text-book scattering theory was used, but $\langle k0|k0+\rangle$ was assumed to be non-zero. Equation (8.30) yields then coupling coefficients of the same order of magnitude as obtained here.)

Deviating conclusions in the literature are based on one or both of the following ways of reasoning, which have been criticized here:

i) Apply standard text-book scattering theory in the zero kinetic energy limit [8.118,119]. The transition matrix elements vanish on the energy shell (see above) and one therefore obtains $\langle k0-|k0+\rangle = 0$.

ii) Ignore the dimensionality of the substrate excitations in calculating the coupling coefficients, i.e., the energy denominator $E_{k0} - E_m$ is replaced by the energetic separation $E_{k0} - E_a$ of two non-coupling gas particle states, the incoming state $|k_z, k_{||}\rangle$ and the localized state $|a(z), k_{||}\rangle$. If the energy of the localized state is not right at threshold, then the energetic separation is of order unity, i.e., independent of the normalization length. The phonon factor $L^{-3/2}$ is consequently ignored in $V_{km}$ and $C_{km}$ becomes of order $L^{-3/2}$. The sticking rate is then of order $L^{-3}$ and the sticking coefficient calculated from (8.31) vanishes in the limit $L \longrightarrow \infty$. In this category (yielding sticking rates of order $L^{-3}$) belong all theories which evaluate the transition rate by taking matrix elements of the gradient of the static potential between localized and delocalized gas particle states [8.112,113,115].

*Brivio* and *Grimley* [8.157] suggested to replace the sticking rate by $2\pi |V_{km}|^2 R_{ph}$ which is two times the negative of the *exact* imaginary part of the self-energy for the gettering Hamiltonian. This is unphysical, because $|k0\rangle$ is not a physical state and because W (or $V_{A-ph}$, respectively) is not a physical transition inducing potential. Also the notion of the "width" (which assumes distribution of the spectral weight over infinitely many continuum states) of the state $|k0\rangle$ becomes doubtful, once the width becomes equal to the energetic separation of two "continuum states".

The "driving force" for the special behaviour at zero kinetic energy is the additional weakening factor $L^{-1}$ which is multiplied to the matrix element (8.43) when going from finite kinetic energies to zero energy. This had the consequence that the coupling between delocalized gas atom states tends to zero in higher order and one is left with the gettering Hamiltonian. The same is going to happen, of course, whenever the gas-surface excitation coupling becomes weak and will necessarily lead to a drastic enhancement of inelasticity.

A striking experimental verification of this is the measurement of the enormously small survival probability ($10^{-4}$) of a metastable helium atom hitting an adsorbed layer of CO molecules [8.33]. The interaction is weak even at thermal energies, because of the very large size of the metastable atom, which makes it turn round at large distances. In the gas phase a metastable He atom colliding with a single CO molecule has a survival probability of ca. 0.5.

An intriguing point is the reason why high inelasticity at zero kinetic energy is special for the gas surface interaction and is not present in gas phase collisions. In order to bear out the central argument assume that there exists a continuum of boson-like excitations in gas phase reactions. In the center-of-mass frame an effective particle moves in a central potential with a repulsive hard core. At the center the delocalized elastic particle wave functions have to tend to zero proportional to $kr$ (where $r$ is now the radial distance) which at zero kinetic energy becomes proportional to $r/L$. The inelastic particle-boson coupling is localized at the center and has then in analogy to (8.42) the form

$$V_{A-boson} = |a(r)\rangle\langle a(r)| \times L^{-3/2} \sum_{k,\ell,q} |\exp(ikx)\rangle g(k,\ell,q)\langle\exp(i\ell x)|(b_q^\dagger + b_q)$$

$$(8.49)$$

The delocalized and localized basis functions are in complete analogy to the surface case given by

$$|k,q\rangle = L^{-3/2}\exp(-ikx)|S_q\rangle \ , \tag{8.50a}$$

$$|m\rangle = |q\rangle = a(r)|S_q\rangle \ . \tag{8.50b}$$

Important are here the normalization constants for the particle part, which are different from the surface case because of the missing parallel translational invariance. The matrix elements are of the following order of magnitude

$$V_{k\ell} = 0(L^{-9/2}) \ , \quad V_{km} = 0(L^{-3}) \quad \text{for finite kinetic energies} \ , \tag{8.51a}$$

$$V_{k\ell} = 0(L^{-11/2}) \ , \quad V_{km} = 0(L^{-4}) \quad \text{for zero kinetic energy} \ . \tag{8.51b}$$

The perturbation mixing coefficients are in the latter case of the following order of magnitude:

$$C_{k\ell} = 0(L^{-3/2}) \ ; \quad C_{km} = 0(L^{-1}) \ . \tag{8.52}$$

*Both* kinds of coefficients tend to zero for $L \longrightarrow \infty$, which means that no inelasticity is possible. This is, of course, a standard text-book result. The surface case is more favourable for inelasticity, because the particle has to localize only in one dimension in order to initialize excitations. Another way of expressing this is to point out that the elastic motion of the particle is effectively one-dimensional for scattering from a flat surface, whereas the inelastic motion is three dimensional. In the gas phase both kinds of motion are effectively three dimensional. For this reason this phenomenon has also been termed a "dimensionality effect" [8.33,163].

The dependence on the dimensionality becomes very obvious, if one investigates the unrealistic (but theoretically sometimes treated) case of one-dimensional gas-particle movement and one-dimensional phonon spectrum. Then the delocalized and localized gas-particle states corresponding to (8.40 and 41) will carry normalization factors $L^{-1/2}$ and $L^0$, respectively. The phonon coupling (8.42) only exhibits a total normalization factor $L^{-3/2}$ (because both the parallel gas particle and phonon movements drop out) and consequently the coupling matrix elements $V_{km}$ will be of order $L^{-1}$ at thermal energies and $L^{-2}$ at zero kinetic energy. The energy denominator $(E_{k0} - E_m)$ is in the one-dimensional case of order $L^{-1}$ and hence $C_{km}$ at zero energy is of order $L^{-1}$ (in contrast to the three-dimensional case where $C_{km}$ was of order unity). Combining this with (8.31) for the sticking coefficient and $n_{coll}$ of order $L^{-2}$ yields sticking coefficients smaller than unity though not necessarily zero [8.161].

The above analysis of the zero kinetic energy problem criticized existing methods and forwarded arguments for an enhanced inelasticity, but did not introduce new methods which allow explicit calculations at zero kinetic energy or – even more difficult – in the intermediate region towards thermal energies. Wave-packet methods should yield reliable results. *Doyen* performed time-dependent wave packet calculations – not using the gettering Hamiltonian but the full Hamiltonian in the low kinetic energy limit [8.164]. The results indicated high probabilities for inelasticity approaching unity.

Wave packet calculations are difficult and expensive, if performed with the necessary accuracy. Therefore a renormalized Lippmann-Schwinger formalism, designed to be capable of handling gettering Hamiltonians, was suggested [8.33,164]. The results were encouraging both in camparison with the wave packet calculations and with experiment.

### 8.4.4 Quantum Partitioning Techniques

The theoretical treatment of dynamical processes at surfaces leads to a complicated many-body problem, which is illustrated in Fig. 8.3. If a gas atom approaches a surface, it experiences first an average potential created by the substrate atoms vibrating around their equilibrium position. When climbing up the repulsive part of the static potential it exerts a force on the surface atoms leading to a displacement of the equilibrium position of the vibrating atoms. During this deformation the gas atom obtains an energetically favourable distance to more surface atoms than is possible in the static case, i.e., the potential

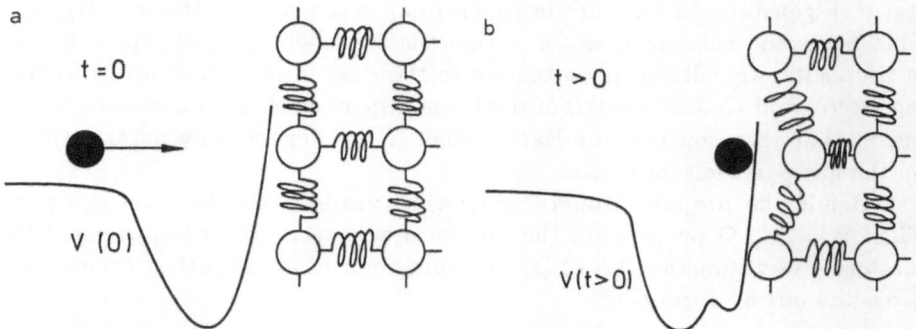

**Fig. 8.3. (a)** Rigid lattice and corresponding static gas atom potential (schematic). **(b)** The local deformation of the surface lattice upon impact of a gas atom and the corresponding phonon adiabatic potential (schematic)

energy of the gas atom is decreased. This happens only for a short period of time. By recoil the gas atom is pushed back into the vacuum or, if it lost enough energy before, it is trapped in the adsorption well. The surface deformation relaxes then.

Energy and momentum transfer occur during the deformation. This local and transient process is therefore the most important but also most difficult part in a theory.

The fundamental idea, which permits a quantum-theoretical solution of acceptable accuracy, is to split off this local problem from the rest of the Hamiltonian and to first investigate it separately. If then a solution of this local deformation problem has been found, it will afterwards be coupled to the translational movement of the gas atom or — to put it in other words — it will be embedded into the continuum of the scattering states.

To describe this formally the Hamiltonian is divided up into three parts:

$$H = H_{nonloc} + H_{loc} + H_{loc-nonloc} \ . \tag{8.53}$$

$H_{nonloc}$ contains the static potential, which does not support bound states. The eigen states of $H_{nonloc}$ are the scattering states of the static potential.

The local Hamiltonian $H_{loc}$ describes the interacting gas atom-surface system, when the gas atom is near the surface and coupling to the elementary excitations of the solid. The eigenstates describe the excitation spectrum with the gas atom adsorbed, i.e., the spectrum will include adsorbate induced modes.

The local and nonlocal Hamiltonians are connected by the interaction term $H_{loc-nonloc}$.

This partitioning was used in [8.165] to study the role of substrate electronic excitations. They use a formulation of reactive scattering due to *O'Malley* [8.166,167] which is based on *Feshbach*'s theory of resonant scattering [8.168]. The eigenstates of $H_{loc}$ (denoted $H_{el}^f$ by *Metiu* and *Gadzuk*) which have the molecular system bound to the surface they call product states. The eigenstates of $H_{nonloc}$ (denoted $H_{el}'$ by *Metiu* and *Gadzuk*) in which the binding of the molecule to the surface is not allowed they call intial states or reactant states. Eigenstates of $H_{loc}$ are in general non-orthogonal to those of $H_{nonloc}$. This has some consequences for a rigorous treatment [8.169]. For the sake of simplicity, we will not pay attention to these technical details in the following. *Metiu* and *Gadzuk* worked in the Born-Oppenheimer approximation which means that they consider the Hamiltonian $H_{el}$ to depend only parametrically on the gas particle coordinates.

Define the projection operator $Q$ which projects on the local subspace. Then $P = 1 - Q$ projects on the non-local subspace. The component of the scattering wave function $|k\beta +\rangle$ (k: gas atom momentum; $\beta$: surface excitations) projected out by $P$ satisfies

$$(PHP + V_{opt} - E_{k\beta})P|k\beta + \rangle = 0 \ . \tag{8.54}$$

The generalized optical potential is here defined as

$$V_{opt} = PHQG^+(E_{k\beta})QHP \ . \tag{8.55}$$

$G^+$ is the Green operator

$$G^+(E) = (E - H + PHQ + QHP)^{-1} \ . \tag{8.56}$$

The optical potential is complex. Its real part has the meaning of a level shift, its imaginary part gives the state $P|k\beta + \rangle$ a finite life time which signals a possible decay of the asymptotic state $|k\beta\rangle$ into eigenstates of $H_{loc}$. Formally transition matrix elements between eigenstates of $H_{loc}$ and $H_{nonloc}$ assume then the form:

$$\langle \ell\alpha|H_{loc-nonloc}|k\beta + \rangle = \langle \ell\alpha|V_{opt}[1 + G^+(E_{k\beta})V_{opt}]|k\beta\rangle \ . \tag{8.57}$$

Here we have used the approximation $PH_{loc-nonloc}P = 0$ with the understanding that $PHQ + QHP = PH_{loc-nonloc}Q + QH_{loc-nonloc}P$. Care is needed when using this matrix element for calculating sticking probabilities, because of the problematics with the definition of the final state for sticking, as discussed in Sect. 8.4.3. A literal transcription of the formalism by *O'Malley* as performed by *Gadzuk* and *Metiu* requires that $PHQ + QHP$ is the potential connecting the asymptotic final states to the residual system.

Under favourable conditions the above transition matrix element might split approximately into a product of several matrix elements which include integration over different degrees of freedom. Factors which are just overlap matrix elements of components of the total wave function are then called Franck-Condon factors in analogy to the related formalism of spectroscopy.

In the case of coupling to electronic substrate excitations the eigenstates of $H_{loc}$ and $H_{nonloc}$ are different from the customary adiabatic states. *Gadzuk* and *Metiu* called them "quasi-adiabatic" states. *Gumhalter* and *Davison* [8.170] followed this line and talk of "diabatic states" in reference to corresponding gas phase work.

*Doyen* and *Grimley* [8.156] applied the partitioning (8.53) to develop a quantum formalism for phonon induced adsorption and desorption. They use a basis set representation of the gas particle so that the gas atom, the phonons and the gas atom-phonon coupling are all conveniently treated in second quantization in an occupation number representation. The basis set contains localized and non-localized functions. The latter are used to define $H_{nonloc}$ and the former to define $H_{loc}$. $H_{loc-nonloc}$ contains both localized and non-localized basis functions, but no phonon operators. A Green's function formalism is used to embed $H_{loc}$ into the continuum of scattering states. For a microcanonical ensemble on the energy shell E the following expression for the desorption rate into the state $|k\beta\rangle$ is obtained:

$$2\pi|T_{desorp}|^2 P(E) = - \sum_{i,j} \langle k\beta|V_{desorp}|i\rangle \langle j|V_{desorp}|k\beta\rangle \ \text{Im}\{G_{ij}(E)\} \ . \tag{8.58}$$

The sum runs over all eigenstates $|i\rangle$, $|j\rangle$ of the local Hamiltonian $H_{loc}$. $V_{desorp}$ is here *not* $H_{loc-nonloc}$ but the coupling of the emitted phonons to the adparticle (Sect. 8.4.3). $G(E)$ is the *exact* Green's operator on the energy shell. In order to get the exact desorption rate we therefore have to evaluate the exact Green's functions $G_{ij}(E)$ in the subspace spanned by the eigenstates of $H_{loc}$. These Green's functions are many-body propagators and contain an embedding self-energy describing the influence of $H_{loc-nonloc}$.

For the transition probabilities the eigenstates of $H_{loc}$ are needed. These cannot be calculated exactly, and *Doyen* and *Grimley* suggested a self-consistent approximation. To outline the idea of this approximation consider a single vibrational adatom state $|v\rangle$ near the surface and a single phonon mode $\alpha$. The adatom-phonon interaction is assumed to be linear in the displacement of this phonon mode:

$$V_{A-ph} = gn_v(\alpha^\dagger + \alpha) \ . \tag{8.59}$$

g is the coupling constant which will be discussed later. $n_v$ is the occupation number operator for the adatom vibrational state. In a self-consistent approximation $V_{A-ph}$ is written as an effective static coupling where the surface atoms move in an averaged static potential determined by the position probability of the gas atom, i.e., $\langle n_v\rangle$. On the other hand, the gas atom moves in an average potential given by the average position $\langle \alpha^\dagger + \alpha\rangle$ of the substrate atoms. This means taking into account the same interaction twice and therefore the product of the expectation values has to be subtracted once

$$V_{A-ph} \longrightarrow V^{sc}_{A-ph} = g(\langle n_v\rangle(\alpha^\dagger + \alpha) + \langle \alpha^\dagger + \alpha\rangle n_v - \langle n_v\rangle\langle \alpha^\dagger + \alpha\rangle) \ . \tag{8.60}$$

This self-consistent version of the local Hamiltonian can be diagonalized by a canonical transformation to a displaced phonon mode $\beta$

$$\beta = \alpha + g\langle n_v\rangle/\omega \tag{8.61}$$

$\omega$ is the frequency of the phonon mode. The result is

$$\omega\alpha^\dagger\alpha + V^{sc}_{A-ph} = \omega\beta^\dagger\beta + g^2|\langle n_v\rangle|^2/\omega - 2g^2\langle n_v\rangle n_v/\omega \ . \tag{8.62}$$

One has obtained a decoupled, displaced phonon mode plus a constant shift of the static potential and a correction to the self-energy of the gas atom. For the gas atom it means a decrease of the effective potential as compared to the static case. One might call this a mass renormalization or − because of the close analogy to the theory of the image force interaction [8.171] − a "phonon image force".

A real solid contains a continuum of phonon modes which apparently renders a straightforward generalisation of the above self-consistent approach impractical. Now the interaction of a gas atom with the surface has local character and it is obviously inefficient to redescribe this local coupling in terms of collective excitations. *Lin* and *Adelman* [8.172] suggested therefore a further partitioning of the local Hamiltonian

$$H_{loc} = H_{prim} + H_{bath} + H_{loc-bath} \ . \tag{8.63}$$

The first term describes the "primary zone" which comprises those surface atoms which directly couple to the gas-particle. The second term is the remainder of the solid and contains the collective excitations. The third part represents the heat bath-solid interaction. These terms can be treated with different levels of approximation according to their different physical significance. *Lin* and *Adelman* proposed to handle the gas-particle interaction with $H_{prim}$ in a close-coupling framework and to treat the rest in a single-phonon approximation.

The above one-phonon mode approximation is now interpreted as describing only $H_{prim}$. $\alpha$ need not mean a single surface atom then. Consider an Einstein solid where the coupling of the gas particle to all phonon modes is known

$$V_{A-ph} = n_v \sum_q g_q(b_q^\dagger + b_q) \ . \tag{8.64}$$

The primary, localized phonon mode can then be defined as

$$\alpha = g^{-1} \sum_q g_q b_q \quad \text{with} \quad g = (\sum_q |g_q|^2)^{1/2} \ . \tag{8.65}$$

$\alpha$ has amplitude on many surface atoms near the point of impact. It can now be demonstrated mathematically [8.173] that it is possible to orthogonalize all the other delocalized phonon modes of the solid in such a way that they do no longer couple to the gas particle. Complete decoupling is only achieved for an Einstein solid. For a realistic phonon spectrum with dispersion one is left with an interaction between the primary mode and the orthogonalized modes. This residual interaction can be handled within the Green's function formalism by introducing a complex self-energy for the primary phonon mode.

Numerical calculations within the described framework have been performed for sticking and desorption [8.174]. In order to evaluate the eigenstates of $H_{loc}$ a set of ca. 20 localized gas atom states was introduced. The primary zone contained a single-phonon mode with a complex self-energy. Out of the self-consistent local states only few show usually a strong "phonon image force" and are important for sticking and desorption. These self-consistent local states are not eigenstates of the total Hamiltonian, not even approximately. They exhibit a considerable width due to coupling to both the phonon continuum and the gas atom scattering continuum. They are therefore transient states and are called 'deformation resonances'.

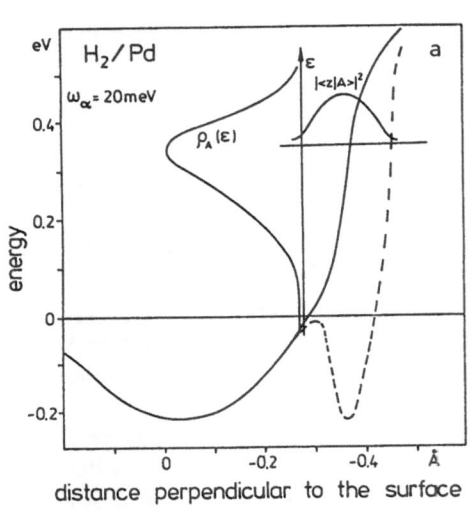

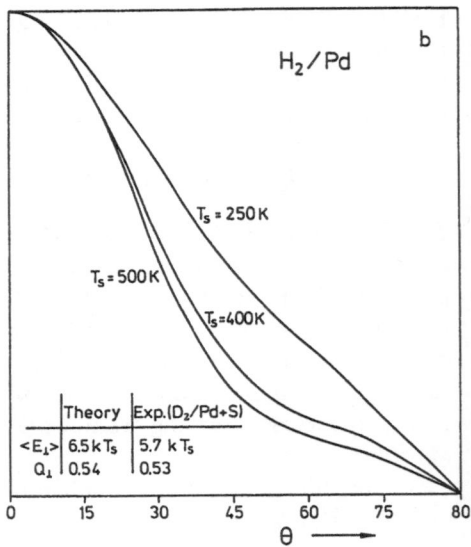

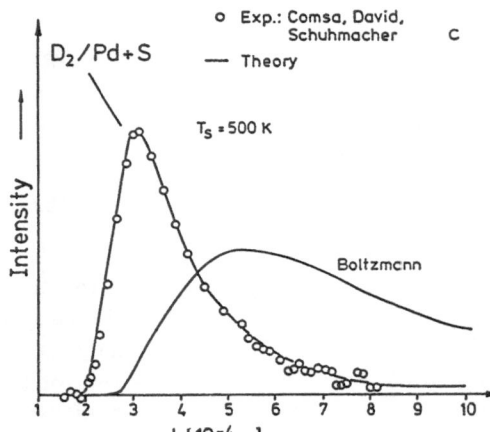

**Fig. 8.4. (a)** Deformation resonance in real space, calculated from the displayed static potential (——) with parameters chosen for $H_2/Pd$. The *dotted curve* is the phonon-adiabatic potential (cf. text). $\omega_\alpha$ is the effective phonon frequency. **(b)** Desorption angular distribution for various substrate temperatures calculated from the deformation resonance of Fig. 8.4a. $\langle E_\perp \rangle$ denotes the mean kinetic energy perpendicular to the surface and $Q_\perp$ the normalized speed ratio for the perpendicular motion. **(c)** Velocity distributions for perpendicular and parallel (glancing) desorption calculated from the deformation resonance of Fig. 8.4a. The *circles* represent the experimental data (perpendicular desorption) of [8.58]

An example – parametrized to describe $H_2$ interacting with palladium – is shown in Fig. 8.4a. It displays the assumed static potential (full curve) and the calculated "phonon-adiabatic" potential. For the latter the gas molecule is kept fixed at the distance given by the abcissa and the corresponding deformation of the surface, and the resulting effective gas atom potential (dotted curve) is evaluated. Within the plotted range of kinetic energies only one self-consistent local state $|A\rangle$ implies a considerable polarisation of the surface. The distribution of this state is exhibited in configurational space and in energy space. It is roughly centered in real space where the phonon-adiabatic potential has its minimum.

The resulting angular and velocity distributions for desorption are given in Figs. 8.4b and c, respectively. Comparison with the experimental data [8.175] is very satisfactory, but the static potential was varied to obtain this fit. The calculated sticking coefficient is 0.03 at an incoming kinetic energy of 45 meV. Experimental values for the initial sticking coefficient (0.13 (Pd (poly)) [8.176], 0.5 (Pd(100)) [8.177], 0.1 (Pd(111)) [8.178]) are for dissociative adsorption and are not relevant for the trapping in the physisorption well to which the theoretical value applies. A sticking probability of ca. 0.15 has been measured for $H_2$ *physisorbed* on low-temperature copper [8.179].

### 8.4.5 Non-Adiabatic Adsorption and Desorption

Phonon inelasticity is commonly treated on adiabatic potential energy surfaces. There are, however, important processes – such as electronic energy exchange, charge transfer and quenching of excited electronic adsorbate states – which occur via transitions between different adiabatic hypersurfaces. The most widely used approach for treating these non-adiabatic effects are the single trajectory approximations. They have been briefly discussed in Sect. 8.3.2, where it was outlined that they have severe limitations when applied to sticking and desorption in the thermal energy range.

A considerable improvement over these crude approaches is given by the "surface-hopping trajectory approach" [8.92]. Here the trajectories are allowed to split into branches, each following a particular potential energy surface. The splitting occurs at "hopping seams" which are taken along surface crossings or avoided crossings. Each branch is assigned a weight, which is computed from the electronic Schrödinger equation or by a semiclassical approximation [8.180]. This scheme works well for gas-phase reactions [8.181]. At metal surfaces the electron-hole pair mechanism provides a continuum of potential energy surfaces and the approach cannot be directly applied without further refinement. *Gadzuk* adapted it to a class of atomic scattering processes at solid surfaces involving a substrate-induced diabatic transition on the incident atom [8.182].

An interesting aspect is that electron-hole pairs of low energy may be "bosonized", i.e., it is possible to define pair excitation operators ("Tomonagons") which approximately obey boson commutation relations [8.183]. Via this procedure it would be possible to utilize the quantum-phonon theories for non-adiabatic effects by just changing the coupling constants and the density of states of the excitations. This approach has not been followed, however, whereas it is quite popular to combine the bosonization with the single-trajectory approximation [8.102].

The exact quantum-mechanical equations for the coupled nuclear-electronic motion were derived more than 30 years ago by Born [8.184]

$$[T_R + E_j(R) - \varepsilon]f_j(R) = - \sum_k C_{kj}(R)f_k(R) . \tag{8.66}$$

This is an infinite system of coupled equations. j and k label the adiabatic potential energy surfaces or, equivalently, the pair excitations. $T_R$ is the kinetic energy of the gas atom. $E_j(R)$ is the total *electronic* energy in the $j^{th}$ electronic state ($j^{th}$ adiabatic potential energy surface) for fixed gas atom distance R. $f_j(R)$ is the nuclear wave function on the $j^{th}$ adiabatic potential energy surface. The coupled equations are derived from the following ansatz for the many-body wave function:

$$W(r, R) = \sum_j f_j(R)g_j(r, R) \ . \tag{8.67}$$

$g_j(r, R)$ is the electronic Born-Oppenheimer wave function depending parametrically on R. The non-adiabatic coupling operators are then given by

$$\begin{aligned} C_{kj}(R) &= \langle g_k|T_R|g_j \rangle - \langle g_k|\nabla_R/m|g_j \rangle \nabla_R \\ &= -B_{kj}(R)/(2m) - (A_{kj}(R)/m)\nabla_R \ . \end{aligned} \tag{8.68}$$

$T_R g_j(r, R) = -\nabla_R^2/(2m)g_j(r, R)$ is a function $h_j(r, R)$ depending on both the electronic coordinates r and the gas atom position R. Hence: $\langle g_k|T_R|g_j \rangle = \int dr\, g_k^*(r, R)h_j(r, R)$.

These equations are difficult to solve, because the index j is continuous and because the evaluation of the coupling quantities $C_{jk}$ implies solving the electronic chemisorption problem at each gas atom distance R.

*Brivio* and *Grimley* were the first to attack this task [8.185] by introducing the following simplifications:

i)    The gas atom motion is one-dimensional.

ii)   The gas atom functions $f_j(R)$ are normalized over a finite length L so that j becomes discrete.

iii)  The chemisorption problem is treated for a one-dimensional chain of metal atoms using the tight-binding approximation and the *Grimley-Pisani* model [8.186]. The excited electronic states are described in Koopmans' approximation, i.e., relaxation effects are neglected.

iv)   Matrix elements of $T_R$ are neglected: $\langle g_k|T_R|g_j \rangle = 0$.

v)    Only transitions between adiabatic potential energy surfaces are taken into account which differ by a single electron-hole pair excitation. This is a consequence of the Koopmans' approximation.

vi)   They evaluate the self-energy of the asymptotic state (gas atom far from the surface) only in lowest order of the non-adiabatic coupling.

Due to the perturbation character of the calculation the results were non-unitary, but the conclusion was that non-adiabatic coupling can lead to large sticking coefficients of the order of unity for hydrogen atoms impinging on a metal surface [8.187].

A similar calculation was performed for He-atoms scattering from a copper surface approximated by a Sommerfeld model [8.188]. The total inelastic probability was $10^{-5}$. This small value is due to the low substrate charge density at the turning point of the He atom. As the excitation probability of electron-hole pairs due to the long-range part of the He-metal potential was found to be also extremely small [8.189], the conclusion is that non-adiabatic effects are not important for rare gas scattering from metal surfaces.

The physical reason for the qualitative different behaviour might perhaps be revealed with the help of the qualitative discussion given by *Norskov* and *Lundqvist* [8.100]. They propose that the affinity level of the adsorbate plays the decisive role. Due to the image potential this level is shifted downwards when the gas particle approaches the surface and could move below the Fermi level becoming then partially filled. If the affinity level is narrow when crossing the Fermi level, the electronic system might not have time to follow the change adiabatically and crossing to a different adiabatic potential-energy curve occurs. However, this picture does not straightforwardly emerge from the calculations by *Brivio* and *Grimley*.

The described calculations have all the difficulty mentioned in Sect. 8.4.3 that the non-adiabatic coupling is not switched off in the asymptotic final (stuck) state as assumed by standard scattering theory. A way out is to formulate the scattering problem for the metal electrons. This point of view was adopted by *Kirson* et al. [8.190,191] and *Crisa* et al. [8.192]. The probability for energy transfer from the gas atom to the metal has then to be obtained by a properly weighted sum over all possible single electron scattering processes. The results are very similar to those obtained with the conventional methods.

An interesting point is to estimate the relative importance of the phonon and electron-hole pair mechanism. All existing theories consider only one of the two channels. If such a theory predicts sticking coefficient unity for say the non-adiabatic coupling, it is still possible that non-adiabaticity will turn out to be negligible, if both mechanisms are included in a unified theory. To see this, consider (8.39) for the definition of the sticking coefficient

$$s = \frac{(1 - r_{el})(k_{ph} + k_{na})}{k_{refl}^{ph} + k_{ph} + k_{refl}^{na} + k_{na}} . \qquad (8.69)$$

The meaning of the symbols is as follows: $r_{el}$ elastic reflection coefficient; $k_{ph}$ phonon-induced sticking rate; $k_{refl}^{ph}$ rate for inelastic reflection induced by phonons; $k_{na}$ non-adiabatic sticking rate; $k_{refl}^{na}$ rate for non-adiabatic inelastic reflection. According to the *Gell-Mann* and *Goldberger* theorem [8.155] the total inelastic rate can be written as the sum of a purely non-adiabatic rate $(k_{na} + k_{refl}^{na})$ and a contribution $(k_{ph} + k_{refl}^{ph})$ which contains both the pure phononic and the mixed phononic-non-adiabatic inelasticity. The sticking probability in a theory which neglects phonons is

$$s_{na} = \frac{(1 - r_{el})k_{na}}{k_{refl}^{na} + k_{na}} . \qquad (8.70)$$

Now assume that $k_{refl}^{na} \ll k_{na}$ and $k_{na} \ll k_{ph}$; $k_{refl}^{na} \ll k_{refl}^{ph}$. Then $s_{na} = (1 - r_{el})$ which can be of the order of unity. The total sticking coefficient is, however, for the assumed relative orders of magnitude solely determined by the phonon induced rates. From these considerations it is even conceivable that a purely non-adiabatic theory predicts a sticking probability near unity, whereas a complete theory would yield a negligible sticking coefficient due to the dominance of the phonon induced rates.

Theoretical attempts have been made to estimate the relative role of phonons and electron-hole pairs by *independently* evaluating phonon rates and non-adiabatic rates [8.193,194]. The calculations are based on a one-dimensional model with a Morse potential for the adsorption well and applying first-order perturbation theory. The electromagnetic field absorption approach was used for the electron-hole pair mechanism. For the considered examples (Ar/W and CO/Cu) the phonon rates were found to be several orders of magnitude larger than the non-adiabatic rates.

*Feibelman* [8.199] pointed out that in the "harpooning" mechanism of *Norskov* and *Lundqvist* [8.100] more energy is transfered, if the local density of states at the Fermi level is lowered. Electronic structure calculations [8.200] have revealed that this might be achieved by introducing low coverages of P, S or Cl atoms. When non-adiabaticity of this special form is dominant, an increase in sticking should be observed.

As another possibility of deciding this question experimentally, *Gadzuk* [8.182] proposed an experiment which is supposed to utilize the similarities of alkali ions and the noble-gas atoms which are adjacent in the periodic table. The similarity concerns the mass as well as the electronic structure which is close shell. Gadzuk expected a similar hard wall-like repulsion at the surface in both cases, although it is quite difficult to believe that the additional charge on the alkali atom should not influence the repulsive part of the potential. The argument continues that for a similar repulsive potential the phonon excitation probabilities should be similar as well and the alkali sticking probability can be estimated by integrating over the noble-gas scattering distribution up to a certain energy. If the integration does not yield the sticking coefficient, then the phonon mechanism cannot be operating and non-adiabatic effects have to be important.

*Robota* et al. [8.53] used the reverse argument in their experiment. Scattering $H_2$ and $D_2$ with equal incoming *velocity* should yield identical scattering distributions, if the electron-hole pair mechanism is operating. This conclusion rests on the Norskov-Lundqvist mechanism according to which the velocity of the incoming particle is the crucial parameter under identical electronic conditions. This velocity has, however, to take into account the acceleration over the adsorption well [8.24]. The well depth can be estimated to be roughly 40 meV [8.195], but it is, of course, unknown at which distance and with which velocity the affinity level crosses the Fermi level (if at all). In the experiment the asymptotic velocities were chosen equal. It is found that a substantially lower fraction of $D_2$ than $H_2$ undergoes elastic scattering. Phonon interaction is expected to depend strongly on the mass of the scattering particle and the conclusion is therefore that the direct inelastic scattering is phonon induced.

The results showed, however, no evidence for phonon-induced sticking into the molecular adsorption well on clean surfaces, in marked contrast to the situation for $H_2$ on copper [8.179].

*Hellsing* gave recently an interesting analysis of the electronic mechanism for desorption [8.196]. In analogy to the classical work by *Montroll* and *Shuler* [8.197] he studied the master equation for an adsorbed gas atom in a truncated harmonic oscillator potential [8.124]. During the desorption process the gas atom climbs up the ladder of bound state levels. If it reaches the continuum, it is desorbed, i.e., recapturing events are neglected. This is a rather simplified picture which has well-known limitations, but is sufficient to demonstrate some concepts. Solving the master equation, *Hellsing* derived an Arrhenius form for the desorption rate constant. The prefactor is given by the transition probability per unit time to the continuum state. (There is only a single continuum state in the model.) Estimating and comparing the prefactors for the electromagnetic field mechanism and the electron-transfer mechanism, he finds that the latter dominates in the case of a hydrogen atom adsorbed on jellium by three orders of magnitude. The conclusion is that the electromagnetic field mechanism (even if corrected with an effective coupling charge) must not be applied to chemisorption systems as done in [8.193]. The same relative orders of magnitude for the two mechanisms were found earlier to apply for vibrational damping of adsorbed molecules [8.198].

## 8.5 Concluding Remarks

The selected problems of Sect. 8.4 all deal with topics which are expected to pose challenges to the scientists in the future. A systematic investigation of constraints in gas-surface interaction is still missing. Only the first steps have been taken and a considerable impact of this approach especially on the interpretation of experimental data appears possible.

Section 8.4.2 revealed that there have been some recent attempts in understanding the long established phenomenon of the compensation effect. Progress in this field might also depend on the availability of more experimental data with reliably and independently evaluated frequency factors and adsorption energies.

It will be obvious to the reader that the sticking problem is far from being resolved and more effort has to be invested here. The electron-hole pair versus phonon problem still awaits its first unified treatment. The discriminating experimental data is more likely to be provided once the theoreticians are able to specify what they need.

The often stated dilemma of the lack of relevant potential energy surfaces will make a judgement of the quality of the dynamical theories difficult. Progress with the potential energy surfaces might be slower because of the enormous computational efforts involved.

New kinds of experiment are likely to be performed. *Reyes* et al. [8.201] suggested an adsorption spectroscopy by polariton-induced desorption (AS-

341

PID). Because one may create nearly monoenergetic surface phonon polaritons, adatom momenta and band structures could be analyzed. More detailed experimental information about vibrational distributions of desorbing molecules and about the role of specific gas particle states in adsorption and desorption processes can be expected for the near future.

Theoretical and experimental progress will be needed in order to gain a quantitative understanding of adsorption and desorption phenomena. On the theoretical side various general and abstract formulations of the basic scattering problem continue to appear [8.202], but application of the existing methods to realistic systems and critical comparison with experiment will most probably lead to a more rapid increase in understandig the elementary steps.

# References

8.1     W. Schommers, P. v. Blanckenhagen (eds.): *Structure and Dynamics of Surfaces I*, Topics Curr. Phys., Vol. 41 (Springer, Berlin, Heidelberg 1986)
8.2     G. Ertl: Ber. Bunsenges. Phys. Chem. **86**, 425 (1982)
        G. Ertl: Surf. Sci. **89**, 525 (1979)
        G. Ertl: Critical. Rev. Solid State Mater. Sci. **10**, 349 (1983)
        G. Ertl: Catal. Rev. Sci. Eng. **21**, 201 (1980)
8.3     R.R. Cavanagh, D.S. King: In *Chemistry and Physics of Solid Surfaces V*, ed. by R. Vanselow, R. Howe, Springer Ser. Chem. Phys., Vol. 35 (Springer, Berlin, Heidelberg 1984) p. 141
8.4     J.A. Barker, D.J. Auerbach: Surf. Sci. Rpts. **4**, 1 (1984)
8.5     M.A. Morris, M. Bowker, D.A. King: In *Chemical Kinetics*, ed. by C.H. Bamford, C.F.H. Tipper, R.G. Compton (Elsevier, New York 1984) p. 1
8.6     N. Garcia, B.A. Barker, K.H. Rieder: Solid State Commun. **45**, 567 (1983)
8.7     J. Harris, A. Liebsch: Phys. Rev. Lett. **49**, 341 (1982)
8.8     R. Schinke, A.C. Luntz: Surf. Sci. **124**, L60 (1983)
8.9     J.K. Norskov, N.D. Lang: Phys. Rev. B **21**, 2136 (1980)
8.10    J.K. Norskov: Phys. Rev. B **26**, 2875 (1982)
8.11    N.J. Stott, E. Zaremba: Phys. Rev. B **22**, 1564 (1980)
8.12    J. Harris, A. Liebsch: J. Phys. C **25**, 2275 (1982)
8.13    D. Drakova, G. Doyen, F. von Trentini: Phys. Rev. B **32**, 6399 (1985)
8.14    Y. Takada, W. Kohn: Phys. Rev. Lett. **54**, 470 (1985)
8.15    J.K. Norskov, A. Houmoller, P. Johansson, B.I. Lundqvist: Phys. Rev. Lett. **46**, 257 (1981)
8.16    D. Tomanek, K.H. Bennemann: Surf. Sci. **127**, L111 (1983)
8.17    A.C. Diebold, G. Wolken: Surf. Sci. **82**, 245 (1979)
8.18    J.C. Tully: In *Many Body Phenomena at Surfaces*, ed. by D. Langreth and H. Suhl (Academic, Orlando, FL 1984) p. 377
8.19    F.O. Goodman: Surf. Sci. **24**, 667 (1971)
8.20    V. Celli, D. Evans: In *Dynamics of Gas Surface Interactions*, ed. by G. Benedek, U. Valbusa, Springer Ser. Chem. Phys., Vol. 21, (Springer, Berlin, Heidelberg 1982)
8.21    A.C. Levi, H. Suhl: Surf. Sci. **88**, 221 (1979)
8.22    K.H. Rieder, W. Stocker: Phys. Rev. B **31**, 3392 (1985)
8.23    J.M. Ziman: *Principles of the Theory of Solids* (Cambridge, Univ. Press, Cambridge 1964) p. 62
8.24    J.L. Beeby: J. Phys. C **4**, L359 (1971)
8.25    G. Armand, J. Lapujoulade, Y. Lejay: Surf. Sci. **63**, 143 (1977)
8.26    G. Armand, J.R. Manson: Surf. Sci. **80**, 532 (1979)
8.27    H.D. Meyer: Surf. Sci. **104**, 117 (1981)
8.28    J. Lapujoulade, Y. Lejay, N. Papanicolaou: Surf. Sci. **90**, 133 (1979)
8.29    J. Lapujoulade, Y. Le Cruer, M. Lefor, Y. Lejay, E. Maurel: Surf. Sci. **103**, L85 (1981)

8.30    J. Lapujoulade, Y. Lejay, G. Armand: Surf. Sci. **95**, 107 (1980)
8.31    H.U. Finzel, H. Frank, H. Hoinkes, M. Luschka, H. Nahr, H. Wilsch, M. Wonka: Surf. Sci. **49**, 577 (1975)
8.32    J.H. Craig, J.T. Dickinson: J. Vacuum Sci. Technol. **10**, 319 (1973)
8.33    H. Conrad, G. Doyen, G. Ertl, J. Küppers, W. Sesselmann: Chem. Phys. Lett. **88**, 281 (1982)
8.34    D.J. Auerbach, C.A. Becker, J.P. Cowin, L. Wharton: Appl. Phys. **14**, 141 (1977)
        J.E. Hurst, C.A. Becker, J.P. Cowin, K.C. Janda, L. Wharton, D.J. Auerbach: Phys. Rev. Lett. **43**, 1175 (1979)
8.35    C. Wang, R. Gomer: Surf. Sci. **84**, 329 (1979)
8.36    M. Sinvani, M.W. Cole, D.L. Goodstein: Phys. Rev. Lett. **51**, 188 (1983)
8.37    L.B. Thomas: In *Rarefied Gas Dynamics*, Vol. 1, ed. by C.L. Brundin (Academic, New York 1967) p. 155
8.38    T. Engel: J. Chem. Phys. **69**, 373 (1978)
8.39    G. Castro: Dissertation University of Munich (1982)
8.40    B.A. Joyce, C.T. Foxon: In *Comprehensive Chemical Kinetics*, Vol. 19, ed. by C.H. Banford, C.F.H. Tipper, and R.G. Compton (Elsevier, Amsterdam 1984) p. 181
8.41    H. Ibach, J.E. Rowe: Surf. Sci. **43**, 481 (1974);
        T. Sakurai, H.D. Hagstrum: Phys. Rev. B **12**, 5349 (1975)
8.42    T. Sakurai, M.J. Cardillo, H.D. Hagstrum: J. Vac. Sci. Technol. **14**, 387 (1977)
8.43    R. Dorn, H. Lüth, H. Ibach: Surf. Sci. **42**, 583 (1974)
8.44    R.J. Madix, M. Boudart: J. Catal. **7**, 240 (1967)
8.45    R.J. Madix, R. Korus: Trans. Faraday Soc. **64**, 2514 (1968)
8.46    B.A. Joyce, J.H. Neave: Surf. Sci. **27**, 499 (1971)
8.47    B.A. Joyce: Surf. Sci. **35**, 1 (1973)
8.48    H. Ibach, K. Horn, R. Dorn, H. Lüth: Surf. Sci. **38**, 433 (1973)
8.49    M.J. Cardillo, G.E. Becker: Phys. Rev. Lett. **40**, 1148 (1978)
8.50    R.L. Palmer, H. Saltsburg, J.N. Smith: J. Chem. Phys. **50**, 4661 (1969)
8.51    R. Sau, R.P. Merrill: Surf. Sci. **34**, 268 (1973)
8.52    S.L. Bernasek, W.J. Siekhaus, G.A. Somorjai: Phys. Rev. Lett. **30**, 1202 (1973)
8.53    H.J. Robota, W. Vielhaber, M.C. Lin, J. Segner, G. Ertl: Surf. Sci. **155**, 101 (1985)
8.54    G. Comsa: Proc. 7th Intern. Vac. Congr. 3rd Intern. Conf. Solid Surfaces, Wien (1977) p. 1317
8.55    E. Bauer, F. Bonczek, H. Poppa, G. Todd: J. Appl. Phys. **45**, 5164 (1974); Surf. Sci. **53**, 87 (1975)
8.56    H. Pfnür, P. Feulner, H.A. Engelhardt, D. Menzel: Chem. Phys. Lett. **59**, 481 (1978)
8.57    G. Comsa, R. David, K.D. Rendulic: Phys. Rev. Lett. **38**, 775 (1977)
8.58    G. Comsa, R. David, B.-J. Schuhmacher: Surf. Sci. **85**, 45 (1979); Surf. Sci. **95**, L210 (1980);
        G. Comsa, R. David: Surf. Sci. **117**, 77 (1982)
8.59    K. Christmann, G. Ertl: Surf. Sci. **60**, 365 (1976);
        R. McCabe, L. Schmidt: Proc. 7th Intern. Vacuum Congr. and 3rd Intern. Conf. Solid Surfaces, Vienna (1977) p. 1201
8.60    J.A. Serri, J.C. Tully, M.J. Cardillo: J. Chem. Phys. **79**, 1530 (1983)
8.61    R.A. Barker, A.M. Horlacher, P.J. Estrup: J. Vac. Sci. Technol. **20**, 536 (1982)
8.62    A. Mödl, H. Robota, J. Segner, W. Vielhaber, M.C. Lin, G. Ertl: J. Chem. Phys. **83**, 4800 (1985)
8.63    J. Segner, H. Robota, W. Vielhaber, G. Ertl, F. Frenkel, J. Häger, W. Krieger, H. Walter: Surf. Sci. **131**, 273 (1982)
8.64    D.A. Mantell, S.B. Ryali, G.L. Haller, J.B. Fenn: J. Chem. Phys. **78**, 4250 (1983);
        R.R. Cavanagh, D.S. King: Phys. Rev. Lett. **47**, 1829 (1981)
8.65    G.D. Kubiak, G.O. Sitz, R.N. Zare: J. Chem. Phys. **83**, 2538 (1985)
8.66    S. Glasstone, K.J. Laidler, H. Eyring: *Theory of Rate Processes* (McGraw-Hill, New York 1941)
8.67    H.A. Kramers. Physica **7**, 284 (1940)
8.68    H. Ibach, W. Erley, H. Wagner: Surf. Sci. **92**, 29 (1980)
8.69    A. Surda, I. Karasova: Report E17-80-343, Communication of the Joint Institute for Nuclear Research, Dubna (1980)
8.70    R.M. Logan, R.E. Stickney: J. Chem. Phys. **44**, 195 (1966)

8.71  R.M. Logan, J.C. Keck: J. Chem. Phys. **49**, 860 (1968)
8.72  J.D. Doll: J. Chem. Phys. **59**, 1038 (1973)
8.73  E.K. Grimmelmann: J.C. Tully, M.J. Cardillo: J. Chem. Phys. **72**, 1039 (1980)
8.74  R.J. Madix, R.A. Korus: J. Phys. Chem. Sol. **29**, 1531 (1968)
8.75  W.L. Nichols, J.H. Weare: J. Chem. Phys. **62**, 3754 (1975); J. Chem. Phys. **63**, 379 (1975)
8.76  Ch. Steinbrüchel: Chem. Phys. Lett. **76**, 58 (1980); Surf. Sci. **115**, 247 (1982)
8.77  J.D. McClure: J. Chem. Phys. **51**, 1687 (1969); **52**, 2712 (1970); **57**, 2810 (1972); **57**, 2823 (1972)
8.78  S.A. Adelman: Adv. Chem. Phys. **44**, 143 (1980)
8.79  J.C. Tully: Acc. Chem. Res. **14**, 188 (1981); Surf. Sci. **111**, 461 (1981)
8.80  J.C. Tully, M. Cardillo: Science **223**, 445 (1984)
8.81  R.W. Zwanzig: J. Chem. Phys. **32**, 1173 (1960)
8.82  F.O. Goodman, H.Y. Wachman: J. Chem. Phys. **46**, 2376 (1967)
8.83  J.A. Barker, D.R. Dion, R.P. Merrill: Surf. Sci. **95**, 15 (1980)
8.84  A. Gelb, M.J. Cardillo: Surf. Sci. **59**, 128 (1976); **64**, 197 (1977)
8.85  M. Balooch, M.J. Cardillo, D. Miller, R.E. Stickney: Surf. Sci. **46**, 358 (1974)
8.86  F.O. Goodman: Surf. Sci. **5**, 283 (1966)
8.87  I. Frenkel: Z. Phys. **26**, 117 (1924)
8.88  G. Armand: Surf. Sci. **9**, 145 (1968)
8.89  P. Jewsbury, J.L. Beeby: J. Phys. C **8**, 1541 (1975)
8.90  S. Holloway, J.L. Beeby: J. Phys. C **8**, 3531 (1975)
8.91  S. Holloway, P. Jewsbury: J. Phys. C **9**, 1907 (1976)
8.92  J.C. Tully, R.K. Preston: J. Chem. Phys. **55**, 562 (1971)
8.93  J.L. Beeby: J. Phys. C **5**, 3438 (1972); J. Phys. C **5**, 3457 (1972)
8.94  B. Bendow, S.C. Ying: Phys. Rev. B **7**, 622 (1973); Phys. Rev. B **7**, 637 (1873)
8.95  A.C. Levi: Nuovo Cimento, **54B**, 357 (1979)
8.96  W. Brenig: Z. Phys. B **36**, 81 (1979)
       J. Böheim, W. Brenig: Z. Phys. B **41**, 243 (1981)
8.97  R.J. Glauber: Phys. Rev. **98**, 1692 (1955)
8.98  J.I. Kaplan, E. Drauglis: Surf. Sci. **36**, 1 (1973)
8.99  E. Müller-Hartmann, T.V. Ramankrishnan, G. Toulouse: Solid State Comm. **9**, 99 (1971); Phys. Rev. B **3**, 1102 (1971)
8.100 J.K. Norskov, B.I. Lundqvist: Surf. Sci. **89**, 251 (1979)
8.101 R. Brako, D.M. Newns: Solid State Commun. **33**, 713 (1980)
8.102 K. Schönhammer, O. Gunnarson: In *Many Body Phenomena at Surfaces*, ed. by D. Langreth, H. Suhl (Academic, New York 1984) p. 421
8.103 E.G. d-Agliano, P. Kumar, W. Schaich, H. Suhl: Phys. Rev. B **11**, 2122 (1975)
       A. Blandin, A. Nourtier, D.W. Hone: J. Phys. Paris **37**, 369 (1976)
       A. Nourtier: J. Physique **38**, 479 (1977)
8.104 J.W. Gadzuk, H. Metiu: Phys. Rev. B **22**, 2603 (1980)
       Z. Crljen, B. Gumhalter: Phys. Rev. B **29**, 6600 (1984)
8.105 G. Drolshagen, E.J. Heller: J. Chem. Phys., **79**, 2072 (1983); Surf. Sci. **139**, 260 (1984)
       P.M. Agrawal, L.M. Raff: J. Chem. Phys. **77**, 3946 (1982)
       C.B. Smith, L.M. Raff, P.M. Agrawal: J. Chem. Phys. **83**, 1411 (1985)
8.106 D. Kumamoto, R. Silbey: J. Chem. Phys. **75**, 5164 (1981)
8.107 K.L. Sebastian: Chem. Phys. Lett. **81**, 14 (1981)
8.108 G. Armand, J. Lapujoulade: Proc. 11th Intern. Symp. Rarefied Gas Dynamics, Cannes (1978) p. 1329
8.109 W. Brenig, H. Kasai, H. Müller: Surf. Sci. **161**, 608 (1985)
8.110 J. Böheim: Surf. Sci. **148**, 463 (1984)
8.111 J.E. Lennard-Jones, A.F. Devonshire, C. Strachan: Proc. Roy. Soc. A **150**, 442; 456 (1935); A **156**, 6; 29; 37 (1936); A **158**, 242; 253; 269 (1937)
8.112 B. Gaffney, J.R. Manson: J. Chem. Phys. **62**, 2508 (1975)
       J. Manson, J. Tompkins: preprint
8.113 N. Cabrera, V. Celli, F.O. Goodman, R. Manson: Surf. Sci. **19**, 67 (1970)
8.114 M.D. Stiles, J.W. Wilkins: Phys. Rev. Lett. **54**, 595 (1985)
8.115 F.O. Goodman: Surf. Sci. **27**, 157 (1971); J. Chem. Phys. **55**, 5742 (1971); Surf. Sci. **111**, 279 (1981)

8.116  N. Garcia, J. Ibanez: J. Chem. Phys. **64**, 4803 (1976)
8.117  F.O. Goodman, N. Garcia: Phys. Rev. B **20**, 813 (1979)
8.118  W. Brenig: Z. Phys. B **36**, 227 (1980)
8.119  J. Böheim, W. Brenig: Z. Phys. B **48**, 43 (1983)
8.120  T.R. Knowles, H. Suhl: Phys. Rev. Lett. **38**, 1417 (1977), Erratum in Phys. Rev. Lett. **40**, 911 (1978)
8.121  Z.W. Gortel, H.J. Kreuzer, D. Spaner: J. Chem. Phys. **72**, 234 (1980)
        E. Goldys, Z.W. Gortel, H. Kreuzer: Surf. Sci. **116**, 33 (1982)
8.122  H.J. Kreuzer, Z.W. Gortel: *Physisorption Kinetics*, Springer Ser. Surf. Sci., Vol. 1 (Springer, Berlin, Heidelberg 1986)
8.123  F.O. Goodman, I. Romero: J. Chem. Phys. **69**, 1086 (1978)
8.124  B.J. Garrison, D.J. Diestler, S.A. Adelman: J. Chem. Phys. **67**, 4317 (1977)
8.125  W.L. Schaich: J. Phys. C **11**, 2519 (1978)
8.126  G. Wolken: J. Chem. Phys. **60**, 2210 (1974)
8.127  D.A. Micha: J. Chem. Phys. **74**, 2054 (1981)
8.128  D.M. Newns: Surf. Sci. **154**, 658 (1985)
8.129  G. Armand, J.R. Manson: Phys. Rev. Lett. **53**, 1112 (1984)
8.130  J.W. Gadzuk, U. Landman, E.J. Kuster, C.L. Cleveland, R.N. Barnett: Phys. Rev. Lett. **49**, 426 (1982)
8.131  U. Landman, G.G. Kleiman, C.L. Cleveland, E. Kuster, R.N. Barnett, J.W. Gadzuk: Phys. Rev. B **29**, 4313 (1984)
8.132  G. Doyen, G. Ertl, H. Robota, J. Segner, W. Vielhaber, F. Frenkel, J. Häger, W. Krieger, H. Walther: J. Vac. Sci. Technol. A **12**, 1269 (1983)
8.133  Y. Alhassid, R.D. Levine: Phys. Rev. A **18**, 89 (1978)
8.134  Y. Alhassid, R.D. Levine: J. Chem. Phys. **67**, 4321 (1977)
8.135  Y. Alhassid, R.D. Levine: Phys. Rev. C **20**, 1775 (1979)
8.136  R.D. Levine, A. Ben-Shaul: In *Chemical and Biochemical Applications of Lasers*, Vol. II (Academic, New York 1977) p. 145
8.137  R.D. Levine, R.B. Berstein: In *Modern Theoretical Chemistry*, Vol. III, ed. by W.H. Miller (Plenum, New York 1976) p. 323
8.138  E. Zamir, R.D. Levine: Chem. Phys. Lett. **104**, 143 (1983)
8.139  R.D. Levine, R. Silbey: J. Chem. Phys. **74**, 4741 (1981)
8.140  H.D. Meyer, R.D. Levine: Chem. Phys. **85**, 189 (1984)
8.141  E.T. Jaynes: Phys. Rev. **106**, 620 (1957)
8.142  G. Doyen: Vacuum **32**, 91 (1982)
8.143  J. Segner, C.T. Campbell, G. Doyen, G. Ertl: Surf. Sci. **138**, 505 (1984)
8.144  H.P. Steinrück, K.D. Rendulic, A. Winkler: Surf. Sci. **154**, 99 (1985)
8.145  F.H. Constable: Proc. Roy. Soc. Lond. Ser. A **108**, 355 (1925)
8.146  A. Galway: Adv. Catalysis **26**, 247 (1977)
8.147  E. Peacock-Lopez, H. Suhl: Phys. Rev. B **26**, 3774 (1982)
8.148  E. Sommer, H.J. Kreuzer: Phys. Rev. Lett. **49**, 61 (1982)
8.149  G. Doyen: In *Vibration at Surfaces*, ed. by A.A. Lucas (Plenum, New York 1982) p. 105
8.150  B.J. McCoy: J. Chem. Phys. **80**, 3627 (1984)
8.151  V.P. Zhdanov: Surf. Sci. **159**, L416 (1985)
8.152  G. Armand: Surf. Sci. **66**, 321 (1977)
8.153  J.L. Beeby, K. Agrawal: Surf. Sci. **122**, 447 (1982)
8.154  M.L. Goldberger, K.M. Watson: *Collision Theory* (Wiley, New York 1964)
8.155  M. Gell-Mann, M.L. Goldberger: Phys. Rev. **91**, 398 (1953)
8.156  G. Doyen, T.B. Grimley: Surf. Sci. **91**, 51 (1980)
8.157  G.P. Brivio, T.B. Grimley: Surf. Sci. **161**, L573 (1985)
8.158  P.A. Wolff: Phys. Rev. **124**, 1030 (1961)
8.159  P.W. Anderson: Phys. Rev. **124**, 41 (1961)
8.160  D.M. Newns: Phys. Rev. **178**, 1123 (1969)
8.161  G. Doyen: Phys. Rev. B **22**, 497 (1980)
8.162  F.O. Goodman, H.Y. Wachman: *Dynamics of Gas Surface Scattering* (Academic, New York 1976)
8.163  G. Doyen: In *Studies in Surface Science and Catalysis*, Vol. 9 (Elsevier, New York 1982) p. 1

8.164  G. Doyen: Surf. Sci. **117**, 85 (1982)

8.165  H. Metiu, J.W. Gadzuk: J. Chem. Phys. **74**, 2641 (1981)

8.166  T.F. O'Malley: Adv. Atomic and Mol. Phys. **7**, 223 (1971)

8.167  T.F. O'Malley: Phys. Rev. **150**, 4 (1966)

8.168  H. Feshbach: Ann.Phys. N.Y. **5**, 357 (1958); **19**, 287 (1962)

8.169  G. Doyen, F. von Trentini, T.B. Grimley: unpublished results

8.170  B. Gumhalter, S.G. Davison: Phys.Rev. B **30**, 3179 (1984)

8.171  A.C. Hewson, D.M. Newns: Jpn. J. Appl. Suppl. **2**, Pt. 2, 121 (1974)

8.172  Y.W. Lin, S.A. Adelman: J. Chem. Phys. **68**, 9 (1978)

8.173  F. Delanaye, A. Lucas, G.D. Mahan: Surf. Sci. **70**, 629 (1978)

8.174  G. Doyen: Habilitation thesis, University of Munich (1981) and unpublished results

8.175  G. Comsa, R. David, B.J. Schumacher: Surf. Sci. **95**, L210 (1980)

8.176  A.W. Aldag, L.D. Schmidt: J. Catal. **22**, 280 (1971))

8.177  J. Behm: Dissertation University of Munich (1980)

8.178  H. Kuipers: Dissertation University of Munich (1980)

8.179  S. Andersson, J. Harris: Phys. Rev. B **27**, 9 (1983)

8.180  W.H. Miller, T.F. George: J. Chem. Phys. **56**, 5637 (1972)

8.181  J.C. Tully: In *Many-Body Phenomena at Surfaces*, ed. by Langreth, H. Suhl (Academic, Orlando, FL 1984) p. 377

8.182  J.W. Gadzuk: Surf. Sci. **118**, 180 (1982)

8.183  S. Tomonaga: Progr. Theor. Phys. **5**, 544 (1950)

8.184  M. Born: Gött. Nachrichten math. phys. Kl. (1951) p. 1

8.185  G.P. Brivio, T.B. Grimley: J. Phys. C **10**, 2351 (1977); Surf. Sci. **89**, 226 (1979)

8.186  T.B. Grimley, C. Pisani: J. Phys. C **7**, 2831 (1980)

8.187  G.P. Brivio, T.B. Grimley: Surf. Sci. **131**, 475 (1983)

8.188  O. Gunnarson, K. Schönhammer: Phys. Rev. B **25**, 2514 (1982)

8.189  W.L. Schaich, J. Harris: J.Phys. F **11**, 65 (1981)

8.190  Z. Kirson, R.B. Gerber, A. Nitzan, Surf. Sci. **124**, 279 (1983)

8.191  Z.Kirson, R.B. Gerber, A. Nitzan, M.A. Ratner: Surf. Sci. **151**, 531 (1985)

8.192  N. Crisa, G. Doyen, F. von Trentini: Surf.Sci. **162**, 120 (1985)

8.193  G.E. Korzeniewski, E. Hood, H. Metiu: J. Vac. Sci. Technol. **20**, 594 (1982); J. Chem. Phys. **80**, 6274 (1984)

8.194  E. Hood, H. Metiu: In *Many-Body Phenomena at Surfaces*, ed. by D. Langreth, H. Suhl (Academic, Orlando, FL 1984) p. 533

8.195  A. Liebsch. J. Harris: Surf. Sci. **130**, L349 (1983)

8.196  B. Hellsing: J. Chem. Phys. **83**, 1371 (1985)

8.197  E.W. Montroll. K.E. Shuler: Adv. Chem. Phys. **1**, 361 (1958)

8.198  B.N.J. Persson, M. Persson: Solid State Commun. **36**, 175 (1980)

8.199  P.J. Feibelman: Surf. Sci. **160**, 139 (1985)

8.200  P.J. Feibelman, D.R. Hamann: Surf. Sci. **149**, 48 (1985)

8.201  J. Reyes, F.O. Goodman, M.W. Cole: Surf. Sci. **151**, 221 (1985)

8.202  B.H. Choi, R.T. Poe: J. Chem. Phys. **83**, 1330 (1985)

# 9. Many-Body Description of Surface Elementary Excitations. Application to Semiconductors

A. Muramatsu and W. Hanke

With 9 Figures

In this chapter a microscopic theory of surface elementary excitations and response functions is summarized. Particular emphasis is placed on surfaces for which the often-used jellium approximation is not valid. Starting from a general formulation in terms of the microscopic Maxwell equations or, formally, of the Bethe-Salpeter equation for the two-particle Green's function, an equation of motion for surface elementary excitations and the related response functions is obtained. This equation is solved in a local wave-function representation. It takes explicitly density fluctuations on a microscopic scale into account (due to the surface profile and the underlying crystal potential). Many-body effects of random-phase (RPA) and electron-hole type are included. Based on this theory a formulation of surface optical response is presented which allows for the inclusion of local-field and excitonic effects. In this way experimentally accessible quantities like the reflectivity can be calculated for a realistic surface, beyond the jellium or the three-medium approximations. Various applications of the present scheme to the special case of an ideal Si(111) surface are reviewed. These applications demonstrate that many-body interactions of excitonic (electron-hole) and (RPA) local-field character play a substantial role in the charge- and spin-density response. In particular, they can lead to instabilities of the ideal paramagnetic surface with respect to spin-density waves with a wavelength corresponding to the observed $(2 \times 1)$ and $(7 \times 7)$ superstructures. Other examples deal with an a-priori calculation of phonons and the electron-phonon interaction of the same surface system.

## 9.1 Background

The concept of elementary excitations provides an extremely useful device for the study of interacting many-particle systems. Landau's idea of quasi-particles [9.1] and the linear-response theory developed by *Pines* and *Nozières* [9.2] and others, proved to be very successful in describing quantum liquids and bulk systems where the homogeneous electron-gas model is applicable, e.g. nearly-free electron metals. However, for most of the systems of current interest, like semiconductors, insulators and transition metals, the inhomogeneity of the electronic charge density is rather important. In the density response of a periodic crystal this fact is reflected in the local-field effects which appear due to the periodicity of the charge distribution [9.3]. Thus, although the external field which couples to the electronic density is macroscopic with a wavelength much larger than the lattice constant, due to Bragg reflections microscopic fields with wavelengths on the scale of the inter-ionic distance are induced.

For periodic bulk systems a practical scheme for calculating the response or the corresponding two-particle Green's function has been devised, which takes the microscopic charge fluctuation of the solid into account. A very essential point of this work has been to demonstrate the intimate interrelation between the localization (for example, of electronic states in transition metals) or the inhomogeneity of the electronic charge distribution and the importance of many-body effects. This interrelation has been substantiated in a wide spectrum of bulk elementary excitations ranging from single- [9.4] and two-particle (optical) [9.5] to collective (plasmon) and phonon excitations as well as broken symmetry behavior, like superconductivity and charge-density wave (CDW) states [9.3].

In the presence of surfaces, the study of many-body effects becomes even more complicated, as demonstrated already by the jellium model, where the breaking of translational invariance in one direction prevents a simple theoretical treatment. Only recently [9.6,7] promising results were obtained for an understanding of surface experiments in nearly-free electron systems, starting from this model. Another source of difficulties is the fact that correlation effects become more important as a result of the charge inhomogeneity inherent to the surface system. For example, it has been shown [9.8] for a jellium slab with infinite barriers that the correlation energy is more dominant at the surface $[\sim 1/(r_s)^{1/2}]$ than in the bulk $(\sim \ln r_s)$ in the high-density limit $(r_s \to 0)$. This implies that a careful study of the many-body effects is needed. In the presence of the atomic lattice, we expect on the ground of our experience in the bulk, that the role of many-body interactions is even more enhanced due to the appearance of local-field effects on top of the non-locality induced by the surface.

We review, in this chapter, a theory aiming at a microscopic description of surface elementary excitations on the same level attained in bulk [9.4,5]. Starting from the lattice structure, and the quantum-mechanical eigenenergies and wave functions of the electronic system, the correlation and response functions are constructed within the frame of quantum statistical mechanics.

In Sect. 9.2 we outline a local treatment of the two-particle Green's function for the surface. Many-body effects are included through the screened electron-hole attraction and its exchange counterpart, the RPA local-field effects. From the Bethe-Salpeter equation we derive an equation of motion for the electron-hole pair in a local representation which allows for a practical determination of the energy spectrum and the amplitudes of the collective excitations. The surface response function is obtained within the same scheme.

In Sect. 9.3 we discuss extensively the optical response of a crystalline surface. To this we follow [9.9] and extend previous work for jellium surfaces [9.6] to the case of a realistic one. With the help of the formalism developed in Sect. 9.2, local-field effects due to charge-density inhomogeneities both parallel and perpendicular to the surface are included. The macroscopic dielectric tensor can then be obtained and the corrections of the classical formulae for reflectivity are presented. The quantities involved are also important for other surface problems like dynamical image forces and surface power absorption [9.6].

Section 9.4 is devoted to the formulation of a microscopic theory of electron-phonon interaction and lattice dynamics on the surface, the main ingredient being the density response function obtained in Sect. 9.2.

Finally, we review in Sect. 9.5 applications of the theory of elementary excitations presented in Sect. 9.2 to the case of an ideal Si(111) surface. These calculations are intended (a) to demonstrate the ability of the formal scheme to handle realistic crystalline surfaces, and (b) to reveal the importance of many-body interactions for semiconductor surfaces.

Results of quantitative calculations of the charge- and spin-density response function of the Si(111) surface establish the importance of including both excitonic (electron-hole) and (RPA) local-field many-body interactions. In particular, they lead to an instability of the ideal paramagnetic surface with respect to spin-density waves (SDW) with wavelength corresponding to the observed $(2 \times 1)$ and $(7 \times 7)$ superstructures. Another example deals with an a-priori calculation of the phonons and the electron-phonon interaction of the same surface system. Various results of the theory such as phonon softening due to the coupling of the charge-density fluctuations to the lattice are summarized and general aspects of the importance of many-body effects for the a-priori determination of surface structures via elementary excitations are discussed.

## 9.2 Elementary Excitations on Surfaces

The most general theoretical framework for a microscopic study of elementary excitations of a quantum-mechanical system, irrespective of its symmetry, is given by the theory of Green's functions [9.1–3,10]. In particular, if we are interested in collective excitations (plasmons, excitons, magnons, etc.), the two-particle Green's function which describes the evolution of an electron-hole pair is the natural quantity whose analytical structure we should study [9.10]. Specifically, it determines completely the dielectric response of the system and its poles correspond to the collective normal modes of the system.

In order to obtain the two-particle Green's function, an integral equation, the Bethe-Salpeter equation [9.3,10], has to be solved. Although a general solution is impossible, a local-orbital treatment of the wave functions allows for a formally exact solution for the two-particle Green's function within the time-dependent screened Hartree-Fock approximation (TDSHFA). This approach proved to be extremely successful in bulk for a quantitative explanation of both collective [9.3,5] and single-particle properties [9.4] of semiconductors.

We summarize in Sect. 9.2.1 and 2 an equivalent treatment of the two-particle Green's function for a surface system. This formalism enables us to obtain an equation of motion for the electron-hole pair which determines the energy spectrum and amplitudes (wave functions) of the collective excitations.

### 9.2.1 Homogeneous Bethe-Salpeter Equations

We discuss in this subsection the homogeneous Bethe-Salpeter equation as a preliminary step for an equation of motion of elementary excitations.

The two-particle Green's function is defined as usual [9.10]

$$G(1, 2; 3, 4) = \langle T_\tau \, \tilde{\psi}(1)\tilde{\psi}(2)\tilde{\psi}^\dagger(3)\tilde{\psi}^\dagger(4) \rangle \ , \tag{9.1}$$

where $1 = (\boldsymbol{r}_1, \tau_1)$, and

$$\tilde{\psi}(1) = \exp[(H - \mu N)\tau_1]\psi(\boldsymbol{r}_1)\exp[-(H - \mu N)\tau_1] \ , \tag{9.2}$$

$\mu$ is the chemical potential, N the particle number and $T_\tau$ the time-ordering operator defined for the temperature Green's function [9.10]. Finally, H is the full Hamiltonian with interparticle interactions. In the following it is always understood that we consider only the limit $T \rightarrow 0$. We suppose the system periodic in directions parallel to the surface.

$G(1, 2; 3, 4)$ can be divided into a free and a bound part [9.1]:

$$G(1, 2; 3, 4) = G(1, 3)G(2, 4) + \delta G(1, 2; 3, 4) \ . \tag{9.3}$$

The free part, given by the first term in (9.3), corresponds to the propagation of two excitations totally independent of one another resulting in a product of two one-particle Green's functions. The information on the collective excitations of the system is then contained in the bound part $\delta G(1, 2; 3, 4)$ which describes the interaction among the single-particle excitations. It satisfies the following Bethe-Salpeter equation [9.11]

$$\begin{aligned}
\delta G(1, 2; 3, 4) = \ &G(2, 2')G(3', 3) \, I(1', 2'; 3', 4') \, G(1, 1')G(4', 4) \\
&+ G(2, 2')G(3', 3) \, I(1', 2'; 3', 4') \, \delta G(1, 4', 1', 4) \ ,
\end{aligned} \tag{9.4}$$

where integrations over repeated arguments are understood. A diagrammatic representation of (9.4) is given in Fig. 9.1. $I(1', 2'; 3', 4')$ is the irreducible electron-hole interaction, irreducible meaning that we cannot obtain a new interaction part by cutting a single interaction line.

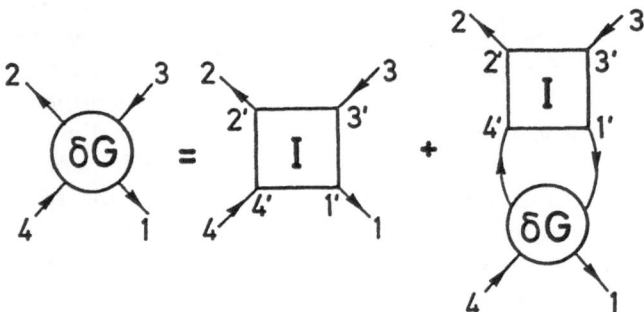

Fig. 9.1. Bethe-Salpeter equation for the bound part $\delta G$

Due to the fact that the poles of the two-particle Green's function determine the energies of the collective excitations, only singular terms in $\delta G(1,2;3,4)$ need to be considered in the vicinity of $\omega = E_{N\lambda q} - E_{N0}$, which we assume to be the excitation energy of an excited state $|N\lambda q\rangle$. The first term in (9.4) is regular (interactions cannot be singular in the normal state) and can therefore be neglected. Then, using the spectral representation of the two-particle Green's function [9.10], the homogeneous Bethe-Salpeter equations are obtained [9.11]:

$$G_{0,\lambda q}(2,3) = G(2,2')G(3',3)\,I(1',2';3',4')\,G_{0\lambda q}(4',1') \quad \text{with} \tag{9.5}$$

$$G_{0\lambda q}(2,3) = \sum_\alpha \frac{\langle N0|\psi(r_2)|N+1,\alpha\rangle\langle N+1,\alpha|\psi^\dagger(r_3)|N\lambda q\rangle}{-\omega_2 + E_{N0} - E_{N+1,\alpha}}$$

$$- \frac{\langle N0|\psi^\dagger(r_3)|N-1,\alpha\rangle\langle N-1,\alpha|\psi(r_2)|N\lambda q\rangle}{-\omega_3 + E_{N0} - E_{N-1,\alpha}} . \tag{9.6}$$

### 9.2.2 Equation of Motion in the Local-Orbital Representation

Once we obtain the homogeneous Bethe-Salpeter equation, which determines the evolution of collective excitations of the system, a "wave function" or amplitude for such excitation should be defined. This can be done [9.11] in a similar way as for the quasi-electron and quasi-hole amplitudes [9.12]:

$$f_{\lambda q}(r_2, r_3) = \langle N0|\psi^\dagger(r_2)\psi(r_3)|N\lambda q\rangle$$

$$= \lim_{\eta \to 0+} \frac{1}{2\pi i} \int_C d\omega_q e^{\omega_2 \eta} G_{0,\lambda q}(r_2, \omega_2; r_3, \omega_2 - \omega) , \tag{9.7}$$

where $f_{\lambda q}$ describes the amplitude of an electron-hole pair for the excited state $|N\lambda q\rangle$. The contour of interaction in (9.7) is taken as in [9.11]. Using (9.5) and the definition (9.7) an integral equation for the amplitude of the collective excitation is obtained [9.11]

$$f_{\lambda q}(r_2, r_3) = \int d^3r_1'\ldots d^3r_4' \sum_{n,n'k}$$

$$\times \frac{\phi_{nk+q}(r_2)\phi_{nk+q}^*(r_2')\phi_{n'k}^*(r_3)\phi_{n'k}(r_3')}{\omega - E_n(k+q) + E_{n'}(k)}$$

$$\times [f_n(k+q) - f_{n'}(k)]\cdot I(r_1', r_2'; r_3', r_4')f_{\lambda q}(r_4', r_1') , \tag{9.8}$$

where $f_n(k)$ is the occupation number of the corresponding state with wave function $\phi_{nk}(r)$. The irreducible electron-hole interaction in (9.8) is restricted to the screened Hartree-Fock approximation (Fig. 9.2):

$$I(1',2';3',4') = \delta(1',3')\delta(2',4')\,V^S(2',3')$$
$$+ \delta(1',4')\delta(2',3')\,V(2',4') . \tag{9.9}$$

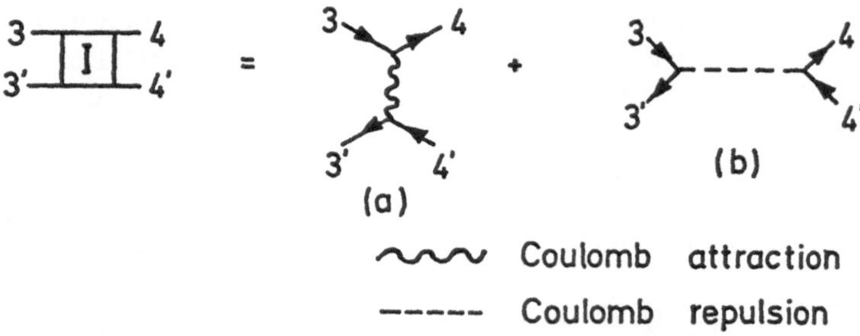

Coulomb attraction

- - - - - Coulomb repulsion

**Fig. 9.2.** The irreducible electron-hole interaction in the TDSHFA (a) screened electron-hole attraction, (b) unscreened exchange

V is the bare Coulomb interaction, whereas $V^S$ corresponds to the screened interaction which in our case is used only in the static approximation [9.4,5].

Following [9.11] we express $f_{\lambda q}$ in terms of the Bloch wave function $\phi_{nk}(r)$ :

$$f_{\lambda q}(r_1, r_2) = \sum_{nn'k} A^{nn'}(k+q, k)\phi_{nk+q}(r_1)\phi^*_{n'k}(r_2) \; . \tag{9.10}$$

We can now introduce a local representation for the surface Bloch wave function and use the formalism of *Hanke* and *Sham* [9.3–5]:

$$\phi_{nk}(r) = \frac{1}{\sqrt{NM}} \sum_{m,\nu} c^m_{n\nu}(k) \sum_{\ell} \exp(ik \cdot R_\ell) \, a^m_\nu(r - R_m - R_\ell) \; , \tag{9.11}$$

where m denotes the m-th layer in a thin slab, M is the total number of layers, $R_\ell$ is a 2-D translation vector, $R_m$ is 3-D basis vector for the 2-D unit cell, $\nu$ is an orbital index, and N is the number of 2-D unit cells. Substituting (9.11 and 10) into the integral equation (9.8) we obtain a matrix equation

$$\sum_{\ell'} \exp(iq \cdot R_{\ell'}) \sum_{\lambda, \lambda', \lambda''} a^m_\nu(r_2 - R_m - R_\lambda - R_{\ell'})$$
$$\times a^{m'*}_{\nu'}(r_3 - R_{m'} - R_{\ell'}) \cdot (X_\lambda + N_{\lambda\lambda'} V^{xc}_{\lambda'\lambda''} X_{\lambda''}) = 0 \; . \tag{9.12}$$

Explicit expressions for $X_\lambda(q)$, $N_{\lambda\lambda'}(q,\omega)$, and $V^{xc}_{\lambda\lambda'}(q)$ can be found in detail in [9.13]. We omit them here for space reasons.

Finally, since (9.12) is valid for any complete local basis, we obtain the following equation of motion for the collective excitations

$$[\overset{\leftrightarrow}{1} + \overset{\leftrightarrow}{N}(q,\omega)\overset{\leftrightarrow}{V}^{xc}(q)] \cdot X(q) = 0 \; . \tag{9.13}$$

The solution of (9.13) has to satisfy the condition

$$\det |\overset{\leftrightarrow}{1} + \overset{\leftrightarrow}{N}(q,\omega)\overset{\leftrightarrow}{V}^{xc}(q)| = 0 \; , \tag{9.14}$$

which determines in a completely general way the energy spectrum and life time of the electronic collective eigenmodes of the surface system. Their amplitude $X(q)$ is given by (9.13). They will provide us with information about the localized nature of those modes and enable us to distinguish the surface localized modes from the extended (bulk) ones.

### 9.2.3 Surface Density-Response Function

We derive here the electronic density-response function for a surface system from the two-particle Green's function discussed in the previous subsection.

Quite generally, the density response function $\chi$ is obtained from the two-particle Green's function as follows [9.5]

$$\chi(1,2) = \sum_{\sigma\sigma'} G_{\sigma\sigma'}(1,2;2^+,1^+) \ , \tag{9.15}$$

where $\sigma$ and $\sigma'$ are spin indices and $\tau_1^+ = \tau_1 + 0^+$.

We can define a polarizability $\tilde{\chi}$ in a similar way. Let us take in the integral equation (9.4) only those diagrams for the irreducible electron-hole interaction $I(1,2;3,4)$, that do not contain contributions like the one in Fig. 9.2.b. This will give all irreducible terms in the two-particle Green's function, the sum of which we call $\tilde{G}$. Then the polarizability is given as:

$$\tilde{\chi}(1,2) = \sum_{\sigma\sigma'} \tilde{G}_{\sigma\sigma'}(1,2;2^+,1^+) \ . \tag{9.16}$$

With this definition, all long-range effects coming from the bare Coulomb interaction are excluded. Then the polarizability is that part of the susceptibility without long-range effects, and the susceptibility is obtained by the following integral equation (Fig. 9.3):

$$\chi(1,2) = \tilde{\chi}(1,2) + \tilde{\chi}(1,1')v(1',2')\chi(2',2) \ , \tag{9.17}$$

where $v(1',2')$ is the unscreened Coulomb interaction.

Due to the fact that the polarizability $\tilde{\chi}$ can be expressed with the help of Bloch wave functions, a local-orbital representation of the wave functions allows us to achieve a separable form for $\tilde{\chi}$ [9.3,5]. In such a representation the Fourier transformed surface polarizability is given by [9.13,14]:

$$\tilde{\chi}(q+G, q+G';z,z') = \sum_{\lambda\lambda'} A_\lambda(q+G,z)\, \tilde{S}_{\lambda\lambda'}(q,\omega)\, A_{\lambda'}^\dagger(q+G,z') \ , \tag{9.18}$$

where $A_\lambda(q+G,z)$ is a generalized charge-density wave

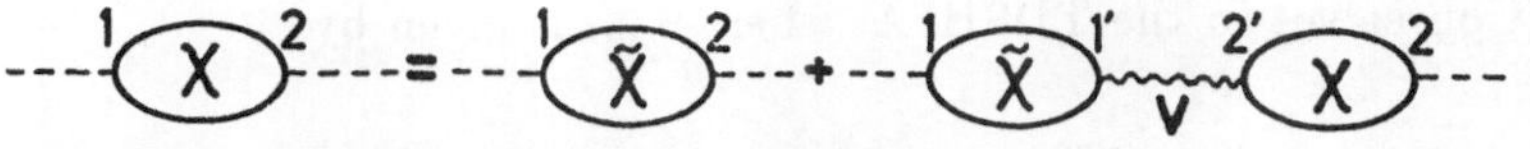

**Fig. 9. 3.** Integral equation for the susceptibility

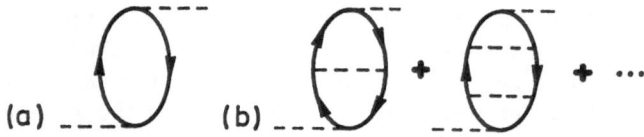

(a)          (b)

**Fig.9.4.** (a) Hartree polarizability. (b) Contribution from the electron-hole attraction in the time-dependent Hartree-Fock approximation

$$A_\lambda(q + G, z) = \int d^2r \; a_\nu^{m*}(r, z - R_m^z) \exp[-i(q + G)\cdot r]$$
$$\times a_{\nu'}^{m'}(r - R_\lambda, z - R_{m'}^z) \; , \tag{9.19}$$

with $G$ a 2-D reciprocal lattice vector and $r$ a 2-D vector parallel to the surface. The index $\lambda$ is the same as the one defined in the previous subsection.

The matrix $\tilde{S}_{\lambda\lambda'}$ is given in the time-dependent Hartree approximation (RPA) by

$$\tilde{S}_{\lambda\lambda'}(q, \omega) = N_{\lambda\lambda'}(q, \omega) \; . \tag{9.20}$$

The corresponding diagram is shown in Fig. 9.4a. In this approximation the polarizability $\tilde{\chi}$ contains only contributions from non-interacting electrons and holes. In the time-dependent screened Hartree-Fock approximation (TDSHFA) we have

$$\tilde{S}_{\lambda\lambda'}(q, \omega) = [N_{\lambda\lambda'}^{-1}(q, \omega) + \tfrac{1}{2}V_{\lambda\lambda'}^s]^{-1} \; . \tag{9.21}$$

The diagrams corresponding to this approximation are shown in Fig. 9.4b, where the interaction lines should be understood as a statically screened Coulomb attraction, as in Fig. 9.2. The separability of the polarizability in (9.18) allows us to solve the integral equation (9.17), to obtain [9.13,14]:

$$\chi(q + G, q + G'; z, z'; \omega) = \sum_{\lambda\lambda'} A_\lambda(q + G, z) S_{\lambda\lambda'}(q, \omega)$$
$$\times A_{\lambda'}^\dagger(q + G', z') \; , \quad \text{where} \tag{9.22}$$

$$S(q, \omega) = [\tilde{S}^{-1} - V^c]^{-1} \quad \text{and} \tag{9.23}$$

$$V_{\lambda\lambda'}^c = \sum_G \int dz \, dz' \, A_\lambda^\dagger(q + G, z) \, v(q + G; |z - z'|) \, A_{\lambda'}(q + G, z') \; . \tag{9.24}$$

The expression (9.24) corresponds to the Coulomb interaction between the generalized charge-density waves $A_\lambda$ and $A_{\lambda'}$ [9.13].

We restrict ourselves to the TDSHFA, where $S_{\lambda\lambda'}$ is given by

$$\overleftrightarrow{S}(q, \omega) = [\overleftrightarrow{N}^{-1}(q, \omega) + \overleftrightarrow{V}^{xc}(q, \omega)]^{-1} \; . \tag{9.25}$$

354

A comparison of (9.14) and (9.25) displays a well-known result, obtained in general from the spectral representation of the density-density correlation function [9.10]. The solutions of the equation of motion (9.13), i.e. the elementary excitations of the system, determine the poles of the density-response function (9.22).

## 9.3 Surface Electromagnetic Response

The general microscopic theory of surface elementary excitations and response function of the previous section is complemented with a discussion of the response of a surface system to an electromagnetic excitation.

A tremendous amount of experimental results from different optical spectroscopic techniques on surfaces has been accumulated in the past years [9.15–18], which were not accompanied by a parallel theoretical development. Theoretical progress has been limited until very recently to two very simplified models for the surface.

On the one hand, the jellium model is used for nearly-free electron metals. Here the electron gas is taken as homogeneous in directions parallel to the surface and the only charge-density variation is in the perpendicular direction, induced by a step-like positive background. Explicit calculations for the surface contribution in ultraviolet photoemission spectroscopy (UPS) were performed with a "jellium" dielectric function $\varepsilon(q, z, z')$ in the RPA [9.6] obtaining remarkably good agreement with experiments for a typical "jellium"-type metal – Al(100). Unfortunately the jellium model is not capable of obtaining the correct surface plasmon dispersion for the Al(111) surface, showing already that effects due to the discrete nature of the lattice become important. Thus, an understanding of the role of the local-field effects due to the charge inhomogeneity of the crystalline lattice is required. We expect from the bulk studies [9.3–5], that this will, in particular, be essential for semiconductor and insulator surfaces.

The opposite limit to the jellium approximation is provided by the extreme tight-binding model. This approximation was and is widely used to study the effect of spatial dispersion of the dielectric function on the reflectivity of semiconductors near exciton absorption lines [9.19,20]. Here, additional boundary conditions (ABC) [9.20] are usually introduced, because the classical ones, i.e. conservation of tangential components of $E$ and $H$, fail to provide sufficient information to compute the reflectivity. In order to avoid this artifact the knowledge of the fully non-local dielectric response function of the system is necessary.

Very recently a theoretical formulation for a macroscopic dielectric tensor for crystal surfaces was presented [9.9], based on the formalism of the previous section. In the following we discuss the optical response along similar lines and at the end will provide a contact with measurable quantities like the reflectivity in a way first introduced in the context of the jellium model [9.6].

### 9.3.1 Surface Macroscopic Dielectric Tensor

In order to discuss the interaction of a many-body system with an electromagnetic field, it will prove convenient to work in the Coulomb gauge, i.e. $\nabla \cdot \boldsymbol{A} = 0$, where $\boldsymbol{A}$ is the vector potential. In this gauge, the Hamiltonian for a system of electrons in the presence of an electromagnetic field can be written as a many-body part $H_0$, the same we used in Sect. 9.2 and a part that describes the coupling with the electromagnetic field only through the current [9.21]

$$H_I = -\tfrac{1}{2} \int d^3r \, \boldsymbol{j}(\boldsymbol{r}) \cdot \boldsymbol{A}(\boldsymbol{r}) \ . \tag{9.26}$$

Quadratic terms in $\boldsymbol{A}$ are neglected because we are interested in the regime of linear response only.

In the frame of linear-response theory [9.10,21], the induced current in the system due to the presence of the vector potential $\boldsymbol{A}$ is given by the conductivity tensor obtained from Kubo's formula [9.21]

$$\boldsymbol{j}^{\text{ind}}(\boldsymbol{r}, \omega) = \int d^3r' \, \overset{\leftrightarrow}{\sigma}(\boldsymbol{r}, \boldsymbol{r}', \omega) \, \boldsymbol{E}_{\text{T}}(\boldsymbol{r}', \omega) \ . \tag{9.27}$$

Longitudinal and transverse fields can be described, in a similar way as in bulk [9.22], with the help of the longitudinal and transverse dyadic defined as

$$\begin{aligned}
&\mathbb{1}_{\text{L}} = \hat{Q}\hat{Q} \ , \\
&\mathbb{1}_{\text{T}} = 1 - \hat{Q}\hat{Q} \ ,
\end{aligned} \tag{9.28}$$

where $\mathbb{1}$ is the unit dyadic. $\hat{Q}$ is a unit vector parallel to the direction of propagation $\boldsymbol{Q}$, given in general by

$$\boldsymbol{Q} = (q_x + G_x, q_y + G_y, k_z) \ . \tag{9.29}$$

Following these definitions we have

$$\boldsymbol{E}_{\text{T}}(\boldsymbol{Q}) = \mathbb{1}_{\text{T}} \cdot \boldsymbol{E}(\boldsymbol{Q}) = \boldsymbol{E}(\boldsymbol{Q}) - \boldsymbol{E}_{\text{L}}(\boldsymbol{Q}) \ . \tag{9.30}$$

Notice that from our choice of gauge, the induced current is not given by the total electric field but only the transverse part, although the conductivity $\overset{\leftrightarrow}{\sigma}$ is the current-current correlation defined as usually. On the other hand, $H_0$ is of such a form that we can calculate $\overset{\leftrightarrow}{\sigma}$ (or any other correlation function) using the formalism developed in Sect. 9.2. The explicit form of $\overset{\leftrightarrow}{\sigma}$ will be given in Sect. 9.3.2, although we will use its properties in the present section, like invertibility, in deriving the macroscopic dielectric tensor.

We have to define now the macroscopic fields. These are fields that result from averaging the microscopic ones over distances of the order of the unit cell, or equivalently, by considering only those Fourier components of the fields for small enough wave vectors, e.g.,

$$E^M(q, k_z) = E(\boldsymbol{q} + \boldsymbol{G}, k_z) \tag{9.31}$$

with $\boldsymbol{G} = 0$ and $k_z < k_c$, where the cutoff value is taken as $\omega/c \ll k_c \ll G$ [9.9]. This is a direct generalization of the notion of macroscopic fields in the bulk [9.22].

An important consequence of the above definition and Maxwell's equations is that the transverse total electric field has negligible microscopic components compared to the macroscopic part [9.9]. This can be easily seen if we combine Maxwell's equations for the electric and magnetic field such that in Fourier space we obtain

$$E_T = -\frac{\omega^2}{c^2} \frac{1}{(|\boldsymbol{q} + \boldsymbol{G}|^2 + k_z^2)} \boldsymbol{D} . \tag{9.32}$$

Then for $\boldsymbol{G} \neq 0$ or $k_z > k_c$, the resulting electric field is much smaller in amplitude than the external one.

In order to proceed further, we consider now the polarization of the system, and a polarizability $\overleftrightarrow{\alpha}$ defined such that

$$P(\boldsymbol{r}, \omega) = \int d^3 r' \overleftrightarrow{\alpha}(\boldsymbol{r}, \boldsymbol{r}'; \omega) E_T(\boldsymbol{r}', \omega) , \tag{9.33}$$

where $\overleftrightarrow{\alpha} = (i/\omega) \overleftrightarrow{\sigma}$.

Recalling that the polarization is related to the total electric field through the displacement field and using the fact that the external field $\boldsymbol{D}$ has no longitudinal components we arrive at

$$E_L = -4\pi P_L , \tag{9.34}$$

that can be written as

$$E_L(\boldsymbol{Q}) = -4\pi \mathbb{1}_L \int dk_z' \overleftrightarrow{\alpha}(\boldsymbol{q} + \boldsymbol{G}, \boldsymbol{q}; k_z, k_z') E_T(\boldsymbol{q}, k_z') , \tag{9.35}$$

where we take explicitly into account the fact that we neglect the microscopic components of $E_T$ [9.9].

Once we arrive at (9.33 and 35), we can look for the macroscopic dielectric tensor. It is defined through the macroscopic total electric field and the external one

$$D(\boldsymbol{q}, k_z; \omega) = \int_{k_z' < k_c} dk_z' \overleftrightarrow{\varepsilon}^M(\boldsymbol{q}; k_z, k_z'; \omega) E^M(\boldsymbol{q}, k_z'; \omega) . \tag{9.36}$$

The macroscopic components of the total electric field can be obtained from (9.30, 35)

$$E^M(\boldsymbol{q}, k_z) = \int_{k_z' < k_c} dk_z' \overleftrightarrow{\gamma}(\boldsymbol{q}; k_z, k_z') E_T(\boldsymbol{q}, k_z') , \quad \text{where} \tag{9.37}$$

357

$$\overleftrightarrow{\gamma}(\boldsymbol{q}; \mathbf{k_z}, \mathbf{k_z'}) = \mathbb{1}\delta(\mathbf{k_z} - \mathbf{k_z'}) - 4\pi \mathbb{1}_\mathrm{L} \cdot \overleftrightarrow{\alpha}(\boldsymbol{q} + \boldsymbol{G}; \mathbf{k_z}, \mathbf{k_z'}) \ . \tag{9.38}$$

If we now assume that we can invert $\overleftrightarrow{\gamma}$, then we immediately obtain the macroscopic dielectric tensor. From (9.37) we have

$$\boldsymbol{E}_\mathrm{T}(\boldsymbol{q}, \mathbf{k_z}) = \int\limits_{\mathbf{k_z'} < \mathbf{k_z}} \mathrm{dk_z'} \, \overleftrightarrow{\gamma}^{-1}(\boldsymbol{q}; \mathbf{k_z}, \mathbf{k_z'}; \omega) \boldsymbol{E}^\mathrm{M}(\boldsymbol{q}, \mathbf{k_z'}) \tag{9.39}$$

and from (9.33 and 36) we finally arrive at

$$\overleftrightarrow{\varepsilon}^\mathrm{M}(\boldsymbol{q}; \mathbf{k_z}, \mathbf{k_z'}; \omega) = \mathbb{1}\delta(\mathbf{k_z} - \mathbf{k_z'}) + 4\pi \int \mathrm{dk_z''} \, \overleftrightarrow{\alpha}(\boldsymbol{q}, \mathbf{k_z}, \mathbf{k_z''})$$
$$\times \overleftrightarrow{\gamma}^{-1}(\boldsymbol{q}, \mathbf{k_z''}, \mathbf{k_z'}; \omega) \ . \tag{9.40}$$

A more tractable form can be obtained for (9.40) following the treatment in [9.9]. Instead of solving for $\overleftrightarrow{\gamma}^{-1}$ explicitly, we can write

$$\int \mathrm{dk_z'' dk_z'''} \, \overleftrightarrow{\alpha}(\boldsymbol{q}; \mathbf{k_z}, \mathbf{k_z''}) \, \overleftrightarrow{\gamma}^{-1}(\boldsymbol{q}; \mathbf{k_z''}, \mathbf{k_z'''}) \boldsymbol{Q}''' \, \mathrm{O}(\boldsymbol{q}; \mathbf{k_z'''}, \mathbf{k_z'})$$
$$= \overleftrightarrow{\alpha}(\boldsymbol{q}; \mathbf{k_z}, \mathbf{k_z'}) \boldsymbol{Q}' \ , \tag{9.41}$$

where as before $\boldsymbol{Q} = (\boldsymbol{q}, \mathbf{k_z})$ and

$$\mathrm{O}(\boldsymbol{q}; \mathbf{k_z}, \mathbf{k_z'}) = \delta(\mathbf{k_z} - \mathbf{k_z'}) - 4\pi \boldsymbol{Q} \overleftrightarrow{\alpha}(\boldsymbol{q}; \mathbf{k_z}, \mathbf{k_z'}) \boldsymbol{Q}' \ . \tag{9.42}$$

Thus, the inversion of a tensor $\overleftrightarrow{\gamma}^{-1}$, which depends on the continuous indices $\mathbf{k_z}$ and $\mathbf{k_z'}$ was simplified now to the inversion of a matrix with the same continuous indices. This last step will be done explicitly in the next section. After this inversion is performed, we can write

$$\int \mathrm{dk_z''} \, \overleftrightarrow{\alpha}(\boldsymbol{q}; \mathbf{k_z}, \mathbf{k_z''}) \, \overleftrightarrow{\gamma}^{-1}(\boldsymbol{q}; \mathbf{k_z''}, \mathbf{k_z'})$$
$$= \overleftrightarrow{\alpha}(\boldsymbol{q}; \mathbf{k_z}, \mathbf{k_z'}) + 4\pi \int \mathrm{dk_z'' dk_z'''} \, \overleftrightarrow{\alpha}(\boldsymbol{q}, \mathbf{k_z}, \mathbf{k_z''}) \boldsymbol{Q}''$$
$$\times \mathrm{O}^{-1}(\boldsymbol{q}; \mathbf{k_z''}, \mathbf{k_z'''}) \boldsymbol{Q}''' \, \overleftrightarrow{\alpha}(\boldsymbol{q}; \mathbf{k_z'''}, \mathbf{k_z'}) \ . \tag{9.43}$$

This last equation together with (9.40) give the final form of the macroscopic dielectric tensor.

### 9.3.2 Microscopic Formulation of the Optical Response

In this subsection we give explicit expressions for the conductivity tensor $\overleftrightarrow{\sigma}$ and the polarizability $\overleftrightarrow{\alpha}$ defined in the previous section. As in Sect. 9.2, we incorporate the information of the band structure and wave functions of the

358

surface system with the help of a local orbital representation of the two-particle Green's function.

According to linear response theory [9.10], the conductivity tensor introduced in (9.27) is given by

$$\sigma_{\alpha\beta}(\mathbf{r}, \mathbf{r}'; \omega) = \frac{1}{\omega} P^{jj}_{\alpha\beta}(\mathbf{r}, \mathbf{r}'; \omega) - i\frac{e^2}{m\omega} \delta_{\alpha\beta} \varrho(\mathbf{r})\delta(\mathbf{r} - \mathbf{r}') \ , \tag{9.44}$$

where

$$P^{jj}_{\alpha\beta}(\mathbf{r}, \mathbf{r}'; \omega) = -\langle T_\tau \tilde{j}_\alpha(\mathbf{r})\tilde{j}_\beta(\mathbf{r}')\rangle_\omega \tag{9.45}$$

with the current operators $\tilde{j}_\alpha(\mathbf{r})$ in the Heisenberg representation. The last term in (9.44) contains the ground-state density.

The current-current correlation function $P^{jj}$ was taken as a time-ordered product, in order to use the diagrammatic expansion of Sect. 9.2. Its retarded version corresponds to the well-known Kubo formula [9.21]. $P^{jj}$ can also be written in terms of a two-particle Green's function [9.10]

$$P^{jj}_{\alpha\beta}(1, 1') = \frac{e^2}{4m^2}(\nabla_1 - \nabla_2)_\alpha (\nabla'_1 - \nabla'_2)_\beta$$

$$\times \sum_{\sigma\sigma'} G_{\sigma\sigma'}(1, 1'; 2', 2)\Bigg|_{\substack{2'=1'+ \\ 2=1+}} \tag{9.46}$$

with the notation introduced in Sect. 9.2.1. By comparing (9.46) with (9.15, 22), we obtain

$$P^{jj}_{\alpha\beta}(\mathbf{q} + \mathbf{G}, \mathbf{q} + \mathbf{G}'; z, z')$$

$$= \frac{e^2}{m^2} \sum_{\lambda\lambda'} K^\alpha_\lambda(\mathbf{q} + \mathbf{G}, z) S_{\lambda\lambda'}(\mathbf{q}, \omega) K^\beta_{\lambda'}(\mathbf{q} + \mathbf{G}; z') \quad \text{with} \tag{9.47}$$

$$K^\alpha_\lambda(\mathbf{q} + \mathbf{G}, z) = \tfrac{1}{2} \int d^2r \ e^{i(\mathbf{q}+\mathbf{G})\cdot\mathbf{r}} \left\{ a^{m*}_\mathbf{r}(\mathbf{r}, z - R^z_m) \right.$$

$$\times \frac{\partial}{\partial r_\alpha} a^{m'}_{\gamma'}(\mathbf{r} - R_\lambda, z - R^z_{m'})$$

$$\left. - \left[\frac{\partial}{\partial r_\alpha} a^{m*}_\gamma(\mathbf{r}, z - R^z_m)\right] a^{m'}_{\gamma'}(\mathbf{r} - R_\lambda, z - R^z_{m'}) \right\} \ . \tag{9.48}$$

Notice that now we have the full $S_{\lambda\lambda'}$, given by (9.25) since we consider the full Green's function in (9.46).

Once the full conductivity was obtained, we concentrate our attention on the different contractions of $\overleftrightarrow{\sigma}$ (or $\overleftrightarrow{\alpha}$) with $\mathbf{Q}$.

359

Let us first consider $Q \cdot \overleftrightarrow{\sigma} \cdot Q$, a response function that gives the longitudinal currents due to longitudinal fields. In the gauge we chose, such a field is entirely determined by an external scalar potential

$$E_L^{ext}(Q) = -iQ \phi^{ext}(Q) \ . \tag{9.49}$$

On the other hand, the longitudinal components of the induced current are related to the induced density via the continuity equation:

$$\omega \varrho(Q,\omega) = Q \cdot j^{ind}(Q,\omega) \ . \tag{9.50}$$

Putting together the last two equations and using $\overleftrightarrow{\alpha} = (i/\omega) \overleftrightarrow{\sigma}$ we obtain

$$Q \cdot \overleftrightarrow{\alpha} \cdot Q = -\frac{1}{|Q||Q'|} \chi(Q,Q') \ , \tag{9.51}$$

where $\chi$ is the density response function discussed in Sect. 9.2.3. Then the matrix O in (9.42) becomes

$$O(q; k_z, k_z') = \delta(k_z - k_z') + \frac{4\pi}{|Q||Q'|} \sum_{\lambda\lambda'} A_\lambda(Q) S_{\lambda\lambda'}(q,\omega) A_{\lambda'}^\dagger(Q') \ . \tag{9.52}$$

For the other contractions present in (9.43) we have to introduce first density-current and current-density correlation functions [9.9,23]

$$P_\alpha^{\varrho j}(r, r'; \omega) = -\langle T_\tau \varrho(r) \tilde{j}_\alpha(r') \rangle_\omega \quad \text{and} \tag{9.53}$$

$$P_\alpha^{j\varrho}(r, r'; \omega) = -\langle T_\tau \tilde{j}_\alpha(r) \varrho(r) \rangle_\omega \ . \tag{9.54}$$

Then, in the present gauge and allowing for an external scalar potential, we can obtain the induced densities and currents as [9.9]:

$$\varrho^{ind}(r) = \frac{1}{i\omega} \int d^3r' \, P_\alpha^{\varrho j}(r, r'; \omega) E_T(r') \quad \text{and} \tag{9.55}$$

$$j_\alpha^{ind}(r) = \int d^3r' \, P_\alpha^{j\varrho}(r, r'; \omega) \phi^{ext}(r') \ . \tag{9.56}$$

In a similar way as (9.47) the correlation functions above can be expressed in a local basis, where they acquire the following form

$$P_\alpha^{\varrho j}(Q, Q') = i\frac{e^2}{m} \sum_{\lambda\lambda'} A_\lambda(Q) S_{\lambda\lambda'}(q,\omega) K_{\lambda'}^{\alpha\dagger}(Q') \ , \tag{9.57}$$

$$P_\alpha^{j\varrho}(Q, Q') = \frac{e^2}{m} \sum_{\lambda\lambda'} K_\lambda^\alpha(Q) S_{\lambda\lambda'}(q,\omega) A_{\lambda'}^\dagger(Q') \ . \tag{9.58}$$

The remaining contractions are finally given by [9.9]

$$Q \cdot \overleftrightarrow{\alpha}(Q, Q') = \frac{1}{\omega Q} \sum_{\lambda \lambda'} A_\lambda(Q) S_{\lambda \lambda'}(q, \omega) K_{\lambda'}^\dagger(Q') \quad \text{and} \tag{9.59}$$

$$\overleftrightarrow{\alpha}(Q, Q') \cdot Q = \frac{1}{\omega Q} \sum_{\lambda \lambda'} K_\lambda(Q) S_{\lambda \lambda'}(q, \omega) A_{\lambda'}^\dagger(Q') . \tag{9.60}$$

Finally, after a little algebra we arrive at a rather simple expression for the macroscopic dielectric tensor

$$\varepsilon_{\alpha \beta}(q; k_z, k_z'; \omega) = 1 + \delta(k_z - k_z') + \frac{4\pi}{\omega^2} \sum_{\lambda \lambda_1 \lambda'} K_\lambda^\alpha(q, k_z) S_{\lambda \lambda_1}(q, \omega)$$

$$\times (1 + \overleftrightarrow{V}_{LR}^c \overleftrightarrow{S})_{\lambda_1 \lambda'}^{-1} K_{\lambda'}^\beta(q, k_z') . \tag{9.61}$$

Notice that in (9.61) only the long-range contribution $V_{LR}^c$ to the Coulomb matrix $V^c$ enters, defined as

$$V_{LR}^c \equiv \int_{k_z < k_c} A^\dagger(q, k_z) \frac{4\pi e^2}{q^2 + k_z^2} A(q, k_z) \, dk_z . \tag{9.62}$$

This is due to the fact that only macroscopic quantities enter in (9.40 and 43). Furthermore, the matrix product in (9.61) is precisely the solution of the Bethe-Salpeter equation without long-range contributions

$$\hat{S} = \tilde{S} + \tilde{S} V_{SR}^c \hat{S} \quad \text{with} \tag{9.63}$$

$$V^c = V_{LR}^c + V_{SR}^c , \tag{9.64}$$

and we recall that the matrix $\tilde{S}$, already defined in (9.18) does not contain any long-range contribution because it corresponds to a sum of irreducible diagrams [9.24]. The relation between $\hat{S}$ and $S$ is given by

$$S = \hat{S} + \hat{S} V_{LR}^c S , \tag{9.65}$$

obtained after inserting (9.63 and 64) into (9.23), or equivalently

$$\hat{S} = S(1 + V_{LR}^c S)^{-1} . \tag{9.66}$$

Then, the macroscopic dielectric tensor is free from non-analyticities associated with the long-range nature of the Coulomb interaction [9.5,9,23,24] and we can safely take the limit $Q \rightarrow 0$ in order to calculate the electromagnetic response.

### 9.3.3 Microscopic Theory of Surface Reflectivity

The purpose of this last subsection on surface electromagnetic response is to formulate corrections to the classical surface reflectivity with the help of the macroscopic dielectric tensor obtained previously.

361

As already mentioned, a large amount of work was done to study the optical properties of jellium surfaces [9.6], but only recently were similar efforts started for real surfaces [9.9,25].

In the following we generalize a development made for jellium surfaces [9.6] with the necessary changes appropriate for real surfaces. For this purpose we will when needed incorporate the results of the previous sections in the general treatment of Maxwell's equations done by *Feibelman* [9.6]. Due to the clarity of this treatment we are not going to repeat it but use directly some results obtained there.

The main difference between a jellium and a real crystal appears in the tensorial form of the dielectric function

$$D_\alpha(z) = \int dz' \sum_\beta \varepsilon_{\alpha\beta}(z, z') E_\beta(z') \ . \tag{9.67}$$

Let us assume that the surface is perpendicular to the z-direction and $\varepsilon_{xy} = \varepsilon_{yx} = 0$. Furthermore, we recall that we are interested in the long-wavelength corrections to the classical results. This implies that we need only to consider the macroscopic fields. The microscopic information is contained in the dielectric tensor derived in Sect. 9.3.1 and 2.

Let us consider first the case of s-polarization. Following [9.6] we have for the parallel components of *E* and *H*

$$E_y(z_b) = E_y(z_v) - iQ(z_b - z_v)H_x(z_v) \quad \text{and} \tag{9.68}$$

$$H_x(z_b) = H_x(z_v) + i\frac{q_x^2}{Q}(z_b - z_v)E_y(z_v)$$

$$+ iQ\left[\int_{z_v}^{z_b} dz' \frac{d}{dz'}D_y(z') - zD_y(z) + z_vD_y(z_v)\right] \ . \tag{9.69}$$

Here $z_b$ and $z_v$ are arbitrary points well inside the bulk and well inside the vacuum, respectively, and $Q = \omega/c$, a small number compared to the inverse of any length characteristic of the system. The above equations give the correction to the classical boundary conditions in first order in Q.

In order to solve the coupled set of equations, we need to obtain $D_y(z)$. The invertibility of $\overset{\leftrightarrow}{\varepsilon}(z, z')$, as shown in Sect. 9.3.2, makes this possible:

$$D_y(z) = \int dz' \varepsilon_{yy}(z, z') E_y(z')$$

$$+ \int dz_1 dz' \varepsilon_{yz}(z_2 z_1) \varepsilon_{zz}^{-1}(z_1, z') D_z(z')$$

$$- \int dz_1 dz_2 dz' \sum_{j=x,y} \varepsilon_{yz}(z, z_1) \varepsilon_{zz}^{-1}(z_1, z_2) \varepsilon_{zj}(z_2, z') E_j(z') \ . \tag{9.70}$$

Recall now that $D_y$ is needed only on zero-th order in Q. This implies that the

362

classical matching conditions can be applied, which for s-polarization lead to $D_z = 0$, $E_x = 0$ and $E_y$ (inside) $= E_y$ (outside). Then we arrive at

$$\int_{z_v}^{z_b} dz'\, z'\, \frac{d}{dz'} D_y(z') = E_y(z_v)(\varepsilon_b - 1)d_\parallel^y(\omega) \ , \tag{9.71}$$

where we defined

$$d_\parallel^y(\omega) = \frac{1}{(\varepsilon_b - 1)} \int_{-\infty}^{\infty} dz\, z\, \frac{d}{dz} \int_{-\infty}^{\infty} dz_1 [\varepsilon_{yy}(z, z_1)$$

$$- \int_{-\infty}^{\infty} dz_2 dz_3 \varepsilon_{yz}(z, z_1)\varepsilon_{zz}^{-1}(z_1, z_2)\varepsilon_{zy}(z_2, z_3)] \ . \tag{9.72}$$

Equation (9.72) gives a generalization for real surfaces of the same quantity introduced by *Feibelman* [9.6] for jellium surfaces. As shown in detail in [9.6] $d_\parallel^y(\omega)$ is not only connected with the surface reflectivity, but will give in general the effective surface location ($\omega$-dependent) and the surface power absorption spectrum.

Let us return now to the calculation of the reflectivity. From now on it continues like in the jellium case. After some tedious algebra, one obtains up to first order in Q,

$$r^{(s)} \equiv \frac{R_y}{E_y^0} = r^{(s),\text{cl.}}[1 + 2iq_z d_\parallel^y(\omega)] \ , \quad \text{where} \tag{9.73}$$

$$r^{(s),\text{cl}} = \frac{1 - q_z'/q_z}{1 + q_z'/q_z} \tag{9.74}$$

is the classical reflectivity. The form (9.73) is the same as the one obtained for jellium surfaces but together with (9.72,61) is now applicable to a real crystalline semiconductor surface.

The reflectivity for p-polarized light can be obtained following the same path as for the previous case. We are not going to repeat the treatment here but refer to *Feibelman*'s work [9.6]

$$r^{(p)} \equiv \frac{R_z}{E_z^0} = r^{(p),\text{cl}} \left\{ 1 - \frac{2iq_z(\varepsilon_b - 1)}{q_z'^2 - \varepsilon_b^2 q_z^2} \right.$$

$$\left. [q_z'^2 d_\parallel^x(\omega) - \varepsilon_b q_x^2 d_\perp(\omega)] \right\} \ , \quad \text{where again} \tag{9.75}$$

$$r^{(p),\text{cl}} = \frac{1 - \frac{q_z'}{q_z \varepsilon_b}}{1 + \frac{q_z'}{q_z \varepsilon_b}} \tag{9.76}$$

is the classical reflectivity. The quantity $d_\parallel^x(\omega)$ is defined as in (9.72) but with indices x instead of y. On the other hand, for $d_\perp(\omega)$ we have

$$d_{\perp}(\omega) = \frac{\varepsilon_b}{\varepsilon_b - 1} \int dz \, dz' \varepsilon_{zz}^{-1}(z, z') \ . \tag{9.77}$$

Furthermore, in order to arrive at (9.75) we used the fact that

$$\varepsilon_{\alpha\beta}(z, z') = \varepsilon_{\beta\alpha}(z', z) \tag{9.78}$$

valid for a basis set of real wave functions, and that $\varepsilon_{\alpha\beta}(z, z')$ is a function with a short-range character in $z$ and $z'$ as a result of (9.61, 66). Then $\varepsilon = \varepsilon_b$ if $z$ or $z'$ equal $z_b$ or 1 for $z_v$.

Expressions (9.73, 75) for the reflectivity, together with the form of $\varepsilon_{\alpha\beta}$ explicitly given in Sect. 9.3.2 complete the microscopic theory of surface optical properties which we developed following the lines of [9.9, 6]. Finally, we would like to bring special attention to the quantities $d_{\parallel}$ obtained in (9.72) and $d_{\perp}$ in (9.77). As already shown for the jellium case [9.6], they play a central role in the determination of image potentials beyond the static approximation. In the present formulation they should allow for the study of particle-surface interactions including the many-body effects on real surfaces. This remains to be done in future work.

## 9.4 Electron-Phonon Interaction and Microscopic Theory of Lattice Dynamics on Surfaces

The study of vibrational excitations on surfaces has recently made significant progress due to the development of elaborate experimental techniques, such as low-energy electron diffraction (LEED) [9.26], characteristic energy-loss [9.27] and scattering of light (He) atoms [9.28]. From the theoretical point of view, the dynamics of crystal surfaces [9.29] have so far been exclusively described by empirical models [9.30]. They employ a bulk parameterization of the force constants to extract the surface vibrational modes. As a result these empirical models work well on surfaces of, for example, alkali halides [9.31] where the electronic charge distribution surrounding the surface ions as well as their geometrical arrangement is practically unchanged compared to the bulk. This is contrasted by the majority of semiconductor and transition-metal surfaces, where the electronic properties are radically different from the bulk. Often this change is accompanied by an instability towards surface reconstruction.

An alternative to the empirical models are local-density calculations of phonons like those already carried out for bulk semiconductors [9.32]. But until now, the numerical inaccuracies and the large amount of atoms involved make this technique rather impractical for surface systems with the present computing power. No such calculation has been performed yet for semiconductor surfaces.

Our lattice-dynamical formulation focuses on the problem of expressing the phonon self-energy in terms of the response of the surface electrons to an external probe. This approach allows a complete solution of the lattice vibrational problem within the adiabatic and harmonic approximation once the potential of the ion cores and a properly defined density-response matrix $\varepsilon^{-1}(\boldsymbol{q} + \boldsymbol{G}, \boldsymbol{q} + \boldsymbol{G}'; z, z')$ are known.

### 9.4.1 Surface Lattice Dynamics

We consider a thin slab with surfaces perpendicular to the z-direction. The normal-mode solutions are of the form

$$u_\alpha(\kappa) = M^{1/2} \xi_\alpha(\kappa, \boldsymbol{q}) \exp(i\boldsymbol{q} \cdot \boldsymbol{R}^0_{\kappa\|}) \exp(-i\omega t) \ , \tag{9.79}$$

where $u_\alpha$ is the $\alpha$-component of the displacement of the ion $\kappa$ with mass M from its equilibrium position $\boldsymbol{R}^0_\kappa = (\boldsymbol{R}^0_{\kappa\|}, \mathrm{R}^0_{\kappa z})$. We remain in the harmonic approximation, where the eigenfrequencies are defined through an eigenvalue equation

$$\sum_{\kappa'\beta} D_{\alpha\beta}(\kappa\kappa', \boldsymbol{q}) \xi^j_\beta(\kappa', \boldsymbol{q}) = \omega^2_j \xi^j_\alpha(\kappa, \boldsymbol{q}) \ . \tag{9.80}$$

The index $j = 1, \ldots, 3N$ classifies the eigenmodes and N is the number of layers in the slab. We assume that there is one atom per layer in the 2-D unit cell in order to simplify the notation. $D_{\alpha\beta}(\kappa\kappa', \boldsymbol{q})$ is the dynamical matrix

$$D_{\alpha\beta}(\kappa\kappa', \boldsymbol{q}) = \frac{1}{M} \sum_{\ell'} \phi_{\alpha\beta} \begin{pmatrix} \kappa\kappa' \\ \ell\ell' \end{pmatrix} \exp[i\boldsymbol{q} \cdot (\boldsymbol{R}_\ell - \boldsymbol{R}_{\ell'})] \tag{9.81}$$

where $\boldsymbol{R}_\ell$ and $\boldsymbol{R}_{\ell'}$ are 2-D translation vectors and $\phi_{\alpha\beta} \begin{pmatrix} \kappa\kappa' \\ \ell\ell' \end{pmatrix}$ are the second-order force constants.

There are two contributions to (9.81). The first one is due to the ion-ion interaction through a repulsive Coulomb potential. This part can be calculated using a modified version of Ewald's method for 2-D lattice sums [9.33] and we are not going to discuss it here. The second part is due to the renormalization of the first contribution by the system of interacting electrons which screen the long-range Coulomb potential giving an effective ion-ion potential. This contribution to the phonon self-energy can be expressed in the harmonic approximation with the help of the density-response function [9.34]

$$\phi^E_{\alpha\beta} \begin{pmatrix} \kappa\kappa' \\ \ell\ell' \end{pmatrix} = \int d^3r \, d^3r' \left[ \frac{\partial w}{\partial r_\alpha} (\boldsymbol{r} - \boldsymbol{R}_\ell - \boldsymbol{R}^0_\kappa) \chi(\boldsymbol{r}, \boldsymbol{r}'; \omega) \right.$$

$$\times \frac{\partial w}{\partial r'_\beta} (\boldsymbol{r}' - \boldsymbol{R}_\ell - \boldsymbol{R}^0_{\kappa'}) - \delta_{\ell\ell'}\delta_{\kappa\kappa'} \sum_{\ell''\kappa''} \frac{\partial w}{\partial r_\alpha} (\boldsymbol{r} - \boldsymbol{R}_\ell - \boldsymbol{R}^0_\kappa)$$

$$\left. \times \chi(\boldsymbol{r}, \boldsymbol{r}'; \omega = 0) \frac{\partial w}{\partial r'_\beta} (\boldsymbol{r}' - \boldsymbol{R}_{\ell''} - \boldsymbol{R}^0_{\kappa''}) \right] \ , \quad \text{where} \tag{9.82}$$

$$w(\boldsymbol{r} - \boldsymbol{R}_\ell - \boldsymbol{R}_\kappa^0) = \frac{Z_\kappa e^2}{|\boldsymbol{r} - \boldsymbol{R}_\kappa^\ell|} \tag{9.83}$$

is the ionic potential experienced by an electron at position $\boldsymbol{r}$ due to an ion at position $\boldsymbol{R}_\kappa^\ell = \boldsymbol{R}_\ell + \boldsymbol{R}_\kappa^0$ with charge $Z_\kappa$. In the adiabatic approximation, the susceptibility $\chi$ has to be calculated in the static limit. The second term in (9.82) is a result of imposing infinitesimal translational invariance.

The part of the dynamical matrix determined by the electronic contribution can be written as [9.35]

$$D_{\alpha\beta}^E(\kappa\kappa', \boldsymbol{q}) = \overline{D}_{\alpha\beta}^E(\kappa\kappa', \boldsymbol{q}) - \overline{D}_{\alpha\beta}^E(\kappa\kappa', 0) \quad \text{with} \tag{9.84}$$

$$\overline{D}_{\alpha\beta}^E(\kappa\kappa', \boldsymbol{q}) = (M_\kappa M_{\kappa'})^{-1/2} \sum_{\lambda\lambda'} F_\alpha^\lambda(\kappa, \boldsymbol{q}) S_{\lambda\lambda'}(\boldsymbol{q}) F_\beta^{\lambda'\dagger}(\kappa', \boldsymbol{q}) \ . \tag{9.85}$$

The force form factor $F_\alpha^\lambda$ corresponds to a force in direction $\alpha$ on the ion $\kappa$ at the site $R_\kappa^0$ due to interaction with a charge-density wave $A_\lambda$. For $\alpha = x, y$ it is given by

$$F_\alpha^\lambda(\kappa, \boldsymbol{q}) = -\mathrm{i} \sum_{\boldsymbol{G}} (\boldsymbol{q} + \boldsymbol{G})_\alpha \exp(\mathrm{i}\boldsymbol{G}\cdot\boldsymbol{R}_{\kappa\|}^0)$$

$$\times \int \mathrm{d}z\, w(\boldsymbol{q} + \boldsymbol{G}, |z - R_{\kappa z}^0|) A_\lambda(\boldsymbol{q} + \boldsymbol{G}, z) \tag{9.86}$$

and for $\alpha = z$,

$$F_z^\lambda(\kappa, \boldsymbol{q}) = -\sum_{\boldsymbol{G}} |\boldsymbol{q} + \boldsymbol{G}| \exp(\mathrm{i}\boldsymbol{G}\cdot\boldsymbol{R}_{\kappa\|}^0)$$

$$\times \int \mathrm{d}z\, \mathrm{sgn}\{z - R_{\kappa z}^0\} w(\boldsymbol{q} + \boldsymbol{G}, z - R_{\kappa z}^0) A_\lambda(\boldsymbol{q} + \boldsymbol{G}, z) \ . \tag{9.87}$$

## 9.5 Applications: Si(111)

The Si(111) surface appears as the best candidate for an application of the theoretical methods developed in the previous sections. It is the most extensively studied surface in recent years [9.36] and continues to be the object of active research [9.37].

Unfortunately a theoretical interpretation of the experimental results is precluded by a determination of the structural properties of the system, which in this case turns out to be rather difficult. In fact, the ideal Si(111) surface, i.e. the one that corresponds to the bulk periodicity, is structurally unstable and displays superstructures with $(2 \times 1)$ and $(7 \times 7)$ reconstruction patterns as seen in LEED measurements [9.36]. Although the periodicities of the reconstructed surfaces has long since been known, the determination of the structure is still the object of continuous efforts both experimentally [9.37] and theoretically [9.38].

We are not going to discuss here the possible structures of reconstructed surface and its experimental consequence [9.37] but rather address the question of the stability of the ideal one against electronic and lattice deformations.

In both cases, electronic and lattice instabilities, we will focus our attention on the role of many-body interactions and their interplay with band-structure effects.

### 9.5.1 Electronic Instabilities in an Ideal Si(111) Slab

Let us first describe the system we used to study the instabilities of an ideal Si(111) surface. We considered an ideal Si(111) slab with 8 layers.

The surface band structure (Fig. 9.5) was obtained in a tight-binding scheme where we considered up to fourth-nearest neighbors. The parameters were determined with a least-square fit to the self-consistent pseudopotential results of *Schlüter* et al. [9.39] and [9.40]. The dangling-bond (DB) band, which appears in the band gap of Fig. 9.5, was fitted to the results of *Appelbaum* and *Hamann* [9.41] in order to describe more accurately the surface states. We obtained good overall agreement for the DB band [9.41] and the band complex of the valence bands [9.39]. For the conduction bands good agreement is achieved for the lowest band [9.40] but unfortunately there is no detailed calculation available for the higher states. The wave functions are expanded in terms of a dangling bond, four bonding and four antibonding orbitals, which are obtained as combinations of $sp^3$ hybrids. Such a description of the wave functions has been shown to be appropriate to investigate the dielectric properties of silicon-bulk taking into account the many-body effects [9.4,5]. Further details concerning technical aspects of the calculation have been given in [9.13].

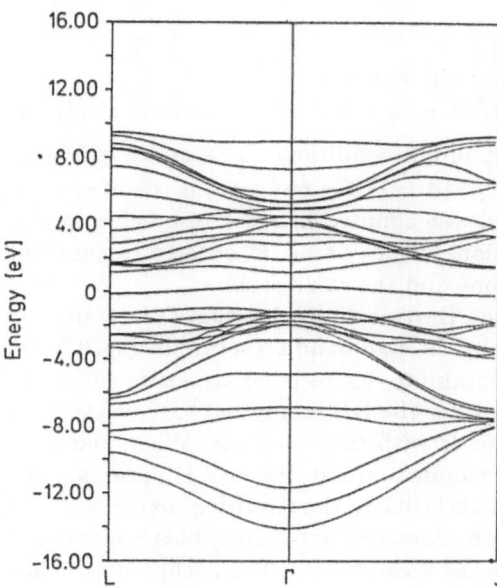

**Fig. 9.5.** Band structure in the tight-binding approximation for an eight-layer slab. The DB band appears in the gap

Once the one-particle properties of the system are determined, we are able to study the elementary excitations of the system, in particular when they become "soft" or trigger an instability of the system. The conditions for such an electronic instability can be formulated in a quantitative form with the help of (9.14). The system will be unstable against a distortion of the original charge-density distribution when such a distortion is possible without delivering energy to the system, i.e. it is incorporated in the ground state. This happens when

$$\det |\overset{\leftrightarrow}{N}{}^{-1}(\boldsymbol{q}, \omega = 0) + \overset{\leftrightarrow}{V}{}^{xc}(\boldsymbol{q})| = 0 \tag{9.88}$$

as a consequence of (9.14). A similar condition is obtained for a spin-density wave instability. In this case we have to consider only triplet states of the electron-hole pair. Then $V^c$ does not contribute because this interaction appears only for singlet states:

$$\det |\overset{\leftrightarrow}{N}{}^{-1}(\boldsymbol{q}, \omega = 0) + \tfrac{1}{2}\overset{\leftrightarrow}{V}{}^{s}(\boldsymbol{q})| = 0 \ , \tag{9.89}$$

where $V^s$ is the screened electron-hole attraction and $V^c$ its exchange counterpart [9.13].

In both cases we observe that an instability is a product of an interplay between band-structure contributions ($\overset{\leftrightarrow}{N}{}^{-1}(\boldsymbol{q}, \omega = 0)$) and many-body effects given by $V^{xc}(\boldsymbol{q})$ or $V^s(\boldsymbol{q})$. In the case we are interested in ferromagnetism in a one orbital system, i.e. q→0, (9.89) reduces to the well-known Stoner criterium [9.21]. This can be easily seen by noticing that in that case

$$\lim_{q \to 0} N(q, \omega = 0) = D(E_F) \ , \tag{9.90}$$

where $D(E_F)$ is the density of states at the Fermi level.

According to (9.25), in order to obtain an instability we have to study the determinant of $\overset{\leftrightarrow}{S}{}^{-1}$ and see when it fulfills conditions (9.88 and 89). Figure 9.6 presents a plot of $(\det S)^{1/d}$ where d (d = 113 in our case) is the dimension of the matrix S [9.13]. The power $1/d$ was adopted in order to be able to plot the different many-body approximations, which result in order of magnitude changes in the determinant of S, on one and the same scale.

The dotted line corresponds to the Hartree approximation ($V^{xc} = 0$), and thus contains only the effect induced by the slab band structure in Fig. 9.5. No structure is observed, ruling out an instability due to band structure effects. In particular, nesting features that appear for the DB band [9.42] are washed out by the interaction of the dangling bonds with the substrate. When the RPA local-field effects (dashed line) are taken into account, the (1 × 1) paramagnetic structure becomes even more stable. This is due to the repulsive character of the interaction $V^c$. A $(1/q)$ divergence is obtained at the $\Gamma$ point. This is a result of the metallic nature of the DB band on the ideal surface. The dashed-dotted line

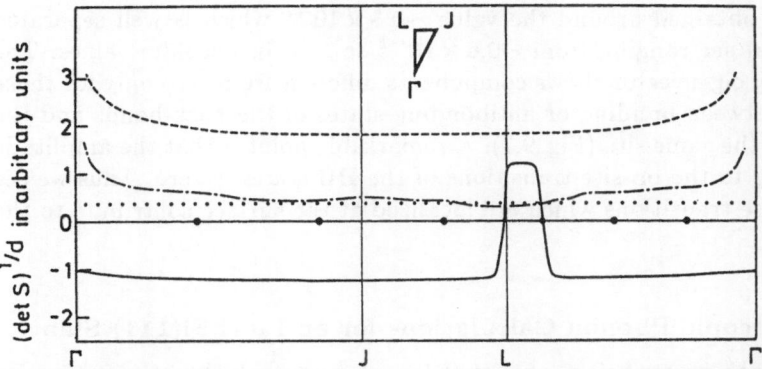

**Fig. 9. 6.** SDW instabilities on the ideal Si(111) slab. The inset shows the irreducible part of the 2-D Brillouin zone. The *dots* on the abscissa correspond to wave vectors for the possible (7 × 7) superstructures along the $\Gamma$-J-L-$\Gamma$ line

finally shows the decisive influence of excitonic effects ($V^s$): the determinant becomes smaller at the zone boundary (J-L line), and even smaller than in the Hartree approximation around the L point. This suggests that the system has a tendency towards a CDW instability in that region.

Next, consider the spin-density fluctuations (full line) in Fig. 9.6. It is evident that the inclusion of the electron-hole attraction $V^s$ leads to two types of instabilities. The determinant goes through zero at wave vectors near to those corresponding to the (2 × 1) and (7 × 7) superstructures, where the first one agrees with pseudopotential local spin-density results [9.43]. The magnetic (7 × 7) superstructure was until now not investigated by other methods to our knowledge.

As a last point we discuss the localization of the magnetic instability. It can be obtained from the amplitude of elementary excitations, as defined in Sect. 9.1. The amplitude becomes especially important for surface systems and allows us to separate extended bulk-like excitations from those localized at the surface. In order to obtain the amplitude "per site" $X_\lambda$ (9.13) we first diagonalize the matrix $\overset{\leftrightarrow}{S}{}^{-1}(q)$ for the wave vector where the determinant goes to zero, and then look for the eigenvector corresponding to the zero eigenvalue. One

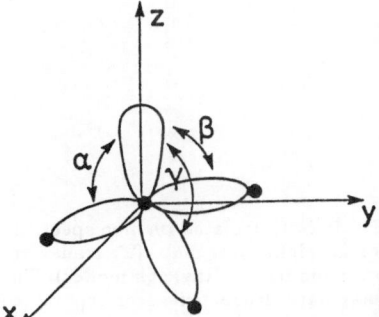

**Fig. 9. 7.** Schematic representation of transitions between DB-orbitals and back-bond orbitals

eigenvalue is obtained around the value $\sim 0.5 \times 10^{-7}$ which is well separated from the total set ranging from $\sim 0.6 \times 10^{-1}$ to $\sim 6$ in absolute values. The corresponding eigenvector shows components different from zero only for those transitions between bonding or antibonding states of the back-bonds and the DB states on the same site (Fig. 9.7). A remarkable point is that the amplitude corresponding to the on-site transitions of the DB states is zero. Thus we see that only these transitions which are localized at the surface contribute to the instability.

### 9.5.2 Microscopic Phonon Calculations for an Ideal Si(111) Slab

According to the prescriptions of Sect. 9.4, we have to calculate the force form factors (9.86 and 87), in addition to the matrix $\overleftrightarrow{S}$ which was already evaluated in Sect. 9.4. Applying the same approximations as for the matrices $\overleftrightarrow{N}$ and $\overleftrightarrow{V}^{xc}$ [9.13], the matrix F has dimension $113 \times 24$, where 24 are the ionic degrees of freedom for a slab with 8 layers.

For the electron-ion interaction, a pseudopotential of the form

$$w(\boldsymbol{r}) = -\frac{e^2}{r}(1 - \beta e^{-\alpha r}) \tag{9.91}$$

was chosen. This pseudopotential corresponds to a frozen-core approximation where the only contributions to the polarizability come from the valence electrons. For further calculational details see [9.35].

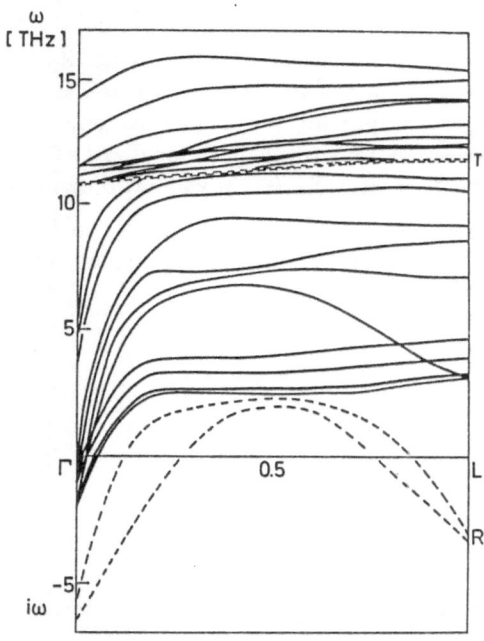

**Fig. 9. 8.** Calculated phonon spectrum for an eight-layer slab. (T: transverse optic modes; R: Rayleigh modes). The imaginary frequencies correspond to the unstable phonons

We present in Fig. 9.8 the phonon spectrum in the $\Gamma$-L direction calculated for an ideal Si(111) slab with 8 layers. The dashed lines correspond to surface modes and the full lines to modes extended through the whole slab. Here we designate as surface modes those modes which are localized in the first and second layers with amplitudes decaying approximately exponentially in directions perpendicular to the surface. We cannot identify resonant modes because the slab is too thin to distinguish them. Due to interactions between both surfaces, the surface modes appear pairwise.

Two kinds of surface modes are obtained. The low-energy modes, characterized with R in Fig. 9.8 appear below the bulk-like modes and are elliptically polarized in the sagittal plane (in our case the x,y-plane). We identify them as Rayleigh modes [9.44]. The other modes are transverse modes, polarized parallel to the surface. Both modes were also obtained by *Zimmermann* [9.45] and *Ludwig* [9.46] with a phenomenological model for a semiinfinite Si crystal.

We can distinguish two different types of instabilities in Fig. 9.8. In the region where $q \leq 1/\ell$, $\ell$ being the thickness of the slab, we obtain phonon instabilities which are related to the fact that we are dealing with an unrelaxed surface. On the other hand, the instabilities of the Rayleigh modes at the border of the BZ are a direct consequence of the coupling to the charge-density wave fluctuations that we discussed in Sect. 9.5.1. The soft phonon at the L point corresponds to a structural instability against $(2 \times 1)$ superstructure. A study of the eigenvectors of this mode shows the displacement pattern of the atoms (Fig. 9.9) which gives rise to a buckled surface. We should remark here that although we obtain an instability towards buckling of the surface, no predictions are intended with respect to the final new configuration of the surface, which until now, is generally agreed to be well described by the $\pi$-bonded chain model of *Pandey* [9.47].

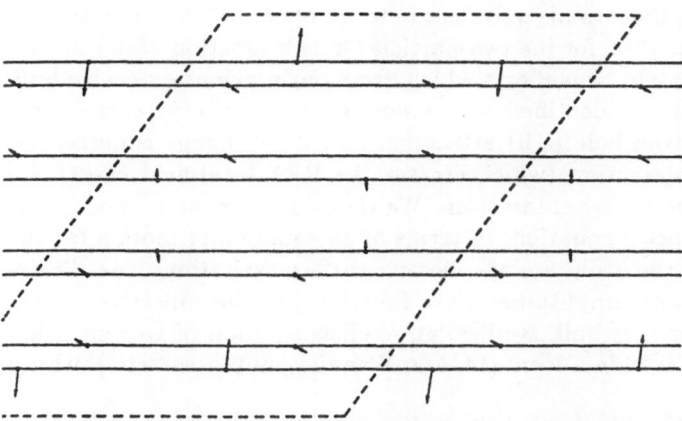

**Fig. 9. 9.** Displacement pattern corresponding to the unstable phonons at the L point of Fig. 9.8

## 9.6 Summary and Perspectives

The purpose of this chapter is to summarize and discuss a recently developed theory of density and spin response of surface systems and its application to elementary excitations. In particular, the theory aims at a microscopic understanding of surface experimental results on electromagnetic response, electronically driven instabilities and electron-phonon interactions on a quantitative level, similar to that achieved for bulk solids [9.3–5]. In the surface theory one starts, like in the complementary bulk studies, from quantum-mechanic wave functions and energies and constructs correlation and response functions according to the standard many-body perturbation theory.

Our discussion begins in Sect. 9.2 with a general theory of surface elementary excitations, which are embedded in the framework of linear response. The theory is general, in that it does not rest on the frequently used jellium or homogeneous electron-gas approximation or even simpler models like the infinite barrier model. Thus, it takes into account the microscopic electronic density fluctuations which, on the one hand, are introduced by the underlying crystal lattice structure and, on the other hand, are created by the very presence of the surface. Previous descriptions of surface response functions and, therefore, also of the extracted elementary excitations either use the jellium simplification or rest on phenomenological and in general macroscopic (i.e., boundary condition) models. These macroscopic models neglect electronic density fluctuations on the scale of the first few surface layers, which are still contained in the jellium model. However, this latter model does not take into account density fluctuations due to the underlying ion cores and thus, in particular, neglects local-field effects in directions parallel to the surface. It is adequate for typical "pseudopotential" metal surfaces, like the alkalis, where the valence-electron density parallel to the surface can approximately be regarded as constant.

Our more general theory, therefore, has to start from the microscopic Maxwell equations or, the formal equivalent in terms of Green's function theory, the Bethe-Salpeter equation for the two-particle Green's function of the surface system. The two many-electron effects which, from our previous extensive bulk studies [9.3–5] have been identified as the dominant many-body corrections, i.e. the screened electron-hole (e-h) attraction and its exchange counterpart, the RPA Coulomb interaction (which creates the RPA local-field effect) are included in the irreducible e-h interaction. We then consider the homogeneous part of the Bethe-Salpeter equation. In terms of an equation of motion for the e-h pair it determines the evolution of collective surface excitations, specifically the energy spectrum and amplitudes (wave functions) of the collective excitations. Like in our previous bulk studies, an explicit solution of this equation of motion can be achieved in a local (LCAO, Wannier, muffin-tin, etc.) orbital basis.

Section 9.3 applies this general, linear-response theory to the optical response. Again emphasis is placed on surfaces, like semiconductor surfaces, for which the jellium approximation is not valid. Our treatment thus extends previous jellium theories [9.6] and, in some respects, follows [9.9], which had already implemented our local-orbital scheme [9.3–5,13,35] to achieve the required inversion of the surface dielectric tensor. In particular, we obtain the optical response in a local-orbital scheme and describe the extraction of measurable quantities such as the reflectivity. No explicit calculation is yet available, which uses this microscopic theory of optical response. However, many interesting applications of this quite general scheme are feasible. One possible application would be the construction of bound excitons including the e-h attraction (which is not necessarily short- or long-ranged in our theory) and its exchange counterpart. Another application of current interest would be the calculation of the optical response for a surface, which has a superstructure, like the $(2 \times 1)$ reconstructed Si(111) surface. Here, accurate experimental determination of optical response is available and could be used to check theoretically proposed reconstruction models, such as the $\pi$-bonded chain model. In principle, however, this would additionally require knowledge of the single-particle electronic states, embedded in the one-particle Green's function. The two-particle propagator of Sect. 9.2 could be used to generate a controlled approximation for the non-local, dynamical self-energy. Within the time-dependent screened HF approximation this program has been carried through already for bulk materials [9.4,5]. It now seems feasible also for surface systems.

Section 9.4 gives a microscopic formulation of the electron-phonon interaction and lattice dynamics of surfaces in the harmonic and adiabatic approximation. The theoretical description of the surface dynamical properties has so far relied exclusively on phenomenological force-constant models, usually extracted from a bulk parametrization of phonon energies. In contrast, our theory starts from the actual microscopic ingredients of the surface system, i.e. quantum mechanic wave-functions and energies. The phonon self-energy is constructed from the non-local density-response or inverse dielectric response function, which we obtain from the formalism of Sect. 9.2. Again many applications of this theory are feasible. They are certainly required in the light of the development of new, powerful experimental techniques for dynamical surface excitations, such as the scattering of light (He) atoms, low-energy electron-diffraction (LEED) and characteristic energy loss.

Section 9.5 finally surveys a few applications to a specific surface system of current interest, the Si(111) unreconstructed (ideal) surface. This surface reconstructs in a $(2 \times 1)$ pattern after cleavage in vacuum, with the $(2 \times 1)$ structure being metastable, and going irreversibly over into a $(7 \times 7)$ superstructure at high temperatures ($\sim 400° C$). With the help of our general formalism of Sect. 9.2 the conditions for the occurrence of an electronic (charge and spin-density) instability at the surface are studied. Again the coupling of such modes to the lattice is obtained through the dielectric response function. Calculations based on non-selfconsistent LCAO wavefunctions and eigenvalues are reviewed which display pronounced many-body effects of RPA and excitonic nature. In particular, a delicate balance of the many-body and the band-structure effects is shown to be responsible for spin-density wave (SDW) instabilities which occur for wave vectors correspondig to a $(2 \times 1)$ and a $(7 \times 7)$ superstructure. A calculation of the phonons of an 8-layer slab of Si(111) reveals that the coupling of the charge-density waves (CDW) to the lattice leads to a soft surface mode with atomic displacements corresponding to a $(2 \times 1)$ reconstruction.

These model applications demonstrate the feasibility of a microscopic calculation of surface collective excitations and their interplay with structural and dynamical properties taking into account the actual surface density profile as well as many-body effects.

# References

9.1   P. Nozières: *Theory of Interacting Fermi Systems* (Benjamin, New York 1964)
9.2   D. Pines, P. Nozières: *Theory of Quantum Liquids* (Benjamin, New York 1966)
9.3   W. Hanke: Adv. Phys. **27**, 287 (1978)
9.4   G. Strinati, H.J. Mattausch, W. Hanke: Phys. Rev. B **25**, 2867 (1982)
9.5   W. Hanke, L.J. Sham: Phys. Rev. B **21**, 4656 (1980)
9.6   P.J. Feibelman: Progr. Surf. Sci. **12**, 287 (1982)
9.7   B.N.J. Persson: Phys. Rev. Lett. **55**, 2957 (1985)
9.8   M. Jonson, G. Srinivasan: Phys. Scr. **10**, 262 (1974)
9.9   R. del Sole, E. Fiorino: Phys. Rev. B **29**, 4631 (1984)
9.10  A.L. Fetter, J.L. Walecka: *Quantum Theory of Many-Particle Sytems* (McGraw Hill, New York 1971)
      E.N. Economou: *Green's Functions in Quantum Physics*, 2nd ed., Springer Ser. Solid-State Sci., Vol. 7 (Springer, Berlin, Heidelberg 1983)
9.11  L.J. Sham, T.M. Rice: Phys. Rev. **144**, 708 (1966)
9.12  L.J.Sham: Phys. Rev. **150**, 720 (1966)
9.13  A. Muramatsu, W. Hanke: Phys. Rev. B **30**, 1911 (1984)
9.14  C. Wu, W. Hanke: Solid State Commun. **23**, 829 (1977)
9.15  H. Lüth, M. Buckel, R. Dorn, M. Liehr, R. Matz: Phys. Rev. **15**, 865 (1977)
9.16  P. Chiaradia, G.Chirotti, S. Nannarone, P. Sassaroli: Solid State Commun. **26**, 813 (1978)
9.17  P.E. Wierenga, A. van Silfhout, M.J. Sparnaay: Surf. Sci. **87**, 43 (1979) and ibid **99**, 59 (1980)
9.18  S. Nannarone, P. Chiaradia, F. Ciccacci, R. Memeo, P. Sassaroli, S. Selci, G. Chiarotti: Solid State Commun. **33**, 593 (1980)
9.19  V.V. Hyzhnyakov, A.A. Maradudin, D.L. Mills: Phys. Rev. B **11**, 3149 (1975)
9.20  A. Stahl, C. Uihlein: Festkörperprobleme **19**, 159 (Vieweg, Braunschweig 1979) and references therein
9.21  G.D. Mahan: *Many-Particle Physics* (Plenum, New York 1981)
9.22  S.L. Adler: Phys. Rev. **126**, 413 (1962)
9.23  H. Ehrenreich: In *The Optical Properties of Solids*, ed. by J. Taue (Academic, New York 1965) p. 106
9.24  A. Ambegaokar, W. Kohn: Phys. Rev. **117**, 423 (1960)
9.25  R. Del Sole, A. Selloni: Phys. Rev. B **30**, 883 (1984)
9.26  See, for example: C.B. Duke, A. Paton, W.K. Ford, A. Kahn, G. Scott: Phys. Rev. B **24**, 3310 (1981)
      M.A. Van Hove, W.H. Weinberg, C.-M. Chan: Low-Energy Electron Diffraction, Springer Ser. Surf. Sci., Vol. 6 (Springer, Berlin, Heidelberg 1986)
9.27  H. Lüth: Physica **118 B**, 810 (1983)
      J.M. Szeftel, S. Lechwald, H. Ibach, T.S. Rahman, J.E. Black, D.L. Mills: Phys.Rev. Lett. **51**, 268 (1983)
9.28  H.D. Mayer, J.P. Toennies: Surf. Sci. **148**, 58 (1984) and references therein
      G. Bruscdeylins, R. Rechsteiner, J.G. Skofronick, J.P. Toennies, G. Benedek, L. Miglio: Phys. Rev. Lett. **54**, 466 (1985)

9.29  For a recent review on surface waves, see A.A. Maradudin: Festkörperprobleme **21**, 15 (Vieweg, Braunschweig 1981)
R.F. Wallis, G.I. Stegeman (eds.): *Electromagnetic Surface Excitations*, Springer Ser. Wave Phen., Vol. 3 (Springer, Berlin, Heidelberg 1986)

9.30  V. Bortolani, G. Santoro, U. Harten, J.P. Toennies: Surf. Sci. **148**, 82 (1984)

9.31  G. Benedek: Surf. Sci. **61**, 603 (1976)
F.W. de Wette, G.P. Alldredge: In *Methods of Computational Physics*, Vol. 15, ed. by G. Gilat et al. (Academic, New York 1976) Chap. 5, p. 163

9.32  See, for example, *Ab Initio Calculation of Phonon Spectra*, ed. by J.T. Devreese, V.E. Van Doren and P.E. Van Camp, (Plenum, New York 1983)

9.33  F.W. de Wette, G.E. Schacher: Phys. Rev. **137 A**, 78 (1965)
S.Y. Tong, A.A. Maradudin: Phys. Rev. **181**, 1318 (1969)

9.34  L.J. Sham: Phys. Rev. **188**, 1431 (1969)

9.35  For more details, A. Muramatsu, W. Hanke: Phys. Rev. B **30**, 1922 (1984)

9.36  For recent review: W. Mönch: Surf. Sci. **86**, 672 (1979);
D.J. Chadi: Surf. Sci. **99**, 1 (1980)

9.37  P. Chiaradia, A. Cricenti, S. Selc, G. Chiarotti: Phys. Rev. Lett. **52**, 1145 (1984) and references therein

9.38  O.H. Nielsen, R.M. Martin, D.J. Chadi, K. Kunc: J. Vac. Sci. Technol. **31**, 714 (1983) and references therein

9.39  M. Schlüter, S.R. Chelikowski, J.G. Louie, M.L. Cohen: Phys. Rev. B **12**, 4200 (1975)

9.40  M. Schlüter, M.L. Cohen: Phys. Rev. B **17**, 716 (1978)

9.41  J.A. Appelbaum, D.R. Hamann: Phys. Rev. B **12**, 1410 (1975)
J.A. Appelbaum, D.R. Hamann: In *Theory of Chemisorption*, ed. by J.R. Smith, Topics Curr. Phys., Vol. 19 (Springer, Berlin, Heidelberg 1980)

9.42  E. Tosatti: Festkörperprobleme **15**, 113 (Vieweg, Braunschweig 1975)

9.43  J.E. Northrup, J. Ihm, M.L. Cohen: Phys. Rev. Lett. **47**, 1910 (1981)

9.44  R.E. Allen, G.P. Alldredge, F.W. de Wette: Phys. Rev. B **4**, 1661 (1971)

9.45  R. Zimmermann: Appl. Phys. **3**, 235 (1974)

9.46  W. Ludwig: Jap. J. Appl. Phys. Suppl. 2, Pt. 2, 879 (1974)

9.47  K.C. Pandey: Phys. Rev. Lett. **47**, 1913 (1981); **49**, 223 (1982)

# Subject Index

# Topics in Current Physics

Founded by Helmut K. V. Lotsch